STUDIES IN APPLIED MECHANICS 1

MECHANICS AND STRENGTH OF MATERIALS

STUDIES IN APPLIED MECHANICS 1

MECHANICS AND STRENGTH OF MATERIALS

BOGDAN SKALMIERSKI
Silesian Polytechnical University
Gliwice, Poland

ELSEVIER SCIENTIFIC PUBLISHING COMPANY
Amsterdam-Oxford-New York
PWN–POLISH SCIENTIFIC PUBLISHERS
Warszawa
1979

Revised translation from the original Polish
Mechanika z wytrzymałością materiałów dla automatyków
Published in 1973 by Państwowe Wydawnictwo Naukowe, Warszawa

Translated by JERZY BACHRACH

Published in co-edition with
PWN—POLISH SCIENTIFIC PUBLISHERS
WARSZAWA

Distribution of this book is being handled by the following publishers

for the U.S.A. and Canada
ELSEVIER/NORTH-HOLLAND, INC.
52 Vanderbilt Avenue, New York, N.Y. 10017

for Albania, Bulgaria, Chinese People's Republic, Czechoslovakia, Cuba, German Democratic Republic, Hungary, Korean People's Democratic Republic, Mongolia, Poland, Rumania, the U.S.R.R., Vietnam, and Yugoslavia
ARS POLONA
Krakowskie Przedmieście 7, 00-068 Warszawa, Poland

for all remaining areas
ELSEVIER SCIENTIFIC PUBLISHING COMPANY
335 Jan van Galenstraat, P.O. Box 211
1000 AE Amsterdam, The Netherlands

Library of Congress Cataloging in Publication Data

Skalmierski, Bogdan.
Mechanics and strength of materials.
(Studies in applied mechanics; 1)
Translation Mechanika z wytrzymałością materiałów dla automatyków.
Bibliography: p.
Includes index.
1. Strength of materials. 2. Mechanics. I. Title. II. Series.
TA405.S46813 620.1′12 78-10909
ISBN 0-444-99793-8

Printed in Poland by D.R.P.

Preface

Mechanics and strength of materials represent today such extensive fields of study that it would be impossible to tackle comprehensively all the problems pertinent to the subject in this modest work. As a result, the author, in order to bring out the salient points, has been obliged to confine himself to essentials. This book has been conceived, therefore, as a concise general introduction to mechanics and strength of materials and in no way is it intended to exhaust the subject matter. The literature cited at the end of this book gives basic sources, to which the reader, wishing to acquire a deeper insight into certain portions of the exposition, is referred. Having understood the basic problems considered in this book, a keen student should be able to go far beyond the scope of this work.

This book is devoted to problems involving mechanics and strength of materials which are in principle two fields of study (though developed upon the same ground); mechanics and strength of materials are thus treated as separate subjects in the syllabuses of academic engineering schools. Mechanics in the traditional sense is the science of the motion of particles and rigid bodies, a fact evidenced by a host of books on the subject.

The idea of transcending this convention is, it seems, justifiable. Since mechanics is the science of motion, studies in this field should cover a wider range of problems than has hitherto been the case using the classical approach. This can be achieved, for example, by the inclusion of problems relating to the mechanics of continuous media and to selected strength problems. By such a complex treatment of the subject, one should be able to more adequately answer the question: what does mechanics in the present sense of the word stand for? 'Mechanics and Strength of Materials' is an unconventional book, in that it treats the two different subjects jointly, an approach which the author has unquestionably derived from his lectures, of the same title, delivered to students in the Department of Automation of the Silesian Polytechnical University. Yet another feature in which the book departs from the classical approach is the mathematical methodology used. Regarding the latter, some comment is required:

(a) The summation convention is fairly consistently used in the book. This method contributes not only to the lucidity of the equations but also to their conciseness; additionally it facilitates transformations. By introducing the theory of geometric objects, emphasis has been placed on the natural link between mechanics and geometry.

(b) The notion of distribution (generalized function) has been introduced. The application of the theory of distribution has allowed a more precise presentation of problems and at the same time a more convenient means of calculation. For example, the beam-deflection problem, when analyzed in the traditional manner, calls for intervallic divisions, the solution being obtained by reference to conformity conditions. By the application of Heaviside's and Dirac's distribution, this tedious procedure can be dispensed with. Furthermore, such classical notions as concentrated force, concentrated moment or particle and the like, themselves bring forth the notion of distribution in a natural way. Reversing the situation, their equivalents in mechanics can be defined.

(c) Stochastic problems are also considered, some aspects of indeterminate mechanical systems being taken into account. This theme has been confined in principle to correlation and spectral analysis. Elements of the theory of departures have also been included.

(d) This book has been so designed as to emphasise that mechanics is a deductive system, a body of theorems of motion, some of them being axioms and others being corollaries deduced from axioms. An effort has also been made to demonstrate a certain universality of the methods peculiar to mechanics by reference to selected problems relating to electrical or electromechanical systems constructed on Lagrange's equations.

To achieve the above objectives, it was found necessary not only to make a suitable selection of problems to be considered but also to find appropriate mathematical means. Since vector analysis is foremost among these means, this branch of mathematics, so important to mechanics, has been recapitulated and set at the beginning of the book.

This book is addressed to all interested in mechanical and allied problems.

B. SKALMIERSKI

Contents

Preface, V

Introduction, 1

Chapter 1. Kinematics, 11

1.1. Motion of a single particle, 11
1.2. Description of the motion of a rigid body, 26
1.3. Relative motion, 40
1.4. Plane motion of a rigid body, 48
1.5. Examples of the determination of velocities and accelarations in the motion of plane mechanisms, 55

Chapter 2. The dynamics of a particle, 69

2.1. Fundamental definitions and theorems, 69
2.2. Motion of a particle, 72
2.3. Centre of mass and centre of gravity of a particle system, 80
2.4. Law of variation of momentum, 86
2.5. Law of variation of angular momentum, 92

Chapter 3. Statics, 95

3.1. Equations of equilibrum, 95
3.2. Couple, 96
3.3. Spatial system of forces. Wrench, 97
3.4. Analytical and graphical methods in the statics of plane systems, 99
Analytical methods, 99
Graphical methods (funicular polygon), 101

3.5. Examples, 106
Triple-jointed system, 106
Free-ends beam, 114
Plane trusses, 117
3.6. The distributive character of transverse loads in simple rods, 121
3.7. The equilibrium of rods loaded with transverse forces, 127
3.8. Friction, 136
Friction on an inclined plane, 138
Bearing friction, 140
Friction of rope against wheel, 141
Problem of friction in the case of a cylinder rolling on plane surface, 142

Chapter 4. The statics of elastic systems, 144

4.1. Hooke's law, 144
4.2. Safety factor, 149
4.3. Statically indeterminate systems, 151
4.4. Problems of simple beam-bending, 155
4.5. Geometric moments of inertia, 159
4.6. Strength calculations of beams, 166
4.7. The equation for the axis of a deflected beam, 169
4.8. Graphical methods of determining deflections of simple beams (Mohr's analogy), 173
4.9. Oblique bending, 176
4.10. Some special problems of bending theory, 177
4.11. Clapeyron's systems, 185
4.12. Buckling problems of axially compressed rods, 199
4.13. Highly curved rods, 203
4.14. Torsion of rods with circular cross-section, 209
4.15. Springs, 212
4.16. Simultaneous bending and torsion of rods with circular cross-section, 217

Chapter 5. The dynamics of rigid bodies, 220

5.1. Moments of inertia of rigid bodies, 220
5.2. The angular momentum of a rigid body in general motion, 226

5.3. Angular momentum in circular motion, 228
5.4. Euler's equations, 230
5.5. The kinetic energy of rigid bodies in general motion, 235

Chapter 6. Dynamics in relative motion, 241

6.1. Differential equation of the motion of a particle in a non-inertial system, 241
6.2. The dynamics of rigid bodies in relative motion, 243

Chapter 7. Fundamentals of analytical mechanics, 249

7.1. Generalized coordinates and degrees of freedom of a mechanical system, 249
7.2. D'Alembert's principle, 255
7.3. Hamilton's principle, 258
7.4. Lagrange equations of the first order, 261
7.5. Lagrange equations of the second order, 265
7.6. Kinetic energy of a system, 272
7.7. Impulsive motion, 272
7.8. Gyroscopic and dissipative forces, 276
7.9. The Lagrange equations for electromechanical systems, 279
7.10. Hamilton's canonical equations, 283
7.11. The total energy of a mechanical system, 284
7.12. Configurational space, 286
7.13. The stability of mechanical systems, 293

Chapter 8. Vibrations of systems with one degree of freedom, 307

8.1. Preliminary discussion, 307
8.2. Free vibrations of harmonic oscillators, 310
8.3. The influence of dissipative forces in the free vibrations of harmonic oscillators, 319
8.4. Forced vibrations of harmonic oscillators, 323
8.5. Vibrations of harmonic oscillators with kinematical input, 332
8.6. Vibrations of harmonic oscillators under periodic input forces, 333
8.7. Vibrations of non-linear systems, 336

Chapter 9. Vibrations of systems with many degrees of freedom, 351

9.1. Preliminary discussion, 351
9.2. Problems of linearization of the equations, 352
9.3. Free vibrations of conservative systems, 356
9.4. Normal coordinates, 362
9.5. Forced vibrations of a system, 363
9.6. Free vibrations of dissipative systems, 364
9.7. Forced vibrations in dissipative systems, 366
9.8. Vibrations of Clapeyron's systems, 367

Chapter 10. Some methods of describing random phenomena in mechanics, 371

10.1. Basic concepts, 371
10.2. Methods of describing stochastic processes, 381
10.3. Stochastic linearization, 387
10.4. Random vibrations of linear systems with one degree of freedom, 389
10.5. Random vibrations of systems with many degrees of freedom, 397
10.6. The problem of departures, 400
10.7. Fokker–Planck–Kolmogorov equations, 415
10.8. Proposal for a method of direct determination of probability density, 420

Bibliography, 425
Subject index, 429

Introduction

Mechanics, being one of the basic branches of theoretical physics, deals with the motion of bodies under the influence of forces acting on them.

A branch of mathematics that we shall be using from the outset is vector analysis; restating the principles of vector algebra in this introduction will therefore serve a useful purpose.

By a *vector* we understand any magnitude that can be represented by an oriented section having an origin (A) and an end (B) (Fig. 0.1). In describing a vector, the following data should be specified:

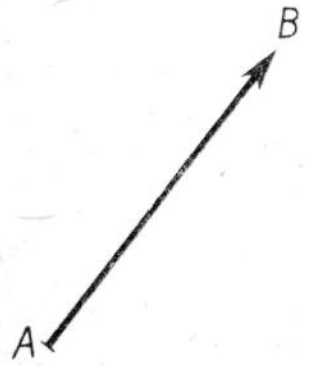

Fig. 0.1

1. The *length* of a section AB, which corresponds to a stated physical quantity, as represented by the vector;
2. The *straight line* defining the direction;
3. The *initial point* or *end* of the vector, or the sense.

In this way, each ordered pair of points A and B defines a vector attached at the point A (origin of the vector). In a special case, a section AA will be regarded as a vector and defined as the zero vector at the point A. In fixing the point of attachment of a vector, i.e., its initial point, it should be noted that its end uniquely specifies a point in space.

Similarly, any point in space uniquely specifies a vector with a fixed end. Thus there is a one-to-one correspondence of vectors starting from a fixed point with points in space. Such a vector is called a *position vector.* It will play an essential role in our considerations.

Two vectors, *AB* and *CD,* will be treated as equivalent and their equivalence denoted by the sign =, on condition that the quadrilateral *ABDC* is a parallelogram.

By a *free vector* we shall understand a class of all equivalent localized vectors. A specimen of a free vector is any localized vector belonging to that class. Just one specimen vector originates from an arbitrary point in space. In fixing the starting point of a free vector we thereby fix the localized vector.

Two vectors are defined as *parallel* if they are lying along two parallel lines.

Two vectors lying in a straight line are called *coaxial.*

The same vector can be represented, depending on the scale used, by parallel or co-axial sections of different length. The number of units corresponding to a given length is called the *modulus.* The vector itself will be denoted by the letters **a**, **b**, **A**, etc., and the corresponding moduli by the letters *a*, *b*, *A*, etc. In some cases the modulus may be denoted by

$$|\mathbf{a}| \quad \text{or} \quad |\mathbf{b}|.$$

Vectors are added by constructing the diagonal of a parallelogram built on the component vectors (Fig. 0.2). We write

$$\mathbf{a}+\mathbf{b} = \mathbf{c}.$$

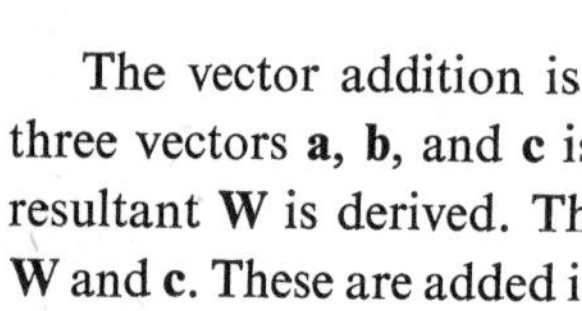

Fig. 0.2

The vector addition is commutative (i.e., $\mathbf{a}+\mathbf{b} = \mathbf{b}+\mathbf{a}$). If the sum of three vectors **a**, **b**, and **c** is to be obtained, first **a** and **b** are added and the resultant **W** is derived. The system is thus reduced to two vectors, namely **W** and **c**. These are added in the same manner as the vectors **a** and **b** (Fig. 0.3) the associative rule being involved here, i.e.,

$$(\mathbf{a}+\mathbf{b})+\mathbf{c} = \mathbf{a}+(\mathbf{b}+\mathbf{c}).$$

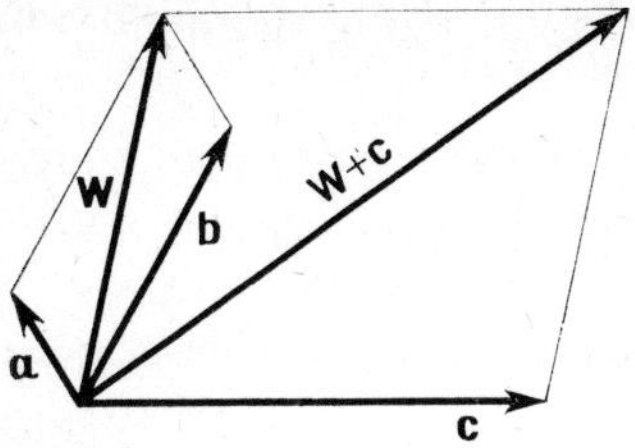

Fig. 0.3

Subtraction of two vectors consists of adding to the minuend the subtrahend with the opposite sense:

$$\mathbf{a}-\mathbf{b} = \mathbf{a}+(-\mathbf{b}). \tag{2}$$

The vectors $-\mathbf{b}$ and $\mathbf{b}$ have the same orientation and absolute value and they differ only in having opposite directions (Fig. 0.4). Any vector can be represented by means of its components. It is convenient to use for

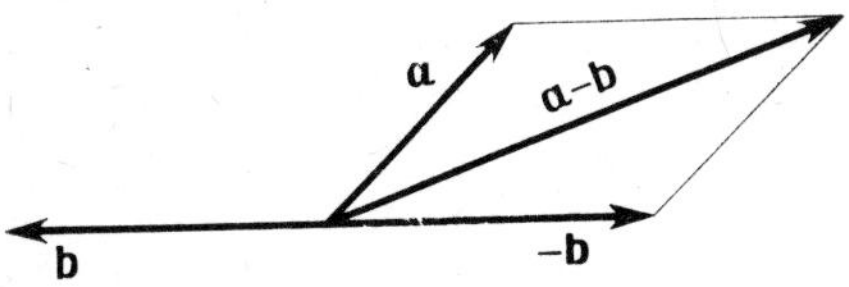

Fig. 0.4

this purpose cartesian coordinates and resolve a vector into three components (in a three-dimensional space) parallel to **x**, **y** and **z**. In space, the resultant of the three vectors is the diagonal of a parallelepiped, in this case a rectangular prism (Fig. 0.5), since $\mathbf{a}_x$, $\mathbf{a}_y$ and $\mathbf{a}_z$ are vectors parallel to the axes of the orthogonal frame of reference $Oxyz$:

$$\mathbf{a} = \mathbf{a}_x+\mathbf{a}_y+\mathbf{a}_z. \tag{3}$$

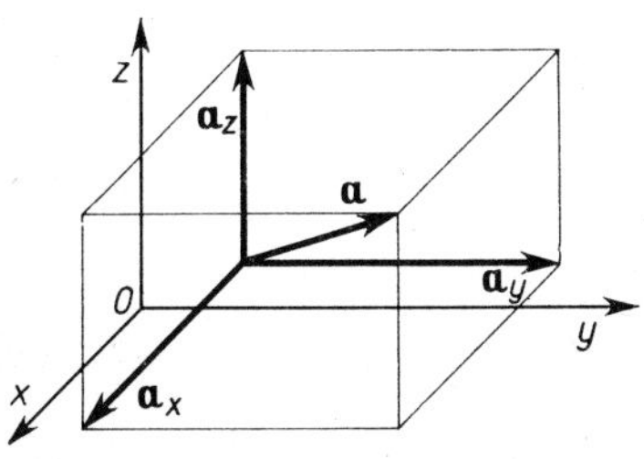

Fig. 0.5

Multiplying the vector by a scalar changes its length and if that number is negative also reverses its sense:

$$\mathbf{b} = \alpha \mathbf{a}. \tag{4}$$

The vector **b** thus derived is a *collinear vector* (parallel or co-axial).

Note that vectors when added and multiplied by a scalar form a linear space, since the following equalities occur here:

$$\left.\begin{aligned}(\mathbf{a}+\mathbf{b})+\mathbf{c} &= \mathbf{a}+(\mathbf{b}+\mathbf{c}),\\ \mathbf{a}+\mathbf{b} &= \mathbf{b}+\mathbf{a},\\ \mathbf{0}+\mathbf{a} &= \mathbf{a},\\ \mathbf{a}+(-\mathbf{a}) &= \mathbf{0},\end{aligned}\right\} \tag{a}$$

and

$$\left.\begin{aligned}\alpha_1(\alpha_2\mathbf{a}) &= (\alpha_1\alpha_2)\mathbf{a},\\ (\alpha_1+\alpha_2)\mathbf{a} &= \alpha_1\mathbf{a}+\alpha_2\mathbf{a},\\ \alpha(\mathbf{a}+\mathbf{b}) &= \alpha\mathbf{a}+\alpha\mathbf{b},\\ 1\cdot\mathbf{a} &= \mathbf{a},\\ (-1)\cdot\mathbf{a} &= -\mathbf{a},\\ 0\cdot\mathbf{a} &= \mathbf{0},\\ \alpha\cdot\mathbf{0} &= \mathbf{0},\end{aligned}\right\} \tag{b}$$

where α, α_1, and α_2 are scalars.

A vector whose modulus is equal to 1 is called a *unit vector.* A unit vector can be derived from any vector by multiplying it by the reciprocal of its length

$$\mathbf{a}^0 = \mathbf{a}\frac{1}{\mathbf{a}}. \tag{5}$$

The unit vectors corresponding to $\mathbf{a}_x$, $\mathbf{a}_y$, $\mathbf{a}_z$ are derived successively in the following manner:

$$\mathbf{i} = \frac{\mathbf{a}_x}{a_x}, \quad j = \frac{\mathbf{a}_y}{a_y}, \quad k = \frac{\mathbf{a}_z}{a_z}. \tag{6}$$

The unit vectors **i, j, k,** like the coordinate axes x, y, and z are at right angles to each other and by virtue of this characteristic they are also called *orthogonal vectors.* Introducing unit vectors into the formula, we have

$$\mathbf{a} = a_x\mathbf{i}+a_y\mathbf{j}+a_z\mathbf{k}. \tag{7}$$

The components a_x, a_y, and a_z are projections of the vector **a** on x, y, and z.

Expressing by components the two vectors **a** and **b**, we can obtain either the sum or the difference by successively adding or subtracting the appropriate projections:

$$\mathbf{a}\pm\mathbf{b} = (a_x\pm b_x)\mathbf{i}+(a_y\pm b_y)\mathbf{j}+(a_z\pm b_z)\mathbf{k}; \tag{8}$$

or in more general terms:

$$\sum_{i=1}^{n}\mathbf{a}_i = \mathbf{i}\sum_{i=1}^{n}a_{xi}+\mathbf{j}\sum_{i=1}^{n}a_{yi}+\mathbf{k}\sum_{i=1}^{n}a_{zi}. \tag{8a}$$

Introducing the following notations

$$\begin{aligned} \mathbf{i} &\equiv \mathbf{e}_1, \quad & \mathbf{j} &\equiv \mathbf{e}_2, \quad & \mathbf{k} &\equiv \mathbf{e}_3, \\ a_x &\equiv a^1, \quad & a_y &\equiv a^2, \quad & a_z &\equiv a^3, \end{aligned} \tag{9}$$

equation (7) can be written in the following form:

$$\mathbf{a} = \sum_{i=1}^{n} a^i\mathbf{e}_i. \tag{7a}$$

We shall use special notation in this book to permit some equations to be written in a shortened form and to be easily transformable. The notations (9) enable the formula (7) to be shortened to (7a). If we also agree that where indices occur twice summation takes place for values of the indices from 1 to 3 depending on the number (1 to 3) of spatial dimensions, then the summation sign in equation (7a) can be omitted and (7a) can now be written in a yet simpler form:

$$\mathbf{a} = a^i\mathbf{e}_i.$$

Moreover, we shall register indices as superscript (a^i) or subscript ($\mathbf{e}_i$).

Vector multiplication. Where two vectors are involved, we distinguish their scalar product and vector product.

(a) The *scalar product* of two vectors **a** and **b**, making an angle α between them, is a scalar calculated from

$$\mathbf{a}\cdot\mathbf{b} \overset{\text{df}}{=} ab\cos\alpha. \tag{10}$$

If vectors **a** and **b** are represented by projections of orthogonal coordinate axes, i.e., if use is made of equation (7), we shall obtain

$$\begin{aligned} \mathbf{a}\cdot\mathbf{b} &= (a_x\mathbf{i}+a_y\mathbf{j}+a_z\mathbf{k})\cdot(b_x\mathbf{i}+b_y\mathbf{j}+b_z\mathbf{k}) \\ &= a_xb_x\mathbf{i}\cdot\mathbf{i}+a_xb_y\mathbf{i}\cdot\mathbf{j}+a_xb_z\mathbf{i}\cdot\mathbf{k}+a_yb_x\mathbf{j}\cdot\mathbf{i}+ \\ &\quad +a_yb_y\mathbf{j}\cdot\mathbf{j}+a_yb_z\mathbf{j}\cdot\mathbf{k}+a_zb_x\mathbf{k}\cdot\mathbf{i}+a_zb_y\mathbf{k}\cdot\mathbf{j}+a_zb_z\mathbf{k}\cdot\mathbf{k}. \end{aligned} \tag{10a}$$

It follows from the definition of scalar product that

$$\begin{aligned} \mathbf{i}\cdot\mathbf{i} &= \mathbf{j}\cdot\mathbf{j} = \mathbf{k}\cdot\mathbf{k} = 1, \\ \mathbf{i}\cdot\mathbf{j} &= \mathbf{j}\cdot\mathbf{i} = \mathbf{k}\cdot\mathbf{i} = \mathbf{i}\cdot\mathbf{k} = \mathbf{k}\cdot\mathbf{j} = \mathbf{j}\cdot\mathbf{k} = 0. \end{aligned} \tag{11}$$

Introducing the notations (9), the relationships (11) can be shortened:

$$\mathbf{e}_m\cdot\mathbf{e}_n = \delta_{mn}, \tag{11a}$$

where δ_{mn} is *Kronecker's symbol*:

$$\delta_{mn} = \begin{cases} 1 & \text{for} \quad m = n, \\ 0 & \text{for} \quad m \neq n. \end{cases} \tag{12}$$

Thus, the scalar product $\mathbf{a}\cdot\mathbf{b}$ can be written as follows:

$$\mathbf{a}\cdot\mathbf{b} = \sum_{m,n=1}^{3} \delta_{mn} a^m b^n \tag{13}$$

or using the summation convention:

$$\mathbf{a}\cdot\mathbf{b} = \delta_{mn} a^m b^n. \tag{13a}$$

Going back to traditional notation, (13a) can be written as

$$\mathbf{a}\cdot\mathbf{b} = a_x b_x + a_y b_y + a_z b_z. \tag{13b}$$

This is the so-called *bilinear product*. Specifically, where $\mathbf{a} = \mathbf{b}$, we have $\mathbf{a}^2 = a_x^2 + a_y^2 + a_z^2$; it is the square of the modulus of the vector $\mathbf{a}$.

The vectors $\mathbf{a}_1, \mathbf{a}_2, \ldots, \mathbf{a}_r$ will be defined as being in *linear dependence* if there are numbers $\beta^1, \beta^2, \ldots, \beta^r$, at least one of which is non-zero, so that

$$\sum_{i=1}^{r} \beta^i \mathbf{a}_i = 0.$$

Note that in equation (7), the vector $\mathbf{a}$ has a linear dependence on vectors $\mathbf{e}_i$ constituting the reference line, since the triplet generates all the vectors in the space under examination.

(b) The *vector product*, $\mathbf{a}\times\mathbf{b} = \mathbf{c}$, of vectors $\mathbf{a}$ and $\mathbf{b}$, including angle α, is a vector with modulus $c = ab\sin\alpha$ and its direction perpendicular to the plane determined by the vectors $\mathbf{a}$ and $\mathbf{b}$, and whose sense is such that the vectors $\mathbf{a}$, $\mathbf{b}$, and $\mathbf{c}$ make a right-handed system, exactly the same as the axial system x, y, z; for example:

$$\mathbf{a}_x \times \mathbf{a}_y = \mathbf{d};$$

the vector $\mathbf{d}$ has the direction of the axis z.

If the senses $\mathbf{a}_x$ and $\mathbf{a}_y$ coincide with those of the unit vectors **i** and **j**, then the sense **d** will coincide with that of the unit vector **k**.

Representing **a** and **b** by projection, we have

$$\begin{aligned}\mathbf{a}\times\mathbf{b} = {} & a_x b_x \mathbf{i}\times\mathbf{i} + a_x b_y \mathbf{i}\times\mathbf{j} + a_x b_z \mathbf{i}\times\mathbf{k} + \\ & + a_y b_x \mathbf{j}\times\mathbf{i} + a_y b_y \mathbf{j}\times\mathbf{j} + a_y b_z \mathbf{j}\times\mathbf{k} + \\ & + a_z b_x \mathbf{k}\times\mathbf{i} + a_z b_y \mathbf{k}\times\mathbf{j} + a_z b_z \mathbf{k}\times\mathbf{k}.\end{aligned} \tag{14}$$

From the definition of vector product we have

$$\begin{aligned}&\mathbf{i}\times\mathbf{i} = \mathbf{j}\times\mathbf{j} = \mathbf{k}\times\mathbf{k} = 0,\\ &\mathbf{i}\times\mathbf{j} = -\mathbf{j}\times\mathbf{i} = \mathbf{k}, \quad \mathbf{k}\times\mathbf{i} = -\mathbf{i}\times\mathbf{k} = \mathbf{j}, \quad \mathbf{j}\times\mathbf{k} = -\mathbf{k}\times\mathbf{j} = \mathbf{i}.\end{aligned} \tag{15}$$

Substituting (15) into (14),

$$\mathbf{a}\times\mathbf{b} = \mathbf{i}(a_y b_z - a_z b_y) - \mathbf{j}(a_x b_z - a_z b_y) + \mathbf{k}(a_x b_y - a_y b_x). \tag{16}$$

It will readily be seen that equation (16) is an expansion of a determinant of the third order

$$\mathbf{a}\times\mathbf{b} = \begin{vmatrix} \mathbf{i} & \mathbf{j} & \mathbf{k} \\ a_x & a_y & a_z \\ b_x & b_y & b_z \end{vmatrix}. \tag{17}$$

Introducing (9), equation (14) can be rewritten. If $\mathbf{a} = a^i\mathbf{e}_i$ and $\mathbf{b} = b^k\mathbf{e}_k$, then

$$\mathbf{a}\times\mathbf{b} = a^i b^k \mathbf{e}_i \times \mathbf{e}_k. \tag{17a}$$

Since $\mathbf{a}\times\mathbf{b} = \mathbf{c}$, the j component is expressed by

$$c_j = (\mathbf{a}\times\mathbf{b})\cdot\mathbf{e}_j = a^i b^k (\mathbf{e}_i\times\mathbf{e}_k)\cdot\mathbf{e}_j = a^i b^k \varepsilon_{jik}. \tag{18}$$

Thus we have

$$\varepsilon_{jik} = (\mathbf{e}_i\times\mathbf{e}_k)\cdot\mathbf{e}_j \tag{19}$$

where ε_{jik} is the so-called *permutation symbol.* It can easily be checked with equation (19) that

$\varepsilon_{jik} = 0$, if two of the indices are equal to each other;

$\varepsilon_{jik} = 1$, if the sequence of j, i, k is a sequence of 1, 2, 3 or its even permutation;

$\varepsilon_{jik} = -1$, if the sequence of j, i, k is an odd permutation of the sequence of 1, 2, 3.

Finally, we have

$$\mathbf{a}\times\mathbf{b} = \delta^{jl} c_j \mathbf{e}_l = \delta^{jl} \varepsilon_{jik} a^i b^k \mathbf{e}_l, \tag{20}$$

where δ^{jl} is Kronecker's symbol.

From the results now derived we can easily see that

$$\mathbf{a}\times\mathbf{b} = -\mathbf{b}\times\mathbf{a}, \tag{21}$$

$$(\alpha\mathbf{a})\times\mathbf{b} = \mathbf{a}\times(\alpha\mathbf{b}) = \alpha(\mathbf{a}\times\mathbf{b}), \tag{22}$$

$$(\mathbf{a}+\mathbf{b})\times\mathbf{c} = \mathbf{a}\times\mathbf{c}+\mathbf{b}\times\mathbf{c}, \tag{23}$$

$$\mathbf{a}\times(\mathbf{b}+\mathbf{c}) = \mathbf{a}\times\mathbf{b}+\mathbf{a}\times\mathbf{c}, \tag{24}$$

$$(\mathbf{a}\times\mathbf{b})\times\mathbf{c} = (\mathbf{a}\cdot\mathbf{c})\mathbf{b}-(\mathbf{c}\cdot\mathbf{b})\mathbf{a}. \tag{25}$$

It is for the reader himself to verify equations (21) to (25).

By introducing a right-handed frame of reference $Oxyz$, along with the triplet of unit vectors $\mathbf{e}_1$, $\mathbf{e}_2$, and $\mathbf{e}_3$, we endow the space with *orientation*. The ordered triplet of vectors, **a**, **b**, and **c** is said to be *oriented positively* or *negatively* when the triple scalar product $(\mathbf{a}\times\mathbf{b})\cdot\mathbf{c}$, which is equal to the volume of the parallelepiped constructed on these vectors, is correspondingly positive or negative.

The right-handed system will be used consistently throughout the book.

One of the essential concepts in mechanics is the moment of the vector **a** about the point O (Fig. 0.6).

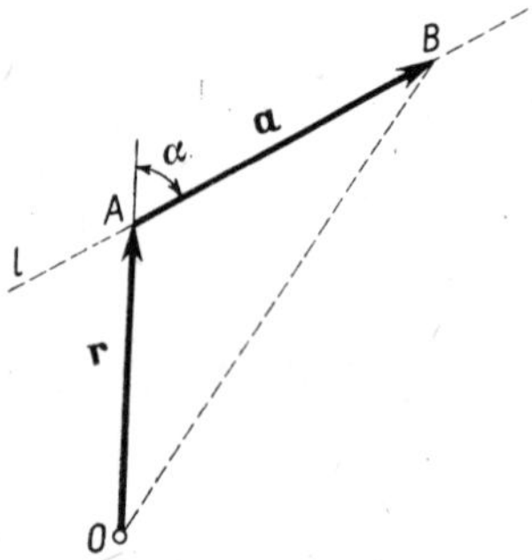

Fig. 0.6

The *moment of the vector* **a**, lying in the line l, is called the vector product

$$\mathbf{M} \overset{\text{df}}{=} \mathbf{r}\times\mathbf{a}. \tag{26}$$

It follows from the definition that equivalent and co-axial vectors have the same moment with respect to the same point in space. For self-evident reasons, the reverse is also true.

We shall now discuss the elements of vector analysis. In mechanics we use vector functions of one or several variables. For any vector function $\mathbf{a} = \mathbf{a}(t)$ in a three-dimensional space there are three corresponding scalar functions $a^1(t)$, $a^2(t)$, and $a^3(t)$. The parameter t usually denotes time in mechanics. There is yet another scalar function associated with the vector function, namely the modulus of the vector **a**:

$$|\mathbf{a}| = \sqrt{[a^1(t)]^2+[a^2(t)]^2+[a^3(t)]^2},$$

which is characterized by the fact that unlike the components $a^1(t)$, $a^2(t)$, and $a^3(t)$ it is independent of the selected frame of reference; it is therefore an invariant.

The relations involved in vector functions are as follows:

$$\lim_{t\to t_0}[\mathbf{a}(t)+\mathbf{b}(t)] = \lim_{t\to t_0}\mathbf{a}(t)+\lim_{t\to t_0}\mathbf{b}(t), \tag{27}$$

$$\lim_{t\to t_0}[\alpha(t)\mathbf{a}(t)] = \lim_{t\to t_0}\alpha(t)\lim_{t\to t_0}\mathbf{a}(t), \tag{28}$$

$$\lim_{t\to t_0}[\mathbf{a}(t)\cdot\mathbf{b}(t)] = \lim_{t\to t_0}\mathbf{a}(t)\cdot\lim_{t\to t_0}\mathbf{b}(t), \tag{29}$$

$$\lim_{t\to t_0}[\mathbf{a}(t)\times\mathbf{b}(t)] = \lim_{t\to t_0}\mathbf{a}(t)\times\lim_{t\to t_0}\mathbf{b}(t). \tag{30}$$

The vector function $\mathbf{a}(t)$ *converges to the limit* $\mathbf{g}$ as $t\to t_0$ if

$$\lim_{t\to t_0}|\mathbf{a}(t)-\mathbf{g}| = 0. \tag{31}$$

This condition can also be written in the following manner:

$$\lim_{t\to t_0}\mathbf{a}(t) = \mathbf{g} \quad \text{or} \quad \mathbf{a}(t)\underset{t\to t_0}{\to}\mathbf{g}.$$

The function $\mathbf{a}(t)$ is limited when its modulus is limited; it is continuous at a point t_0, provided that for any $\varepsilon > 0$ there exists a number $\delta > 0$, where $|t-t_0| < \delta$, such that $|\mathbf{a}(t)-\mathbf{a}(t_0)| < \varepsilon$. Using the symbols of mathematical logic, this condition can be written as follows:

$$\bigwedge_{\varepsilon>0}\bigvee_{\delta>0}|t-t_0| < \delta \Rightarrow |\mathbf{a}(t)-\mathbf{a}(t_0)| < \varepsilon. \tag{32}$$

The symbols $\bigwedge$ and $\bigvee$ are termed *quantificators, general* and *particular.* The clause $\bigwedge_{\varepsilon>0} f(\varepsilon)$ should read: for any $\varepsilon > 0$, the function $f(\varepsilon)$ is satisfied; $\bigvee_{\delta>0} f(\delta)$ implies that there exists such a $\delta > 0$ that satisfies the function $f(\delta)$. The sign $\Rightarrow$ is the implication sign. The clause $a\Rightarrow b$ should be read: if a, then b.

If the clause (32) is valid for any t_0, the vector function $\mathbf{a}(t)$ can be defined in short as continuous.

The *derivative* of the vector function, analogous to the derivative of the scalar function, is defined as the limit of the differential quotient. It will be denoted by $\lim \mathbf{a}(t)$ or $\mathrm{d}\mathbf{a}(t)/\mathrm{d}t$, or in brief $\dot{\mathbf{a}}$ or $\mathrm{d}\mathbf{a}/\mathrm{d}t$:

$$\left.\frac{\mathrm{d}\mathbf{a}}{\mathrm{d}t}\right|_{t=t_0} = \dot{\mathbf{a}}(t_0) = \lim_{t\to t_0} \frac{\mathbf{a}(t)-\mathbf{a}(t_0)}{t-t_0}. \tag{33}$$

The presence of a derivative thus defined means that a given function is differentiable at a point t_0.

The function $\mathbf{a}(t)$ is said to be *differentiable* if a derivative is obtainable for any point t_0.

Starting out with the definition of the derivative of a vector function, we shall find, without difficulty, that the following equalities occur:

$$\frac{\mathrm{d}}{\mathrm{d}t}(\mathbf{a}+\mathbf{b}) = \dot{\mathbf{a}}+\dot{\mathbf{b}}, \tag{34}$$

$$\frac{\mathrm{d}}{\mathrm{d}t}(\alpha\mathbf{a}) = \dot{\alpha}\mathbf{a}+\alpha\cdot\dot{\mathbf{a}}, \tag{35}$$

$$\frac{\mathrm{d}}{\mathrm{d}t}(\mathbf{a}\cdot\mathbf{b}) = \dot{\mathbf{a}}\cdot\mathbf{b}+\mathbf{a}\cdot\dot{\mathbf{b}}, \tag{36}$$

$$\frac{\mathrm{d}}{\mathrm{d}t}(\mathbf{a}\times\mathbf{b}) = \dot{\mathbf{a}}\times\mathbf{b}+\mathbf{a}\times\dot{\mathbf{b}}. \tag{37}$$

The equations (34) to (37) will be used quite frequently in our exposition.

As in the case of a scalar function, derivatives of a higher order than the first will be applicable. They are denoted successively

$$\ddot{\mathbf{a}},\ \dddot{\mathbf{a}},\ \mathbf{a}^{(4)},\ \mathbf{a}^{(5)},\ \ldots,\ \mathbf{a}^{(n)}$$

or correspondingly

$$\frac{\mathrm{d}^2\mathbf{a}}{\mathrm{d}t^2},\ \frac{\mathrm{d}^3\mathbf{a}}{\mathrm{d}t^3},\ \frac{\mathrm{d}^4\mathbf{a}}{\mathrm{d}t^4},\ \frac{\mathrm{d}^5\mathbf{a}}{\mathrm{d}t^5},\ \ldots,\ \frac{\mathrm{d}^n\mathbf{a}}{\mathrm{d}t^n}.$$

It is also true that the function $\mathbf{a}(t)$ is of the C_n class if it has a continuous derivative of the nth order.

CHAPTER 1

Kinematics

The primary concepts in mechanics are space and time. Regarding space, we shall assume that it is Euclidean (plane), therefore an orthogonal frame of reference can be applied to it.

Furthermore, it will be assumed that space is *uniform* and *isotropic*, i.e., there are no distinct loci nor preferred directions in it.

With regard to time, it will be recognized as *uniform*, i.e., preferred instants are non-existent in it.

Some parts of this book deal with problems within the purview of classical mechanics, therefore the following two hypotheses have been accepted:

1. *Time is absolute, i.e. it runs identically in all frames of reference* (*moving and fixed*).

2. *The distance between two arbitrary points in space, irrespective of the frame of reference in which it is measured, is identical* (*it is an invariant*).

Kinematics is a science of motion but it is not concerned with the causes liable to induce or disturb motion.

By *motion* we shall understand changes in time in the position of a body referred to a system treated as stationary.

The *position* of a body or *configuration* is an area of Euclidean space, in which the particles of that body have been mapped one-to-one and continuously. This mapping we shall call *homeomorphism*. Accordingly, motion is the change of mapping in time. Note that the concept of motion is relative, since it depends on the adopted frame of reference.

1.1 Motion of a single particle

We assume a right-handed frame of reference $Oxyz$ in a space (Fig. 1.1), embracing a triplet of unit vectors, $\mathbf{i}$, $\mathbf{j}$, and $\mathbf{k}$, which correspond to x, y, and z. In a space endowed with such orientation we shall consider the motion of a single particle determined by a position vector $\mathbf{r}(t)$. This vector, being the function of time t, can be written as follows:

$$\mathbf{r}(t) = x(t)\mathbf{i}+y(t)\mathbf{j}+z(t)\mathbf{k} \tag{1}$$

or more concisely, using the summation convention,

$$\mathbf{r}(t) = x^i(t)\mathbf{e}_i. \tag{1a}$$

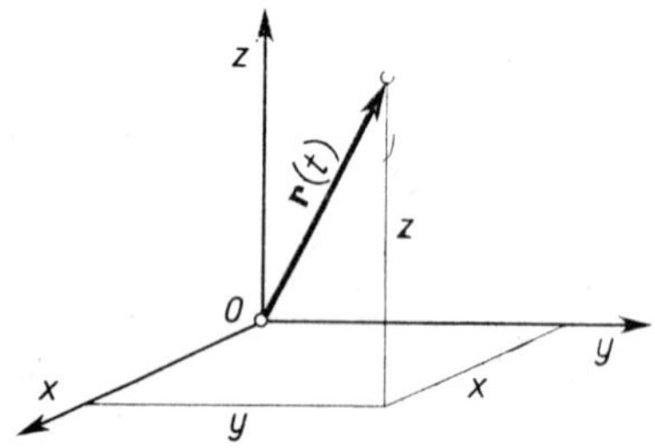

Fig. 1.1

Knowing the functions $x(t)$, $y(t)$, and $z(t)$, we are in fact in possession of all the necessary information about the motion of a particle since we have not only its path but also its speed and acceleration. On a plane surface, the motion will be considered in the Oxy reference frame. The position vector has then only two components

$$x = x(t), \qquad y = y(t). \tag{2}$$

We may often be interested in the equation of the path of motion written as

$$F(x, y) = 0. \tag{2a}$$

To derive (2a) from equations (2), the parameter t should be eliminated from the latter.

Two simple examples will illustrate how we should proceed to derive the equation of the path (2a) from (2).

Example 1. A position vector has the components

$$x = x_0\sin(\omega t+\alpha), \qquad y = y_0\cos(\omega t+\beta). \tag{a}$$

Find from (a) the equation of trajectory (2a).

To this end, the following substitution should be made:

$$\omega t+\alpha = \varphi$$

so that

$$\omega t+\beta = \varphi+\gamma,$$

where

$$\gamma = \beta - \alpha.$$

This type of substitution will greatly facilitate further transformations. Introducing the relationships into (a), we have

$$x = x_0 \sin\varphi, \quad y = y_0 \sin(\varphi + \gamma).$$

Hence

$$\sin\varphi = \frac{x}{x_0}, \quad y = y_0(\sin\varphi\cos\gamma + \cos\varphi\sin\gamma)$$

or, eliminating the angle φ which is dependent on time,

$$y = y_0\left(\frac{x}{x_0}\cos\gamma + \sin\gamma\sqrt{1-\left(\frac{x}{x_0}\right)^2}\right).$$

Transforming further, we obtain

$$\frac{y}{y_0} = \frac{x}{x_0}\cos\gamma + \sin\gamma\sqrt{1-\left(\frac{x}{x_0}\right)^2}.$$

Transferring the first term, on the right-hand side of the equation, to the left and then removing irrationality, we have

$$\left(\frac{y}{y_0} - \frac{x}{x_0}\cos\gamma\right)^2 = \sin^2\gamma\left(1 - \frac{x^2}{x_0^2}\right)$$

or

$$\frac{y^2}{y_0^2} + \frac{x^2}{x_0^2}\cos^2\gamma + \frac{x^2}{x_0^2}\sin^2\gamma - \frac{2xy}{x_0 y_0}\cos\gamma = \sin^2\gamma.$$

Remembering that $\gamma = \beta - \alpha$, the final result will be

$$\frac{y^2}{y_0^2} + \frac{x^2}{x_0^2} - 2\frac{xy}{x_0 y_0}\cos(\beta - \alpha) = \sin^2(\beta - \alpha).$$

Example 2. The system given in Fig. 1.2 consists of two rods of equal length: $OA = AB = l$, connected with an articulated joint at A. The rod OA can rotate about the stationary point O, whereas B moves in a straight horizontal line which is the axis x. Determine the motion of particle P of the rod AB, which is βl distant from A, where $0 \leqslant \beta \leqslant 1$.

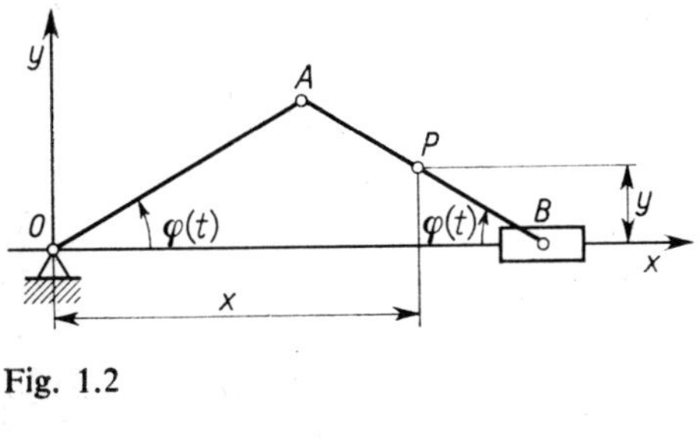

Fig. 1.2

It will be seen from the illustration that

$$x = l\cos\varphi(t)+\beta l\cos\varphi(t) = l(1+\beta)\cos\varphi(t),$$
$$y = l\sin\varphi(t)-\beta l\sin\varphi(t) = l(1-\beta)\sin\varphi(t).$$

Hence

$$\cos\varphi(t) = \frac{x}{l(1+\beta)}, \qquad \sin\varphi(t) = \frac{y}{l(1-\beta)}.$$

But $\cos^2\varphi(t)+\sin^2\varphi(t) = 1$. Therefore

$$\frac{x^2}{l^2(1+\beta)^2}+\frac{y^2}{l^2(1-\beta)^2} = 1.$$

Thus, we find that the particle P will be moving along an elliptic path.

Let us now go back to the concept of velocity. Consider two points A and B of a trajectory (Fig. 1.3). A position vector $\mathbf{r}(t_A)$ is assigned to A and a vector $\mathbf{r}(t_B)$ to B. The *velocity* is the limit of the following differential quotient:

$$\mathbf{v} \stackrel{\mathrm{df}}{=} \lim_{t_B \to t_A} \frac{\mathbf{r}(t_B)-\mathbf{r}(t_A)}{t_B-t_A} = \frac{\mathrm{d}\mathbf{r}}{\mathrm{d}t}. \qquad (3)$$

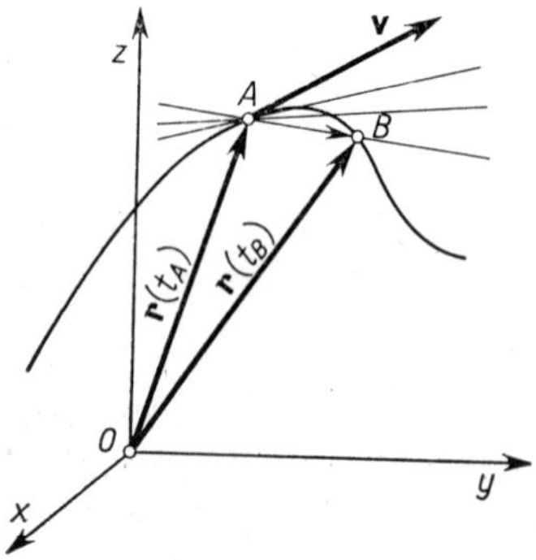

Fig. 1.3

As B approaches A, the chord determined by the two points approximates to the tangent. The velocity $\mathbf{v}$, it will be seen, is a vector contiguous to the trace.

Differentiating (1) we obtain the velocity vector

$$\mathbf{v} = v_x\mathbf{i}+v_y\mathbf{j}+v_z\mathbf{k}, \qquad (4)$$

where

$$v_x = \frac{dx(t)}{dt} = \dot{x}(t), \quad v_y = \frac{dy(t)}{dt} = \dot{y}(t), \quad v_z = \frac{dz(t)}{dt} = \dot{z}(t) \quad (5)$$

or using the summation convention

$$\mathbf{v} = \dot{x}^i(t)\cdot\mathbf{e}_i. \quad (4a)$$

If $\mathbf{v}(t)$ is treated as a position vector, its end will be moving in a trace called the *hodograph*.

Example 3. Determine the amplitude x_0 of the free vibration frequency of a particle if it is known that with a deflection x_1 it had velocity v_1 and with a deflection x_2, velocity v_2 (Fig. 1.4).

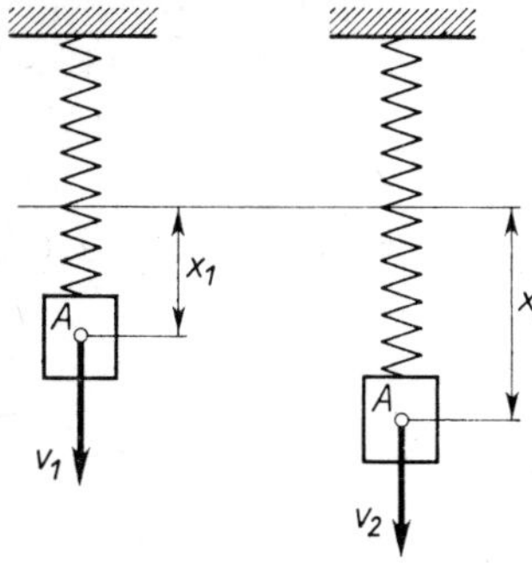

Fig. 1.4

First of all the relationship between the instantaneous position x and the instantaneous velocity $\mathbf{v} = \dot{x}$ should be found. The considered motion is rectilinear:

$$x = x_0 \sin\omega t. \quad (a)$$

Differentiating (a) with respect to time, we have

$$\dot{x} = x_0\,\omega\cos\omega t = x_0\,\omega\sqrt{1-\sin^2\omega t}. \quad (b)$$

calculating $\sin\omega t$ from (a) and then substituting into (b), we find the position-velocity relationship:

$$\dot{x} = x_0\,\omega\sqrt{1-\frac{x^2}{x_0^2}}. \quad (c)$$

We have thus obtained

$$v_1 = x_0\,\omega\sqrt{1-\frac{x_1^2}{x_0^2}}, \quad v_2 = x_0\,\omega\sqrt{1-\frac{x_2^2}{x_0^2}}. \quad (d)$$

Solving equations (d) for x_0 and ω, we have

$$x_0 = \sqrt{\frac{v_1^2 x_2^2 - v_2^2 x_1^2}{v_1^2 - v_2^2}}, \quad \omega = \sqrt{\frac{v_1^2 - v_2^2}{x_2^2 - x_1^2}}. \quad (e)$$

Example 4. Calculate the velocity of the particle P of the connecting-rod described in Example 2 if the rod AO moves according to the equation $\varphi(t) = \omega t$, where $\omega = \text{const}$.

The equations of motion are as follows:

$$x = l(1+\beta)\cos\omega t, \quad y = l(1-\beta)\sin\omega t. \tag{a}$$

Differentiating (a), we obtain

$$v_x = -l(1+\beta)\,\omega\sin\omega t, \quad v_y = l(1-\beta)\,\omega\cos\omega t.$$

The absolute value will be

$$v = \sqrt{v_x^2+v_y^2} = \sqrt{1+\beta^2-2\beta\cos 2\omega t}\, l\omega.$$

In giving the definition of the velocity vector we have used the notion of position vector $\mathbf{r}$. If vector $\mathbf{v}$ is treated as a position vector, its end will trace a hodograph. The velocity with which the end of vector $\mathbf{v}$ is moving in the hodograph is called *acceleration*; it is a vector contiguous to the hodograph (Fig. 1.5). Acceleration is defined as

$$\mathbf{a} \overset{\mathrm{df}}{=} \frac{\mathrm{d}\mathbf{v}}{\mathrm{d}t} = \frac{\mathrm{d}^2\mathbf{r}}{\mathrm{d}t^2}. \tag{6}$$

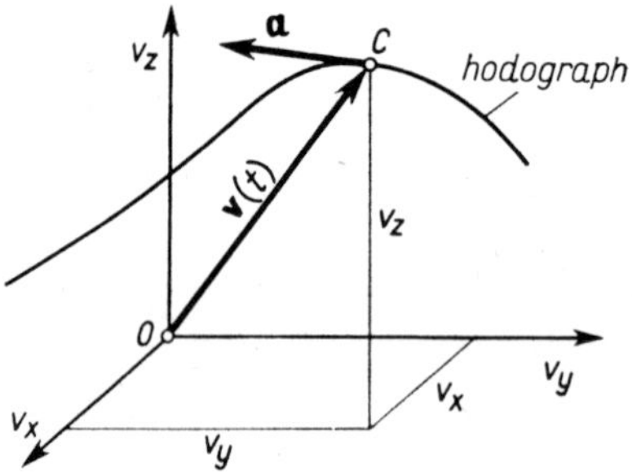

Fig. 1.5

The acceleration vector can be represented as follows:

$$\mathbf{a} = a_x\mathbf{i}+a_y\mathbf{j}+a_z\mathbf{k}, \tag{7}$$

where

$$a_x = \frac{\mathrm{d}v_x}{\mathrm{d}t} = \frac{\mathrm{d}^2x}{\mathrm{d}t^2}, \quad a_y = \frac{\mathrm{d}v_y}{\mathrm{d}t} = \frac{\mathrm{d}^2y}{\mathrm{d}t^2}, \quad a_z = \frac{\mathrm{d}v_z}{\mathrm{d}t} = \frac{\mathrm{d}^2z}{\mathrm{d}t^2} \tag{8}$$

or writing the same using the summation convention

$$\mathbf{a} = \ddot{x}^i\mathbf{e}_i = a^i\mathbf{e}_i. \tag{7a}$$

In general, vectors **a** and **v** are not parallel, consequently the acceleration vector has two components, one contiguous to the trajectory, $\mathbf{a}^t$, and the other normal to it, $\mathbf{a}^n$ (Fig. 1.6). So we can write

$$\mathbf{a} = \mathbf{a}^t + \mathbf{a}^n. \tag{9}$$

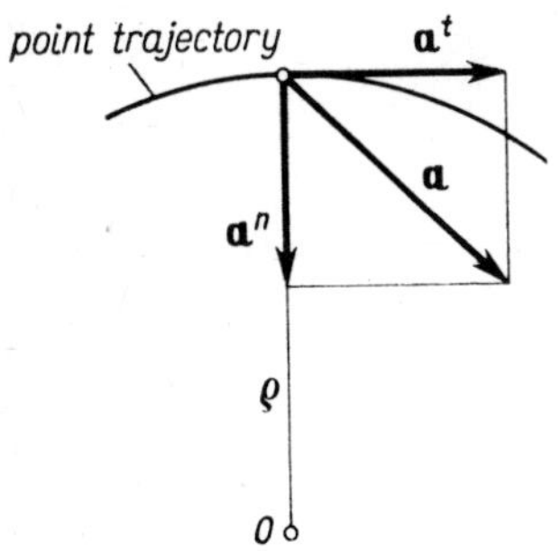

Fig. 1.6

We shall now find these two components. To this end, the unit vector $\boldsymbol{\tau}$ is subordinated to the direction of the tangential acceleration and the unit vector **n** to the direction of the normal acceleration. Both the vectors lie on a plane strictly tangential to the trajectory. The velocity **v** can be expressed as follows:

$$\mathbf{v} = v\boldsymbol{\tau}. \tag{10}$$

Differentiating (10), we obtain

$$\mathbf{a} = \frac{\mathrm{d}\mathbf{v}}{\mathrm{d}t} = \frac{\mathrm{d}v}{\mathrm{d}t}\boldsymbol{\tau} + v\frac{\mathrm{d}\boldsymbol{\tau}}{\mathrm{d}t}. \tag{11}$$

Now

$$\frac{\mathrm{d}\boldsymbol{\tau}}{\mathrm{d}t} = \lim_{\Delta t \to 0} \frac{\Delta\boldsymbol{\tau}}{\Delta t}.$$

But by virtue of the similarity of $\triangle AOB$ and $\triangle CBD$ (Fig. 1.7),

$$\Delta\boldsymbol{\tau} = \frac{AB}{\varrho} \quad \text{since} \quad \boldsymbol{\tau} = 1$$

where ϱ is the radius of curvature of the trajectory of the particle. Regarding the limit, we can write

$$\lim_{\Delta t \to 0} \frac{\Delta\boldsymbol{\tau}}{\Delta t} = \lim_{\Delta t \to 0} \mathbf{n}' \frac{AB}{\varrho \Delta t} = \mathbf{n}\frac{v}{\varrho}.$$

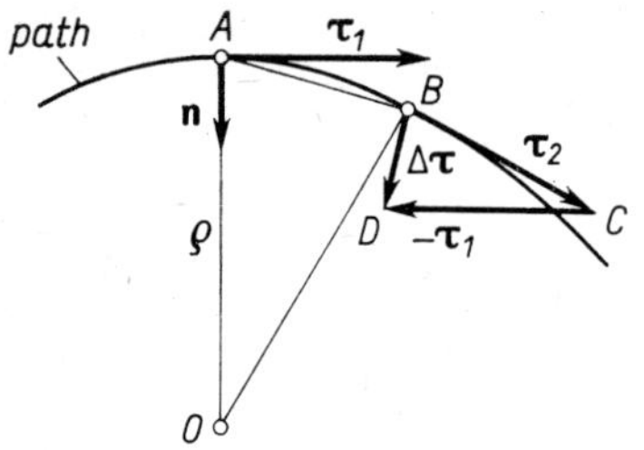

Fig. 1.7

Hence

$$\mathbf{a} = \frac{d\mathbf{v}}{dt} = \frac{dv}{dt}\boldsymbol{\tau} + \frac{v^2}{\varrho}\mathbf{n}, \tag{12}$$

Example 5. A particle P is moving in a circular path of diameter $d = 100$ cm, in such a way that the position vector $\mathbf{r}$ rotates at constant angular velocity $\omega = \pi \mathrm{s}^{-1}$.

The origin of the vector $\mathbf{r}$ is one of the points on the path (Fig. 1.8). Calculate the velocity and acceleration of P.

We begin by establishing the position vector $\mathbf{r}$

$$\mathbf{r} = x\mathbf{i} + y\mathbf{j},$$

where

$$x = r\cos\omega t, \qquad y = r\sin\omega t.$$

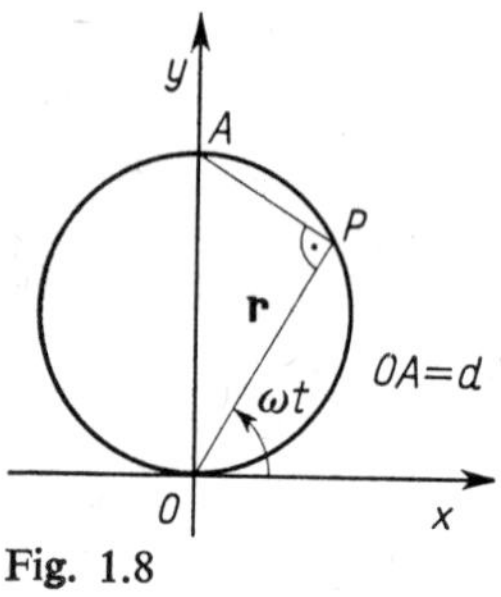

Fig. 1.8

As we can easily see, $\triangle OPA$ (Fig. 1.8) is a right-angled triangle, therefore

$$r = d\sin\omega t.$$

We can thus write

$$x = d\sin\omega t\cos\omega t, \qquad y = d\sin^2\omega t.$$

Differentiating, we obtain the velocity components

$$v_x = d(\omega\cos^2\omega t - \omega\sin^2\omega t) = d\omega\cos 2\omega t = 100\pi\cos 2\pi t,$$

$$v_y = 2d\omega\cos\omega t\sin\omega t = d\omega\sin 2\omega t = 100\pi\sin 2\pi t.$$

Differentiating again, we get the acceleration:

$$a_x = -2d\omega^2 \sin 2\omega t = -200\pi^2 \sin 2\pi t,$$
$$a_y = 2d\omega^2 \cos 2\omega t = 200\pi^2 \cos 2\pi t.$$

Example 6. A particle is moving along the arc of a circle of radius $\varrho = 50$ cm, in accordance with the equation $s = 40 \sin \frac{1}{2} \pi t$. Specify the tangential and normal acceleration of this particle. Tangential acceleration

$$a^t = \frac{dv}{dt} = \frac{d}{dt}\left(\frac{ds}{dt}\right) = -10\pi^2 \sin \tfrac{1}{2} \pi t.$$

Normal acceleration

$$a^n = \frac{v^2}{\varrho} = \frac{1}{\varrho}\left(\frac{ds}{dt}\right)^2 = \frac{400\pi^2}{50} \cos^2 \frac{\pi}{2} t = 8\pi^2 \cos^2 \frac{\pi}{2} t.$$

Total acceleration can be derived from

$$a = \sqrt{(a^t)^2 + (a^n)^2} = \sqrt{100\pi^4 \sin^2 \tfrac{1}{2} \pi t + 64\pi^4 \cos^4 \tfrac{1}{2} \pi t}.$$

Example 7. A particle is moving along a helical curve in accordance with

$$x = r\cos\omega t, \quad y = r\sin\omega t, \quad z = v_z t.$$

Calculate the radius of curvature of the helical path.

Remembering that normal acceleration is expressed by

$$a^n = \frac{v^2}{\varrho},$$

it is possible, knowing $\mathbf{a}^n$ and $\mathbf{v}$, to find the radius of curvature ϱ:

$$\varrho = \frac{v^2}{a^n}.$$

The velocity components are expressed by

$$v_x = -r\omega \sin\omega t, \quad v_y = r\omega \cos\omega t, \quad v_z = v_z,$$

and the square of the velocity is:

$$v^2 = v_x^2 + v_y^2 + v_z^2 = r^2\omega^2 + v_z^2.$$

Since $\mathbf{v}$ = const, the tangential acceleration is

$$a^t = 0.$$

Differentiating the equations of motion once more, we find the components of acceleration

$$a_x = -r\omega^2 \cos \omega t, \quad a_y = -\omega^2 r \sin \omega t, \quad a_z = 0.$$

Hence

$$a = \sqrt{a_x^2 + a_y^2 + a_z^2} = r\omega^2.$$

The specified acceleration is a normal acceleration because $a^t = 0$. Therefore

$$\varrho = \frac{r^2\omega^2 + v_z^2}{r\omega^2} = r + \frac{v_z^2}{r\omega^2}.$$

Example 8. Derive the equation for the curvature of a two-dimensional curve, $y = y(x)$.

We should first make it clear that the curvature is equal to the ratio of normal acceleration a^n to the square of the velocity of a point moving along the curve under consideration. Therefore,

$$\frac{1}{\varrho} = \frac{a^n}{v^2}. \tag{a}$$

We can readily see that

$$a^n = |\mathbf{v}\times\mathbf{a}|\frac{1}{v} = \frac{1}{v}(a_y v_x - a_x v_y) = \frac{\ddot{y}\dot{x}-\ddot{x}\dot{y}}{\sqrt{\dot{x}^2+\dot{y}^2}}. \tag{b}$$

Substituting (b) into (a), we have

$$\frac{1}{\varrho} = \frac{\ddot{y}\dot{x}-\ddot{x}\dot{y}}{(\dot{x}^2+\dot{y}^2)^{3/2}} = \frac{\ddot{y}\dot{x}-\ddot{x}\dot{y}}{\dot{x}^3}\,\frac{1}{(1+\dot{y}^2/\dot{x}^2)^{3/2}}. \tag{c}$$

Note that

$$y' = \frac{dy}{dx} = \frac{dy}{dt}\frac{dt}{dx} = \frac{dy/dt}{dx/dt} = \frac{\dot{y}}{\dot{x}}, \tag{d}$$

and

$$y'' = \frac{d^2y}{dx^2} = \frac{d}{dx}\left(\frac{\dot{y}}{\dot{x}}\right) = \frac{d}{dt}\left(\frac{\dot{y}}{\dot{x}}\right)\frac{1}{\dot{x}} = \frac{\ddot{y}\dot{x}-\ddot{x}\dot{y}}{\dot{x}^3}. \tag{e}$$

Considering (d) and (e) in (c), we obtain

$$\frac{1}{\varrho} = \frac{y''}{(1+y'^2)^{3/2}}. \tag{f}$$

Equation (f) will be applicable in the theory of simple beam-bending.

Let us now examine the motion of a particle in curvilinear coordinates associated with orthogonal coordinates in the following relationship:

$$\mathbf{r} = \mathbf{r}(q^1, q^2, q^3). \tag{13}$$

We shall call the coordinates q^i *generalized coordinates.* Solving (13) with respect to the coordinates q^1, q^2, and q^3, we have

$$q^1 = f^1(x, y, z), \quad q^2 = f^2(x, y, z), \quad q^3 = f^3(x, y, z). \tag{14}$$

The function $q^1 = \text{const}$ represents a surface on which only q^2 and q^3 change. Fixing the value of two of the three curvilinear coordinates, we obtain the line of changes in the third one. Differentiating (13), we obtain the relationship between velocity in the old and the new system:

$$\mathbf{v} = \frac{\partial\mathbf{r}}{\partial q^1}\dot{q}^1 + \frac{\partial\mathbf{r}}{\partial q^2}\dot{q}^2 + \frac{\partial\mathbf{r}}{\partial q^3}\dot{q}^3. \tag{15}$$

Using the summation convention, we can write (15) as

$$\mathbf{v} = \frac{\partial \mathbf{r}}{\partial q^i}\dot{q}^i. \tag{16}$$

The vector equation (15) is equivalent to three scalar equations:

$$v_x = \dot{x} = \frac{\partial x}{\partial q^i}\dot{q}^i, \quad v_y = \dot{y} = \frac{\partial y}{\partial q^i}\dot{q}^i, \quad v_z = \dot{z} = \frac{\partial z}{\partial q^i}\dot{q}^i. \tag{17}$$

It is more convenient to introduce the following notations:

$$x \equiv x^1, \quad y \equiv x^2, \quad z \equiv x^3,$$

since then the relationship (15) can be expressed in one equation:

$$\dot{x}^k = \frac{\partial x^k}{\partial q^i}\dot{q}^i. \tag{18}$$

Denoting the derivatives $\partial x^k/\partial q^i$ by A_i^k, we have

$$\dot{x}^k = A_i^k \dot{q}^i. \tag{19}$$

If the matrix A is not singular, then equation (19) can be solved in relation to the generalized velocities $\dot{q}^i$:

$$\dot{q}^i = B_k^i \dot{x}^k, \tag{20}$$

where the matrix $\boldsymbol{B}$ is inverse of $\boldsymbol{A}$.

Example 9. Introducing a polar system (Fig. 1.9) in the plane of motion, find the velocity components $\dot{r}$ and $\dot{\varphi}$. From the drawing we have the relations

$$x = r\cos\varphi, \quad y = r\sin\varphi, \quad z = 0. \tag{a}$$

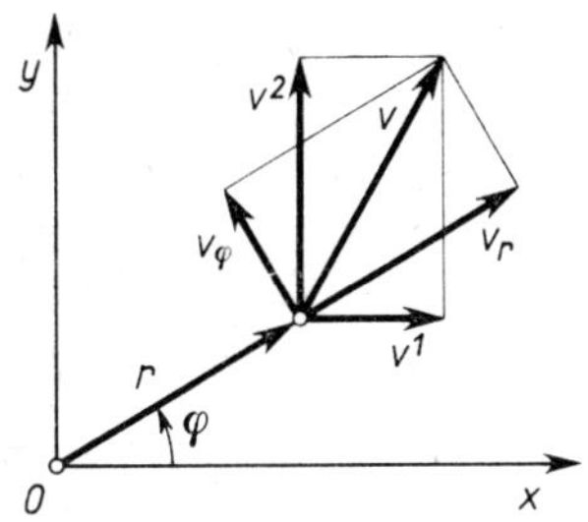

Fig. 1.9

These are equations (13). The system (19) for this special case will take the following form:

$$\dot{x}^1 = \frac{\partial x}{\partial r}\dot{r} + \frac{\partial x}{\partial \varphi}\dot{\varphi}, \quad \dot{x}^2 = \frac{\partial y}{\partial r}\dot{r} + \frac{\partial y}{\partial \varphi}\dot{\varphi}. \tag{b}$$

We identify matrix A as

$$A = \begin{bmatrix} \dfrac{\partial x}{\partial r} & \dfrac{\partial x}{\partial \varphi} \\ \dfrac{\partial y}{\partial r} & \dfrac{\partial y}{\partial \varphi} \end{bmatrix} = \begin{bmatrix} \cos\varphi & -r\sin\varphi \\ \sin\varphi & r\cos\varphi \end{bmatrix}$$

and matrix B as

$$B = \begin{bmatrix} \cos\varphi & \sin\varphi \\ -\dfrac{1}{r}\sin\varphi & \dfrac{1}{r}\cos\varphi \end{bmatrix};$$

therefore

$$\dot{r} = v^1\cos\varphi + v^2\sin\varphi, \quad \dot{\varphi} = \frac{1}{r}(-v^1\sin\varphi + v^2\cos\varphi). \tag{c}$$

The equations obtained have evident geometrical interpretation. The generalized velocity $\dot{r} = v_r$ has the direction of the radius vector r and for that reason it is called *radial velocity*. The product $\dot{\varphi}r = v_\varphi$ is a transverse component and is oriented perpendicularly to the direction of v_r.

Both components are shown in Fig. 1.9, the radial v_r which has the direction of the radius-vector r, and the transverse v_φ at right angles to it.

Note further that the generalized velocities $\dot{r}$ and $\dot{\varphi}$ lack dimensional compatibility, being a natural consequence of introducing new coordinates. The usefulness of the formulae derived here will be demonstrated in the following example.

Example 10. Find the equation for the path of a particle in polar coordinates (r, φ) if the ratio of radial to transverse velocity is known to be a function of angle φ, which is $\alpha = \alpha(\varphi)$, and initial conditions are: $r = r_0$ and $\varphi = 0$.

First of all, we write the conditions of the problem in analytical form:

$$\frac{v_r}{v_\varphi} = \alpha(\varphi). \tag{a}$$

Using $\dot{r} = v_r$ and $\dot{\varphi}r = v_\varphi$ from the previous example, we can rewrite (a) as

$$\frac{\dot{r}}{r\dot{\varphi}} = \alpha(\varphi). \tag{b}$$

Hence

$$\frac{dr}{r} = \alpha(\varphi)d\varphi. \tag{c}$$

Integrating, we get

$$\int_{r_0}^{r} \frac{dr}{r} = \int_{0}^{\varphi} \alpha(\varphi)d\varphi$$

or after performing the necessary operations and elementary transformations

$$r = r_0 \exp\left(\int_0^{\varphi} \alpha(\varphi) d\varphi\right). \tag{d}$$

The question may now be asked: How on the basis of the given components of the generalized velocities $\dot{q}^i$ can the value of the modulus $|\mathbf{v}| = v$ be determined? We know that

$$v^2 = \delta_{ik} \dot{x}^i \dot{x}^k, \tag{21}$$

but

$$\dot{x}^i = \frac{\partial x^i}{\partial q^l} \dot{q}^l,$$

therefore

$$v^2 = \delta_{ik} \frac{\partial x^i}{\partial q^l} \frac{\partial x^k}{\partial q^r} \dot{q}^l \dot{q}^r. \tag{22}$$

Introducing

$$g_{lr} = \delta_{ik} \frac{\partial x^i}{\partial q^l} \frac{\partial x^k}{\partial q^r}, \tag{23}$$

we now have

$$v^2 = g_{lr} \dot{q}^l \dot{q}^r. \tag{24}$$

Knowing generalized velocities $\dot{r}$ and $\dot{\varphi}$ from Example 9, we can calculate the value of velocity of the particle. From the relationship (a) in this example we have

$$g_{11} = \frac{\partial x}{\partial r} \frac{\partial x}{\partial r} + \frac{\partial y}{\partial r} \frac{\partial y}{\partial r} = \cos^2 \varphi + \sin^2 \varphi = 1,$$

$$g_{12} = \frac{\partial x}{\partial r} \frac{\partial x}{\partial \varphi} + \frac{\partial y}{\partial r} \frac{\partial y}{\partial \varphi} = -r \cos \varphi \sin \varphi + r \sin \varphi \cos \varphi = 0,$$

$$g_{21} = g_{12} = 0,$$

$$g_{22} = \frac{\partial x}{\partial \varphi} \frac{\partial x}{\partial \varphi} + \frac{\partial y}{\partial \varphi} \frac{\partial y}{\partial \varphi} = r^2 \sin^2 \varphi + r^2 \cos^2 \varphi = r^2,$$

therefore

$$v^2 = g_{11} \dot{q}^1 \dot{q}^1 + 2 g_{12} \dot{q}^1 \dot{q}^2 + g_{22} \dot{q}^2 \dot{q}^2.$$

Since

$$\dot{q}^1 \equiv \dot{r} \quad \text{and} \quad \dot{q}^2 \equiv \dot{\varphi},$$

we have

$$v^2 = g_{11}\dot{r}^2 + 2g_{12}\dot{r}\dot{\varphi} + g_{22}\dot{\varphi}^2 = \dot{r}^2 + r^2\dot{\varphi}^2.$$

It follows that the square of the velocity is equal to the sum of the squares of the velocities $v_r^2 + v_\varphi^2$, which results from the orthogonal nature of the system (r, φ).

We shall regard a given coordinate system as orthogonal when the matrix g_{ik} is diagonal.

As we have already noted, generalized velocities usually lack compatible dimensions. This is due to the fact that the transformation rule (20) cannot ensure compatibility on account of the different denomination of generalized coordinates. A coordinate of this type can be not only the length of arc but equally well an angle or a ratio of certain characteristic magnitudes giving the position of a point in space. In order to rid oneself of this inconvenience, the velocity $\mathbf{v}$ should be projected on the directions of the generalized coordinates at a given point.

This requirement does not present any particular difficulty, considering that a unit vector having that direction can be defined in the following manner:

$$\mathbf{e}_j = \frac{1}{\left|\dfrac{\partial \mathbf{r}}{\partial q^j}\right|}\frac{\partial \mathbf{r}}{\partial q^j}, \tag{25}$$

where

$$\left|\frac{\partial \mathbf{r}}{\partial q^j}\right| = \sqrt{\sum_{i=1}^{3}\left(\frac{\partial x^i}{\partial q^j}\right)^2}. \tag{25a}$$

Thus the component velocities of these new coordinates are

$$v_j = \mathbf{v}\cdot\mathbf{e}_j. \tag{26}$$

All v_j components of vector $\mathbf{v}$ have compatible dimensions here.

Let us now consider the question of accelerations in curvilinear systems. The acceleration components in the new generalized coordinates will be derived by scalar multiplication of the acceleration by the unit vector $\mathbf{e}_j$ as described by equation (25). Therefore

$$a_j = \mathbf{a}\cdot\frac{\partial \mathbf{r}}{\partial q^j}\frac{1}{\left|\dfrac{\partial \mathbf{r}}{\partial q^j}\right|}. \tag{27}$$

We shall transform the right-hand expression to an invariable form with respect to the transformation. Note for this purpose that

$$\mathbf{a}\cdot\frac{\partial \mathbf{r}}{\partial q^j} = \dot{\mathbf{v}}\cdot\frac{\partial \mathbf{r}}{\partial q^j} = \frac{\mathrm{d}}{\mathrm{d}t}\left(\mathbf{v}\cdot\frac{\partial \mathbf{r}}{\partial q^j}\right) - \mathbf{v}\cdot\frac{\mathrm{d}}{\mathrm{d}t}\left(\frac{\partial \mathbf{r}}{\partial q^j}\right). \tag{28}$$

But

$$\dot{\mathbf{r}} = \frac{\partial \mathbf{r}}{\partial q^r}\dot{q}^r,$$

therefore

$$\frac{\partial \dot{\mathbf{r}}}{\partial \dot{q}^j} = \frac{\partial \mathbf{r}}{\partial q^j}. \tag{29}$$

Next

$$\frac{\mathrm{d}}{\mathrm{d}t}\left(\frac{\partial \mathbf{r}}{\partial q^j}\right) = \frac{\partial^2 \mathbf{r}}{\partial q^j \partial q^r}\dot{q}^r,$$

also

$$\frac{\partial \dot{\mathbf{r}}}{\partial q^j} = \frac{\partial^2 \mathbf{r}}{\partial q^j \partial q^r}\dot{q}^r;$$

hence

$$\frac{\mathrm{d}}{\mathrm{d}t}\left(\frac{\partial \mathbf{r}}{\partial q^j}\right) = \frac{\partial \dot{\mathbf{r}}}{\partial q^j}. \tag{30}$$

Substituting (29) and (30) into (28), we obtain

$$\begin{aligned} a_j &= \left[\frac{\mathrm{d}}{\mathrm{d}t}\left(\mathbf{v}\cdot\frac{\partial \mathbf{v}}{\partial \dot{q}^j}\right) - \mathbf{v}\cdot\frac{\partial \mathbf{v}}{\partial q^j}\right]\frac{1}{\left|\dfrac{\partial \mathbf{r}}{\partial q^j}\right|} \\ &= \left\{\frac{\mathrm{d}}{\mathrm{d}t}\left[\frac{\partial}{\partial \dot{q}^j}\left(\frac{1}{2}v^2\right)\right] - \frac{\partial}{\partial q^j}\left(\frac{1}{2}v^2\right)\right\}\frac{1}{\left|\dfrac{\partial \mathbf{r}}{\partial q^j}\right|}. \end{aligned} \tag{31}$$

Equation (31) constitutes the required invariant form for the determination of the components of the generalized accelerations.

As an example, we take a cylindrical system where

$$x = r\cos\varphi, \quad y = r\sin\varphi, \quad z = z,$$

$(q^1 \equiv r,\ q^2 \equiv \varphi,\ q^3 \equiv z)$ and

$$v_r = \dot{r}, \quad v_\varphi = r\dot{\varphi}, \quad v_z = \dot{z};$$

we have

$$v^2 = \dot{r}^2 + r^2\dot{\varphi}^2 + \dot{z}^2.$$

Using equation (31),

$$a_r = \ddot{r} - r\dot{\varphi}^2, \quad a_\varphi = 2\dot{r}\dot{\varphi} + r\ddot{\varphi}, \quad a_z = \ddot{z}.$$

Correspondingly in a spherical system (Fig. 1.10) where

$$x = r\sin\theta\cos\varphi, \quad y = r\sin\theta\sin\varphi, \quad z = r\cos\theta$$

$(q^1 \equiv r,\ q^2 \equiv \theta,\ q^3 \equiv \varphi)$, there is

$$v_r = \dot{r}, \quad v_\theta = r\dot{\theta}, \quad v_\varphi = r\dot{\varphi}\sin\theta;$$

hence

$$v^2 = \dot{r}^2 + r^2\dot{\theta}^2 + r^2\dot{\varphi}^2\sin^2\theta.$$

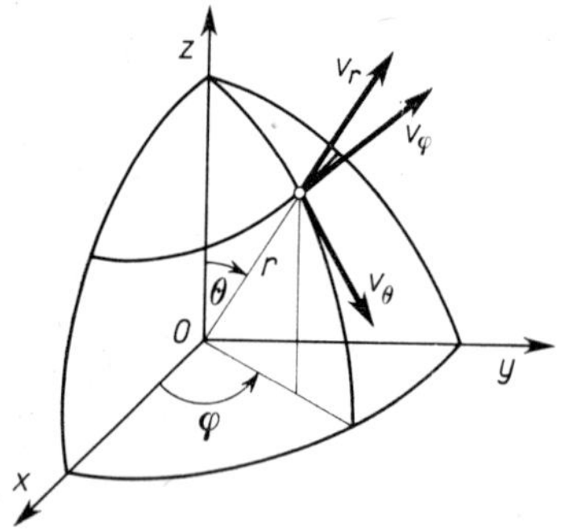

Fig. 1.10

The acceleration components are therefore equal to

$$a_r = \ddot{r} - r\dot{\theta}^2 - r\dot{\varphi}^2\sin^2\theta,$$

$$a_\theta = \frac{1}{r}\left[\frac{\mathrm{d}}{\mathrm{d}t}(r^2\dot{\theta}) - r^2\dot{\varphi}^2\sin\theta\cos\theta\right],$$

$$a_\varphi = \frac{1}{r\sin\theta}\frac{\mathrm{d}}{\mathrm{d}t}(r^2\dot{\varphi}\sin^2\theta).$$

1.2 Description of the motion of a rigid body

The discussion up to now has been concerned with the kinematics of a particle. In describing the motion of a rigid body we shall deal with an infinite set of particles. These will fill the entire space. Thus, a stated velocity and acceleration will correspond to any point in space. We assume that the functions describing these vector fields are differentiable with respect to space and time variables.

Let

$$v^i = v^i(x^1, x^2, x^3, t) \tag{1}$$

denote component of velocity having the direction of the axis x^i at an instant t. Since the function (1) is differentiable, the variation, neglecting time variation, will be

$$\delta v^i = \frac{\partial v^i}{\partial x^k}\,\delta x^k.* \tag{2}$$

The increase in velocity through a displacement $\delta \mathbf{x}$ at the given instant t is therefore the product of the matrix

$$\boldsymbol{T} = \begin{bmatrix} \dfrac{\partial v^1}{\partial x^1} & \dfrac{\partial v^1}{\partial x^2} & \dfrac{\partial v^1}{\partial x^3} \\ \dfrac{\partial v^2}{\partial x^1} & \dfrac{\partial v^2}{\partial x^2} & \dfrac{\partial v^2}{\partial x^3} \\ \dfrac{\partial v^3}{\partial x^1} & \dfrac{\partial v^3}{\partial x^2} & \dfrac{\partial v^3}{\partial x^3} \end{bmatrix}$$

and the vector

$$\delta \mathbf{x} = \begin{bmatrix} \delta x^1 \\ \delta x^2 \\ \delta x^3 \end{bmatrix},$$

which in matrix notation is expressed by

$$\delta \mathbf{v} = \boldsymbol{T}\delta \mathbf{x}. \tag{2a}$$

The matrix $\boldsymbol{T}$ is called the *matrix of velocity gradient*. Since the elements of the matrix are of tensor nature, it will subsequently be referred to as the *tensor of velocity gradient*.

Basic information concerning the tensor calculus will be given in Chapter 7 with reference to configurational space. The knowledge of the elements of this theory is not required for the present. All we need to know is the summation convention.

---

* Readers unfamiliar with the variation calculus should see references [30, 39, 41, 47].

Since we are dealing here with objects in the cartesian coordinate system, the indices can be written discretionally as superscripts or subscripts; this does not alter the character of a given magnitude in any way. For example:

$$T^i_k = T_{ik} \quad \text{(though formally it should be } T^i_k\,\delta_{ir} = T_{kr}),$$
$$v^i = v_i \quad \text{(in this case } v^i = \delta^{ir} v_r).$$

We can break up $\boldsymbol{T}$ into symmetrical and antisymmetrical parts, according to the following equation:

$$T_{ik} = T_{(ik)} + T_{[ik]}, \tag{3}$$

where

$$T_{(ik)} = \frac{1}{2}\left(\frac{\partial v_i}{\partial x^k} + \frac{\partial v_k}{\partial x^i}\right), \tag{3a}$$

$$T_{[ik]} = \frac{1}{2}\left(\frac{\partial v_i}{\partial x^k} - \frac{\partial v_k}{\partial x^i}\right). \tag{3b}$$

Assuming that we are dealing with a perfectly rigid body, i.e., such that it is not deformable, we have

$$\delta\mathbf{v}\cdot\delta\mathbf{r} = 0. \tag{4}$$

In this way, the possibility of an increase in velocity in the direction of a near neighbouring point is excluded, admitting only the angular motion of the vector

$$\delta\mathbf{r} = \delta x^1\mathbf{i} + \delta x^2\mathbf{j} + \delta x^3\mathbf{k}.$$

Thus for a rigid body only such velocities that do not contradict equation (4) are possible. Writing them in the summation convention, with the use of (2), we obtain

$$\frac{\partial v_i}{\partial x^k}\,\delta x^k \delta x^i = 0 \tag{4a}$$

or

$$T_{ik}\,\delta x^k \delta x^i = 0. \tag{4b}$$

Considering (3), we have

$$T_{(ik)}\,\delta x^i \delta x^k + T_{[ik]}\,\delta x^i \delta x^k = 0. \tag{4c}$$

Note that

$$T_{[ik]}\delta x^i \delta x^k \equiv 0 \quad (T_{[ik]} = -T_{[ki]}).$$

Therefore, the condition (4) is satisfied when all the elements of the symmetric part of matrix $\boldsymbol{T}$ are absent,

$$T_{(ik)} = 0. \tag{5}$$

Thus, the possible vector field of velocity for a rigid body is a field such that the tensor of velocity gradient is antisymmetrical. The residual symmetric part occurs in the general case where the body is subjected to strain, and for this reason it has been named the *velocity tensor of strain*

$$T_{(ik)} = \dot{\varepsilon}_{ik} = \frac{1}{2}\left(\frac{\partial v_i}{\partial x^k} + \frac{\partial v_k}{\partial x^i}\right). \tag{6}$$

The aim we set ourselves now is to find the vector field of velocity associated with the motion of a rigid body. The problem amounts to solving a system of six independent partial differential equations consequent upon condition (5):

$$\begin{gathered}
\frac{\partial v_1}{\partial x^1} = 0, \qquad \frac{\partial v_2}{\partial x^2} = 0, \qquad \frac{\partial v_3}{\partial x^3} = 0, \\
\frac{\partial v_1}{\partial x^2} = -\frac{\partial v_2}{\partial x^1}, \qquad \frac{\partial v_2}{\partial x^3} = -\frac{\partial v_3}{\partial x^2}, \qquad \frac{\partial v_3}{\partial x^1} = -\frac{\partial v_1}{\partial x^3}.
\end{gathered} \tag{7}$$

From the first three equations of the system we infer that

$$v_1 = v_1(x^2, x^3, t), \qquad v_2 = v_2(x^3, x^1, t), \qquad v_3 = v_3(x^1, x^2, t), \tag{8}$$

which can be shortened to

$$v_i = v_i(x^j, x^k, t) \qquad (i \neq j,\ j \neq k,\ k \neq i). \tag{9}$$

Making use of the remaining three equations, we note, taking (9) into account, the following relationship

$$\frac{\partial}{\partial x^k}\left(\frac{\partial v_i}{\partial x^k}\right) = -\frac{\partial}{\partial x^i}\left(\frac{\partial v_k}{\partial x^k}\right) = 0.$$

Hence we conclude that:

$$\frac{\partial v_i}{\partial x^k} = \psi_{ik}(x^j, t).$$

Differentiating with respect to the remaining variable x^j, we have

$$\frac{\partial^2 v_i}{\partial x^j \partial x^k} = \frac{\partial \psi_{ik}}{\partial x^j}.$$

Interchanging j with k we obtain

$$\frac{\partial^2 v_i}{\partial x^k \partial x^j} = \frac{\partial \psi_{ij}}{\partial x^k}.$$

Hence

$$\frac{\partial \psi_{ik}}{\partial x^j} = \frac{\partial \psi_{ij}}{\partial x^k}. \tag{10}$$

Remembering that i, j, and k are different terms of the sequence 1, 2, 3, it follows from (10) that

$$\frac{\partial \psi_{ij}}{\partial x^k} = \lambda_{ijk}(t) \tag{11}$$

and

$$\lambda_{ikj} = \lambda_{ijk}. \tag{11a}$$

Integrating (11), we find

$$\psi_{ij}(x^k, t) = \lambda_{ijk} x^k - \omega_{ij}(t). \tag{12}$$

Next, we form the variation δv_i (see (2)):

$$\delta v_i = \frac{\partial v_i}{\partial x^j} \delta x^j = \psi_{ij} \delta x^j = (\lambda_{ijk} x^k - \omega_{ij}) \delta x^j$$

or

$$\delta v_i = \delta(\tfrac{1}{2}\lambda_{ijk} x^j x^k - \omega_{ij} x^j + v_{io}).$$

Hence

$$v_i = \tfrac{1}{2}\lambda_{ijk} x^j x^k - \omega_{ij} x^j + v_{io}. \tag{13}$$

The individual functions of time t occurring in equation (13) should now be determined and for this purpose it is necessary to go back to the group of equations (7), from which we find that

$$\psi_{ij} = -\psi_{ji}.$$

Therefore,

$$\lambda_{ijk} x^k - \omega_{ij} \equiv -\lambda_{jik} x^k + \omega_{ij};$$

hence we have

$$\lambda_{ijk} + \lambda_{jik} = 0 \tag{14}$$

and

$$\omega_{ij} = -\omega_{ij}. \tag{15}$$

From (14) and (11a) we infer that all the functions λ_{ijk} are equal to zero whereas

$$\omega_{ij} = -T_{[ij]}. \tag{16}$$

Thus the vector field involved in the motion of a rigid body is described by

$$v_i = v_{io} - \omega_{ij} x^j. \tag{17}$$

Consequently, we can say that the general motion of a rigid body is a *superposition* of two componential motions, one of which is translational and the other, circular about a fixed point.

For translatory motion, all components of the matrix T_{ij} are equal to zero. Consequently, every point has the same velocity

$$v_i = v_{io}. \tag{18}$$

This is the simplest case of motion where all the particles of a body cover equal distances in arbitrary small infinitesimal time intervals, which follows from (18) since

$$\mathrm{d}r = v_0 \mathrm{d}t, \tag{18a}$$

or the infinitesimal displacements of all points of the body are all exactly the same. Since angular motion is not involved, the sections connecting any two points of the body remain parallel to their initial position (parallel translation). An example of such motion is the motion of element AB of a crank-rocker system, in which the crank and the rocker are of equal length (Fig. 1.11).

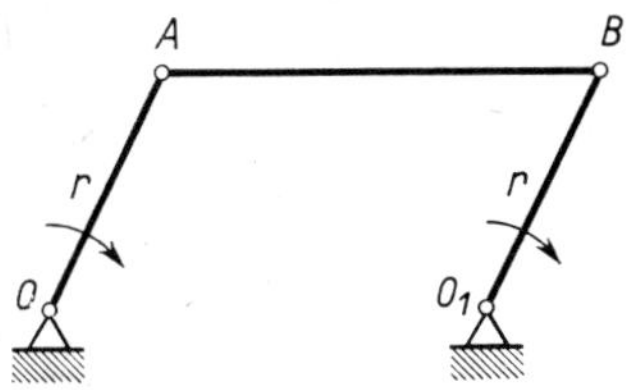

Fig. 1.11

We shall now consider a different type of motion, namely one in which a single particle of a body remains permanently at rest. This corresponds to the case of a velocity vector field described by the equation

$$v_i = -\omega_{ik} x^k. \tag{19}$$

This is the case of gyration about a fixed point (all points are moving on the surfaces of concentric spheres). Writing in detail equation (19), we have

$$\begin{aligned} v_1 &= -\omega_{12}x^2-\omega_{13}x^3, \\ v_2 &= -\omega_{23}x^3-\omega_{21}x^1, \\ v_3 &= -\omega_{31}x^1-\omega_{32}x^2. \end{aligned} \tag{19a}$$

In the vector form, it will be

$$\mathbf{v} = \begin{vmatrix} \mathbf{i} & \mathbf{j} & \mathbf{k} \\ \omega_1 & \omega_2 & \omega_3 \\ x^1 & x^2 & x^3 \end{vmatrix} = \boldsymbol{\omega}\times\mathbf{r}, \tag{20}$$

where the components of the vector $\boldsymbol{\omega}$ can be derived from the components of the antisymmetrical parts of the tensor of velocity gradient, according to the following equations:

$$\omega_1 = -\omega_{32} = \omega_{23}, \qquad \omega_2 = -\omega_{13} = \omega_{31}, \qquad \omega_3 = -\omega_{21} = \omega_{12} \tag{21}$$

or, using the permutation symbol ε_{rik}, we can write (21) in shortened form

$$\omega_r = \tfrac{1}{2}\sum_{i=1}^{3}\sum_{k=1}^{3} \varepsilon_{rik}\omega_{ik}. \tag{21a}$$

The vector $\boldsymbol{\omega}$ we shall call *angular velocity vector* since as one can easily see it expresses the derivative of the angle of instantaneous rotation with respect to time

$$\boldsymbol{\omega} = \lim_{\Delta t\to 0}\frac{\Delta\boldsymbol{\varphi}}{\Delta t} = \dot{\boldsymbol{\varphi}}. \tag{22}$$

Indeed,

$$\omega_{ik} = -\frac{\partial v_i}{\partial x^k} = \frac{\mathrm{d}\varphi_{ik}}{\mathrm{d}t} = \pm\frac{\mathrm{d}\varphi_j}{\mathrm{d}t} = \pm\omega_j \qquad (i \neq j \neq k).$$

The *plus* sign is valid when i, j, k is an even permutation of the sequence of indices 1, 2, 3, and the *minus* sign when an odd permutation is involved. This shows that the interpretation of the vector $\boldsymbol{\omega}$ according to equation (22) is correct.

The vector of the angular displacement $\mathrm{d}\boldsymbol{\varphi}$ is perpendicular to the instantaneous plane of rotation, and its sense we fix according to the right-handed system (the *right-hand screw rule*).

Thus in rotation about a fixed point, the velocity of any point of a rigid body, whose position is defined by the vector **r**, is equal to the vector product $\boldsymbol{\omega} \times \mathbf{r}$, so we can write

$$\mathbf{v} = \boldsymbol{\omega} \times \mathbf{r} = \frac{d\boldsymbol{\varphi}}{dt} \times \mathbf{r}. \tag{23}$$

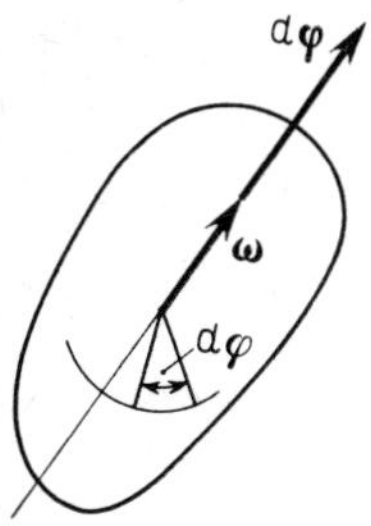

Fig. 1.12

Just as for translatory motion the characteristic magnitudes are velocity and acceleration, so for rotation about a fixed point we have respectively, angular velocity and angular acceleration, the latter being the derivative of the vector $\boldsymbol{\omega}$ with respect to time:

$$\boldsymbol{\epsilon} \overset{\text{df}}{=} \frac{d\boldsymbol{\omega}}{dt}. \tag{24}$$

If the direction of angular velocity remains unchanged, a special case of rotation about a fixed point occurs, namely angular motion.

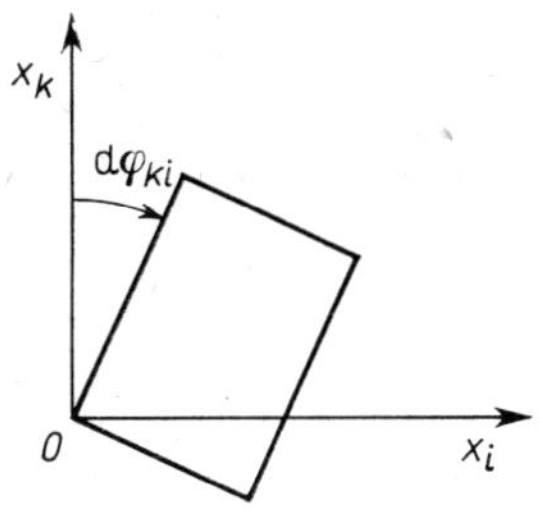

Fig. 1.13

In rotation about a fixed point, the configuration of a rigid body can be determined by introducing three angles of rotation ψ, ϑ and φ, called *Euler's angles* (Fig. 1.14), corresponding to three degrees of freedom.

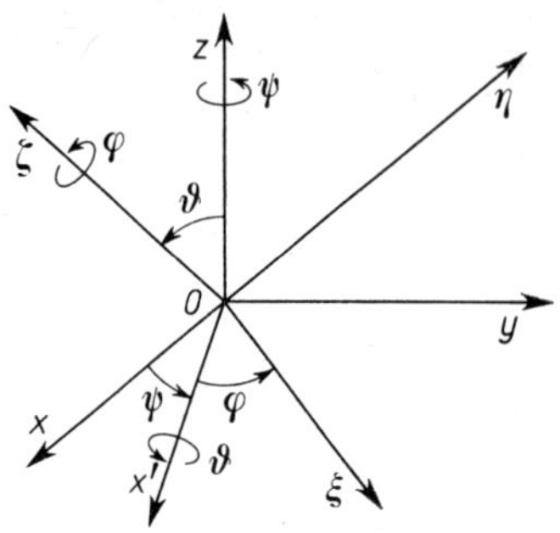

Fig. 1.14

Let us introduce an orthogonal frame of reference $Oxyz$. Turning the frame about the axis z by an angle ψ (Fig. 1.15), we have

$$\mathbf{i}' = \mathbf{i}\cos\psi + \mathbf{j}\sin\psi, \quad \mathbf{j}' = -\mathbf{i}\sin\psi + \mathbf{j}\cos\psi. \tag{25}$$

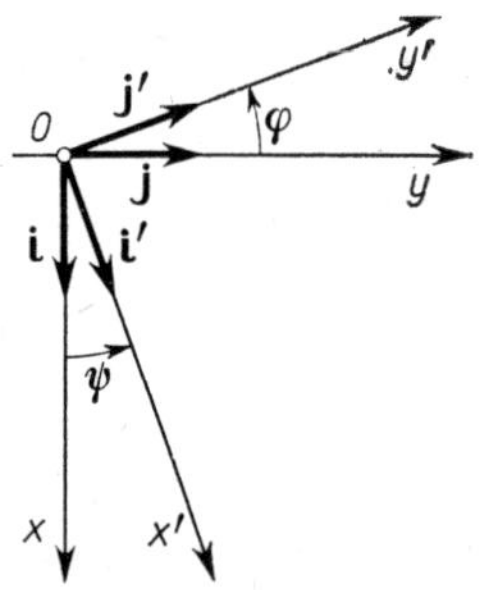

Fig. 1.15

Next we turn the frame of reference about the axis x' by an angle ϑ (Fig. 1.16).

$$\mathbf{j}'' = \mathbf{j}'\cos\vartheta + \mathbf{k}\sin\vartheta, \quad \mathbf{e}_\zeta = \mathbf{k}\cos\vartheta - \mathbf{j}'\sin\vartheta. \tag{26}$$

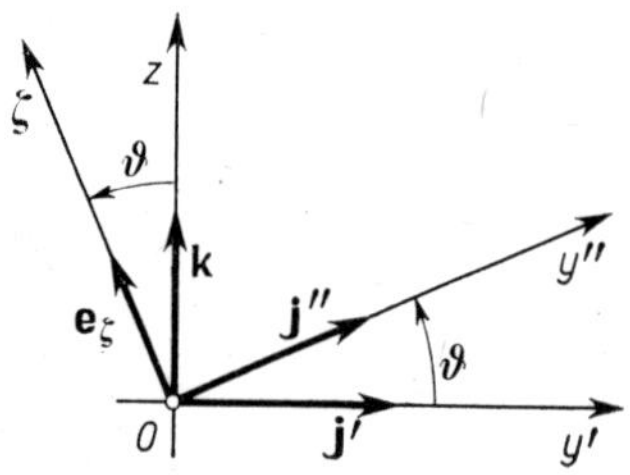

Fig. 1.16

Continuing, we turn the frame about the axis ζ by an angle φ (Fig. 1.17):

$$\mathbf{e}_\xi = \mathbf{i}'\cos\varphi + \mathbf{j}''\sin\varphi, \quad \mathbf{e}_\eta = \mathbf{j}''\cos\varphi - \mathbf{i}'\sin\varphi. \tag{27}$$

From (25), (26), and (27) we have

$$\begin{aligned} \mathbf{e}_\xi &= \mathbf{i}(\cos\psi\cos\varphi - \sin\psi\cos\vartheta\sin\varphi) + \\ &\quad + \mathbf{j}(\sin\psi\cos\varphi + \cos\psi\cos\vartheta\sin\varphi) + \mathbf{k}\sin\vartheta\sin\varphi, \\ \mathbf{e}_\eta &= -\mathbf{i}(\cos\psi\sin\varphi + \sin\psi\cos\vartheta\cos\varphi) - \\ &\quad - \mathbf{j}(\sin\psi\sin\varphi - \cos\psi\cos\vartheta\cos\varphi) + \mathbf{k}\sin\vartheta\cos\varphi, \\ \mathbf{e}_\zeta &= \mathbf{i}\sin\psi\sin\vartheta - \mathbf{j}\cos\psi\sin\vartheta + \mathbf{k}\cos\vartheta, \end{aligned} \tag{28}$$

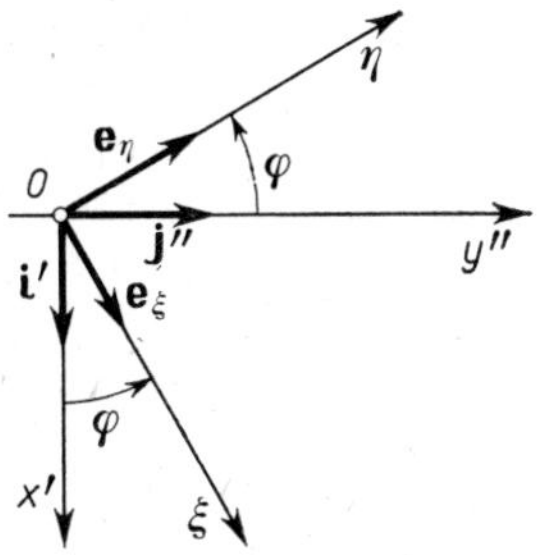

Fig. 1.17

where $\mathbf{e}_\xi$, $\mathbf{e}_\eta$, and $\mathbf{e}_\zeta$ constitute the basic vectors in the $\Omega\xi\eta\zeta$ frame. To find the angles between the axes it is enough to scalar multiply the unit vectors of the $\Omega\xi\eta\zeta$ frame by the unit vectors of the $Oxyz$ frame:

$$\begin{aligned} &\cos(x,\xi) = \mathbf{i}\cdot\mathbf{e}_\xi = A^1_{1'}, \quad \cos(y,\xi) = \mathbf{j}\cdot\mathbf{e}_\xi = A^2_{1'}, \quad \cos(z,\xi) = \mathbf{k}\cdot\mathbf{e}_\xi = A^3_{1'}, \\ &\cos(x,\eta) = \mathbf{i}\cdot\mathbf{e}_\eta = A^1_{2'}, \quad \cos(y,\eta) = \mathbf{j}\cdot\mathbf{e}_\eta = A^2_{2'}, \quad \cos(z,\eta) = \mathbf{k}\cdot\mathbf{e}_\eta = A^3_{2'}, \\ &\cos(x,\zeta) = \mathbf{i}\cdot\mathbf{e}_\zeta = A^1_{3'}, \quad \cos(y,\zeta) = \mathbf{j}\cdot\mathbf{e}_\zeta = A^2_{3'}, \quad \cos(z,\zeta) = \mathbf{k}\cdot\mathbf{e}_\zeta = A^3_{3'}. \end{aligned} \tag{29}$$

Equations (28) can be shortened as follows:

$$\mathbf{e}_{j'} = A^k_{j'}\mathbf{e}_k \quad (j' = 1, 2, 3). \tag{30}$$

Naturally, $\mathbf{e}_1 \equiv \mathbf{i}$, $\mathbf{e}_2 \equiv \mathbf{j}$, $\mathbf{e}_3 \equiv \mathbf{k}$, $\mathbf{e}_{1'} \equiv \mathbf{e}_\xi$, $\mathbf{e}_{2'} \equiv \mathbf{e}_\eta$, $\mathbf{e}_{3'} \equiv \mathbf{e}_\zeta$.

As we have just noted, transitions from one position to another can be obtained by three angular displacements ψ, ϑ, and φ, respectively. Consequently, the vector $\boldsymbol{\omega}$ can be treated as the sum of three components. Therefore, we can write

$$\boldsymbol{\omega} = \dot{\psi}\mathbf{k} + \dot{\vartheta}\mathbf{i}' + \dot{\varphi}\mathbf{e}_\zeta. \tag{31}$$

The projections of the vector $\boldsymbol{\omega}$ on the respective axes of the $Oxyz$ system we write in the following form:

$$\begin{aligned}\omega_x &= \boldsymbol{\omega}\cdot\mathbf{i} = \dot{\vartheta}\cos\psi+\dot{\varphi}A^1_{3'} = \dot{\vartheta}\cos\psi+\dot{\varphi}\sin\psi\sin\vartheta,\\ \omega_y &= \boldsymbol{\omega}\cdot\mathbf{j} = \dot{\vartheta}\sin\psi+\dot{\varphi}A^2_{3'} = \dot{\vartheta}\sin\psi-\dot{\varphi}\cos\psi\sin\vartheta,\\ \omega_z &= \boldsymbol{\omega}\cdot\mathbf{k} = \dot{\psi}+\dot{\varphi}A^3_{3'} = \dot{\psi}+\dot{\varphi}\cos\vartheta\,.\end{aligned} \tag{32}$$

In a similar way we can find the projections of the vector $\boldsymbol{\omega}$ on the axes of a system associated with a moving rigid body, i.e., a system in motion

$$\begin{aligned}\omega_\xi &= \dot{\psi}\sin\vartheta\sin\varphi+\dot{\vartheta}\cos\varphi,\\ \omega_\eta &= \dot{\psi}\sin\vartheta\cos\varphi-\dot{\vartheta}\sin\varphi,\\ \omega_\zeta &= \dot{\psi}\cos\vartheta+\dot{\varphi}.\end{aligned} \tag{33}$$

It can be proved that

$$\begin{aligned}\omega_\xi &= \dot{A}^1_{2'}A^1_{3'}+\dot{A}^2_{2'}A^2_{3'}+\dot{A}^3_{2'}A^3_{3'} = -(A^1_{2'}\dot{A}^1_{3'}+A^2_{2'}\dot{A}^2_{3'}+A^3_{2'}\dot{A}^3_{3'}),\\ \omega_\eta &= \dot{A}^1_{3'}A^1_{1'}+\dot{A}^2_{3'}A^2_{1'}+\dot{A}^3_{3'}A^3_{1'} = -(A^1_{3'}\dot{A}^1_{1'}+A^2_{3'}\dot{A}^2_{1'}+A^3_{3'}\dot{A}^3_{1'}),\\ \omega_\zeta &= \dot{A}^1_{1'}A^1_{2'}+\dot{A}^2_{1'}A^2_{2'}+\dot{A}^3_{1'}A^3_{2'} = -(A^1_{1'}\dot{A}^1_{2'}+A^2_{1'}\dot{A}^2_{2'}+A^3_{1'}\dot{A}^3_{2'}).\end{aligned} \tag{34}$$

The proof is left to the reader.

The square of the angular velocity we find using the equations previously derived:

$$\begin{aligned}\omega^2 &= \omega_\xi^2+\omega_\eta^2+\omega_\zeta^2\\ &= \dot{\psi}^2\sin^2\vartheta\sin^2\varphi+2\dot{\psi}\dot{\vartheta}\sin\vartheta\sin\varphi\cos\varphi+\dot{\vartheta}^2\cos^2\varphi+\\ &\quad+\dot{\psi}^2\sin^2\vartheta\cos^2\varphi-2\dot{\psi}\dot{\vartheta}\sin\vartheta\cos\varphi\sin\varphi+\dot{\vartheta}^2\sin^2\varphi+\\ &\quad+\dot{\psi}^2\cos^2\vartheta+2\dot{\psi}\dot{\varphi}\cos\vartheta+\dot{\varphi}^2\\ &= \dot{\psi}^2+\dot{\vartheta}^2+\dot{\varphi}^2+2\dot{\psi}\dot{\varphi}\cos\vartheta.\end{aligned} \tag{35}$$

Since $\mathbf{v} = \boldsymbol{\omega}\times\mathbf{r}$, all points in a straight line having the direction of the vector $\boldsymbol{\omega}$ are at rest; this is the so-called *instantaneous rotational axis.* This straight line, permanently passing through the reference origin and continuously changing its position in space, form a certain coincoid, called the *axode*. The axode is, therefore, the geometric locus of instantaneous axes of rotation.

The equation of instantaneous axes of rotation will be defined by the condition of parallelism of the vectors $\boldsymbol{\omega}$ and $\mathbf{r}$ in a fixed and in a moving system. The required conditions are:

$$\frac{x}{\omega_x} = \frac{y}{\omega_y} = \frac{z}{\omega_z}, \tag{36}$$

$$\frac{\xi}{\omega_\xi} = \frac{\eta}{\omega_\eta} = \frac{\zeta}{\omega_\zeta}. \tag{37}$$

Eliminating time from the above equations, we obtain the equations of coincoids. These are correspondingly the equations of fixed and moving axode. During motion, the moving axode rolls on the fixed axode. The generating tangent lines are successively the instantaneous axes of rotation.

Acceleration in the case of rotation about a fixed point we find by differentiating the expression $\mathbf{v} = \boldsymbol{\omega} \times \mathbf{r}$. Therefore,

$$\mathbf{a} = \boldsymbol{\epsilon} \times \mathbf{r} + \boldsymbol{\omega} \times (\boldsymbol{\omega} \times \mathbf{r}). \tag{38}$$

The general motion of a rigid body, which we shall now consider, is a superposition of two motions: translation and rotation about a fixed point. If the body is rigidly bound to the axial system ξ, η, ζ with its origin at Ω, then the vector defining the position of an arbitrary particle A of the body is the sum (Fig. 1.18) of the position vector $\mathbf{r}_\Omega$, the origin of the moving system and the position vector $\boldsymbol{\rho}$ of the particle A in the moving system:

$$\mathbf{r} = \mathbf{r}_\Omega + \boldsymbol{\rho}. \tag{39}$$

Differentiating (39) we have

$$\frac{\mathrm{d}\mathbf{r}}{\mathrm{d}t} = \frac{\mathrm{d}\mathbf{r}_\Omega}{\mathrm{d}t} + \frac{\mathrm{d}\boldsymbol{\rho}}{\mathrm{d}t}. \tag{40}$$

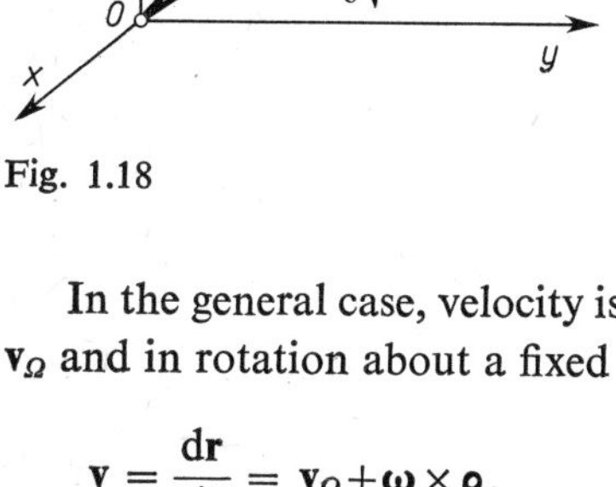

Fig. 1.18

In the general case, velocity is the sum of velocities in translatory motion $\mathbf{v}_\Omega$ and in rotation about a fixed point, therefore $\mathrm{d}\boldsymbol{\rho}/\mathrm{d}t = \boldsymbol{\omega} \times \boldsymbol{\rho}$. Thus

$$\mathbf{v} = \frac{\mathrm{d}\mathbf{r}}{\mathrm{d}t} = \mathbf{v}_\Omega + \boldsymbol{\omega} \times \boldsymbol{\rho}, \tag{41}$$

which is consistent with (17).

We directly infer from (41) that acceleration in general motion is expressed by

$$\mathbf{a} = \frac{d\mathbf{v}_\Omega}{dt} + \boldsymbol{\epsilon}\times\boldsymbol{\rho} + \boldsymbol{\omega}\times(\boldsymbol{\omega}\times\boldsymbol{\rho}). \tag{42}$$

This shows that acceleration in general motion is, like velocity, the sum of the acceleration in translation and the acceleration in rotation about a fixed point. Using identity (25) given in the Introduction, we can write

$$\boldsymbol{\omega}\times(\boldsymbol{\omega}\times\boldsymbol{\rho}) = \boldsymbol{\omega}(\boldsymbol{\omega}\cdot\boldsymbol{\rho}) - \boldsymbol{\rho}\omega^2.$$

Note further that the choice of the origin of the moving system is arbitrary. This may raise the question of whether the vectors $\boldsymbol{\omega}$ and $\boldsymbol{\epsilon}$ depend in fact on the choice of that point. The answer is in the negative, which can easily be proved.

Let us assume that on passing from point Ω to point Ω_1 the angular velocity $\boldsymbol{\omega}$ changes to $\boldsymbol{\omega}_1$ (Fig. 1.19). Then

$$\mathbf{v}_A = \mathbf{v}_\Omega + \boldsymbol{\omega}\times\boldsymbol{\rho} = \mathbf{v}_{\Omega_1} + \boldsymbol{\omega}_1\times\boldsymbol{\rho}_1, \tag{43}$$

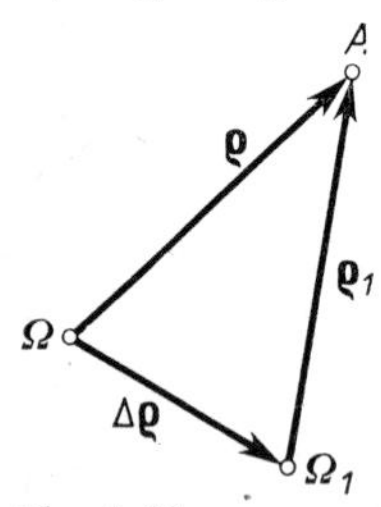

Fig. 1.19

but the velocity

$$\mathbf{v}_{\Omega_1} = \mathbf{v}_\Omega + \boldsymbol{\omega}\times\Delta\boldsymbol{\rho} = \mathbf{v}_\Omega + \boldsymbol{\omega}\times(\boldsymbol{\rho}-\boldsymbol{\rho}_1). \tag{44}$$

Considering (44) in equation (43), we have

$$\mathbf{v}_\Omega + \boldsymbol{\omega}\times\boldsymbol{\rho} = \mathbf{v}_\Omega + \boldsymbol{\omega}\times(\boldsymbol{\rho}-\boldsymbol{\rho}_1) + \boldsymbol{\omega}_1\times\boldsymbol{\rho}_1, \tag{45}$$

which after simplification gives

$$(\boldsymbol{\omega}_1 - \boldsymbol{\omega})\times\boldsymbol{\rho}_1 = \mathbf{0}. \tag{46}$$

Owing to the discretional character of the vector $\boldsymbol{\rho}_1$,

$$\boldsymbol{\omega}_1 \equiv \boldsymbol{\omega}. \tag{47}$$

Consequently, the vector $\boldsymbol{\omega}$ does not change on transition from one point of a rigid body to another. Similarly, it can also be proved that $\boldsymbol{\epsilon}$ is independent of the choice of a point in space. The proof is analogous.

We shall now find a straight line whose points have velocities in directions consistent with the direction of the angular velocity $\boldsymbol{\omega}$. Along this line, called the *instantaneous helical axis*, the vector $\boldsymbol{\omega}$ will coincide with the vector $\mathbf{v}$; therefore, the following relationship is legitimate for its points

$$\boldsymbol{\omega} \times \mathbf{v} = \mathbf{0}. \tag{48}$$

By virtue of (41)

$$\boldsymbol{\omega} \times (\mathbf{v}_\Omega + \boldsymbol{\omega} \times \boldsymbol{\rho}_S) = \mathbf{0}. \tag{49}$$

Equation (49) can be written in a somewhat different form, namely,

$$\boldsymbol{\omega} \times \mathbf{v}_\Omega + \boldsymbol{\omega}(\boldsymbol{\omega} \cdot \boldsymbol{\rho}_S) - \omega^2 \boldsymbol{\rho}_S = \mathbf{0}.$$

The identity (25) given in the Introduction has been used here.

Selecting a point S in the instantaneous helical axis s in such a way that $\boldsymbol{\omega} \cdot \boldsymbol{\rho}_S = 0$ (i.e., selecting $\boldsymbol{\rho}_S$ at right angles to $\boldsymbol{\omega}$), we find the position vector of point S (Fig. 1.20) or the vector $\boldsymbol{\rho}_S$,

$$\boldsymbol{\rho}_S = \frac{\boldsymbol{\omega} \times \mathbf{v}_\Omega}{\omega^2}. \tag{50}$$

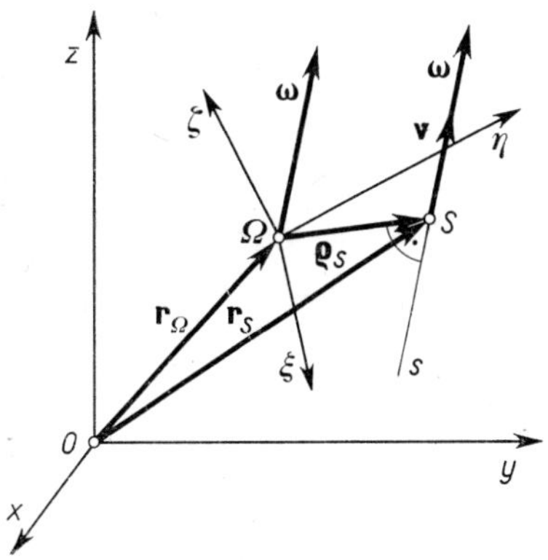

Fig. 1.20

The position of an arbitrary point on the straight line s is given by the vector

$$\boldsymbol{\rho} = \boldsymbol{\rho}_S + \alpha\boldsymbol{\omega} \tag{51}$$

in a moving system, and by the vector

$$\mathbf{r} = \mathbf{r}_S + \alpha\boldsymbol{\omega} \tag{52}$$

in a fixed system. Considering the coordinates of the vectors and the constancy of the coefficient α, we obtain

$$\frac{\xi-\xi_S}{\omega_\xi}=\frac{\eta-\eta_S}{\omega_\eta}=\frac{\zeta-\zeta_S}{\omega_\zeta}, \tag{51a}$$

$$\frac{x-x_S}{\omega_x}=\frac{y-y_S}{\omega_y}=\frac{z-z_S}{\omega_z}. \tag{52a}$$

Equations (51) and (52) are equations of helical axes. Eliminating time from (51a) and (52), we obtain the equations of the moving and the fixed axodes. If in general motion there exists a helical axis, then this motion is sometimes called *helical motion.*

1.3 Relative motion

Relative motion is a most essential kinematical problem. We shall now consider how, knowing the characteristics of a particle moving in a frame of reference as well as the characteristics of that frame moving against some background frame assumed to be at rest, we can deduce the motion of a particle against that latter frame. In other words, the problem amounts to finding the absolute motion of the particle at given velocity and acceleration in the moving frame, knowing also the general motion of the frame against the background space. Let us begin with the determination of velocity. We introduce two reference frames, one in motion $\xi\eta\zeta$ with origin Ω and the other fixed xyz with origin O. The elementary (infinitesimal) displacement $d\mathbf{r}$ of point A (Fig. 1.21) in the $Oxyz$ frame is the result of the general motion of the frame and the relative motion of A. Thus we can write

$$d\mathbf{r} = \mathbf{v}_\Omega dt+\omega\times\boldsymbol{\rho}\,dt+\check{d}\boldsymbol{\rho}, \tag{1}$$

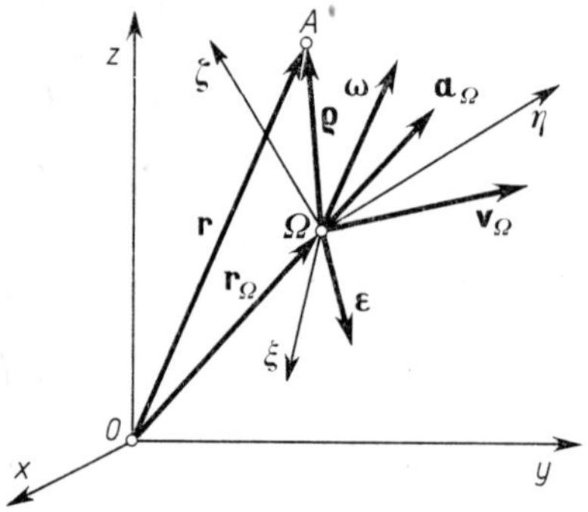

Fig. 1.21

where $\check{d}\boldsymbol{\rho}$ denotes the elemental displacement of A in the $\Omega\xi\eta\zeta$ frame. Dividing (1) by dt we obtain the sought relationship between the velocities:

$$\mathbf{v} = \mathbf{v}_\Omega + \boldsymbol{\omega} \times \boldsymbol{\rho} + \frac{\check{d}\boldsymbol{\rho}}{dt}. \tag{2}$$

It follows that in relative motion velocity is equal to the sum of the velocity which a particle would have if it were rigidly attached to a moving frame (known as the *velocity of a moving space* or *velocity of transportation*) and the relative velocity $\check{d}\boldsymbol{\rho}/dt = \mathbf{v}_r$

$$\mathbf{v} = \mathbf{v}_{tr} + \mathbf{v}_r. \tag{3}$$

Note that the relationship just derived conceals a dependence between the derivative of the vector in the fixed frame and the corresponding derivative in the moving frame of reference. To find the transformation formula for the derivative of a vector, it is necessary to use the relationship shown in (2), which defines the velocities of points A and B constituting the origin and the end of the vector $\mathbf{c}$ (Fig. 1.22)

$$\frac{d\mathbf{r}_A}{dt} = \frac{d\mathbf{r}_\Omega}{dt} + \boldsymbol{\omega} \times \boldsymbol{\rho}_A + \frac{\check{d}\boldsymbol{\rho}_A}{dt}, \tag{4}$$

$$\frac{d\mathbf{r}_B}{dt} = \frac{d\mathbf{r}_\Omega}{dt} + \boldsymbol{\omega} \times \boldsymbol{\rho}_B + \frac{\check{d}\boldsymbol{\rho}_B}{dt}. \tag{5}$$

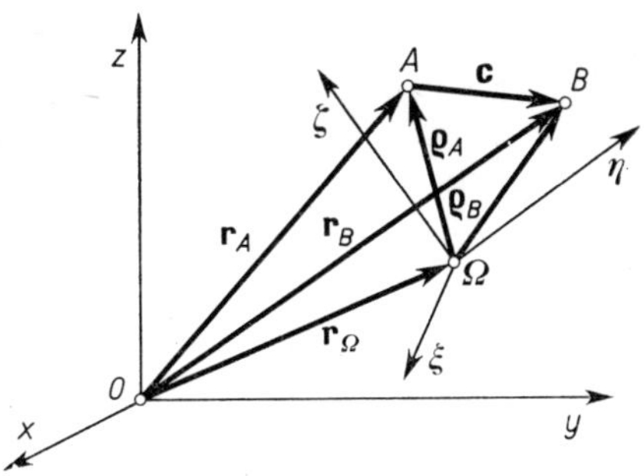

Fig. 1.22

Subtracting (4) from (5) we obtain

$$\frac{d}{dt}(\mathbf{r}_B - \mathbf{r}_A) = \boldsymbol{\omega} \times (\boldsymbol{\rho}_B - \boldsymbol{\rho}_A) + \frac{\check{d}}{dt}(\boldsymbol{\rho}_B - \boldsymbol{\rho}_A).$$

It will be seen directly from Fig. 1.22 that

$$\mathbf{r}_B - \mathbf{r}_A = \boldsymbol{\rho}_B - \boldsymbol{\rho}_A = \mathbf{c}.$$

Therefore

$$\frac{d\mathbf{c}}{dt} = \frac{\check{d}\mathbf{c}}{dt} + \boldsymbol{\omega} \times \mathbf{c}. \tag{6}$$

and this is the sought transformation formula.

To explain the meaning of the derivative in the moving frame of reference and to show how it differs from the corresponding derivative in the fixed frame, it should be appreciated that the same vector can be expressed in both frames. In the moving frame it is

$$\boldsymbol{\rho} = \varrho^{i'} \mathbf{e}'_i$$

and in the fixed frame

$$\boldsymbol{\rho} = \varrho^{i} \mathbf{e}_i .$$

Let us calculate the derivative in the fixed $Oxyz$ frame in both the cases:

$$\frac{d\boldsymbol{\rho}}{dt} = \frac{d}{dt}(\varrho^{i'} \mathbf{e}'_i) = \frac{d\varrho^{i'}}{dt} \mathbf{e}'_i + \varrho^{i'} \frac{d\mathbf{e}'_i}{dt},$$

$$\frac{d\boldsymbol{\rho}}{dt} = \frac{d}{dt}(\varrho^{i} \mathbf{e}_i) = \frac{d\varrho^{i}}{dt} \mathbf{e}_i ,$$

whereas in the moving $\Omega\xi\eta\zeta$ frame

$$\frac{\check{d}\boldsymbol{\rho}}{dt} = \frac{d}{dt}(\varrho^{i'} \mathbf{e}'_i) = \frac{d\varrho^{i'}}{dt} \mathbf{e}'_i .$$

The acceleration in relative motion we find by differentiating (2):

$$\frac{d^2\mathbf{r}}{dt^2} = \frac{d^2\mathbf{r}_\Omega}{dt^2} + \frac{d\boldsymbol{\omega}}{dt} \times \boldsymbol{\rho} + \boldsymbol{\omega} \times \frac{d\boldsymbol{\rho}}{dt} + \frac{d}{dt}\left(\frac{\check{d}\boldsymbol{\rho}}{dt}\right) \tag{7}$$

where $d^2\mathbf{r}_\Omega/dt^2 = \mathbf{a}_\Omega$ is the acceleration of the frame $\Omega\xi\eta\zeta$ in translatory motion. In keeping with the transformation formula

$$\frac{d\boldsymbol{\omega}}{dt} = \frac{\check{d}\boldsymbol{\omega}}{dt} + \boldsymbol{\omega} \times \boldsymbol{\omega} = \frac{\check{d}\boldsymbol{\omega}}{dt} = \boldsymbol{\epsilon}, \tag{a}$$

$$\frac{d\boldsymbol{\rho}}{dt} = \frac{\check{d}\boldsymbol{\rho}}{dt} + \boldsymbol{\omega} \times \boldsymbol{\rho} = \mathbf{v}_r + \boldsymbol{\omega} \times \boldsymbol{\rho}, \tag{b}$$

$$\frac{d}{dt}\left(\frac{\check{d}\boldsymbol{\rho}}{dt}\right) = \frac{\check{d}^2\boldsymbol{\rho}}{dt^2} + \boldsymbol{\omega} \times \frac{\check{d}\boldsymbol{\rho}}{dt} = \mathbf{a}_r + \boldsymbol{\omega} \times \mathbf{v}_r . \tag{c}$$

Considering (a), (b), and (c) in (7), we have

$$\mathbf{a} = \mathbf{a}_\Omega + \boldsymbol{\epsilon} \times \boldsymbol{\rho} + \boldsymbol{\omega} \times (\boldsymbol{\omega} \times \boldsymbol{\rho}) + 2\boldsymbol{\omega} \times \mathbf{v}_r + \mathbf{a}_r . \tag{8}$$

It will be seen that acceleration in relative motion consists of acceleration of transportation

$$\mathbf{a}_{\mathrm{tr}} = \mathbf{a}_{\Omega}+\boldsymbol{\epsilon}\times\boldsymbol{\rho}+\boldsymbol{\omega}\times(\boldsymbol{\omega}\times\boldsymbol{\rho}),$$

Coriolis accelereration

$$\mathbf{a}_{\mathrm{Cor}} = 2\boldsymbol{\omega}\times\mathbf{v}_r$$

and relative acceleration

$$\mathbf{a}_r = \frac{\check{\mathrm{d}}^2\boldsymbol{\rho}}{\mathrm{d}t^2}.$$

Therefore,

$$\mathbf{a} = \mathbf{a}_{\mathrm{tr}}+\mathbf{a}_{\mathrm{Cor}}+\mathbf{a}_r. \tag{9}$$

Example 1. Determine the acceleration of a particle moving in a circle according to the equation $s = \alpha\sin\varkappa t$, where s = length of arc, α—constant value of amplitude. The circle rotates about its diameter with constant angular velocity ω = const.

The problem is solved in the following manner.

(1) We calculate the acceleration of transportation

$$\mathbf{a}_{\mathrm{tr}} = \mathbf{a}_{\Omega}+\boldsymbol{\epsilon}\times\boldsymbol{\rho}+\boldsymbol{\omega}\times(\boldsymbol{\omega}\times\boldsymbol{\rho}).$$

In the considered case, transportation is an angular motion. Since we have set up the frames of reference (Fig. 1.23) in such a way that their origins coincide so

$$\mathbf{a}_{\Omega} = \mathbf{0},$$

also

$$\boldsymbol{\epsilon} = \mathbf{0},$$

therefore

$$\mathbf{a}_{\mathrm{tr}} = \boldsymbol{\omega}\times(\boldsymbol{\omega}\times\boldsymbol{\rho}) = \boldsymbol{\omega}(\boldsymbol{\omega}\cdot\boldsymbol{\rho})-\omega^2\boldsymbol{\rho}.$$

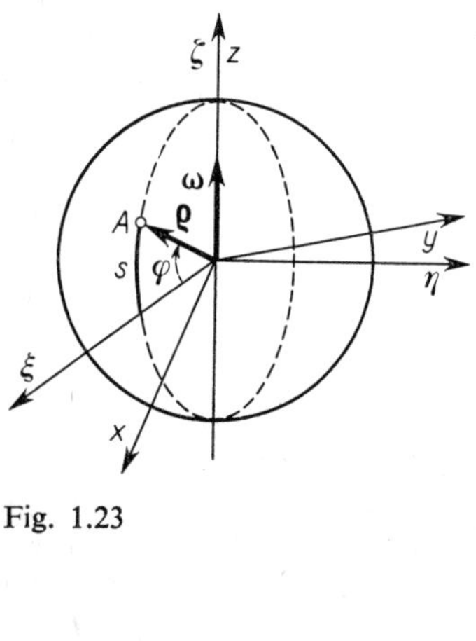

Fig. 1.23

The circle spins together with the $\xi\eta\zeta$ frame of reference, wherein the position vector of the particle under examination can be expressed in the form

$$\boldsymbol{\rho} = \mathbf{i}'\xi+\mathbf{k}'\zeta$$

but

$$\boldsymbol{\omega} = \mathbf{k}'\omega .$$

Therefore,

$$\mathbf{a}_{tr} = \mathbf{k}'\omega^2\zeta-\omega^2(\mathbf{i}'\xi+\mathbf{k}'\zeta) = -\omega^2\xi\mathbf{i}',$$

$$\xi = R\cos\varphi = R\cos\left(\frac{1}{R}\alpha\sin\varkappa t\right).$$

Consequently, the expression for the acceleration of transportation is as follows:

$$\mathbf{a}_{tr} = -\mathbf{i}'\omega^2\cos\left(\frac{1}{R}\alpha\sin\varkappa t\right).$$

(2) Next, we calculate the relative acceleration

$$\mathbf{a}_r = \frac{\check{d}^2\boldsymbol{\rho}}{dt^2} = \mathbf{i}'\frac{d^2\xi}{dt^2}+\mathbf{k}'\frac{d^2\zeta}{dt^2} =$$

$$= -\mathbf{i}'\left\{\cos\left[\frac{1}{R}\alpha\sin\varkappa t\right]\frac{1}{R}\alpha^2\varkappa^2\cos^2\varkappa t-\sin\left[\frac{\alpha}{R}\sin\varkappa t\right]\alpha\varkappa^2\sin\varkappa t\right\} -$$

$$-\mathbf{k}'\left\{\sin\left[\frac{\alpha}{R}\sin\varkappa t\right]\frac{1}{R}\alpha^2\varkappa^2\cos^2\varkappa t+\cos\left[\frac{\alpha}{R}\sin\varkappa t\right]\alpha\varkappa^2\sin\varkappa t\right\}.$$

(3) Finally, the Coriolis acceleration:

$$\mathbf{a}_{Cor} = 2\boldsymbol{\omega}\times\mathbf{v}_r,$$

$$\mathbf{v}_r = \frac{\check{d}\boldsymbol{\rho}}{dt} = -\mathbf{i}'\sin\left[\frac{\alpha}{R}\sin\varkappa t\right]\alpha\varkappa\cos\varkappa t+\mathbf{k}'\cos\left[\frac{\alpha}{R}\sin\varkappa t\right]\alpha\varkappa\cos\varkappa t.$$

Therefore,

$$\mathbf{a}_{Cor} = \begin{vmatrix} \mathbf{i}' & \mathbf{j}' & \mathbf{k}' \\ 0 & 0 & \omega \\ v_\xi & 0 & v_\zeta \end{vmatrix} = \mathbf{j}'\omega v_\xi = -\mathbf{j}'\omega\alpha\varkappa\cos\varkappa t\sin\left[\frac{1}{R}\alpha\sin\varkappa t\right].$$

Total acceleration:

$$\mathbf{a} = \mathbf{a}_{tr}+\mathbf{a}_r+\mathbf{a}_{Cor} = a_\xi\,\mathbf{i}'+a_\eta\,\mathbf{j}'+a_\zeta\,\mathbf{k}',$$

where

$$a_\xi = -\left(\omega^2R+\frac{\alpha^2\varkappa^2}{R}\cos^2\varkappa t\right)\cos\left(\frac{1}{R}\alpha\sin\varkappa t\right)+\alpha\varkappa^2\sin\varkappa t\sin\left(\frac{1}{R}\alpha\sin\varkappa t\right),$$

$$a_\eta = -\omega\alpha\varkappa\cos\varkappa t\sin\left(\frac{1}{R}\alpha\sin\varkappa t\right),$$

$$a_\zeta = -\frac{\alpha^2\varkappa^2}{R}\cos^2\varkappa t\sin\left(\frac{\alpha}{R}\sin\varkappa t\right)-\alpha\varkappa^2\sin\varkappa t\cos\left(\frac{1}{R}\alpha\sin\varkappa t\right).$$

We shall now consider the relative motion of a rigid body. For this purpose let us imagine a rigid solid associated with a frame of reference $\Omega_1\xi_1\eta_1\zeta_1$ which is in general motion in relation to another frame $\Omega_2\xi_2\eta_2\zeta_2$ likewise in general, though a different motion, in relation to a frame $Oxyz$ assumed to be fixed (Fig. 1.24). The motion of the $\Omega_2\xi_2\eta_2\zeta_2$ frame in relation to the $Oxyz$ frame we shall call *transportation* and the motion of the $\Omega_1\xi_1\eta_1\zeta_1$ frame and of the rigid body along with it we shall call *relative motion.*

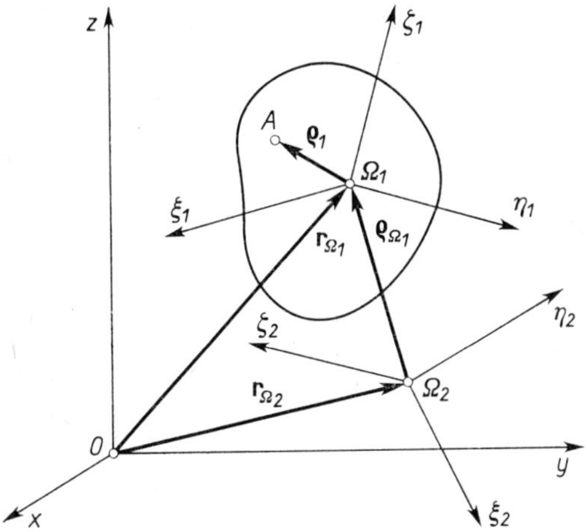

Fig. 1.24

Let the $\Omega_1\xi_1\eta_1\zeta_1$ frame have an angular velocity in absolute motion, i.e., in relation to $Oxyz$, equal ω and the $\Omega_2\xi_2\eta_2\zeta_2$ frame in the same absolute motion an angular velocity $\boldsymbol{\omega}_{tr}$ (angular velocity of transportation). The relative motion of the body, i.e., the motion of the $\Omega_1\xi_1\eta_1\zeta_1$ frame in relation to the $\Omega_2\xi_2\eta_2\zeta_2$ frame, is characterized by the velocity $\mathbf{v}_{\Omega,tr}$ and acceleration $\mathbf{a}_{\Omega,r}$ of point Ω_1 and also by the relative angular velocity $\boldsymbol{\omega}_r$ and the relative angular acceleration $\boldsymbol{\epsilon}_r$. The velocities and accelerations of an arbitrary point of a body in relative motion are expressed by

$$\mathbf{v}_r = \mathbf{v}_{\Omega_1 r}+\boldsymbol{\omega}_r\times\boldsymbol{\rho}_1, \tag{10}$$

$$\mathbf{a}_r = \mathbf{a}_{\Omega_1 r}+\boldsymbol{\epsilon}_r\times\boldsymbol{\rho}_1+\boldsymbol{\omega}_r\times(\boldsymbol{\omega}_r\times\boldsymbol{\rho}_1). \tag{11}$$

Transportation is the motion of a body rigidly fixed to the $\Omega_2\xi_2\eta_2\zeta_2$ frame of reference. The velocity and acceleration of transportation are expressed as follows:

$$\mathbf{v}_{tr} = \mathbf{v}_{\Omega_2}+\boldsymbol{\omega}_{tr}\times(\boldsymbol{\rho}_{\Omega_1}+\boldsymbol{\rho}_1), \tag{12}$$

$$\mathbf{a}_{tr} = \mathbf{a}_{\Omega_2}+\boldsymbol{\epsilon}_{tr}\times(\boldsymbol{\rho}_{\Omega_1}+\boldsymbol{\rho}_1)+\boldsymbol{\omega}_{tr}\times[\boldsymbol{\omega}_{tr}\times(\boldsymbol{\rho}_{\Omega_1}\times\boldsymbol{\rho}_1)]. \tag{13}$$

Knowing $\mathbf{v}_r$, $\mathbf{v}_{tr}$, $\mathbf{a}_r$, and $\mathbf{a}_{tr}$, we can find the absolute velocity and acceleration:

$$\mathbf{v} = \mathbf{v}_{tr}+\mathbf{v}_r = \mathbf{v}_{\Omega_2}+\mathbf{v}_{\Omega_1 r}+\boldsymbol{\omega}_{tr}\times(\boldsymbol{\rho}_{\Omega_1}+\boldsymbol{\rho}_1)+\boldsymbol{\omega}_r\times\boldsymbol{\rho}_1, \tag{14}$$

$$\begin{aligned}\mathbf{a} &= \mathbf{a}_{tr}+\mathbf{a}_r+\mathbf{a}_{Cor} = \mathbf{a}_{\Omega_2}+\mathbf{a}_{\Omega_1 r}+\boldsymbol{\epsilon}_{tr}\times(\boldsymbol{\rho}_{\Omega_1}+\boldsymbol{\rho}_1)+\\ &\quad+\boldsymbol{\omega}_r\times(\boldsymbol{\omega}_r\times\boldsymbol{\rho}_1)+\boldsymbol{\epsilon}_r\times\boldsymbol{\rho}_1+\boldsymbol{\omega}_{tr}\times[\boldsymbol{\omega}_{tr}\times(\boldsymbol{\rho}_{\Omega_1}+\boldsymbol{\rho}_1)]+\\ &\quad+2\boldsymbol{\omega}_{tr}\times(v_{\Omega_1 r}+\boldsymbol{\omega}_r\times\boldsymbol{\rho}_1).\end{aligned} \tag{15}$$

The question to be answered next is: In what way are the individual angular magnitudes related to each other? ($\boldsymbol{\omega}$, $\boldsymbol{\epsilon}$, $\boldsymbol{\omega}_{tr}$, $\boldsymbol{\epsilon}_{tr}$, $\boldsymbol{\omega}_r$, $\boldsymbol{\epsilon}_r$.)

Now

$$\mathbf{v} = \mathbf{v}_{\Omega_1}+\boldsymbol{\omega}\times\boldsymbol{\rho}_1.$$

Therefore

$$\mathbf{v}_{\Omega_1}+\boldsymbol{\omega}\times\boldsymbol{\rho}_1 = \mathbf{v}_{\Omega_2}+\mathbf{v}_{\Omega_1 r}+\boldsymbol{\omega}_{tr}\times(\boldsymbol{\rho}_{\Omega_1}+\boldsymbol{\rho}_1)+\boldsymbol{\omega}_r\times\boldsymbol{\rho}_1. \tag{16}$$

But

$$\mathbf{v}_{\Omega_1} = \mathbf{v}_{\Omega_2}+\mathbf{v}_{\Omega_1 r}+\boldsymbol{\omega}_{tr}\times\boldsymbol{\rho}_{\Omega_1}. \tag{17}$$

Considering (17) in (16), we have

$$(\boldsymbol{\omega}-\boldsymbol{\omega}_{tr}-\boldsymbol{\omega}_r)\times\boldsymbol{\rho}_1 = \mathbf{0}. \tag{18}$$

Since the vector $\boldsymbol{\rho}_1$ is arbitrary, the following equality occurs:

$$\boldsymbol{\omega} = \boldsymbol{\omega}_{tr}+\boldsymbol{\omega}_r. \tag{19}$$

This means that the angular velocity of a body in absolute motion is the sum of angular velocity of transportation $\boldsymbol{\omega}_{tr}$ and angular velocity of relative motion $\boldsymbol{\omega}_r$.

We shall now examine angular accelerations.

The absolute acceleration of an arbitrary point of a body is

$$\mathbf{a} = \mathbf{a}_{\Omega_1}+\boldsymbol{\epsilon}\times\boldsymbol{\rho}_1+\boldsymbol{\omega}\times(\boldsymbol{\omega}\times\boldsymbol{\rho}_1). \tag{20}$$

Comparing the right-hand sides of (20) and (15) we obtain

$$\begin{aligned}&\mathbf{a}_{\Omega_1}+\boldsymbol{\epsilon}\times\boldsymbol{\rho}_1+\boldsymbol{\omega}\times(\boldsymbol{\omega}\times\boldsymbol{\rho}_1)\\ &\quad= \mathbf{a}_{\Omega_2}+\mathbf{a}_{\Omega_1 r}+\boldsymbol{\epsilon}_{tr}\times(\boldsymbol{\rho}_{\Omega_1}+\boldsymbol{\rho}_1)+\boldsymbol{\omega}_r\times(\boldsymbol{\omega}_r\times\boldsymbol{\rho}_1)+\boldsymbol{\epsilon}_r\times\boldsymbol{\rho}_1+\\ &\quad+\boldsymbol{\omega}_{tr}\times[\boldsymbol{\omega}_{tr}\times(\boldsymbol{\rho}_{\Omega_1}+\boldsymbol{\rho}_1)]+2\boldsymbol{\omega}_{tr}\times(\mathbf{v}_{\Omega_1 r}+\boldsymbol{\omega}_r\times\boldsymbol{\rho}_1).\end{aligned}$$

Considering that

$$\mathbf{a}_{\Omega_1} = \mathbf{a}_{\Omega_2}+\mathbf{a}_{\Omega_1 r}+\boldsymbol{\omega}_{tr}\times(\boldsymbol{\omega}_{tr}\times\boldsymbol{\rho}_{\Omega_1})+2\boldsymbol{\omega}_{tr}\times\mathbf{v}_{\Omega_1 r}+\boldsymbol{\epsilon}_{tr}\times\boldsymbol{\rho}_{\Omega_1},$$

we can write

$$\boldsymbol{\epsilon}\times\boldsymbol{\rho}_1+\boldsymbol{\omega}\times(\boldsymbol{\omega}\times\boldsymbol{\rho}_1) = \boldsymbol{\epsilon}_{tr}\times\boldsymbol{\rho}_1+\boldsymbol{\omega}_r\times(\boldsymbol{\omega}_r\times\boldsymbol{\rho}_1)+\boldsymbol{\epsilon}_r\times\boldsymbol{\rho}_1+ +\boldsymbol{\omega}_{tr}\times(\boldsymbol{\omega}_{tr}\times\boldsymbol{\rho}_1)+2\boldsymbol{\omega}_{tr}\times(\boldsymbol{\omega}_r\times\boldsymbol{\rho}_1).$$

Considering result (19) on the left-hand side, we have

$$\boldsymbol{\epsilon}\times\boldsymbol{\rho}_1+\boldsymbol{\omega}_{tr}\times(\boldsymbol{\omega}_{tr}\times\boldsymbol{\rho}_1)+\boldsymbol{\omega}_r\times(\boldsymbol{\omega}_{tr}\times\boldsymbol{\rho}_1)+\boldsymbol{\omega}_{tr}\times(\boldsymbol{\omega}_r\times\boldsymbol{\rho}_1)+ +\boldsymbol{\omega}_r\times(\boldsymbol{\omega}_r\times\boldsymbol{\rho}_1)$$
$$= \boldsymbol{\epsilon}_{tr}\times\boldsymbol{\rho}_1+\boldsymbol{\omega}_r\times(\boldsymbol{\omega}_r\times\boldsymbol{\rho}_1)+\boldsymbol{\epsilon}_r\times\boldsymbol{\rho}_1+\boldsymbol{\omega}_{tr}\times(\boldsymbol{\omega}_{tr}\times\boldsymbol{\rho}_1)+2\boldsymbol{\omega}_{tr}\times(\boldsymbol{\omega}_r\times\boldsymbol{\rho}_1),$$

which when simplified gives

$$\boldsymbol{\epsilon}\times\boldsymbol{\rho}_1+\boldsymbol{\omega}_r\times(\boldsymbol{\omega}_{tr}\times\boldsymbol{\rho}_1) = \boldsymbol{\epsilon}_{tr}\times\boldsymbol{\rho}_1+\boldsymbol{\epsilon}_r\times\boldsymbol{\rho}_1+\boldsymbol{\omega}_{tr}\times(\boldsymbol{\omega}_r\times\boldsymbol{\rho}_1).$$

Note that by use of the identity (25) given in the Introduction, we obtain

$$\boldsymbol{\omega}_r\times(\boldsymbol{\omega}_{tr}\times\boldsymbol{\rho}_1)-\boldsymbol{\omega}_{tr}\times(\boldsymbol{\omega}_r\times\boldsymbol{\rho}_1)$$
$$= \boldsymbol{\omega}_{tr}(\boldsymbol{\omega}_r\boldsymbol{\rho}_1)-\boldsymbol{\rho}_1(\boldsymbol{\omega}_r\boldsymbol{\omega}_{tr})-\boldsymbol{\omega}_r(\boldsymbol{\omega}_{tr}\boldsymbol{\rho}_1)+\boldsymbol{\rho}_1(\boldsymbol{\omega}_{tr}\boldsymbol{\omega}_r)$$
$$= \boldsymbol{\rho}_1\times(\boldsymbol{\omega}_{tr}\times\boldsymbol{\omega}_r).$$

Therefore we have

$$(\boldsymbol{\epsilon}-\boldsymbol{\epsilon}_{tr}-\boldsymbol{\epsilon}_r-\boldsymbol{\omega}_{tr}\times\boldsymbol{\omega}_r)\times\boldsymbol{\rho}_1 = 0. \tag{21}$$

Since relation (21) occurs for any non-zero vector $\boldsymbol{\rho}_1$, we have

$$\boldsymbol{\epsilon} = \boldsymbol{\epsilon}_{tr}+\boldsymbol{\epsilon}_r+\boldsymbol{\omega}_{tr}\times\boldsymbol{\omega}_r. \tag{22}$$

The result (22) can also be obtained by formal differentiation of the angular velocity as expressed by equation (19).

$$\boldsymbol{\epsilon} = \frac{d\boldsymbol{\omega}_{tr}}{dt}+\frac{d\boldsymbol{\omega}_r}{dt}.$$

Using (6), we have

$$\boldsymbol{\epsilon} = \boldsymbol{\omega}_{tr}\times\boldsymbol{\omega}_{tr}+\frac{\check{d}\boldsymbol{\omega}_{tr}}{dt}+\boldsymbol{\omega}_{tr}\times\boldsymbol{\omega}_r+\frac{\check{d}\boldsymbol{\omega}_r}{dt}$$
$$= \frac{\check{d}\boldsymbol{\omega}_{tr}}{dt}+\boldsymbol{\omega}_{tr}\times\boldsymbol{\omega}_r+\boldsymbol{\omega}_r\times\boldsymbol{\omega}_r+\frac{\check{\check{d}}\boldsymbol{\omega}_r}{dt}$$
$$= \frac{\check{d}\boldsymbol{\omega}_{tr}}{dt}+\frac{\check{\check{d}}\boldsymbol{\omega}_r}{dt}+\boldsymbol{\omega}_{tr}\times\boldsymbol{\omega}_r.$$

The above result accords with the result (22).

In this way, we have established that the absolute angular acceleration $\boldsymbol{\epsilon}$ is equal to the sum of the angular acceleration of transportation, the relative angular acceleration, and the product of $\boldsymbol{\omega}_{tr}\times\boldsymbol{\omega}_r$.

1.4 Plane motion of a rigid body

Plane motion of a rigid body is such that any particle of the body moves in one plane only. It is easy to visualize that this is possible only when the individual planes run parallel to each other. In plane motion, all points lying in straight lines at right angles to these planes move in exactly the same way. Consequently, it will be sufficient to examine the motion of points in just one plane.

We shall introduce two plane frames of reference Oxy and $\Omega\xi\eta$ (Fig. 1.25). The latter is in motion and rigidly attached to the moving body. The velocity and acceleration of any particle of the body is expressed by

$$\mathbf{v} = \mathbf{v}_\Omega + \boldsymbol{\omega} \times \boldsymbol{\rho} \tag{1}$$

and

$$\mathbf{a} = \mathbf{a}_\Omega + \boldsymbol{\epsilon} \times \boldsymbol{\rho} + \boldsymbol{\omega} \times (\boldsymbol{\omega} \times \boldsymbol{\rho}) \tag{2}$$

or, using identity (25) from the Introduction, we have

$$\mathbf{a} = \mathbf{a}_\Omega + \boldsymbol{\epsilon} \times \boldsymbol{\rho} - \boldsymbol{\rho}\omega^2. \tag{2a}$$

We shall first deal with velocity.

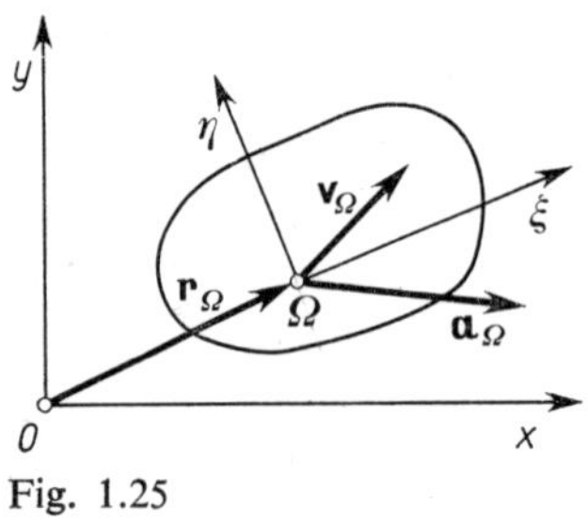

Fig. 1.25

Note that the velocity of an arbitrary particle of a body in plane motion is the geometric sum of the velocity of another arbitrary particle and the velocity which the body would have if it remained in angular motion about that particle; plane motion is a superposition of translatory and angular motion as a special case of rotation about a fixed point. Passing from A to B (Fig. 1.26), we have

$$\mathbf{v}_B = \mathbf{v}_A + \mathbf{v}_{BA}, \tag{3}$$

where $\mathbf{v}_A$ is a velocity in translatory motion, and $\mathbf{v}_{BA} = \boldsymbol{\omega} \times \boldsymbol{\rho}_{AB}$ velocity in angular motion with its centre at point A. The vector $\mathbf{v}_{BA}$ is at right angles to $\boldsymbol{\rho}_{AB}$.

This last statement is relevant to the graphical methods used in the kinematics of plane motion. For it has to be remembered that if the vector $\mathbf{v}_A$ is moved to B and a straight line at right angles to AB passes through its end, somewhere in that line must be the velocity vector of point B. Of course, to find $\mathbf{v}_B$ a certain additional condition is needed, e.g. the direction or the angular velocity $\boldsymbol{\omega}$.

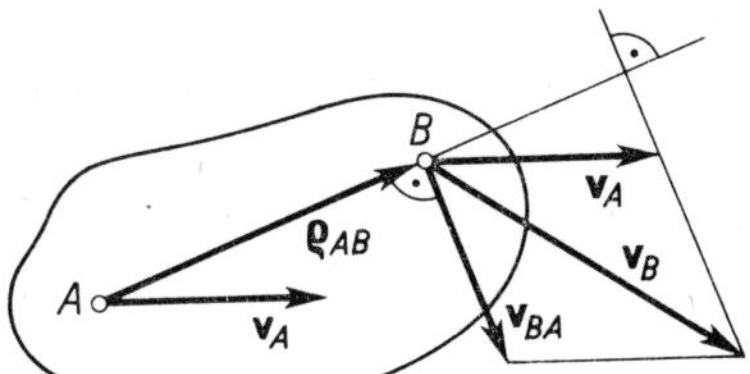

Fig. 1.26

We shall now look for such a point Q, called the *instantaneous centre of velocities*, at which velocity is equal to zero. Let us assume that such a point exists and its position is defined by the vector

$$\boldsymbol{\rho}_Q = \alpha_Q \mathbf{v}_A + \beta_Q \boldsymbol{\omega} \times \mathbf{v}_A. \tag{4}$$

Using equations (1) and (4), we write

$$\mathbf{v}_Q = \mathbf{v}_A + \alpha_Q \boldsymbol{\omega} \times \mathbf{v}_A + \beta_Q \boldsymbol{\omega} \times (\boldsymbol{\omega} \times \mathbf{v}_A) = 0. \tag{5}$$

Since

$$\boldsymbol{\omega} \times (\boldsymbol{\omega} \times \mathbf{v}_A) = \boldsymbol{\omega}(\boldsymbol{\omega} \cdot \mathbf{v}_A) - \mathbf{v}_A \omega^2$$

and in plane motion $\boldsymbol{\omega} \cdot \mathbf{v}_A = 0$ (perpendicular vectors),

$$\mathbf{v}_A(1 - \beta_Q \omega^2) + \alpha_Q \boldsymbol{\omega} \times \mathbf{v}_A = 0. \tag{6}$$

For the above equation to be satisfied

$$1 - \beta_Q \omega^2 = 0, \qquad \alpha_Q = 0; \tag{7}$$

hence

$$\boldsymbol{\rho}_Q = \frac{\boldsymbol{\omega} \times \mathbf{v}_A}{\omega^2}. \tag{8}$$

This shows that the vector $\boldsymbol{\rho}_Q$ is at right angles to v_A. If from any two points straight lines at right angles to their velocities are drawn, then the point of intersection of these lines will be the instantaneous centre of velocities (Fig. 1.27). The geometric locus of instantaneous centres of velocities in a fixed frame of reference is called the *fixed centroide*, and correspondingly in a moving frame, the *moving centroide*.

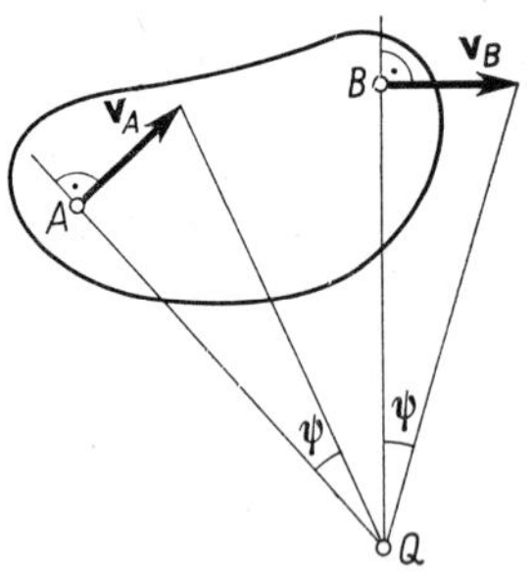

Fig. 1.27

Example 1. A rod AB (Fig. 1.28) is in plane motion so that the point A is sliding along the axis x, whereas C is the point at which the rod is contiguous to an edge situated at height h. Find the equation for the fixed centroide and for the moving centroide.

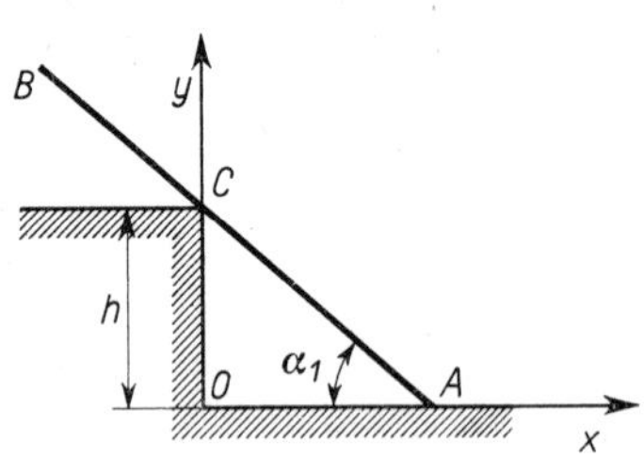

Fig. 1.28

First, a frame of reference is introduced and a variable parameter, the angle α_1, is chosen. The point Q is the centre of instantaneous velocities (Fig. 1.29). In the considered position it is simultaneously a point common to the fixed and moving centroide. In a fixed frame of reference, the coordinates of this point we define by the following relationships:

$$x = h\cot\alpha_1, \quad y = h + x\cot\alpha_1 = h(1+\cot^2\alpha_1).$$

Eliminating parameter α_1 from the equations, we have

$$y = h\left(1+\frac{x^2}{h^2}\right).$$

In moving coordinates

$$\eta = \frac{h}{\sin\alpha_1}, \quad \xi = \eta\cot\alpha_1 = \frac{h\cos\alpha_1}{\sin^2\alpha_1}.$$

Eliminating the parameter α_1 from the frame as in the previous case, we obtain the equation of the moving centroide

$$\eta^2 - h^2\eta^2 - h^2\xi^2 = 0.$$

During motion, the moving centroide is rolling on the fixed centroide.

Problem. Solve graphically the problem considered in Example 1 (Fig. 1.29).

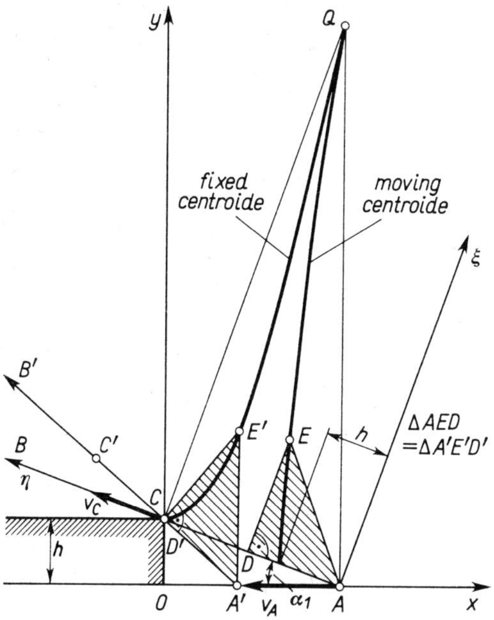

Fig. 1.29

Acceleration in plane motion is obtained from (2). Analogously to velocity, the acceleration of any particle B of a body in plane motion is the sum of the acceleration of point A in translatory motion and the acceleration of point B moving in a circle with centre at point A. Therefore,

$$\mathbf{a}_B = \mathbf{a}_A + \mathbf{a}_{BA}. \tag{9}$$

The acceleration $\mathbf{a}_{BA}$ consists of the tangential acceleration $\mathbf{a}_{BA}^t = \boldsymbol{\varepsilon} \times \boldsymbol{\rho}_{AB}$ an the normal acceleration $\mathbf{a}_{BA}^n = -\omega^2 \boldsymbol{\rho}_{AB}$. Equation (9) can be written in the following form:

$$\mathbf{a}_B = \mathbf{a}_A + \mathbf{a}_{BA}^n + \mathbf{a}_{BA}^t, \tag{10}$$

where $\mathbf{a}_{BA}^n$ has the direction of the radius $\boldsymbol{\rho}_{AB}$ and $\mathbf{a}_{AB}^t$ has its direction at right angles to it (Fig. 1.30).

We shall now look for a point P, called the *instantaneous centre of acceleration* whose acceleration is equal to zero. We assume that such a point exists and that its position is defined by the vector

$$\boldsymbol{\rho}_P = \alpha_P \mathbf{a}_A + \beta_P \boldsymbol{\epsilon} \times \mathbf{a}_A. \tag{11}$$

In accordance with (2)

$$\mathbf{a}_P = \mathbf{a}_A + \boldsymbol{\epsilon} \times \boldsymbol{\rho}_P - \omega^2 \boldsymbol{\rho}_P = 0. \tag{12}$$

Considering (11) in (12), we have

$$\mathbf{a}_A + \alpha_P \boldsymbol{\epsilon} \times \mathbf{a}_A + \beta_P \boldsymbol{\epsilon} \times (\boldsymbol{\epsilon} \times \mathbf{a}_A) - \alpha_P \omega^2 \mathbf{a}_A - \beta_P \omega^2 \boldsymbol{\epsilon} \times \mathbf{a}_A = 0.$$

Simplifying, we obtain

$$\mathbf{a}_A(1 - \alpha_P \omega^2 - \beta_P \varepsilon^2) + (\alpha_P - \beta_P \omega^2) \boldsymbol{\epsilon} \times \mathbf{a}_A = 0, \tag{13}$$

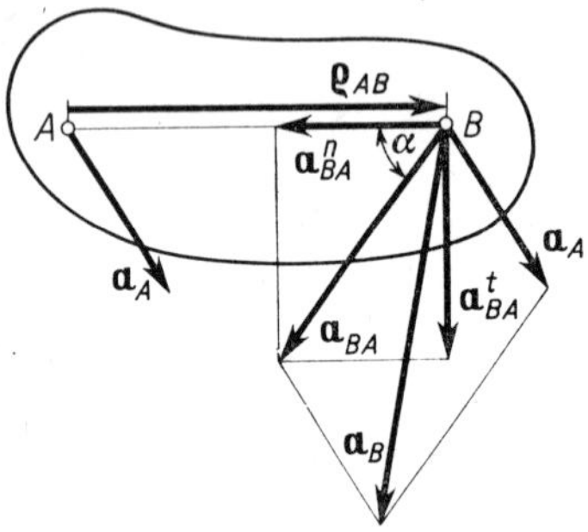

Fig. 1.30

since in plane motion $\boldsymbol{\epsilon} \mathbf{a}_A = 0$. By virtue of (13)

$$1 - \alpha_P \omega^2 - \beta_P \varepsilon^2 = 0, \qquad \alpha_P - \beta_P \omega^2 = 0; \tag{13a}$$

hence

$$\alpha_P = \frac{\omega^2}{\omega^4 + \varepsilon^2}, \qquad \beta_P = \frac{1}{\omega^4 + \varepsilon^2}. \tag{13b}$$

Therefore,

$$\boldsymbol{\rho}_P = \frac{1}{\omega^4 + \varepsilon^2} (\omega^2 \mathbf{a}_A + \boldsymbol{\epsilon} \times \mathbf{a}_A). \tag{14}$$

The vector $\boldsymbol{\rho}_P$ defines the position of the instantaneous centre of accelerations. Calculating the modulus, we obtain

$$\rho_P = \frac{a_A}{\sqrt{\omega^4 + \varepsilon^2}}. \tag{15}$$

The vector $\boldsymbol{\rho}_P$ is attached at point A and it makes an angle α with the vector $\mathbf{a}_A$. This angle we find from the following relationship:

$$\cos\alpha = \frac{\boldsymbol{\rho}_P}{\varrho_P}\,\frac{\mathbf{a}_P}{a_A} = \frac{\omega^2}{\sqrt{\omega^4+\varepsilon^2}}. \tag{16}$$

The tangent of this angle, as it will readily be seen, is expressed by a simple equation, namely

$$\tan\alpha = \frac{\varepsilon}{\omega^2}. \tag{17}$$

Knowing the position of the instantaneous centre of accelerations, the determination of accelerations of other points of the body is greatly facilitated. For, starting out with P, the acceleration is derived from the following relationship:

$$\mathbf{a} = \boldsymbol{\epsilon}\times\boldsymbol{\rho}-\omega^2\boldsymbol{\rho} = \mathbf{a}^t+\mathbf{a}^n, \tag{18}$$

therefore it is exactly the same as in angular motion.

Note further that the vector $\boldsymbol{\rho}_P$ varies with time.

Example 2. Let us go back to Example 1. Let the point A have a velocity $\mathbf{v}_A$ and acceleration $\mathbf{a}_A$. The question to be answered is this: what will be the acceleration of the particles B and C of the rod AB (Fig. 1.31)?

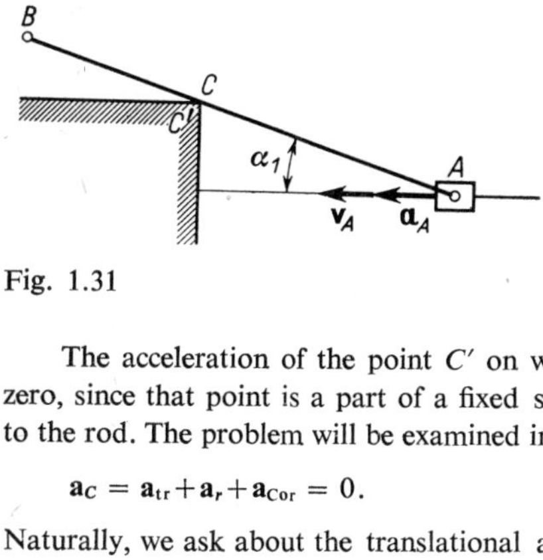

Fig. 1.31

The acceleration of the point C' on which the rod is sliding is obviously equal to zero, since that point is a part of a fixed support. However, it is not stationary relative to the rod. The problem will be examined in the aspect to relative motion:

$$\mathbf{a}_C = \mathbf{a}_{tr}+\mathbf{a}_r+\mathbf{a}_{Cor} = 0. \tag{a}$$

Naturally, we ask about the translational acceleration, since it is the acceleration of the particle C of the rod. It can be resolved into the acceleration in the direction of the rod axis $\mathbf{a}_{tr}^n$ and the acceleration at right angles to that direction $\mathbf{a}_{tr}^t$. Equation (a) can be rewritten in the following form:

$$\mathbf{a}_{tr}^t+\mathbf{a}_{tr}^n+\mathbf{a}_r+2\omega\times\mathbf{v}_r = 0. \tag{b}$$

Note that for equation (b) to be satisfied, it is necessary and sufficient to have

$$\mathbf{a}_r+\mathbf{a}_{tr}^n = 0, \quad \mathbf{a}_{tr}^t+2\boldsymbol{\omega}\times\mathbf{v}_r = 0. \tag{c}$$

It follows that vectors at right angles to each other add up to zero only when each of them is simultaneously equal to zero. From Fig. 1.32 we have

$$v_{CA} = v_A \sin\alpha_1 . \tag{d}$$

But

$$\omega = \frac{v_{CA}}{r_{CA}} = \frac{v_A \sin^2\alpha_1}{h} . \tag{e}$$

Further,

$$a^n_{tr} = -\frac{v^2_{CA}}{r_{CA}} + a_A \cos\alpha_1 = -\frac{v^2_A \sin^3\alpha_1}{h} + a_A \cos\alpha_1 , \tag{f}$$

$$a^t_{tr} = 2\omega v_C = 2\frac{v^2_A \sin^2\alpha_1}{h} \cos\alpha_1 . \tag{g}$$

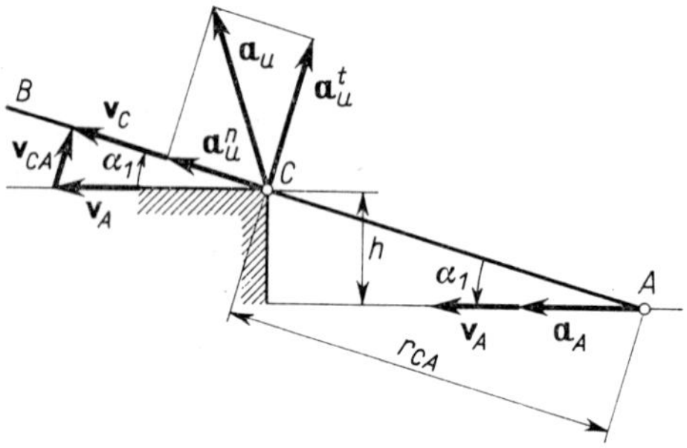

Fig. 1.32

The next step is to determine the instantaneous centre of accelerations. Note first that

$$a^t_{tr} = a^t_A + a^t_{CA} . \tag{h}$$

Hence

$$a^t_{CA} = a^t_{tr} - a^t_A = 2\frac{v^2_A \sin^2\alpha_1}{h} \cos\alpha_1 + a_A \sin\alpha_1 . \tag{i}$$

Further

$$\varepsilon = \frac{a^t_{CA}}{r_{CA}} = \frac{a^t_{AC} \sin\alpha_1}{h} = 2\frac{v^2_A \sin^3\alpha_1}{h^2} \cos\alpha_1 + \frac{a_A \sin^2\alpha_1}{h} . \tag{j}$$

We can nov find the position vector of the instantaneous centre of accelerations ϱ_P, originating from both A and C. Knowing the acceleration of the two points, we calculate the angle α given by (17):

$$\tan\alpha = \frac{\varepsilon}{\omega^2} = 2\cot\alpha_1 + \frac{a_A h}{v^2_A \sin^2\alpha_1} . \tag{k}$$

By virtue of (15), the following relationship is legitimate:

$$\frac{a_A}{PA} = \frac{a_C}{PC} = \frac{a_B}{PB} = \sqrt{\varepsilon^2 + \omega^4}, \tag{l}$$

which justifies the construction of accelerations given in Fig. 1.33.

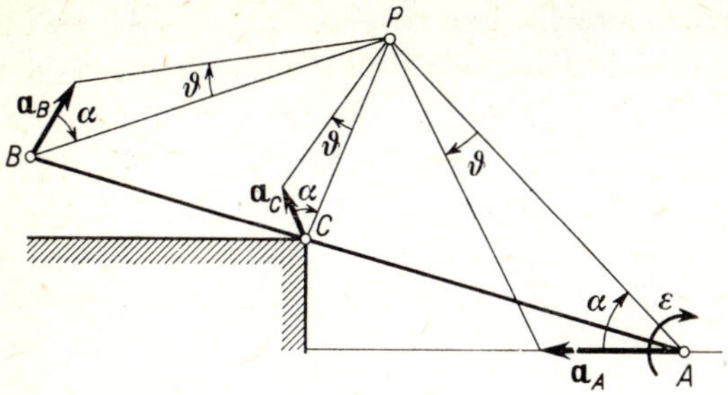

Fig. 1.33

1.5 Examples of the determination of velocities and accelarations in the motion of plane mechanisms

We begin with the kinematics of a crank-slide system (Fig. 1.34). It is a triple-hinged system, in which the point C is in motion. The system has one degree of freedom.

We shall examine the method by which the velocity and acceleration of this point can be determined. Let us assume that the angular velocity

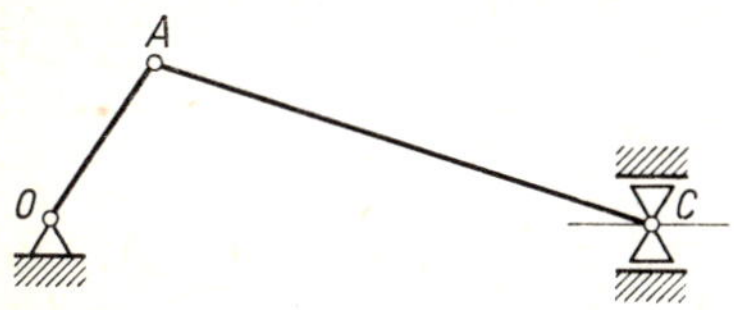

Fig. 1.34

of the crank CA is constant $\omega = 2\pi n$, where n denotes the number of revolutions in unit time. To find the velocity we shall start out with the fact that the velocity projections of two points of a rigid body on a straight line connecting these points are exactly the same. The construction of the velocity vector of C is shown in Fig. 1.35. Moving the vector $\mathbf{v}_A$ to C and

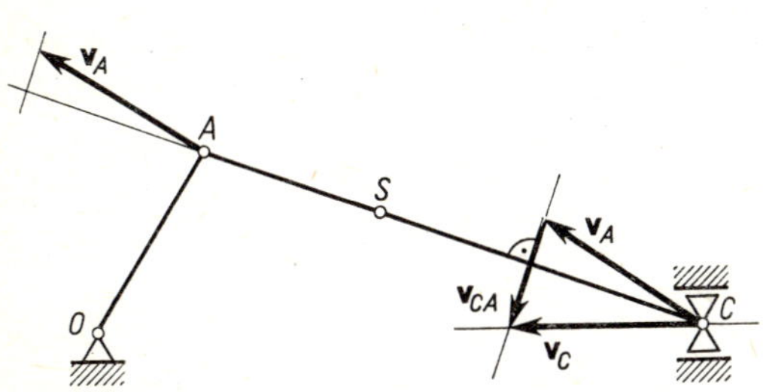

Fig. 1.35

then drawing through its end a straight line perpendicular to AC to the point where it intersects the straight line specifying the direction in which C moves, we determine the velocity $\mathbf{v}_C$. To find the velocity of any point S in the section AC, a polygon of velocities is constructed (Fig. 1.36).

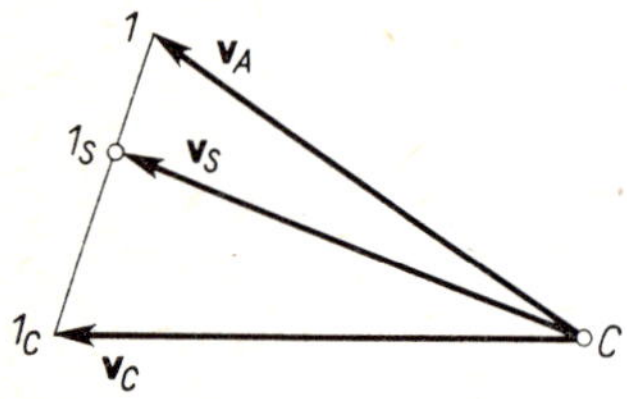

Fig. 1.36

The point 1_S divides the section $(1\text{–}1_C)$ in the same ratio as does the point S the section AC. The section $(1\text{–}1_S)$ is proportional to the angular velocity about point A, which checks out our construction. The velocity scale (μ_v) chosen should be such that the length of the velocity vector at point A is equal to the length of OA in the drawing. Therefore,

$$v_A = \mu_v C1; \tag{1}$$

hence

$$\mu_v = \frac{v_A}{C1} = \frac{\omega r}{OA}. \tag{2}$$

Taking a particular case, we assume: number of revolutions, $n = 200$ r.p.m., $r = 20$ cm, $01 = 2$ cm. Substituting, we have

$$\mu_v = \frac{2\pi \cdot 200 \cdot 20}{2} = 0.21 \text{ ms}^{-1}/\text{mm}.$$

A section 1 mm long corresponds, therefore, to a velocity of 0.21 m/s.

We shall now determine the accelerations. The motion of the coupler AC, being plane motion, is a superposition of translatory and angular motions. It can be treated, therefore, as a resultant motion comprised of translatory motion with the velocity of point A and angular motion about this point. The acceleration of point C is the vector sum of accelerations:

$$\mathbf{a}_C = \mathbf{a}_A + \mathbf{a}_{CA}. \tag{3}$$

We shall first construct the acceleration of the point A. On account of the constant angular velocity of the crank, it is a normal acceleration.

The construction of normal acceleration is shown in Fig. 1.37. By elementary geometry we have

$$AB:AO = AC:AB.$$

Hence

$$\frac{(AB)^2}{AO} = AC. \tag{4}$$

If the section AB is proportional in the scale μ_v to the velocity and the section OA is proportional in the construction scale μ_k to the real distance between A and B, then

$$AB = \frac{v}{\mu_v}, \tag{5}$$

$$AO = \frac{L_{AO}}{\mu_k}. \tag{6}$$

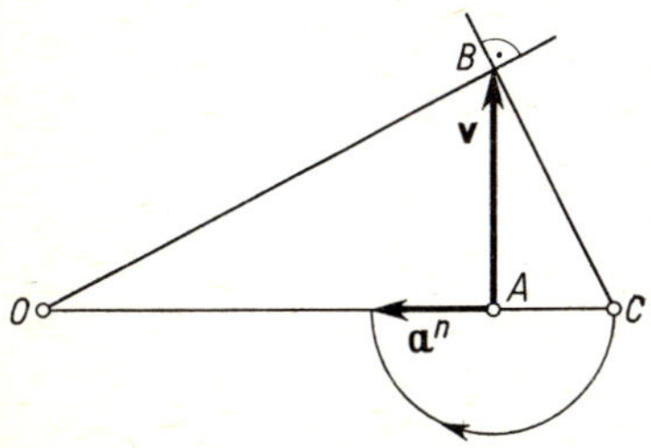

Fig. 1.37

Considering (5) and (6) in (4), we have

$$\frac{v^2}{L_{AO}} \frac{\mu_k}{\mu_v^2} = AC$$

or, after a transformation,

$$\frac{v^2}{L_{AO}} = AC \frac{\mu_v^2}{\mu_k} = a^n.$$

Evidently, the section AC is proportional to the normal acceleration, and moreover the acceleration scale coefficient is dependent on μ_v and μ_k, and is expressed by

$$\mu_a = \frac{\mu_v^2}{\mu_k}. \tag{7}$$

In Fig. 1.38 we make use of the construction of normal acceleration twice, firstly to determine the acceleration of point A and secondly to determine the acceleration $\mathbf{a}_{CA}^n$.

The operations required to find the acceleration of point C are carried out in the following order:

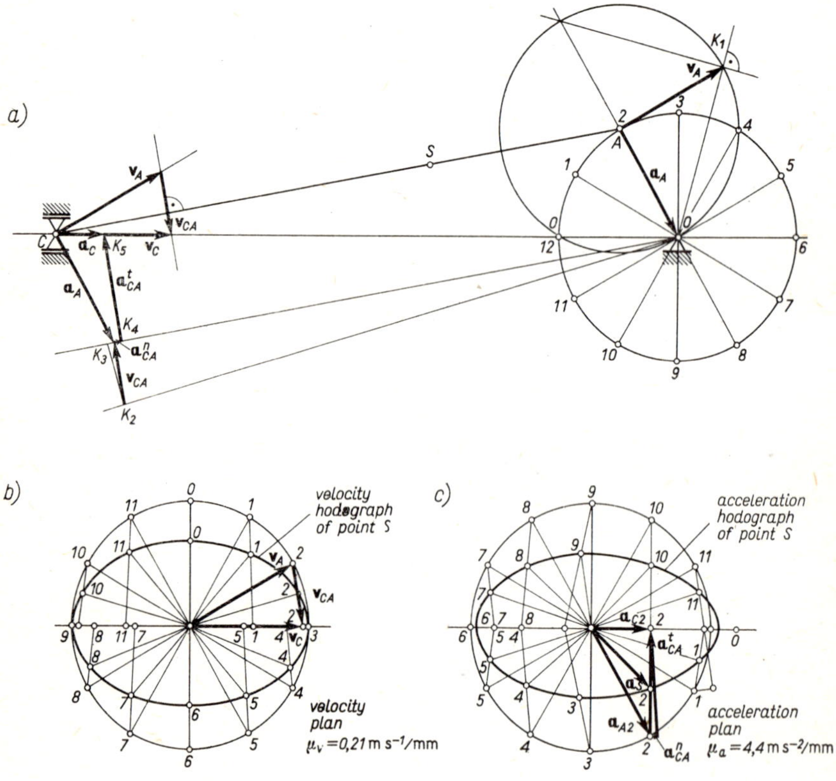

Fig. 1.38

(1) We construct the acceleration of point A.

(2) We translate the acceleration of point A to point C.

(3) We construct the acceleration $\mathbf{a}_{CA}^n = v_{CA}^2/L_{CA}$.

(4) Through the end of the vector $\mathbf{a}_{CA}^n$ we draw straight line perpendicular to it.

(5) We find the point of intersection of the line just named with the straight line giving the direction of motion of C. The section connecting

this point with C is proportional to the acceleration of C. On account of the adopted construction of normal acceleration, the scale coefficient $\mu_a = \mu_v^2/\mu_k$ is valid.

Now

$$\mu_v = 0.21\ \mathrm{ms^{-1}/mm} \quad \text{and} \quad \mu_k = 10,$$

therefore

$$\mu_a = \left(\frac{0.21\ \mathrm{ms^{-1}}}{\mathrm{mm}}\right)^2 \cdot \frac{1}{10} = \frac{4.4\ \mathrm{ms^{-2}}}{\mathrm{mm}}.$$

Consequently, a section 1 mm long corresponds to an acceleration of 4.4 m/s².

In conclusion, we shall examine the angular acceleration of the coupler AC. We may calculate it from

$$\varepsilon = \frac{a^t_{CA}}{L_{CA}}. \tag{8}$$

It follows that the sections proportional to the value of accelerations a^t_{CA} are also proportional to ε. The scale coefficient we calculate as follows:

$$\varepsilon = \frac{(K_4K_5)\cdot\mu_a}{CA\cdot\mu_k}; \quad \text{hence} \quad \mu_\varepsilon = \frac{\mu_a}{L_{CA}}. \tag{9}$$

We now come to another example. We shall consider the crank-yoke system shown in Fig. 1.39. The crank OA rotates with constant angular velocity $\boldsymbol{\omega}$. The problem is to find the angular velocity and acceleration of the yoke O_1B.

Note that knowing the angular velocity of the crank and its length, we know at the same time the velocity and acceleration of point A. Considering the problem as a problem of relative motion, we find that the velocity of this point is the sum of velocity of transportation and relative velocity:

$$\mathbf{v}_A = \mathbf{v}_{tr} + \mathbf{v}_r. \tag{10}$$

How the velocity $\mathbf{v}_A$ is resolved is shown in Fig. 1.39. Knowing $\mathbf{v}_{tr}$, we can calculate the angular velocity of the yoke:

$$\boldsymbol{\omega} = \frac{\mathbf{v}_{tr}}{L_{O_1A}}.$$

With regard to accelerations, we note that

$$\mathbf{a}_A = \mathbf{a}_{tr} + \mathbf{a}_r + \mathbf{a}_{Cor}, \tag{11}$$

where $\mathbf{a}_{tr}$ is acceleration of transportation, $\mathbf{a}_r$ relative acceleration, $\mathbf{a}_{Cor}$ the Coriolis acceleration.

Acceleration of transportation is equal to the sum of accelerations: normal $\mathbf{a}_{tr}^n$ having the direction of the rod axis O_1B, and tangential $\mathbf{a}_{tr}^t$. The former we can find by the graphical method, using the construction of normal acceleration, or else its value can be determined from

$$a_{tr}^n = \frac{v_{tr}^2}{L_{O_1A}} = \omega^2 L_{O_1A}. \tag{12}$$

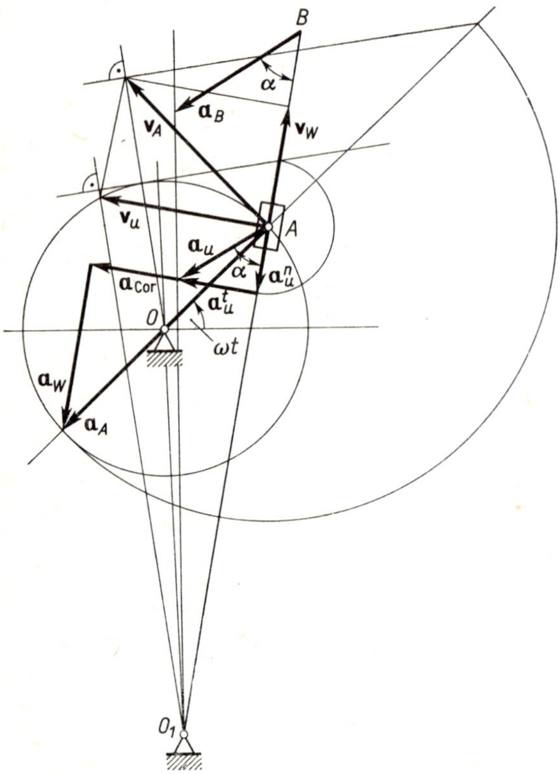

Fig. 1.39

The acceleration $\mathbf{a}_r$ has likewise the direction of the rod O_1B. Therefore the component of the acceleration of point A in the direction of O_1B is

$$\mathbf{a}_{A1} = \mathbf{a}_{tr}^n + \mathbf{a}_r; \tag{13}$$

hence

$$\mathbf{a}_{tr}^n = \mathbf{a}_{A1} - \mathbf{a}_r. \tag{13a}$$

The other component (Fig. 1.39) of the vector $\mathbf{a}_A$:

$$\mathbf{a}_{A2} = \mathbf{a}_{tr}^t + \mathbf{a}_{Cor}. \tag{13b}$$

From (13b) we calculate $\mathbf{a}_{tr}^t = \mathbf{a}_{A2} - \mathbf{a}_{Cor}$.

The Coriolis acceleration $\mathbf{a}_{Cor}$ we find making use of

$$a_{Cor} = 2\frac{v_{tr}}{L_{O_1A}}v_r. \tag{14}$$

Thus we have determined the acceleration of transportation, i.e., the acceleration of a point of the rod O_1B tangential at a given instant to point A of the crank. O_1 is the instantaneous centre of accelerations; therefore, the angular acceleration of the yoke O_1B we calculate by dividing a_{tr}^t by the instantaneous distance O_1A. The construction of the acceleration of point B is also indicated in Fig. 1.39.

Problem. Find the velocity and acceleration of points A and B of the crank-rocker mechanism shown in Fig. 1.40, if it is known that the angular velocity of the crank is constant and equal $\omega = 5 s^{-1}$, crank length $r = 35$ cm, and $L_{AB} = 90$ cm, $R_{BO_1} = 80$ cm and $L_{OB} = 100$ cm.

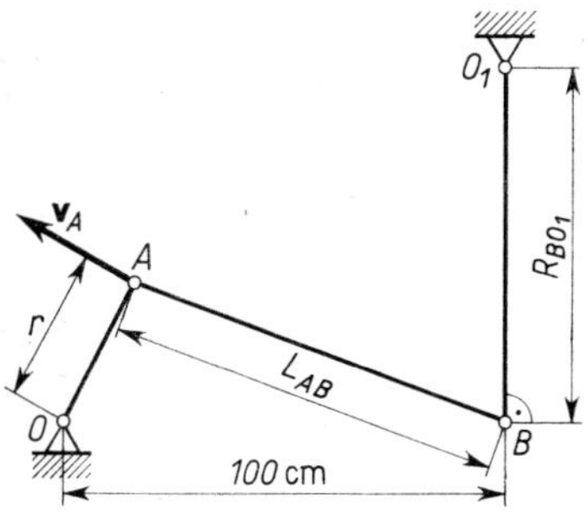

Fig. 1.40

In conclusion we shall consider an example, which is an illustration of apparent difficulty in solving the problem of velocities and accelerations. The system to be examined is given in Fig. 1.41. It consists of three cranks O_1A_1, O_2A_2, O_3A_3 moving at constant angular velocities ω_1, ω_2, ω_3, and in this way setting in motion through intermediary rods A_1B_1, A_2B_2, A_3B_3 the quadrilateral $B_1B_2B_3B_4$. The rod B_4C is also hinged at B_4. The point C is moving along a guide bar in the shape of a circular arc with radius R. The problem is to find the velocities and accelerations of the point C.

The above problem is somewhat more involved than those previously discussed. But it will prove easy to solve by recourse to a certain expedient, used for the first time by Assur. To obtain the velocities and accelerations of B_4, it is necessary beforehand to find the velocity and acceleration of the

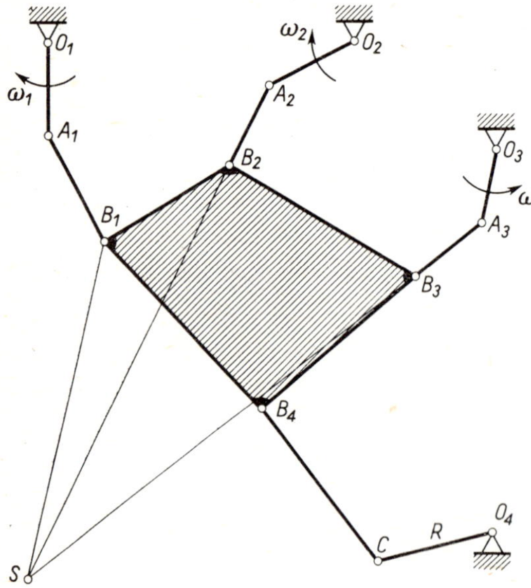

Fig. 1.41

intermediary point S at the intersection of the extended axes of the rods A_2B_2 and A_3B_3 (Fig. 1.41) (but equally well S can be assumed to be the point of intersection of the axes of the rods A_1B_1 and A_2B_3 or A_1B_1 and A_3B_3). This is known as *Assur's point.* The velocity of S we find by reference to the following relationships:

$$\begin{aligned} \mathbf{v}_S &= \mathbf{v}_{B_2} + \mathbf{v}_{SB_2} = \mathbf{v}_{A_2} + \mathbf{v}_{B_2A_2} + \mathbf{v}_{SB_2}, \\ \mathbf{v}_S &= \mathbf{v}_{B_3} + \mathbf{v}_{SB_3} = \mathbf{v}_{A_3} + \mathbf{v}_{B_3A_3} + \mathbf{v}_{SB_3}. \end{aligned} \tag{15}$$

Since the points A_2B_2S and A_3B_3S are correspondingly in the same straight lines, vector $\mathbf{v}_{B_2A}$ has the same direction as the vector $\mathbf{v}_{SB_2}$, and correspondingly the vector $\mathbf{v}_{B_3A_3}$ has the same direction as the vector $\mathbf{v}_{SB_3}$. These are at the same time the directions of the straight lines at right angles to the sections SA_2 and SA_3, respectively. Therefore, we can write

$$\begin{aligned} \mathbf{v}_S &= \mathbf{v}_{A_2} + \mathbf{v}_{SA_2}, \\ \mathbf{v}_S &= \mathbf{v}_{A_3} + \mathbf{v}_{SA_3}, \end{aligned} \tag{16}$$

where

$$\mathbf{v}_{SA_2} = \mathbf{v}_{B_2A_2} + \mathbf{v}_{SB_2} \perp \overline{A_2S},$$

$$\mathbf{v}_{SA_3} = \mathbf{v}_{B_3A_2} + \mathbf{v}_{SB_3} \perp \overline{A_3S}$$

(the symbol $\perp$ denotes perpendicularity).

Figure 1.42 shows the method of successive determination of the velocity of point S, next of point B_1 and lastly of points B_4 and C. To find the velocity of point S, we "translate parallelly" the velocities of points A_2 and A_3 to point S. Through the ends of the vectors with origin at point S we plot two straight lines, one through the end of the vector $\mathbf{v}_{A_2}$ at right angles to the section A_2S and the other through the end of the vector

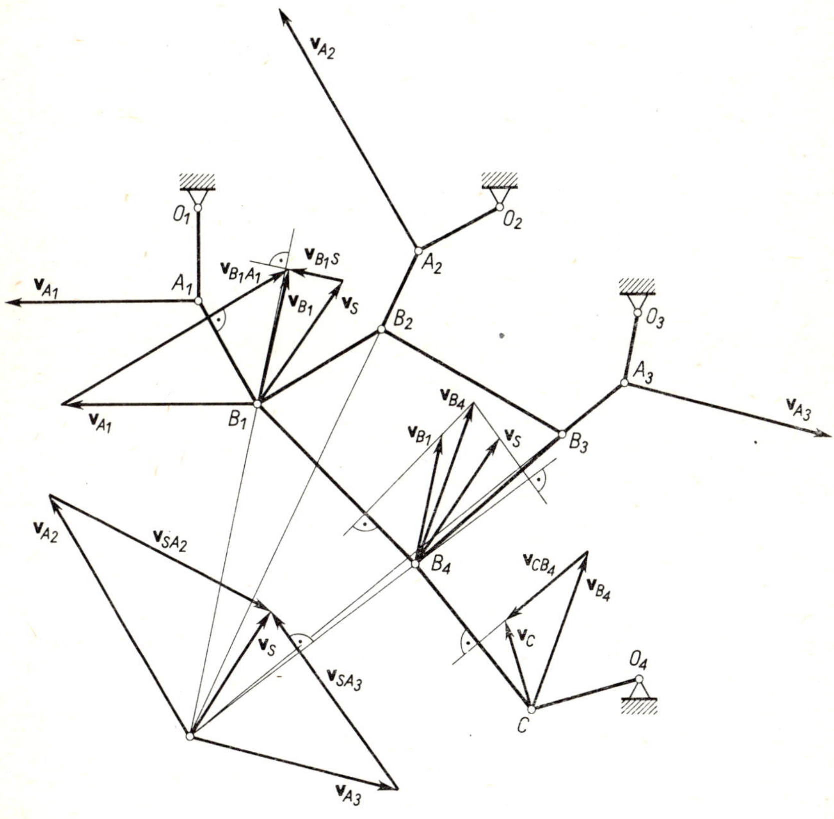

Fig. 1.42

$\mathbf{v}_{A_3}$ at right angles to the section A_3S. The vectors $\mathbf{v}_{SA_2}$ and $\mathbf{v}_{SA_3}$ are respectively in these lines, and so is the end of the velocity vector of point S. Since the end of the vector $\mathbf{v}_S$ is simultaneously placed in both lines, it cannot be situated anywhere else but at their intersection as shown in Fig. 1.42. Having the velocity of point S and knowing the velocity of point A_1, we find, in an analogous manner to the finding of S, the velocity of point B_1. We "translate parallelly" the vectors $\mathbf{v}_{A_1}$ and $\mathbf{v}_S$ to point B_1 and plot next, through the end of the vector $\mathbf{v}_{A_1}$, a straight line at right angles to A_1B_1 and another through the end of the vector $\mathbf{v}_S$ at right angles to SB_1. The point of intersection of these lines is the end of the velocity vector of the point B_1.

The same procedure is followed to find the velocity of B_4, making use of the known velocities of the points S and B_1. Having found the vector $\mathbf{v}_{B_4}$, we can now set out to determine the velocity of C.

If the point O_4 were in motion at speed $\mathbf{v}_{O_4}$, the method of velocity determination for point C would be exactly the same as for S, B_1 or B_4. However, a special case may often prove harder to solve than a more general one. We can then make use of limit property.

Since the velocity of O_4 is equal to zero, the possible velocity of C is one in a direction at right angles to O_4C. Consequently, to solve the problem, it is necessary to "translate parallelly" vector $\mathbf{v}_{B_4}$ to the point C and draw through its end a straight line at right angles to B_4C as far as the intersection with the line perpendicular to the section O_4C. The point of intersection of the two lines gives the end of the velocity vector of point C. The construction of the velocity of C (Fig. 1.42) can be simplified in a way by translating the vectors to a single point as shown in Fig. 1.43.

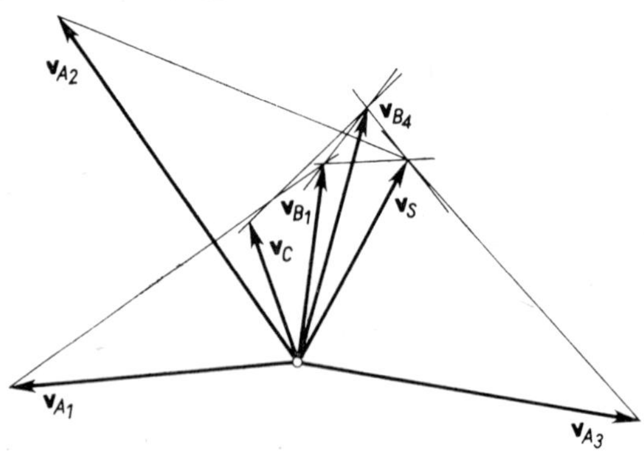

Fig. 1.43

We shall now construct the acceleration of the point C. It is determined in stages, using intermediate points, as in the case of velocity, and in an analogous manner we can therefore write the equations for the relationships between the accelerations of individual points of the system

$$\begin{aligned}\mathbf{a}_S &= \mathbf{a}_{B_2}+\mathbf{a}_{SB_2} = \mathbf{a}_{A_2}+\mathbf{a}_{B_2A_2}+\mathbf{a}_{SB_2}\\ &= \mathbf{a}_{A_2}+\mathbf{a}^n_{B_2A_2}+\mathbf{a}^n_{SB_2}+\mathbf{a}^t_{B_2A_2}+\mathbf{a}^t_{SB_2},\\ \mathbf{a}_S &= \mathbf{a}_{A_3}+a^n_{B_3A_3}+a^n_{SB_3}+a^t_{B_3A_3}+a^t_{SB_3};\end{aligned} \tag{17}$$

we observe here the parallelism of the respective vectors which we shall write down as follows:

$$\mathbf{a}^n_{B_iA_i} \parallel \mathbf{a}^n_{SB_i} \quad \text{for} \quad i = 2, 3, \tag{17a}$$

and

$$\mathbf{a}^t_{B_iA_i} \parallel \mathbf{a}^t_{SB_i}.$$

We can, therefore, write

$$\mathbf{a}_S = \mathbf{a}_{A_i}+\mathbf{a}^n_{SA_i}+\mathbf{a}^t_{SA_i} \qquad (i = 1, 2), \tag{18}$$

where

$$\mathbf{a}^n_{SA_i} = \mathbf{a}^n_{B_iA_i}+\mathbf{a}^n_{SB_i} \perp \overline{A_iS} \qquad (i = 1, 2), \tag{19}$$

$$\mathbf{a}^t_{SA_i} = \mathbf{a}^t_{B_iA_i}+\mathbf{a}^t_{SB_i} \perp \overline{A_iS} \qquad (i = 1, 2). \tag{20}$$

To find the accelerations $\mathbf{a}_{A_i}$, $\mathbf{a}^n_{B_iA_i}$, $\mathbf{a}^n_{SB_i}$ and $\mathbf{a}^n_{CB_4}$, the construction of normal acceleration can be used but then, it will be remembered, the scale coefficient, as defined by equation (7), is valid. Alternatively, one can apply the analytical method, making use of the following relationships:

$$a_{A_i} = \omega^2 L_{O_iA}, \tag{21a}$$

$$a^n_{B_iA_i} = \frac{v^2_{B_iA_i}}{L_{A_iB_i}}, \tag{21b}$$

$$a^n_{SB_i} = \frac{v^2_{SB_i}}{L_{B_iS}}, \tag{21c}$$

$$a^n_{CB_4} = \frac{v^2_{CB_4}}{L_{B_4C}}, \tag{21d}$$

$$a^n_C = \frac{v^2_C}{L_{O_4C}}. \tag{21e}$$

The vectors $\mathbf{v}_{B_iA_i}$, $\mathbf{v}_{SB_i}$, $\mathbf{v}_{CB}$, and $\mathbf{v}_C$ are found from the solution of the first part of the velocity determination problem.

By virtue of (19) and (17a),

$$a_{SA_i} = \frac{v_{B_iA_i}^2}{L_{A_iB_i}} + \frac{v_{SB_i}^2}{L_{B_iS}}. \tag{22}$$

To determine the acceleration of point S, the vectors $\mathbf{a}_{A_2}$ and $\mathbf{a}_{A_3}$ should be "translated parallelly" to the point S. Next the acceleration $\mathbf{a}_{SA_2}^n$ should be added to $\mathbf{a}_{A_2}$ and correspondingly $\mathbf{a}_{SA_3}^n$ to $\mathbf{a}_{A_3}$. Further, straight lines should be drawn through the ends of vectors $\mathbf{a}_{SA_2}^n$ and $\mathbf{a}_{SA_3}^n$ correspondingly at right angles to SA_2 and SA_3. The point of intersection of these lines gives the end of the acceleration vector of point S. Once $\mathbf{a}_S$ has been determined, we can proceed with finding the acceleration of point B_1, since

$$\mathbf{a}_{B_1} = \mathbf{a}_{A_1} + \mathbf{a}_{B_1A_1}^n + \mathbf{a}_{B_1A_1}^t \tag{23}$$

and

$$\mathbf{a}_{B_1} = \mathbf{a}_S + \mathbf{a}_{B_1S}^n + \mathbf{a}_{B_1S}^t. \tag{24}$$

Making the point B_1 the origin of the vectors $\mathbf{a}_{A_1}$ and $\mathbf{a}_S$, we add to them $\mathbf{a}_{B_1A_1}^n$ and $\mathbf{a}_{B_1S}^n$, respectively, and then plot straight lines, in which are the vectors $\mathbf{a}_{B_1A_1}^t$ and $\mathbf{a}_{B_1S}^t$. The point of intersection of these lines defines the end of the vector $\mathbf{a}_{B_1}$ with origin at point B_1.

By analogous procedure we can now determine the vector $\mathbf{a}_{B_4}$, since we have

$$\mathbf{a}_{B_4} = \mathbf{a}_{B_1} + \mathbf{a}_{B_4B_1}^n + \mathbf{a}_{B_4B_1}^t \tag{25}$$

and

$$\mathbf{a}_{B_4} = \mathbf{a}_S + \mathbf{a}_{B_4S}^n + \mathbf{a}_{B_4S}^t. \tag{26}$$

As previously, we "translate parallelly" the vectors $\mathbf{a}_{B_1}$ and $\mathbf{a}_S$ to the point B_4 and add respectively the vectors $\mathbf{a}_{B_4B_1}^n$ and $\mathbf{a}_{B_4S}^n$, and then plot straight lines at right angles to the last two vectors, passing through the ends of the vectors $\mathbf{a}_{B_4B_1}^n$ and $\mathbf{a}_{B_4S}^n$. The point of intersection in this case gives the end of the acceleration vector $\mathbf{a}_{B_4}$ with origin at B_4.

The described method of determining the acceleration of B is not the only one available; in this particular case it would be much simpler to use the method of the instantaneous centre of accelerations.

The determination of the instantaneous centre of accelerations consists of finding the angle between the vector $\mathbf{a}_{B_1S}$ and a straight line passing through the points B_1 and S. Next, a straight line is drawn through B_1

at an angle α to the vector $\mathbf{a}_{B_1}$ (α is laid off to conform to the direction of the angular acceleration $\boldsymbol{\epsilon}$ of the quadrilateral $B_1 B_2 B_3 B_4$). A straight line is also plotted at the same angle from point S to $\mathbf{a}_S$ (see Fig. 1.44). The point of intersection P of these lines is the instantaneous centre of accelerations. After finding P, the acceleration of B_4 can be determined.

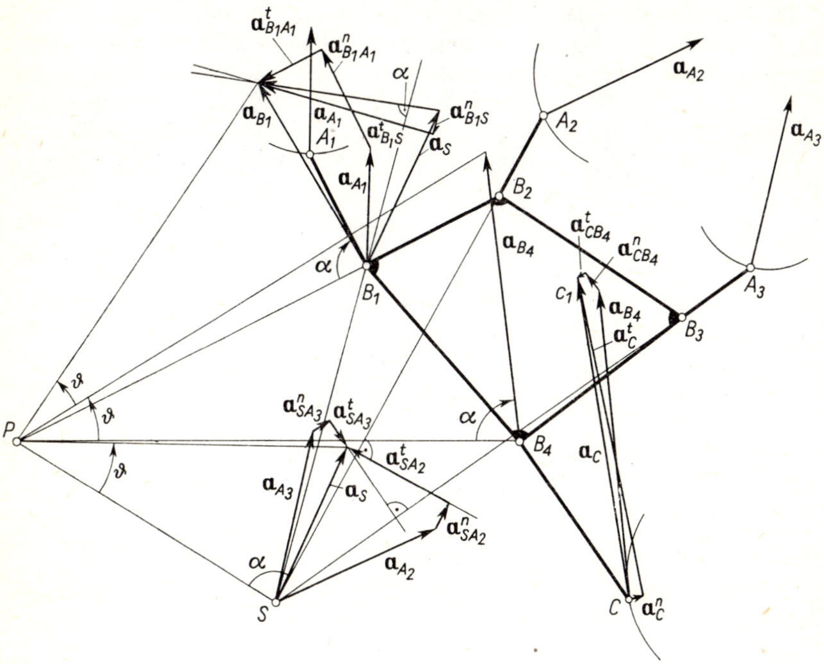

Fig. 1.44

The final part of the problem consists of finding the acceleration of the point C.

$$\mathbf{a}_C = \mathbf{a}_{B_4} + \mathbf{a}^n_{CB_4} + \mathbf{a}^t_{CB_4}, \tag{27}$$

$$\mathbf{a}_C = \mathbf{a}_{O_4} + \mathbf{a}^n_{CO_4} + \mathbf{a}^t_{CO_1} = \mathbf{a}^n_C + \mathbf{a}^t_C, \tag{28}$$

since $\mathbf{a}O_4 \equiv 0$.

We begin with the determination of the normal acceleration of point C, using either the graphical or analytical method. Next, we plot a straight line at right angles to CO_4, somewhere along which is the end of the vector $\mathbf{a}_C$. To find its exact position we "translate parallelly" the vector

$\mathbf{a}_{B_4}$ to the point C and add to it $\mathbf{a}^n_{OB_4}$, to which we construct a perpendicular intersecting the perpendicular to CO_4 at point C. That point of intersection is the end of the acceleration vector of the point C.

As in the case of velocity determination for C, all the vectors could be translated to a single point and a drawing made for accelerations, analogous to Fig. 1.43. The reader is advised to sketch such a drawing as an exercise.

Problem. Determine the velocities and accelerations of points A, B, C and D of the mechanism shown in Fig. 1.45 if the angular velocity of the crank O_1A is known to be constant, $\omega = 5\ s^{-1}$.

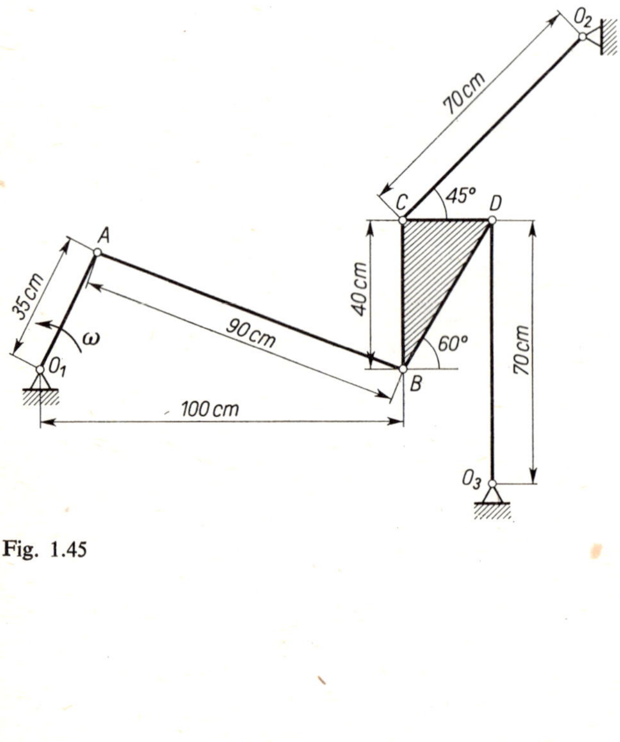

Fig. 1.45

CHAPTER 2

The Dynamics of a Particle

2.1 Fundamental definitions and theorems

In this chapter we shall discuss the dynamic aspects of a particle in motion. We begin with the basic laws of dynamics.

AXIOM 1 (Newton's second law). *If a force* **P** *acts on a particle, the acceleration thus produced is proportional to that force, which can be written as follows*:

$$\mathbf{P} = m\mathbf{a}, \tag{1}$$

where m is the mass of the particle.

We shall treat mass as a primary concept.

The force **P** should be regarded as the resultant of the forces acting on the particle, that is,

$$\mathbf{P} = \sum_{i=1}^{n} \mathbf{P}_i. \tag{1a}$$

The cited law brings into association three basic concepts (force, mass and motion) of mechanics. It is valid in inertial systems (see Chapter 5). Under the SI system, the unit of force is the newton:

$$1\ \mathrm{kg\ m\ s^{-2}} = 1\ \mathrm{N}.$$

For a unit of force we can also take the force with which the earth attracts 1 kg of mass:

$$1\ \mathrm{kgf} = 1\ \mathrm{kg} \cdot g,$$

where $g = 9.80665\ \mathrm{m\ s^{-2}}$ and is the normal value of acceleration of gravity. The equality sign is valid between inert and heavy mass.

AXIOM 2 (Newton's third law). *If a particle A acts on another particle B with force* $\mathbf{P}_{AB}$, *then simultaneously B acts on A with force* $\mathbf{P}_{BA}$ *of equal absolute value but with an opposite sense, i.e.*

$$\mathbf{P}_{AB} + \mathbf{P}_{BA} = 0. \tag{2}$$

This is known as the *law of action and reaction.*

We shall now introduce the definition of work. By *work* one should understand a process in which resistance is being overcome along a certain route. This definition is, however, imprecise, and for that reason it is better to define work in concise mathematical notation:

Work analytically formulated is a curvilinear integral

$$W \overset{\mathrm{df}}{=} \int_A^B \mathbf{P}\cdot \mathrm{d}\mathbf{s}. \tag{3}$$

If under the integral (3) a total differential occurs, then the forces doing the work are said to have *potential V.* A decrease in potential is tantamount to an increase in work:

$$-\delta V = \delta W.$$

Therefore

$$P_i = -\frac{\partial V}{\partial x^i}. \tag{4}$$

Potential forces, as it will easily be seen, act in the direction of the maximum drop of the potential. It can easily be verified that the work of a force in the field of potential force is independent of the route. Indeed,

$$W = -\int_A^B \mathrm{d}V = V_A - V_B. \tag{5}$$

The SI unit of work (Système International d'Unités) is joule

$$1\ \mathrm{J} = 1\ \mathrm{Nm}$$

and under the weight system, 1 kG m.

We shall now consider the concept of energy. If a system is in a state in which it is capable of doing work, it is said to have an *accumulated energy.* This is illustrated by the wound-up spring of a watch, water moving a turbine or a falling body, and the like. The examples we have just given refer only to mechanical energy, ignoring, e.g., heat, nuclear or chemical energy, and so forth. Since energy is measured by the work done, it is expressed in units of work.

We shall now discuss the concept of kinetic energy of motion. If an elementary system (a particle) is to be given some velocity, it has to be subjected to the action of a force **P**.

If the work of force **P** has been wholly used up to impart velocity to a mass m, then from (1) and (3) we have

$$W = \int_A^B m\mathbf{a}\,\mathrm{d}\mathbf{s} = \int_A^B m a_i \,\mathrm{d}x^i = \int_A^B m\frac{\mathrm{d}v_i}{\mathrm{d}t}\mathrm{d}x^i$$

$$= \int_A^B m\frac{\mathrm{d}x^i}{\mathrm{d}t}\mathrm{d}v_i = \int_A^B m v^i \mathrm{d}v_i = m\frac{v^i v_i}{2}\bigg|_A^B. \tag{6}$$

By the summation convention, we obtain

$$v^i v_i = v_1^2+v_2^2+v_3^2 = v^2,$$

mass m = const, therefore

$$W = \tfrac{1}{2}m\underset{B}{v^2}-\tfrac{1}{2}m\underset{A}{v^2}. \tag{6a}$$

Assuming that the velocity at a point A was zero and $\mathbf{v}_B = \mathbf{v}$, we obtain

$$W = E = \tfrac{1}{2}mv^2. \tag{7}$$

The above equation expresses the kinetic energy of a particle having a mass m and velocity $\mathbf{v}$. The kinetic energy, as we can see, increases proportionally to the square of velocity.

In the case of a system of n particles, the kinetic energy is expressed by

$$E = \tfrac{1}{2}\sum_{k=1}^{n} m_k v_k^2. \tag{7a}$$

Comparing the right-hand sides of (5) and (6a), we have

$$E_B-E_A = -(V_B-V_A), \tag{8}$$

or

$$E_B+V_B = E_A+V_A.$$

It follows that for a system on which only the forces of the potential field act, the sum of the kinetic and the potential energy remains constant, that is

$$E+V = \text{const}. \tag{8a}$$

Systems like that we shall call *conservative* and the law (8a), the *law of conservation of mechanical energy*.

To conclude, we shall quote the definition of power. By *instantaneous power* of a force we shall understand

$$N \overset{\mathrm{df}}{=} \frac{\mathrm{d}W}{\mathrm{d}t} = \frac{\mathbf{P}\cdot\mathrm{d}\mathbf{s}}{\mathrm{d}t} = \mathbf{P}\cdot\frac{\mathrm{d}\mathbf{s}}{\mathrm{d}t} = \mathbf{P}\cdot\mathbf{v}. \tag{9}$$

Thus instantaneous power, being a derivative of work, is equal to the scalar product of force **P** and velocity **v**. The unit of power in SI is the watt:

$$\frac{1\ \mathrm{J}}{1\ \mathrm{s}} = \frac{1\ \mathrm{N}\cdot 1\ \mathrm{m}}{1\ \mathrm{s}} = 1\ \mathrm{W},$$

and in the gravitational system, 1 kG m s^{-1} or horse power (hp):

$$1\ \mathrm{hp} = \frac{75\ \mathrm{k\ Gm}}{1\ \mathrm{s}} = 735.49875\ \mathrm{W}.$$

2.2 Motion of a particle

In dynamics we are confronted with two basic problems. One involves the determination of forces at work from the description of the motion of a system and its properties. The second problem is the reverse of the first; with the loads and properties of a system given, establish its kinematics. The basis of discussion in both cases is Newton's second law.

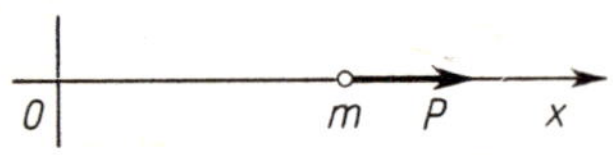

Fig. 2.1

In dealing with the problem of the rectilinear motion of a particle, the procedure can be confined to a scalar equation of the form

$$m\ddot{x} = P. \tag{1}$$

In keeping with what we have said at the beginning, we can determine from equation (1) either the acceleration, velocity and position of the mass or the force **P** when the acceleration is given.

Acceleration we assume to be positive if the sense of the vector accords with the positive sense of the axis x: in the same way we determine the sign of the force **P** (Fig. 2.1). These facts are most essential in laying out differential equations of motions.

We shall now consider the case where force **P**, being the function of time $P = f(t)$, is known. Then we have

$$m\ddot{x} = f(t). \tag{2}$$

Integrating (2),

$$m\dot{x} = C + \int_0^t f(\beta)\,\mathrm{d}\beta. \tag{3}$$

The next integration gives

$$mx = C_1 + Ct + \int_0^t \int_0^\alpha f(\beta) \mathrm{d}\beta \mathrm{d}\alpha. \tag{4}$$

Example 1. A force

$$P = f(t) = \sum_{n=0}^{r} a_n t^n,$$

where a_n are constant coefficients, acts on a particle. Find the equation of motion if in the initial instant $t = 0$, $v_0 = \dot{x}_0 = 0$.

Making use of equation (3), we write

$$m\dot{x} = \sum_{n=0}^{r} \frac{a_n}{n+1} t^{n+1}. \tag{a}$$

Next, using equation (4) we obtain

$$mx = \sum_{n=0}^{r} \frac{a_n}{(n+1)(n+2)} t^{n+2}. \tag{b}$$

The constants C_1 and C_2 are equal to zero by virtue of the initial conditions. If the force remains constant, then

$$m\dot{x} = a_0 t; \quad \text{hence} \quad mx = a_0 \frac{t^2}{2}. \tag{c}$$

Continuing, we shall examine the case where force **P** is the function of velocity, i.e., $\mathbf{P} = f(\dot{x})$. Then

$$m\ddot{x} = f(\dot{x}). \tag{5}$$

Note that $\ddot{x} = \dfrac{\mathrm{d}\dot{x}}{\mathrm{d}t}$; therefore we can write

$$m \frac{\mathrm{d}\dot{x}}{f(\dot{x})} = \mathrm{d}t. \tag{6}$$

Integrating (6) we obtain

$$\int \frac{m\mathrm{d}\dot{x}}{f(\dot{x})} + C = t. \tag{7}$$

The constant C we find from the initial condition ($t = 0$, $\dot{x} = \dot{x}_0$). Denoting the left-hand side of equation (7) by $\varphi(\dot{x})$, we have

$$\varphi(\dot{x}) = t. \tag{7a}$$

In the problem under examination, having obtained (7a), it is necessary to find the inverse function to $\varphi(\dot{x})$:

$$\dot{x} = \psi(t). \tag{7b}$$

Then

$$x = \int_0^t \psi(\beta)\,\mathrm{d}\beta + C_1. \tag{8}$$

The constant C_1 we find from the other initial condition, i.e., from the condition $x(0) = x_0$.

Example 2. Find the equation of velocity $\dot{x}(t)$ and the equation of motion $x(t)$ of a particle falling under the force of gravity, making allowance for the resistance of a medium which is proportional to square velocity. Assume at the same time that the motion is taking place near the earth, so that the gravitational field can be regarded as constant.

In the particular case, the equation of motion will be (Fig. 2.2)

$$m\ddot{x} = mg - k^2 mg\dot{x}^2, \tag{a}$$

where k is the coefficient of resistance of the medium and g the acceleration of gravity. Integrating equation (a) by the described method, we obtain

$$\int \frac{m\,\mathrm{d}\dot{x}}{f(\dot{x})} = \int \frac{\mathrm{d}\dot{x}}{g(1-k^2\dot{x}^2)} = \frac{1}{2gk}\left(\int \frac{\mathrm{d}\dot{x}}{1/k+\dot{x}} + \int \frac{\mathrm{d}\dot{x}}{1/k-\dot{x}}\right) = \frac{1}{2gk}\ln\left|\frac{1/k+\dot{x}}{1/k-\dot{x}}\right|. \tag{b}$$

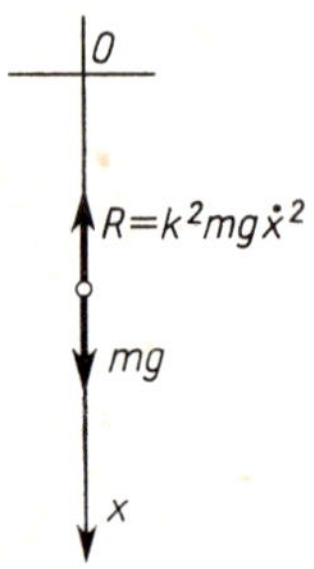

Fig. 2.2

Assuming that the particle was initially at rest we find that $C = 0$. Note also that if velocity falls off to zero and $\dot{x} < 1/k$ is continuously true, we can neglect in (b) the absolute value sign of the expression under the logarithm. Therefore

$$\frac{1}{2gk}\ln\frac{1/k+\dot{x}}{1/k-\dot{x}} = t. \tag{c}$$

In this way we have determined the function $\varphi(\dot{x})$. We now find the inverse function $\psi(t)$. It is easily seen to be

$$\varphi(t) = \dot{x} = \frac{1}{k}\tanh gkt. \tag{d}$$

It is evident from the above result that the velocity of the particle tends to the limit value $1/k$ (Fig. 2.3).

To find the equation of motion, we shall make use of the relationship (8)

$$x = \frac{1}{k}\int_0^t \tanh gk\beta \, d\beta + C_1 = \frac{1}{k^2 g}\ln(\cosh gkt) + C_1 .$$

Assuming that for $t = 0$ there is $x = 0$, we determine $C_1 = 0$. So finally we have

$$x = \frac{1}{k^2 g}\ln(\cosh gkt). \tag{f}$$

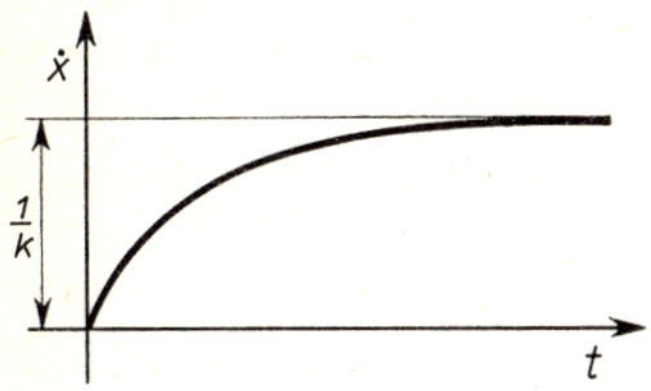

Fig. 2.3

To continue our considerations, we shall now deal with the case where force **P** is a function of position. Here the equation of particle motion is

$$m\ddot{x} = f(x). \tag{9}$$

Note that

$$\ddot{x} = \frac{d\dot{x}}{dt} = \frac{d\dot{x}}{dx}\frac{dx}{dt} = \frac{d\dot{x}}{dx}\dot{x}. \tag{10}$$

Consequently,

$$m\dot{x}\,d\dot{x} = f(x)\,dx. \tag{11}$$

Integrating equation (11) we have

$$\tfrac{1}{2}m\dot{x}^2 + F(x) = C, \tag{12}$$

where

$$F(x) = -\int f(x)\,dx \tag{13}$$

is the potential energy of the system.

Equation (12) expresses in the particular case the law of conservation of mechanical energy. A system whose total energy is of constant magnitude is defined as *conservative*, as it has been shown in the preceding section. If equation (12) describes a closed curve in the system, we deal with the case of vibrations, i.e., with periodically recurrent motion (Fig. 2.4). Finding $\dot{x}$ from equation (12) and then integrating, we obtain

$$\int_0^x \frac{\mathrm{d}x}{\sqrt{2\frac{1}{m}[C-F(x)]}} + C_1 = \pm t. \tag{14}$$

The constants we find from the initial conditions. The constant C is determined by putting $x(0) = x_0$ and $\dot{x}(0) = v_0$ in equation (12). Whence

$$C = \tfrac{1}{2}mv_0^2 + F(x_0). \tag{15}$$

The constant C_1 we find from equation (14), considering the condition that for $t = 0$ we have $x = x_0$:

$$C_1 = -\int_0^{x_0} \frac{\mathrm{d}x}{\sqrt{2\frac{1}{m}[C-F(x)]}}. \tag{16}$$

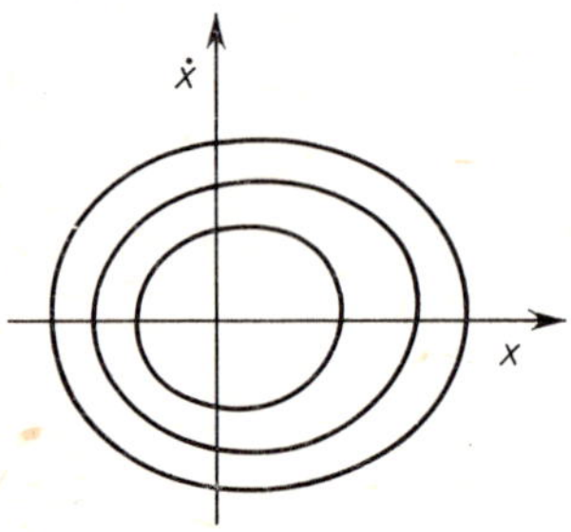

Fig. 2.4

We can see therefore that, taking into account the initial conditions, equation (14) can be written in the following form:

$$\int_{x_0}^x \frac{\mathrm{d}x}{\sqrt{2\frac{1}{m}[C-F(x)]}} = \pm t. \tag{17}$$

This brings our discussion of rectilinear motion to an end and we shall now consider problems relating to curvilinear motion of a particle.

Examining the curvilinear motion of a particle in generalized coordinates q, we can write

$$ma_j = \mathbf{P}\cdot\mathbf{e}_j. \tag{18}$$

The above equation of motion, after considering equation (31) from Section 1.1 for acceleration in curvilinear systems, will change to the following form:

$$\frac{m}{\left|\dfrac{\partial \mathbf{r}}{\partial q^j}\right|}\left\{\frac{\mathrm{d}}{\mathrm{d}t}\left[\frac{\partial}{\partial \dot{q}^j}\left(\frac{1}{2}v^2\right)\right]-\frac{\partial}{\partial q^j}\left(\frac{1}{2}v^2\right)\right\} = P_j, \tag{19}$$

where

$$P_j = \mathbf{P}\cdot\frac{\partial \mathbf{r}}{\partial q^j}\cdot\frac{1}{\left|\dfrac{\partial \mathbf{r}}{\partial q^j}\right|}.$$

(a) In a cylindrical frame of reference we have

$$m(\ddot{r}-r\dot{\varphi}^2) = P_r, \quad m(2\dot{r}\dot{\varphi}+r\ddot{\varphi}) = P_\varphi, \quad m\ddot{z} = P_z. \tag{20}$$

(b) In a spherical frame (Fig. 1.10)

$$\begin{gathered} m(\ddot{r}-r\dot{\theta}^2-r\dot{\varphi}^2\sin^2\theta) = P_r, \\ \frac{1}{2}m\left(\frac{\mathrm{d}}{\mathrm{d}t}(r^2\dot{\theta})-r^2\dot{\varphi}^2\sin\theta\cos\theta\right) = P_\theta, \\ m\frac{1}{r\sin\theta}\frac{\mathrm{d}}{\mathrm{d}t}(r^2\dot{\varphi}\sin^2\theta) = P_\varphi. \end{gathered} \tag{21}$$

Note that an orthogonal frame of reference constitutes a special curvilinear frame in which

$$m\ddot{x} = P_x, \quad m\ddot{y} = P_y, \quad m\ddot{z} = P_z, \tag{22}$$

which is consequential from the general equations of motion (1) considered in Section 2.1.

Example 3. We shall now derive the differential equation of motion for a particle in a polar frame of reference, where only a central force, i.e., one in the direction of the radius, acts on this particle.

This means that $P_\varphi = 0$. Therefore, from (20) we have

$$2\dot{r}\dot{\varphi}+r\ddot{\varphi} = 0; \tag{a}$$

hence

$$\frac{\mathrm{d}}{\mathrm{d}t}(r^2\dot{\varphi}) = 0. \tag{b}$$

Consequently,

$$r^2\dot{\varphi} = \text{const} = C. \tag{c}$$

Equation (c) is completed by taking into account the first of the relationships (20):

$$m(\ddot{r}-r\dot{\varphi}^2) = P_r. \tag{d}$$

We transform the expression in parentheses

$$\ddot{r}-r\dot{\varphi}^2 = \frac{\mathrm{d}}{\mathrm{d}t}\left(\frac{\mathrm{d}r}{\mathrm{d}\varphi}\frac{\mathrm{d}\varphi}{\mathrm{d}t}\right)-r\left(\frac{\mathrm{d}\varphi}{\mathrm{d}t}\right)^2 = \left[\frac{\mathrm{d}}{\mathrm{d}\varphi}\left(\frac{\mathrm{d}r}{\mathrm{d}\varphi}\frac{\mathrm{d}\varphi}{\mathrm{d}t}\right)-r\frac{\mathrm{d}\varphi}{\mathrm{d}t}\right]\frac{\mathrm{d}\varphi}{\mathrm{d}t}. \tag{e}$$

But from (c)

$$\frac{\mathrm{d}\varphi}{\mathrm{d}t} = \frac{C}{r^2},$$

therefore

$$\ddot{r}-r\dot{\varphi}^2 = \left[\frac{\mathrm{d}}{\mathrm{d}\varphi}\left(\frac{\mathrm{d}r}{\mathrm{d}\varphi}\frac{C}{r^2}\right)-r\frac{C}{r^2}\right]\frac{C}{r^2}.$$

Consequently, equation (d) will change to the following form:

$$\frac{mC^2}{r^2}\left[\frac{\mathrm{d}^2}{\mathrm{d}\varphi^2}\left(\frac{1}{r}\right)+\frac{1}{r}\right] = -P_r. \tag{f}$$

Equation (f) is *Binet's equation*. According to this equation, the central force acting on the particle can be determined, with the equation of trajectory given; it may also serve as the basis of solving the inverse problem.

Let the particle considered be moving along the spiral of Archimedes:

$$r = \alpha\varphi$$

where α = const. Then

$$P_r = -\frac{mc^2}{r^3}\left(1+\frac{2\alpha^2}{r^2}\right).$$

The path in which the particle is moving under the general force of gravity given by the equation

$$P_r = -\frac{\alpha' m}{r^2}$$

is defined by the integral of the following equation:

$$\frac{\mathrm{d}^2}{\mathrm{d}\varphi^2}\left(\frac{1}{r}\right)+\frac{1}{r} = \frac{\alpha'}{C^2}.$$

Note that the general solution of this differential equation is the function

$$\frac{1}{r} = \frac{\alpha'}{C^2}(1+A\cos(\varphi+\varepsilon))$$

where A and ε are the integration constants. Hence

$$r = \frac{C^2}{\alpha'(1+A\cos(\varphi+\varepsilon))}.$$

It follows that the particle is moving along a conical curve with its focus at the centre of force.

Example 4. Consider the motion of a particle in a curve expressed by

$$r = r(\varphi). \tag{a}$$

A force T tangential to the trajectory acts on the particle (Fig. 2.5).

Since the particle is moving in a strictly defined path, it has one degree or freedom, thus the frame of reference (20) can be reduced to a single equation, making use of the relationship (a). Note first that

$$\dot{r} = \frac{\mathrm{d}r}{\mathrm{d}\varphi}\dot{\varphi}, \quad \ddot{r} = \frac{\mathrm{d}^2 r}{\mathrm{d}\varphi^2}\dot{\varphi}^2 + \frac{\mathrm{d}r}{\mathrm{d}\varphi}\ddot{\varphi}. \tag{b}$$

Allowing for (b) in (20), we have

$$m\left(\frac{\mathrm{d}^2 r}{\mathrm{d}\varphi^2}\dot{\varphi}^2 + \frac{\mathrm{d}r}{\mathrm{d}\varphi}\ddot{\varphi} - r\dot{\varphi}^2\right) = P_r, \quad m\left(2\frac{\mathrm{d}r}{\mathrm{d}\varphi}\dot{\varphi}^2 + r\ddot{\varphi}\right) = P_\varphi. \tag{c}$$

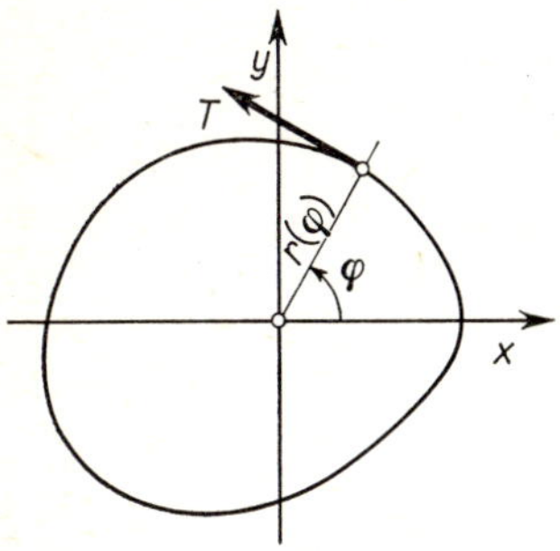

Fig. 2.5

The force T we find by forming the scalar product of vector $\mathbf{P} = P_r\mathbf{e}_r + P_\varphi\mathbf{e}_\varphi$ and the unit vector $\mathbf{e}_s$ tangential to the trajectory (Fig. 2.5). Since

$$\mathbf{e}_s = \frac{v_r\mathbf{e}_r + v_\varphi\mathbf{e}_\varphi}{\sqrt{v_r^2 + v_\varphi^2}},$$

we have

$$T = \mathbf{P}\cdot\mathbf{e}_s = \frac{v_r P_r + v_\varphi P_\varphi}{\sqrt{v_r^2 + v_\varphi^2}}, \tag{d}$$

$$v_r = \dot{r} = \frac{\mathrm{d}r}{\mathrm{d}\varphi}\dot{\varphi}, \quad v_\varphi = r\dot{\varphi}.$$

Therefore

$$T\sqrt{r^2 + \left(\frac{\mathrm{d}r}{\mathrm{d}\varphi}\right)^2} = \frac{\mathrm{d}r}{\mathrm{d}\varphi}P_r + rP_\varphi. \tag{e}$$

If we consider in (e) the left-hand sides of equations (c), we obtain

$$T\sqrt{r^2 + \left(\frac{\mathrm{d}r}{\mathrm{d}\varphi}\right)^2} = m\left\{\left[r^2 + \left(\frac{\mathrm{d}r}{\mathrm{d}\varphi}\right)^2\right]\ddot{\varphi} + \frac{\mathrm{d}r}{\mathrm{d}\varphi}\left(\frac{\mathrm{d}^2 r}{\mathrm{d}\varphi^2} + r\right)\dot{\varphi}^2\right\}. \tag{f}$$

Equation (f) is usually very difficult to integrate. If the curve is a circle, the derived equation is much simpler:

$$Tr = mr^2\varphi. \tag{f$_1$}$$

In the case of the spiral of Archimedes $r = \alpha\varphi$, we shall have

$$T\alpha\sqrt{1+\varphi^2} = m[(1+\varphi^2)\alpha^2\ddot{\varphi} + \alpha^2\varphi\dot{\varphi}^2]. \tag{g}$$

2.3 Centre of mass and centre of gravity of a particle system

Let us examine now a system consisting of n particles. The position of a particle i having a mass m_i is given by position vector $\mathbf{r}_i$.

By the *position vector of the centre of mass* $\mathbf{r}_S$ of this system we shall understand a vector defined in the following manner:

$$\mathbf{r}_S \overset{\text{df}}{=} \frac{\sum_i m_i \mathbf{r}_i}{\sum_i m_i}. \tag{1}$$

The product $\mathbf{r}_S \sum_i m_i \overset{\text{df}}{=} \mathbf{s}$ will be called the *statical moment* of the system.

It follows from definition (1) that with two masses m_1 and m_2 involved, their centre is located somewhere in the straight line connecting them. For, if we assume $\mathbf{r}_1 = 0$, then

$$\mathbf{r}_S = \frac{m_2}{m_1+m_2}\mathbf{r}_2. \tag{1a}$$

Since $\mathbf{r}_S$ and $\mathbf{r}_i$ can be expressed by coordinates, by definition (1) we have

$$(x_S\mathbf{i}+y_S\mathbf{j}+z_S\mathbf{k})\sum_i m_i = \sum_i (x_i\mathbf{i}+y_i\mathbf{j}+z_i\mathbf{k})m_i, \tag{2}$$

which is equivalent to the following equations:

$$\begin{aligned} &x_S \sum_{i=1}^{n} m_i = \sum_{i=1}^{n} x_i m_i, \quad y_S \sum_{i=1}^{n} m_i = \sum_{i=1}^{n} y_i m_i, \\ &z_S \sum_{i=1}^{n} m_i = \sum_{i=1}^{n} z_i m_i. \end{aligned} \tag{3}$$

Equation (2) gives the substance of the theorem on statical moments. The coordinates x_S, y_S, and z_S give the position of the centre of mass.

The statical moment of a system of n masses is the product of the coordinate of the centre of gravity and the mass of the whole system.

For a spatially continuous system the mass distribution is characterized by the function $\mu(x, y, z)$ which, for each of the particles considered, represents the ratio of mass to the corresponding volume:

$$\mu(x, y, z) \overset{\text{df}}{=} \frac{\mathrm{d}m}{\mathrm{d}V}; \tag{4}$$

hence

$$\mathrm{d}m = \mu(x, y, z)\mathrm{d}V.$$

This being the case, the theorem on statical moments (2) is represented by

$$\begin{aligned} x_S \int_V \mu(x, y, z)\mathrm{d}V &= \int_V x\mu(x, y, z)\mathrm{d}V, \\ y_S \int_V \mu(x, y, z)\mathrm{d}V &= \int_V y\mu(x, y, z)\mathrm{d}V, \\ z_S \int_V \mu(x, y, z)\mathrm{d}V &= \int_V z\mu(x, y, z)\mathrm{d}V, \end{aligned} \tag{5}$$

where V is the area taken up by the mass. A body is homogeneous when μ = const. For such bodies relationships (5) will be greatly simplified because the integrals on the left-hand side of the equations will then be the body volume, whereas the right-hand sides will be the so-called *geometrical statical moments*:

$$x_S V = \int_V x\mathrm{d}V, \quad y_S V = \int_V y\mathrm{d}V, \quad z_S V = \int_V z\mathrm{d}V. \tag{6}$$

In a two-dimensional problem areas correspond to volumes. This will be of relevance especially with reference to strength of materials. The centre of gravity (centroid) in a uniform gravitational field (roughly the conditions surrounding us) coincides with the centre of mass. When a mass system is introduced into a uniform gravitational field, the masses involved are loaded with forces of gravity $Q_i = m_i g$, where g stands for acceleration of gravity. The system is orientated in such a way that the vector of these forces assumes the form: $\mathbf{Q}_i = \mathbf{i}m_i g$. Next we make the sum of the vector products as the sum of the moments of the forces of gravity $\mathbf{Q}_i$ about the reference origin:

$$M = \sum_{i=1}^{n} \mathbf{r}_i \times \mathbf{Q}_i = \sum_{i=1}^{n} \begin{vmatrix} \mathbf{i} & \mathbf{j} & \mathbf{k} \\ x_i & y_i & z_i \\ m_i g & 0 & 0 \end{vmatrix} = \mathbf{j}g \sum_{i=1}^{n} z_i m_i - \mathbf{k}g \sum_{i=1}^{n} y_i m_i. \tag{7}$$

On the other hand, the resultant of the forces of gravity $\mathbf{Q}_i$ is

$$\mathbf{Q} = \mathbf{i}g \sum_{i=1}^{n} m_i, \tag{8}$$

and its moment

$$\mathbf{M}_q = \mathbf{r}_q \times \mathbf{Q} = \begin{vmatrix} \mathbf{i} & \mathbf{j} & \mathbf{k} \\ x_q & y_q & z_q \\ Q & 0 & 0 \end{vmatrix} = \mathbf{j}gz_q \sum_{i=1}^{n} m_i - \mathbf{k}gy_q \sum_{i=1}^{n} m_i. \tag{9}$$

Obviously, the moment of the resultant force $\mathbf{Q}$ is equal to the sum of moments of the individual forces, that is,

$$\mathbf{M}_q = \mathbf{M},$$

therefore,

$$\mathbf{j}gz_q \sum_{i=1}^{n} m_i - \mathbf{k}gy_q \sum_{i=1}^{n} m_i = \mathbf{j}g \sum_{i=1}^{n} z_i m_i - \mathbf{k}g \sum_{i=1}^{n} y_i m_i; \tag{10}$$

hence

$$z_q \sum_{i=1}^{n} m_i = \sum_{i=1}^{n} z_i m_i, \tag{11}$$

$$y_q \sum_{i=1}^{n} m_i = \sum_{i=1}^{n} y_i m_i. \tag{12}$$

On comparing the above expression with the relationships (3), we find that

$$z_q = z_S \quad \text{and} \quad y_q = y_S;$$

similarly, it can be proved that $x_q = x_S$. All that needs to be done is to turn the system round, so that the field vector will be parallel to the axis y or z, and to repeat the entire reasoning over again.

In this way, we have proved that in a uniform gravitational field the centre of mass S coincides with the centre of gravity q.

The next problem considered is that of finding the position vector of the centre of mass in plane homogeneous figures. In this case, the area of integration is two-dimensional, and instead of the volume V we have the area A with only the first two equations (6) being applicable.

$$x_S A = \int_A x \mathrm{d}A, \qquad y_S A = \int_A y \mathrm{d}A. \tag{13}$$

The right-hand sides of the above equations will be called *geometrical statical moments.*

The position vector of the centre of gravity of a plane figure can be found from (13):

$$\mathbf{r}_S = \frac{1}{A} \int_A \mathbf{r} \mathrm{d}A, \tag{14}$$

where

$$\mathbf{r}_S = x_S \mathbf{i} + y_S \mathbf{j}, \qquad \mathbf{r} = x\mathbf{i} + y\mathbf{j}. \tag{14a}$$

With the integral on the right-hand side of (14) endowed with the property of denumerable additivity we can write

$$\mathbf{r}_S = \frac{1}{A} \sum_{n=1}^{k} \int_{A_n} \mathbf{r}\, \mathrm{d}A. \tag{15}$$

According to (14)

$$\int_{A_n} \mathbf{r}\, \mathrm{d}A = A_n \mathbf{r}_n. \tag{16}$$

Substituting (16) into (15), we have

$$\mathbf{r}_S = \sum_{n=1}^{k} \frac{A_n \mathbf{r}_n}{A}. \tag{17}$$

For Fig. 2.6, the position vector of centre of mass has the form

$$\mathbf{r}_S = \frac{A_1}{A_1 + A_2} \mathbf{r}_1 + \frac{A_2}{A_1 + A_2} \mathbf{r}_2. \tag{17a}$$

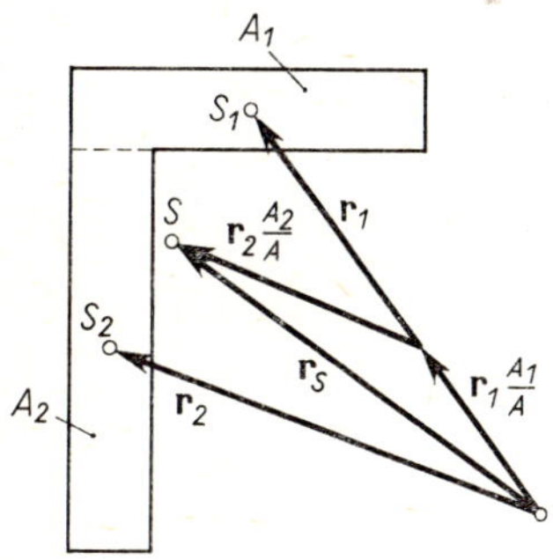

Fig. 2.6

It follows that if a plane figure is divisible into parts and the position of the centre of mass of each part is known, it is always possible to determine the centre of mass of the figure as a whole by way of simple descriptive geometrical construction.

The position of the centre of mass of a plane figure can be determined in another way by making use of the proven fact that in a uniform gravitational field the centre of mass coincides with the centre of gravity. So our uniform plane figure is brought into a gravitational field, i.e., parallel vectors, proportional in length to the componential fields, are attached at their centres of gravity. Next, the straight line is plotted on which the resultant lies. Somewhere on that line is also the centre of mass. To find its exact position the figure should be turned by an arbitrary angle and analogously the straight line in which the resultant of the new parallel vector system lies is determined. The point of intersection of the two lines will give the exact position of the centre of masa of the system (Fig. 2.7).

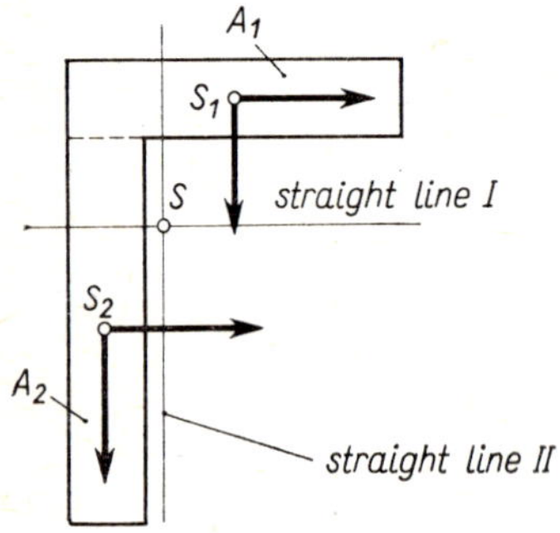

Fig. 2.7

The described graphical methods are very convenient. But for more accurate results, one should revert to analytical methods, making use of the equations for the position vector of the centre of mass of a system.

Example 1. Find the position of the centre of mass of a triangle of base a and height h (Fig. 2.8), assuming that the mass is uniformly distributed (μ = const).

We have

$$y_S = \frac{1}{\frac{1}{2}ah}\int_0^h yb(y)\,\mathrm{d}y, \tag{a}$$

where

$$b(y) = C_1y+C_2,$$

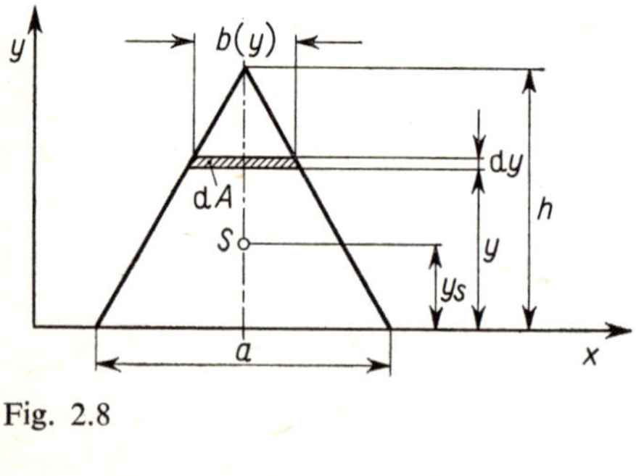

Fig. 2.8

because it is a section parallel to the base limited by straight lines. The constants C_1 and C_2 are determined from the following conditions:

$$b(0) = a = C_2, \quad b(h) = 0 = C_1 h + C_2.$$

Hence

$$C_2 = a, \quad C_1 = -\frac{a}{h}.$$

Therefore

$$b(y) = -\frac{a}{h}y + a. \tag{b}$$

Substituting (b) into (a) we obtain

$$y_S = \frac{2}{ah}\int_0^h \left(ay - \frac{a}{h}y^2\right) \mathrm{d}y = \frac{2}{ah}\left(a\frac{h^2}{2} - \frac{ah^2}{3}\right) = \frac{1}{3}h.$$

So the centre of gravity of the triangle is at $h/3$ above the base.

Example 2. Calculate the coordinates of the centroid of a circular arc of radius R and of the corresponding circular sector (Figs. 2.9 and 2.10).

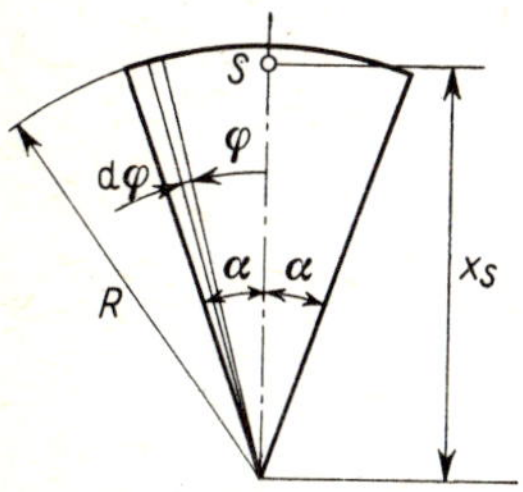

Fig. 2.9

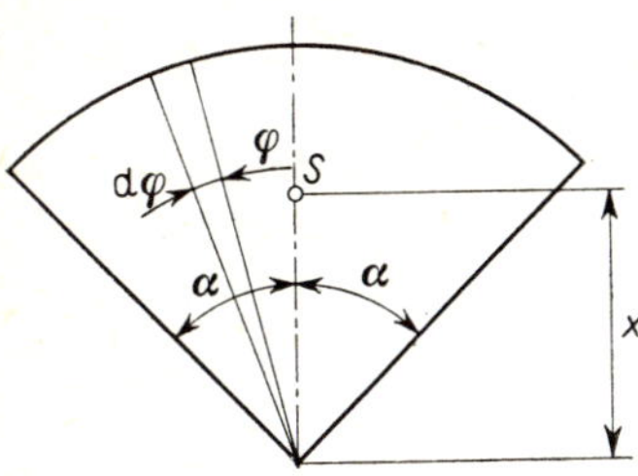

Fig. 2.10

The solution for the circular arc is as follows. We deal here with the case where the mass is continuously distributed along a curve. We assume that it is distributed uniformly, i.e., μ = const. Using the theorem on statical moments, we have

$$x_S = \frac{1}{2R\alpha}\int_{-\alpha}^{\alpha} x(\varphi) R \mathrm{d}\varphi = \frac{1}{2R\alpha}\int_{-\alpha}^{\alpha} R^2 \cos\varphi \, \mathrm{d}\varphi = R\frac{\sin\alpha}{\alpha}.$$

To find the solution for the sector, we write

$$x_S = \frac{1}{\alpha R^2} \int_{-\alpha}^{\alpha} \frac{2}{3} R\cos\varphi \frac{1}{2} RR\mathrm{d}\varphi = \frac{2R}{3\alpha} \sin\alpha .$$

Obviously for the semicircle it is

$$x_S = \frac{4R}{3\pi}.$$

2.4 Law of variation of momentum

We shall consider now a system consisting of n particles. By the *momentum* $\mathbf{p}$ of the system we understand the sum of the products of the masses and their velocities:

$$\mathbf{p} \overset{\mathrm{df}}{=} \sum_{i=1}^{n} m_i \mathbf{v}_i . \tag{1}$$

Note that if the masses m_i are not subject to change with time, the relationship between the statical moments and the momentum $\mathbf{p}$ is such as between a function and its derivative. Therefore we have

$$\mathbf{p} = \frac{\mathrm{d}\mathbf{s}}{\mathrm{d}t}.$$

Hence it follows that

$$\mathbf{p} = \mathbf{v}_S \sum_i m_i = \sum_i m_i \mathbf{v}_i . \tag{2}$$

Thus we see that the momentum of the system is the product of the velocity $\mathbf{v}_S$ of the centre of mass of the system and its sum $\sum_i m_i$. We shall now perform several operations on (2), leading to one of the major theorems of mechanics, the *law of variation of momentum*. First, we shall differentiate (2) with respect to time. We obtain

$$\frac{\mathrm{d}\mathbf{p}}{\mathrm{d}t} = \sum_i m_i \frac{\mathrm{d}\mathbf{v}_i}{\mathrm{d}t}. \tag{3}$$

In keeping with Newton's third law, the sum on the right-hand side of equation (3) is the sum of the forces acting on the system. These forces can be divided into two groups, namely external forces $\mathbf{P}_{i(1)}$ and internal force $\mathbf{P}_{i(2)}$ (forces of interaction of particles). So we can write:

$$\frac{\mathrm{d}\mathbf{p}}{\mathrm{d}t} = \sum_{i=1}^{n} (\mathbf{P}_{i(1)} + \mathbf{P}_{i(2)}). \tag{3a}$$

Since

$$\sum_i (\mathbf{P}_{i(1)}+\mathbf{P}_{i(2)}) = \sum_i \mathbf{P}_{i(1)} + \sum_i \mathbf{P}_{i(2)},$$

we have

$$\frac{d\mathbf{p}}{dt} = \sum_i \mathbf{P}_{i(1)} + \sum_i \mathbf{P}_{i(2)}. \tag{4}$$

The sum $\sum_i \mathbf{P}_{i(2)}$, by Newton's third law, as it will be demonstrated in a moment, is equal to zero.

If the ith particle acts on the jth particle with a force $\mathbf{P}_{ij}$, then the jth particle reacts on the ith particle with a force $\mathbf{P}_{ji}$. The resultant of the external forces on the ith particle is

$$\mathbf{P}_{i(2)} = \sum_{j=1}^{n} \mathbf{P}_{ji}. \tag{5}$$

Therefore

$$\sum_i \mathbf{P}_{i(2)} = \sum_{i,j} \mathbf{P}_{ji} = \sum_{i,j} \mathbf{P}_{ij} = \tfrac{1}{2} \sum_{i,j} (\mathbf{P}_{ij}+\mathbf{P}_{ji}) = 0. \tag{6}$$

Since the sum $\sum_i \mathbf{P}_{i(1)}$ is the resultant of the external forces on the system, while the sum of the internal forces is equal to zero, we can write equation (4) in the final form as follows:

$$\frac{d\mathbf{p}}{dt} = \mathbf{P}, \tag{7}$$

where

$$\mathbf{P} = \sum_i \mathbf{P}_{i(1)}. \tag{7a}$$

Equation (7) expresses the *law of variation of momentum.*

If the sum of the external forces on the system is equal to zero, the momentum of the system remains invariable, and with constant mass, the velocity of the centre of mass v_S remains invariable. Differentiating (2) we obtain

$$\mathbf{a}_S \sum_i m_i = \frac{d\mathbf{p}}{dt} = \mathbf{P}. \tag{8}$$

From equations (2) and (8) we infer that the centre of mass behaves as if the entire mass of the system were concentrated in it. Equations (2) and (8) embody the principle of motion of centre of mass.

Example 1. A rod AB rests with one end against a smooth horizontal plane at an angle α. Two masses m_A and m_B are concentrated at the points A and B. The length of the rod is l. Write the equation for the path of fall of point B when that end of the rod is released.

Two forces act on the falling rod, namely the force of gravity Q and the reaction of the smooth plane surface R. They have the same orientation but opposite directions (Fig. 2.11). Introducing a frame of reference Oxy the following equations can be written in accordance with the law of variation of momentum

$$(m_A+m_B)\ddot{x}_S = 0, \qquad (m_A+m_B)\ddot{y}_S = R-Q. \tag{a}$$

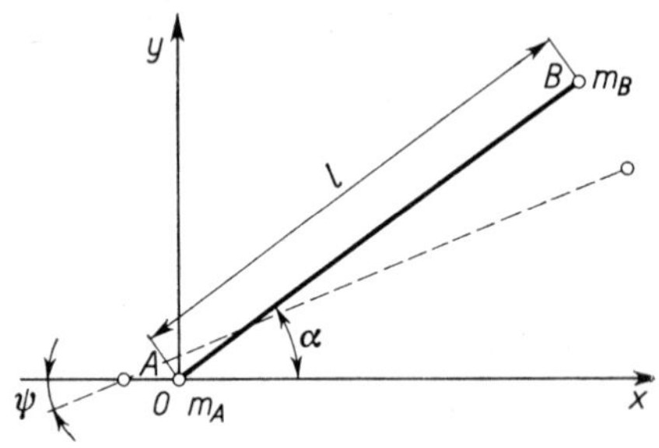

Fig. 2.11

If, initially, velocity $\dot{x}_S = 0$, then in the absence of a force in the direction of the axis x, it will continue to be equal to zero. Hence it is to be inferred that the centre of mass will be moving in a straight line parallel to the axis y. The distance between the centre of mass and the point A will be

$$r_S = \frac{m_B}{m_A+m_B} l. \tag{b}$$

Initially, point A coincided with the reference origin, therefore the coordinates of point B can be determined in the following manner:

$$x_B = r_S \cos\alpha + (l-r_S)\cos\psi, \quad y_B = l\sin\varphi. \tag{c}$$

Eliminating parameter ψ from the system, we obtain

$$\frac{(x_B-r_S\cos\alpha)^2}{(l-r_S)^2} + \frac{y_B^2}{l^2} = 1. \tag{d}$$

Example 2. We shall now consider the case of the motion of a system with variable mass.

Let the momentum of the system at a stated instant be

$$\mathbf{p} = m\mathbf{v}_S. \tag{a}$$

Suppose that after a time dt the system has changed its mass by dm. If mass is added to the system, then $dm > 0$ and if substracted from it, then $dm < 0$. The velocity of the added or subtracted mass dm is $\mathbf{v}_1$. Consequently, at an instant t the total momentum of the system can be calculated as the sum of the momentum given by equation (a) and the momentum of the mass dm which will join the mass m in an instant:

$$\mathbf{p}_1 = m\mathbf{v}_S + dm\mathbf{v}_1. \tag{b}$$

After the time dt, the mass of the system increased by dm and its velocity by $d\mathbf{v}_S$. Hence, the momentum of the system will take the following value:

$$\mathbf{p}_2 = (m+dm)(\mathbf{v}_S+d\mathbf{v}_S). \tag{c}$$

Substracting $\mathbf{p}_1$ from $\mathbf{p}_2$, we obtain the increase of momentum $d\mathbf{p}$:

$$d\mathbf{p} = md\mathbf{v}_S + dm(\mathbf{v}_S - \mathbf{v}_1). \tag{d}$$

Note also that

$$\mathbf{v}_S - \mathbf{v}_1 = -\mathbf{v}_r, \tag{e}$$

where $\mathbf{v}_r$ is the relative velocity of the mass dm with reference to the mass m. Dividing equation (d) by dt and taking (7) into account, we write

$$\mathbf{P} = m\frac{d\mathbf{v}_S}{dt} - \frac{dm}{dt}\mathbf{v}_r. \tag{f}$$

For $dm < 0$, that is, when the mass considered is subtracted from the system, the momentum at instant t is described by equation (a) and after time dt by

$$\mathbf{p}_2 = (m+dm)(\mathbf{v}_S+d\mathbf{v}_S) - dm\mathbf{v}_1. \tag{g}$$

Subtracting (a) from (g) we obtain (d), consequently, the dynamic effects attending the addition and subtracting of the mass are described by one and the same differential equation. Note that from equation (f) we can calculate, for example, the thrust of a jet engine. For this case, putting $\mathbf{v}_S = \text{const}$ in the equation, we obtain

$$\mathbf{P} = -\frac{dm}{dt}\mathbf{v}_r, \tag{h}$$

where dm/dt is the flow rate of the mass of gases passing through the jet.

If the same flux be sent at right angles to a fixed wall, the reaction of the wall will be of the same magnitude as that force. In the case a rocket is moving vertically upward with a constant outlet velocity of the gas $\mathbf{v}_r$ and with a known combustion rate $m = m_0 f(t)$ (m_0 = constant) and when besides the force of gravity mg air resistance $R(x, \dot{x})$ depending on position and velocity acts, equation (f) will take the following form:

$$\ddot{x} = -g - \frac{\dot{f}(t)}{f(t)}\mathbf{v}_r - \frac{R(x, \dot{x})}{m_0 f(t)}. \tag{i}$$

Example 3. Considering the collision of two masses, we shall make use of the law of variation of momentum. Suppose two masses m_1 and m_2 are moving in a straight line with velocities v_1 and v_2. If $v_1 < v_2$, m_2 will catch up with m_1 (Fig. 2.12) and collision

Fig. 2.12

will take place. During collision, only internal forces are involved. Hence it is to be inferred that the momentum of the mass system considered will not change. If initially the momentum of the system is $p_1 = m_1 v_1 + m_2 v_2$, and after the collision, $p_2 = m_1 v_1^* + m_2 v_2^*$, then because of the constant momentum, the following relation must occur:

$$m_1 v_1 + m_2 v_2 = m_1 v_1^* + m_2 v_2^*. \tag{a}$$

It may happen that after contact two masses will unite; this will be the case of collision of perfectly plastic masses. Then we shall consider in the equation an additional condition

$$v_1^* = v_2^* = v. \tag{b}$$

By virtue of this condition we can write

$$v = \frac{m_1 v_1 + m_2 v_2}{m_1 + m_2}. \tag{c}$$

A certain loss of the kinetic energy of the system takes place here. That loss which follows from the condition (b), is

$$E_{\text{loss}} = \frac{1}{2}\left[m_1 v_1^2 + m_2 v_2^2 - (m_1 v_1 + m_2 v_2)^2 \frac{1}{m_1 + m_2}\right] = \frac{1}{2}\,\frac{m_1 m_2}{m_1 + m_2}(v_2 - v_1)^2. \tag{d}$$

In the case of perfectly elastic impact no loss of energy occurs, i.e.,

$$\tfrac{1}{2}(m_1 v_1^2 + m_2 v_2^2 - m_1 v_1^{*2} - m_2 v_2^{*2}) = 0. \tag{e}$$

Solving the system of equations (a) and (e) with respect to v_1^* and v_2^*, we obtain

$$v_1^* = v \pm \frac{m_2}{m_1 + m_2}(v_1 - v_2), \tag{f}$$

$$v_2^* = v \pm \frac{m_1}{m_1 + m_2}(v_1 - v_2), \tag{g}$$

where v is given by equation (c).

The question now to be answered is which sign, plus or minus, is to be used for equations (f) and (g). This can be resolved in the following manner. If m_1 collided with a infinitely large stationary mass m_2, then according to (f)

$$v_1^* = \lim_{m_2 \to \infty} \frac{\pm m_2}{m_1 + m_2} v_1 = \pm v_1,$$

because the first member of the equation, with $m_2 \to \infty$ and $v_2 = 0$, tends to zero. The choice of the sign in this particular case is obvious, since we know that a body, after a perfectly elastic rebound, will be moving with a velocity having a reversed sense, therefore the minus sign is valid. Reversing the situation, i.e., taking m_1 as the infinitely large stationary mass, we find, following exactly the same reasoning, that the plus sign is valid in equation (g). Thus

$$v_1^* = v - \frac{m_2}{m_1 + m_2}(v_1 - v_2), \tag{f$_1$}$$

$$v_2^* = v + \frac{m_1}{m_1 + m_2}(v_1 - v_2). \tag{g$_1$}$$

We have considered two extreme cases. The question to be answered is: how can we calculate the velocities after impact in case of elastic-plastic impact? It will readily be seen that the results (f) and (g) differ from the result (c) in that they have an additional member. Multiplying this member by the number k whose value varies in the interval [0, 1], we have

$$v_1^* = v - k\frac{m_2}{m_1+m_2}(v_1-v_2), \tag{h}$$

$$v_2^* = v + k\frac{m_1}{m_1+m_2}(v_1-v_2). \tag{i}$$

The number k will be found in literature under the name of the *coefficient of restitution.*

Note that, putting $k = 0$ in equations (h) and (i), we obtain the expression for velocity in the case of perfectly inelastic collision and, putting $k = 1$, we obtain the expression for velocity in the case of perfectly elastic collision.

Example 4. Another example of the application of the law of variation of momentum is the dynamical effect of a fluid flux. A fluid flowing, e.g., through a curved conduit, changes its momentum. This is due to the action of the conduit walls on the flux.

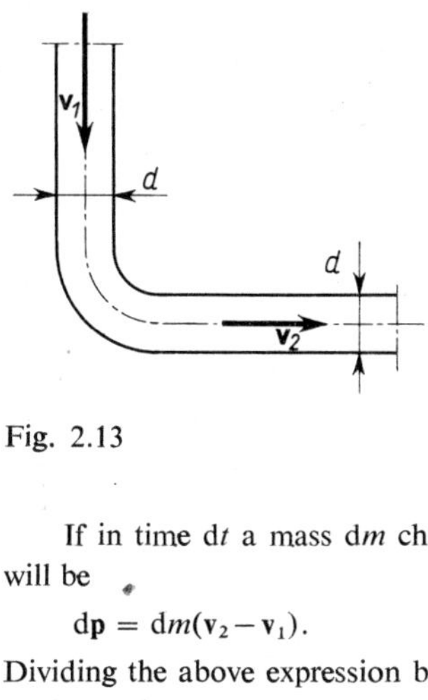

Fig. 2.13

If in time dt a mass dm changes its velocity by $\mathbf{v}_2-\mathbf{v}_1$, the change of momentum will be

$$\mathbf{dp} = dm(\mathbf{v}_2-\mathbf{v}_1). \tag{a}$$

Dividing the above expression by dt, we obtain

$$\frac{\mathbf{dp}}{dt} = \frac{dm}{dt}(\mathbf{v}_1-\mathbf{v}_2). \tag{b}$$

The expression (b) represents the force with which the conduit acts on the flux. Obviously, the action of the flux on the conduit is the reverse (the law of action and reaction). To determine the action of the fluid flowing at the rate of $M = dm/dt$, on the conduit bend (Fig. 2.13), we write

$$\mathbf{P} = M(\mathbf{v}_2-\mathbf{v}_1). \tag{c}$$

The value of this force is

$$\mathbf{P} = Mv\sqrt{2},$$

with $M = \mu Av$, where $A = \pi d^2/4$ is the section of the conduit, and μ density of fluid.

2.5 Law of variation of angular momentum

Let us consider a system of particles. We taken an arbitrary point in space. Making that point the initial point of the position vectors of the masses, we can determine the moment of momentum. By the *moment of momentum* we shall understand the angular momentum of a single particle. The *angular momentum* of a system of particles is called the sum of the moments of momentum of all particles:

$$\mathbf{K} \stackrel{\text{df}}{=} \sum_i \mathbf{r}_i \times m_i \mathbf{v}_i . \tag{1}$$

Differentiating (1) with respect to time, we obtain

$$\frac{\mathrm{d}\mathbf{K}}{\mathrm{d}t} = \sum_i \left(\mathbf{v}_i \times m_i \mathbf{v}_i + \mathbf{r}_i \times m_i \frac{\mathrm{d}\mathbf{v}_i}{\mathrm{d}t} \right); \tag{2}$$

the first product included under the summation sign is equal to zero and the other is the sum of the moments of the forces on the particles, therefore

$$\frac{\mathrm{d}\mathbf{K}}{\mathrm{d}t} = \sum_i \mathbf{r}_i \times (\mathbf{P}_{i(1)} + \mathbf{P}_{i(2)}). \tag{3}$$

The sum $\sum_i \mathbf{r}_i \times \mathbf{P}_{i(2)}$ can be transformed as follows:

$$\begin{aligned}
\sum_i \mathbf{r}_i \times \mathbf{P}_{i(2)} &= \sum_{i,j} \mathbf{r}_i \times \mathbf{P}_{ij} = \sum_{i,j} \mathbf{r}_j \times \mathbf{P}_{ji} \\
&= \frac{1}{2} \sum_{i,j} (\mathbf{r}_i \times \mathbf{P}_{ij} + \mathbf{r}_j \times \mathbf{P}_{ji}) \\
&= \frac{1}{2} \sum_{i,j} \mathbf{r}_i \times (\mathbf{P}_{ij} + \mathbf{P}_{ji}) = 0.
\end{aligned}$$

Consequently, the following equality is legitimate:

$$\frac{\mathrm{d}\mathbf{K}}{\mathrm{d}t} = \sum_i \mathbf{r}_i \times \mathbf{P}_{i(1)}. \tag{4}$$

Equation (4) embodies the *law of variation of angular momentum* in terms of which the derivative of the angular momentum with respect to time is the sum of the moments of the external forces on the system.

Example 1. A mass m has been attached to the end of a thread passing through a hole in a smooth plane horizontal surface (Fig. 2.14). Find the motion of the mass if it is known that initially the distance between the mass and the hole was l, the velocity of the thread passing through the hole $u = \text{const}$, and the transverse velocity in the initial instant w. Also calculate the tension of the string.

It is evident that the moment of the forces acting on the particle at the point at which the hole has been placed is equal to zero. Consequently,

$$\frac{d\mathbf{K}}{dt} = 0; \tag{a}$$

hence it follows that

$$\mathbf{K} = \text{const}. \tag{b}$$

The relationship (b) embodies the law of conservation of angular momentum which will be used in our problem. If for any instant t the angular momentum remains constant, then the following equation will be valid:

$$lmw = rmr\dot{\varphi} \tag{c}$$

which expresses the law of conservation of angular momentum.

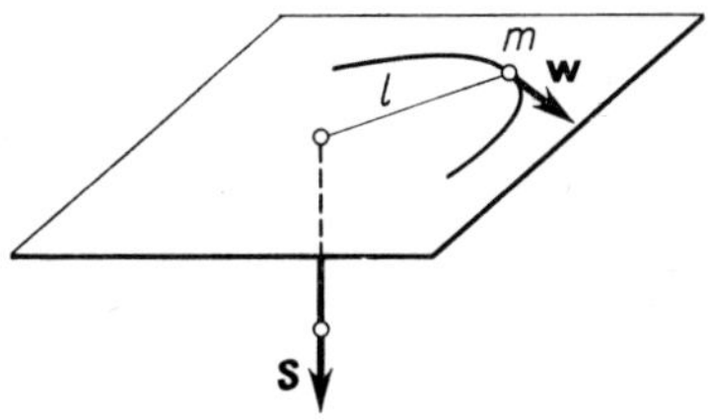

Fig. 2.14

Also note that the radial velocity is constant, i.e.,

$$\frac{dr}{dt} = -u. \tag{d}$$

From (c) and (d) we can write

$$-\frac{u}{lw}\, d\varphi = \frac{dr}{r^2}. \tag{e}$$

Integrating

$$-\frac{u}{lw}\int_0^{\varphi} d\varphi = \int_l^r \frac{dr}{r^2},$$

we obtain

$$-\frac{u\varphi}{lw} = \frac{1}{l} - \frac{1}{r}. \tag{f}$$

Since $r = l - ut$,

$$\varphi = \frac{wt}{l-ut}. \tag{g}$$

The equation of the path $r = r(\varphi)$ is described in implicit form by the relationship (f); after transformations we obtain

$$r = \frac{lw}{u\varphi + w}. \tag{h}$$

We calculate the tension of the thread using the first of equations (20), where P_r is the force acting on the particle.

$$m(\ddot{r} - r\dot{\varphi}^2) = P_r. \tag{i}$$

In our case

$$\ddot{r} = 0, \quad r\dot{\varphi}^2 = \frac{w^2 l^2}{(l-ut)^3}.$$

The action of the particle on the thread is the magnitude looked for and which according to the law of action and reactions

$$S = -P_r = m\frac{w^2 l^2}{(l-ut)^3}. \tag{j}$$

CHAPTER 3

Statics

3.1 Equations of equilibrum

We shall now examine the conditions which are satisfied by forces when a body remains at rest. If a body is at rest both the linear momentum **p** and the angular momentum **K** are simultaneously equal to zero. On this basis we can tell that its being at rest entails the disappearance of derivatives

$$\frac{\mathrm{d}\mathbf{p}}{\mathrm{d}t} = 0 \tag{1}$$

and

$$\frac{\mathrm{d}\mathbf{K}}{\mathrm{d}t} = 0. \tag{2}$$

By the principles of linear and angular momentum, the disappearance of the derivatives (1) and (2) brings about the disappearance of the sum of the forces and of the moments of external forces acting on the body.

Thus in the considered case, the system of vector equations, called the *equations of equilibrium*, is satisfied. This system of equations is the basis for the solution of all statical problems:

$$\sum_i \mathbf{P}_{i(1)} = 0, \tag{3}$$

$$\sum_i \mathbf{r}_i \times \mathbf{P}_{i(1)} = 0. \tag{4}$$

Equations (3) and (4) give the fundamental theorem on equilibrium of forces. A body will remain in equilibrium when such conditions of support are satified that ensure the standstill of the body.

In this situation, the applied forces, together with the reactions, satisfy the conditions of equilibrium, i.e., satisfy equations (3) and (4). In other words:

If a body remains in equilibrium, then the sum of external forces and the sum of moments of these forces about an arbitrary point in space are both simultaneously equal to zero.

The converse is not true. Note should be taken of the linear character of equations (3) and (4), by virtue of which in solving them use can be made of superposition of solutions.

3.2 Couple

Let us imagine two forces in two parallel straight lines l and m (Fig. 3.1). If the line l together with the force is translated until the two lines superpose, then the lengths of the two vectors can be compared. When these lengths are identical but the vectors have opposite senses, we are dealing with what is called a *couple*.

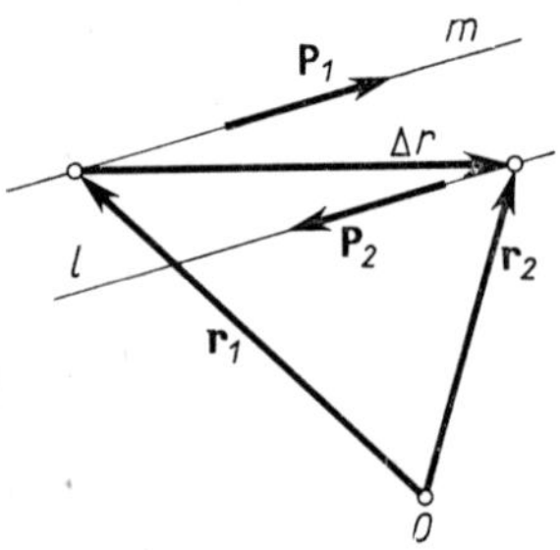

Fig. 3.1

The sum of projections of the forces making a couple can easily be seen to be equal to zero, since

$$\mathbf{P}_1 = P_{x1}\mathbf{i}+P_{y1}\mathbf{j}, \quad \mathbf{P}_2 = P_{x2}\mathbf{i}+P_{y2}\mathbf{j} \tag{1}$$

and

$$P_{x1} = -P_{x2}, \quad P_{y1} = -P_{y2}. \tag{2}$$

Now we calculate the moment

$$M_o = \sum_{i=1}^{2} (P_{yi}x_i - P_{xi}y_i) = P_{y1}x_1 - P_{x1}y_1 + P_{y2}x_2 - P_{x2}y_2 . \tag{3}$$

Considering (2), we have

$$M_o = P_{y1}(x_1 - x_2) - P_{x1}(y_1 - y_2). \tag{4}$$

Evidently, the moment M_o calculated according to (4) is independent of the reference origin but is dependent on the distance between l and m.

If in the plane of l and m we translate the forces making the couple, leaving the distance between them unchanged, the value of the general moment M_o calculated for an arbitrary position of these forces will always be the same.

It can easily be proved that the moment of the couple about any point in space is constant. In fact, $\mathbf{P}_2 = \mathbf{P}_1$, therefore

$$\mathbf{M}_o = \mathbf{r}_2 \times \mathbf{P}_2 + \mathbf{r}_1 \times \mathbf{P}_1 = (\mathbf{r}_2 - \mathbf{r}_1) \times \mathbf{P}_2. \tag{5}$$

But as we can see in Fig. 3.1, $\mathbf{r}_2 - \mathbf{r}_1 = \Delta\mathbf{r}$ does not depend on the choice of point O. It follows that the moment of a couple depends only on the magnitude of the forces involved and the distance between the lines of action.

3.3 Spatial system of forces. Wrench

We shall begin our considerations with the problem of parallel translation of a force. Let the vector of a force $\mathbf{P}$ in a straight line l be given. The question to be answered is this: What is the mechanical magnitude that will appear with parallel translation of the force $\mathbf{P}$ to another straight line m parallel to l? To find that magnitude we take any point A of the line m and attach to it two forces $\mathbf{P}$ and $-\mathbf{P}$ (Fig. 3.2) whose sum, as one

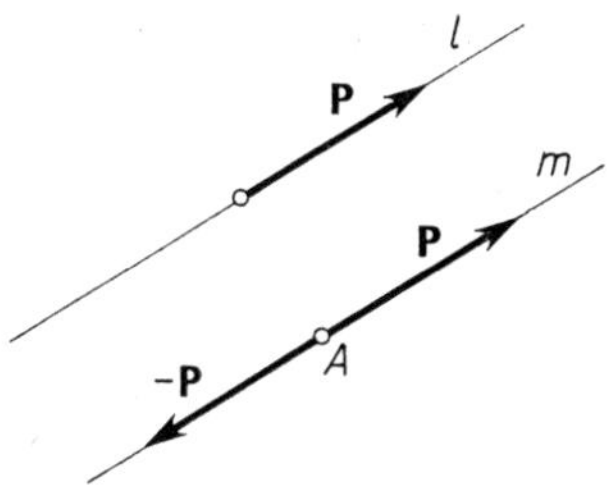

Fig. 3.2

can see, is equal to zero. By this addition we have obtained a new set of three forces, with two of them making a couple; the remaining one is the existing force $\mathbf{P}$ in the straight line m. With parallel translation, the new equivalent system consists of a single force and of the moment of a couple of forces. The force acting on a gear-wheel at the point of contact of the wheels, giving not only a torque but also a reaction on the bearings serves as an example here.

We shall now consider a more general case. We know that two groups of forces are acting on a body remaining in equilibrium, namely active forces and passive forces (reactions). The two groups together make a system of external forces, which, as we know from statics, is reduced to zero (by virtue of being in equilibrium).

The question, we may now ask, is: Which is the simplest system that the active forces can be reduced to? In other words, which is the simplest system equivalent to the given system of active forces that will produce the same effect in the form of reaction? If rectangular coordinates $Oxyz$ be introduced and then all the forces involved be translated parallelly to the reference origin, we shall obtain a concurrent system of forces and a sum of moments of couples of forces (parallel translation of force).

Let the resultant of this concurrent system of forces be called the *resultant vector* $\mathbf{P}$ and the resultant of moments, the *principal moment* $\mathbf{M}_o$:

$$\mathbf{P} = \mathbf{i} \sum_i P_{xi} + \mathbf{j} \sum_i P_{yi} + \mathbf{k} \sum_i P_{zi}, \tag{1}$$

$$\mathbf{M}_o = \sum_i \mathbf{r}_i \times \mathbf{P}_i = \mathbf{i} \sum_i M_{xi} + \mathbf{j} \sum_i M_{yi} + \mathbf{k} \sum_i M_{zi}. \tag{2}$$

We can now find a straight line l parallel to $\mathbf{P}$ such that as a result of the parallel translation the new moment $\mathbf{M}'$ will be parallel to $\mathbf{P}$. We shall call this line the *central axis* of the system (Fig. 3.3) and the set of vectors $\mathbf{M}$ and $\mathbf{P}$, the *wrench*. The equation of central axis follows from the condition of parallelism

$$\mathbf{M}' = \alpha \mathbf{P}$$

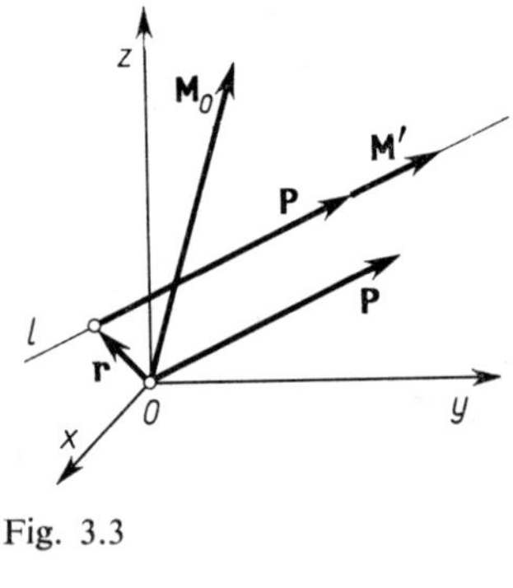

Fig. 3.3

or

$$\frac{M_x'}{P_x} = \frac{M_y'}{P_y} = \frac{M_z'}{P_z}. \tag{3}$$

On the other hand,

$$\mathbf{M}' = -\mathbf{r}\times\mathbf{P}+\mathbf{M}_o.$$

The minus sign before the product $\mathbf{r}\times\mathbf{P}$ can easily be justified if the fact be considered that the moment of an arbitrary force $\mathbf{P}$ about any point in space is

$$\mathbf{M}_A = \mathbf{r}_{BA}\times\mathbf{P},$$

where

$$\mathbf{r}_{BA} = \mathbf{r}_B-\mathbf{r}_A.$$

If the point B (Fig. 3.4) coincides with the reference origin, then

$$\mathbf{M}_A = -\mathbf{r}_A\times\mathbf{P}. \tag{4}$$

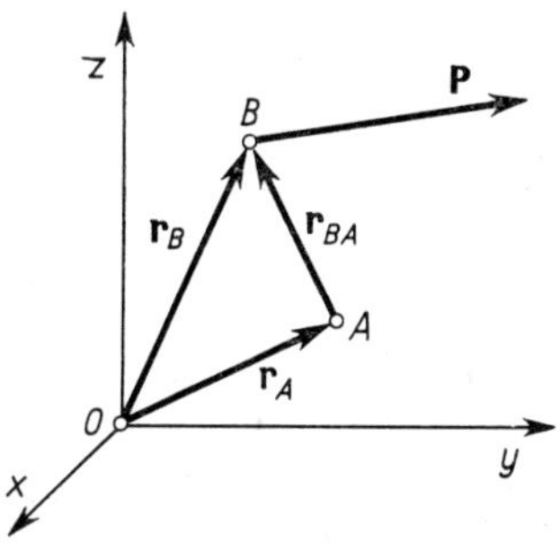

Fig. 3.4

3.4 Analytical and graphical methods in the statics of plane systems

We are often confronted in engineering practice with two-dimensional problems. To extend the reasoning to three-dimensional (spatial) problems presents no particular difficulty.

We shall now consider the analytical and graphical methods of determining the resultant of forces and the moments in a plane system of forces.

Analytical methods

Let a system of n forces be given. Each of the forces can be written in the following form (Fig. 3.5):

$$\mathbf{P}_i = P_{xi}\mathbf{i}+P_{yi}\mathbf{j}. \tag{1}$$

The resultant **P** of n forces is

$$\mathbf{P} = \sum_{i=1}^{n} \mathbf{P}_i = \mathbf{i}\sum_{i=1}^{n} P_{xi} + \mathbf{j}\sum_{i=1}^{n} P_{yi}\,. \tag{2}$$

If

$$\mathbf{P} = P_x\mathbf{i} + P_y\mathbf{j}, \tag{3}$$

then from equations (2) and (3) it follows that

$$P_x = \sum_{i=1}^{n} P_{xi}, \qquad P_y = \sum_{i=1}^{n} P_{yi}\,. \tag{4}$$

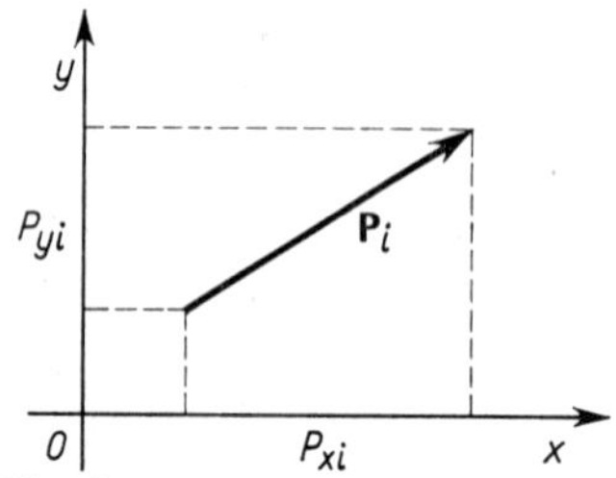

Fig. 3.5

If the position vectors correspond to the initial points of the vectors of the forces

$$\mathbf{r}_i = x_i\mathbf{i} + y_i\mathbf{j}, \tag{5}$$

then the moment of the force $\mathbf{P}_i$ about the reference origin is

$$\mathbf{M}_{oi} = \begin{vmatrix} \mathbf{i} & \mathbf{j} & \mathbf{k} \\ x_i & y_i & 0 \\ P_{xi} & P_{yi} & 0 \end{vmatrix} = \mathbf{k}(P_{xi}x_i - P_{xi}y_i). \tag{6}$$

The general moment $\mathbf{M}_o$ is the sum of moments $\mathbf{M}_{oi}$,

$$\mathbf{M}_o = \sum_{i=1}^{n} \mathbf{M}_{oi} = \mathbf{k}\sum_{i=1}^{n} (P_{yi}x_i - P_{xi}y_i). \tag{7}$$

If the plane system considered is in equilibrium, then both the resultant vector **P** and the general moment $\mathbf{M}_o$ vanish; consequently, from (4) and (7) we can write

$$\sum_{i=1}^{n} P_{xi} = 0, \qquad \sum_{i=1}^{n} P_{yi} = 0,$$
$$\sum_{i=1}^{n} (P_{yi}x_i - P_{xi}y_i) = 0. \tag{8}$$

Graphical methods (funicular polygon)

Any force $\mathbf{P}_i$ can be treated as the resultant of, let us say, forces $\mathbf{P}_{i1}$ and $\mathbf{P}_{i2}$.

The system of forces is given in Fig. 3.6. It shows the disposition of forces $\mathbf{P}_i$ in the appropriate straight lines. Figure 3.6 must conform only to the so-called *construction scale* expressed by the coefficient μ_k, i.e., the ratio of the real length to the length in the drawing. The forces $\mathbf{P}_{i1}$ and $\mathbf{P}_{i2}$ summated at point A, give the resultant $\mathbf{P}_i$.

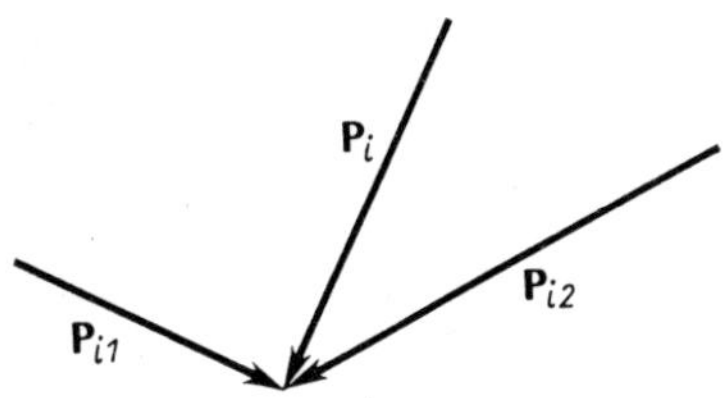

Fig. 3.6

The quantitative relations according to the parallelogram rule are given in Fig. 3.7. This drawing is called the *polygon of forces.* The scale of forces valid here is represented by the scale coefficient μ_p, expressing the ratio of the force to the vector length in the drawing. It should be noted that corresponding to the point A in the plan of forces is the field of the polygon of forces (triangle of forces).

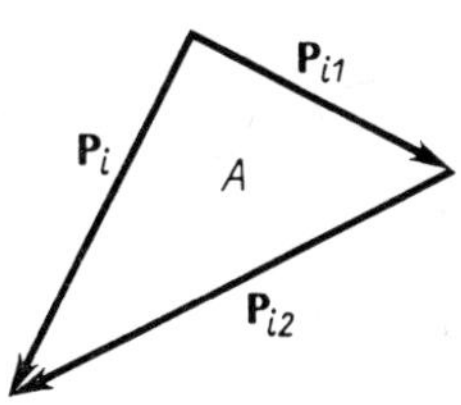

Fig. 3.7

After these introductory remarks we can now proceed with the consideration of the case of two forces $\mathbf{P}_1$ and $\mathbf{P}_2$. As in the preceding case, we resolve the two forces into components in such a manner that

$$\mathbf{P}_1 = \mathbf{P}_{11} + \mathbf{P}_{12}, \tag{9}$$

$$\mathbf{P}_2 = \mathbf{P}_{21} + \mathbf{P}_{22}, \tag{10}$$

and

$$\mathbf{P}_{12} = -\mathbf{P}_{21}. \tag{11}$$

Figure 3.8 gives an exact interpretation of equations (9), (10), and (11).

The result of our operation is such that we have ultimately obtained the resultant $\mathbf{P}$ of the forces $\mathbf{P}_1$ and $\mathbf{P}_2$ as the resultant of the forces $\mathbf{P}_{11}$ and $\mathbf{P}_{22}$.

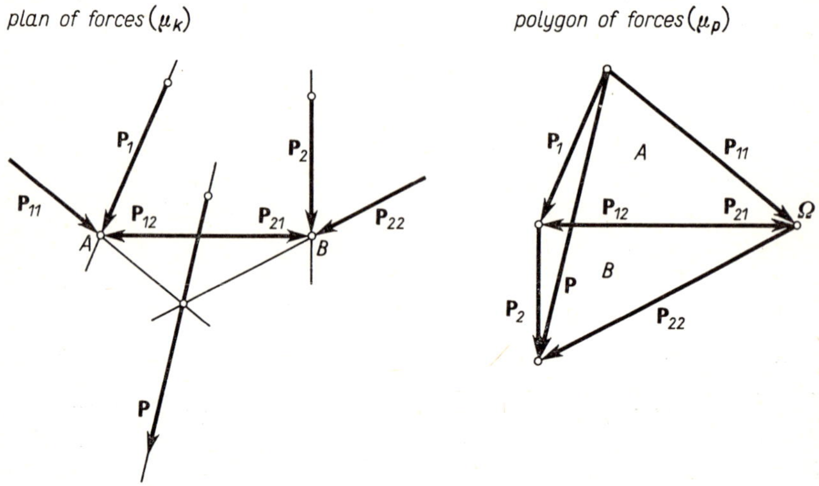

Fig. 3.8

Let us now examine the case of three forces $\mathbf{P}_1$, $\mathbf{P}_2$, $\mathbf{P}_3$:

$$\mathbf{P} = \mathbf{P}_1+\mathbf{P}_2+\mathbf{P}_3 = \mathbf{P}_{11}+\mathbf{P}_{12}+\mathbf{P}_{21}+\mathbf{P}_{22}+\mathbf{P}_{31}+\mathbf{P}_{32}. \tag{12}$$

Since

$$\mathbf{P}_{12} = -\mathbf{P}_{21}, \quad \mathbf{P}_{22} = -\mathbf{P}_{31},$$

therefore by virtue of (12)

$$\mathbf{P} = \mathbf{P}_{11}+\mathbf{P}_{32}. \tag{13}$$

Obviously, in the case of three forces, the problem is reduced to finding the resultant of two forces, according to (13) (Fig. 3.9). We can easily infer that proceeding as in the case of n forces, we obtain the resultant

$$\mathbf{P} = \sum_{i=1}^{n} \mathbf{P}_i = \sum_{i=1}^{n} (\mathbf{P}_{i1}+\mathbf{P}_{i2}), \tag{14}$$

$$\mathbf{P}_{i2} = -\mathbf{P}_{(i+1)1}, \tag{15}$$

whence

$$\mathbf{P} = \sum_{i=1}^{n} \mathbf{P}_i = \mathbf{P}_{11} + \mathbf{P}_{n2}. \tag{16}$$

So we can see that the resultant of an arbitrary plane system of forces we find as the sum of two forces, in keeping with (16).

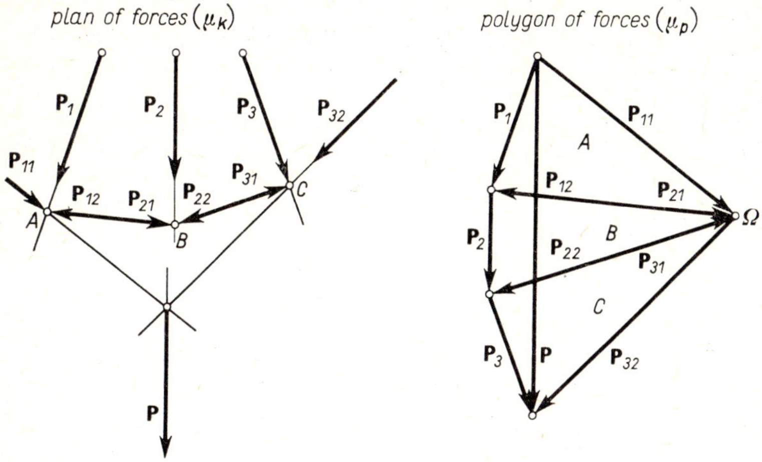

Fig. 3.9

Let us now consider how to determine the moment of force **P** about an arbitrary point *O*. We shall make use of the funicular polygon construction. The force **P** will be treated as the resultant of two forces. The moment of the force **P** about the point *O* is (Fig. 3.10)

$$\mathbf{M} = \mathbf{r} \times \mathbf{P}.$$

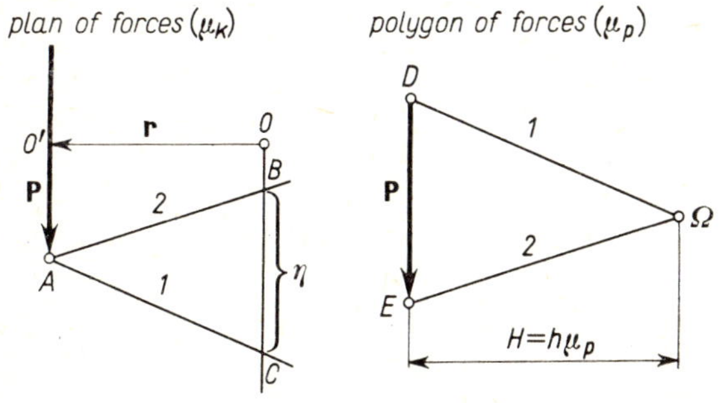

Fig. 3.10

Considering that the angle between **r** and **P** is $\pi/2$,

$$M = rP. \tag{17}$$

In the plan of forces we must conform to the construction scale, therefore $r = OO' \cdot \mu_k$. The value of **P** can be defined as the product of the section DE and the scale coefficient μ_p. In the light of the above,

$$M = OO' \cdot DE \cdot \mu_k \mu_p. \tag{18}$$

We shall consider the similarity of ΔABC and $\Delta DE\Omega$. The height-base ratio in both triangles is the same, therefore

$$\frac{O'O}{BC} = \frac{h}{DE}, \qquad BC = \eta,$$

whence

$$OO' \cdot DE = \eta h. \tag{19}$$

Substituting (19) into (18) we have

$$M = \eta h \mu_k \mu_p. \tag{20}$$

It will easily be seen that the moment of the force **P** about point O is proportional to η in the $h\mu_k\mu_p$ scale. The moment of several forces about an arbitrary point can be determined by making use of the law of superposition.

The moment of the forces $\mathbf{P}_1$ and $\mathbf{P}_2$ about point O (Fig. 3.11) is

$$M = (\eta_1 h_1 + \eta_2 h_2)\mu_k \mu_p. \tag{21}$$

In the case of n forces the result obtained can be generalized:

$$M = \mu_k \mu_p \sum_i \eta_i h_i, \tag{21a}$$

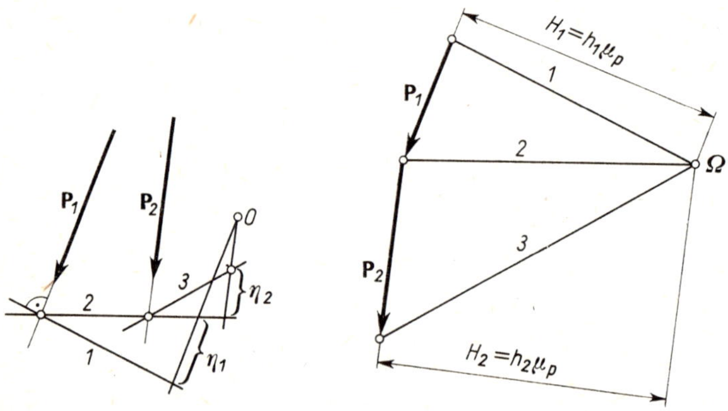

Fig. 3.11

where η_i may assume plus or minus values, depending on whether the force rotates about the point O clockwise or anticlokwise. What we have just said is a convention which we can modify according to the problem considered.

If a set of forces remains in equilibrium, then:
firstly,

$$\mathbf{P}_{11}+\mathbf{P}_{n2} = 0; \tag{22}$$

secondly,

$$\sum_i \eta_i h_i = 0. \tag{23}$$

Take a set of three forces in which equilibrium is maintained (Fig. 3.12). Obviously, $\mathbf{P}_{11}+\mathbf{P}_{32} = 0$ and

$$h\eta_1 - h\eta_2 + h\eta_3 = 0, \quad \eta_1+\eta_3 = \eta_2.$$

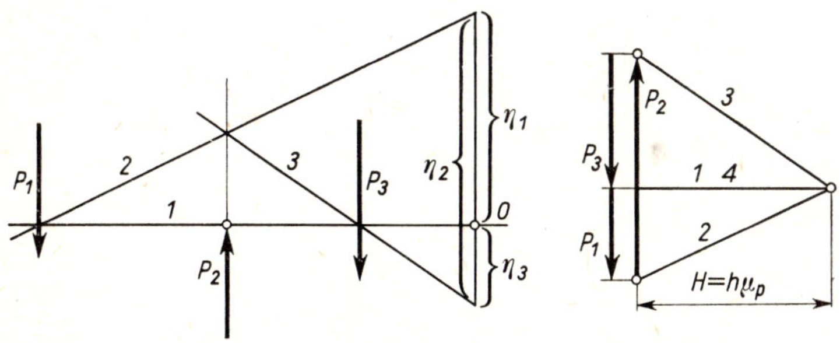

Fig. 3.12

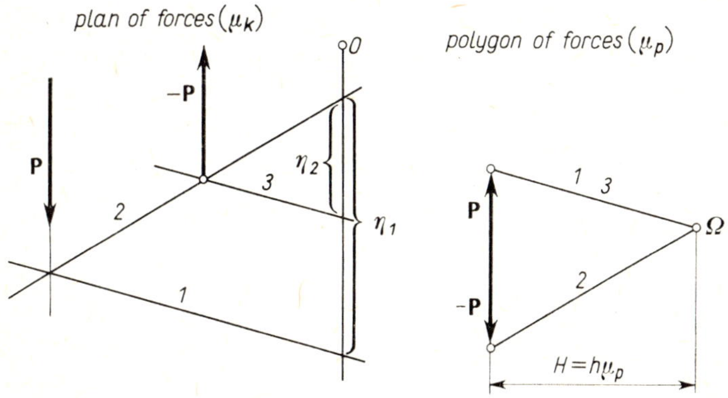

Fig. 3.13

Considering a couple of forces, we find that condition (22) is satisfied but (23) is not. Consequently, such a set of forces is not in equilibrium since

$$\eta_1 - \eta_2 \neq 0,$$

as indicated in Fig. 3.13.

3.5 Examples

Triple-jointed system

A triple-jointed system is shown in Fig. 3.14. It consists of elements *I* and *II* hinged at *C* and connected to the base by articulations *A* and *B*.

It is characteristic of an articulation that it allows mutual movements of the parts (rotation about the articulation) at the same time as not allow-

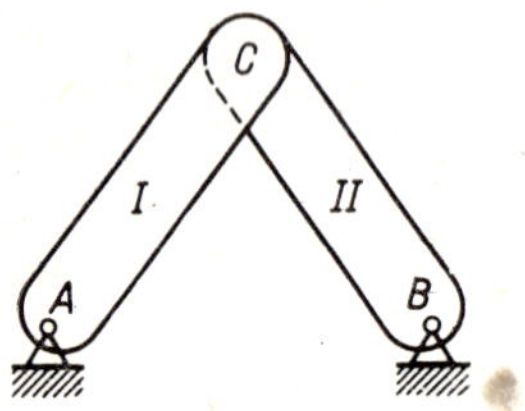

Fig. 3.14

ing displacements against each other. An example of an articulated joint is a door hinge or a short thread connecting the ends of rods, and the like. If there is no friction in the articulation, it is said to be *smooth*.

In our considerations we shall deal only with smooth articulations. Before we proceed with solving the considered system, let us examine a rod which is loaded with one force. It is connected to the base by an articulation (Fig. 3.15). Equilibrium is possible only when the force has the direction of the rod axis. Unless this condition is satisfied, the rod turns about the articulation (the force **P** gives a moment about point *O* other than zero).

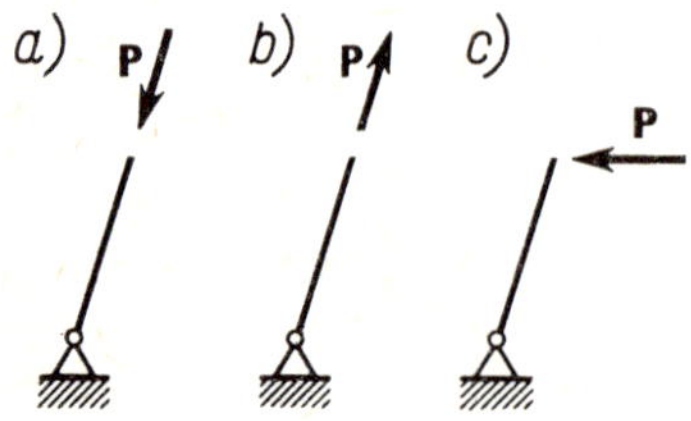

Fig. 3.15

The situation would be the same if we replaced the rod by a body which can also rotate about the articulation (Fig. 3.16). If the body remains in equilibrium, the set of forces is reduced to a resultant which passes through the articulation.

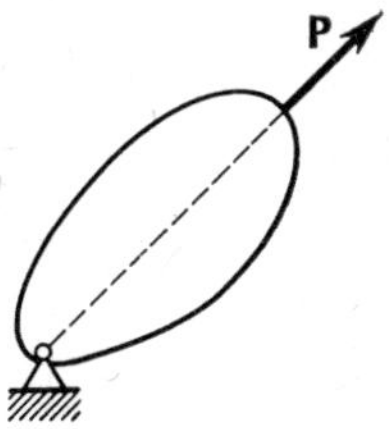

Fig. 3.16

This fact is used in finding the position of the centre of gravity of a body. For this purpose, it should be suspended from an articulation. Its position will be such that the centre of gravity will be in a vertical passing through the articulation. Suspending the body in this manner from two different points, we determine graphically two straight lines whose intersection gives the position or the centre of gravity.

Returning to the triple-jointed system, we shall inquire into the ways of solving it. The problem amounts to finding the reaction. In this particular case the reaction is the force with which the base acts via the articulation on an element of the system. The simplest problem will be to find the reaction forces when a force **P**, applied to the articulation C, is acting on the system (Fig. 3.17). The force may be a resultant of several forces applied at this point. The rods AC and BC can transmit only axial forces. These forces are simultaneously the reactions at points A and B.

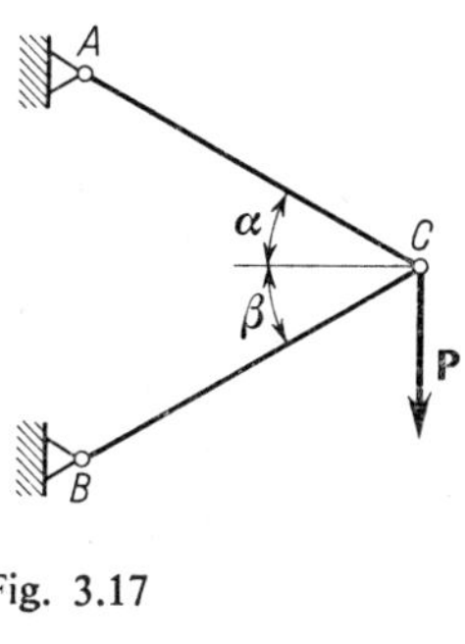

Fig. 3.17

We single out the point C and inquire into its equilibrium. The sum of projections of the forces on the x-axis (Fig. 3.18) is

$$-R_A \cos\alpha - R_B \cos\beta = 0. \tag{a}$$

Similarly the sum of projections of the forces on the y-axis is

$$R_A \sin\alpha - R_B \sin\beta - P = 0. \tag{b}$$

The sum of moments of the forces about point C is similarly equal to zero. From equation (a) we have

$$R_A = -R_B \frac{\cos\beta}{\cos\alpha}. \tag{c}$$

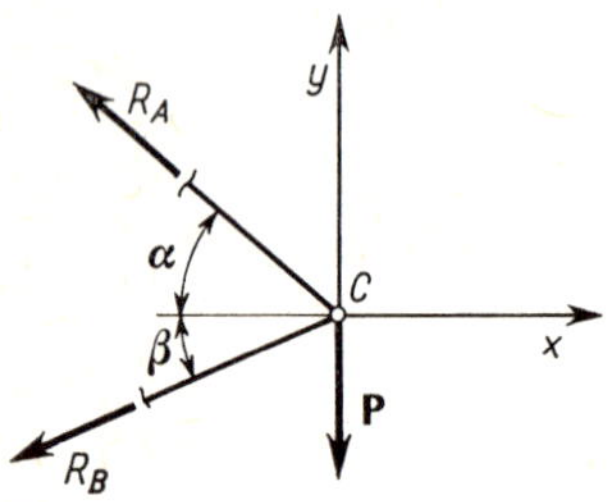

Fig. 3.18

Substituting (c) into (b) we obtain

$$-R_B \frac{\cos\beta}{\cos\alpha} \sin\alpha - R_B \sin\beta - P = 0, \tag{d}$$

whence

$$R_B = -\frac{P\cos\alpha}{\sin(\alpha+\beta)}. \tag{e}$$

Substituting next (e) into (c), we calculate the reaction R_A:

$$R_A = \frac{P\cos\beta}{\sin(\alpha+\beta)}. \tag{f}$$

The expression (e) contains the minus sign, which is a proof that the sense of the force R_B is opposite to the assumed one.

The same problem can be solved graphically. We plot to scale the triangle of mutually equilibrate forces (Fig. 3.19). From the triangle of the forces involved we obtain the substantiation of equations (e) and (f) (use should be made of the Sinus Theorem).

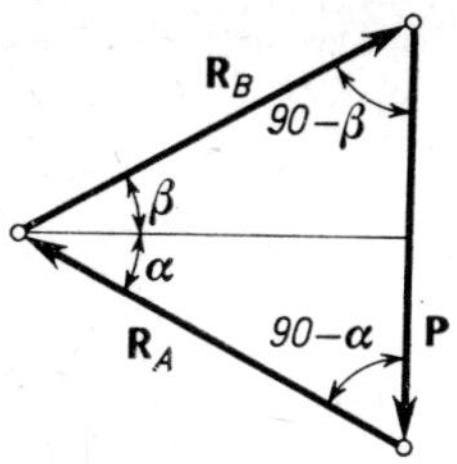

Fig. 3.19

We shall now discuss the case where a resultant force $\mathbf{P}_1$ (Fig. 3.20) acts on member *AC*. Note first that the reaction at point *B* has the direction of the straight line *CB*, since this is the only possible direction of the force for equilibrium to be maintained in this part of the system (the rod *CB*).

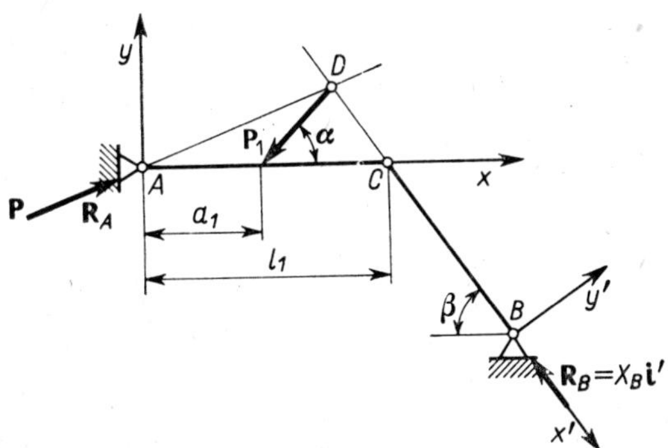

Fig. 3.20

The resultant of forces $\mathbf{R}_B$ and $\mathbf{P}$ passes through the point *D*. $\mathbf{R}_A$ is equilibrated by the force $\mathbf{P}_1+\mathbf{R}_B$ in the straight line determined by points *A* and *D*. The graphical solution is presented in Fig. 3.21 and it is the graphic interpretation of the equations of equilibrium for this special case.

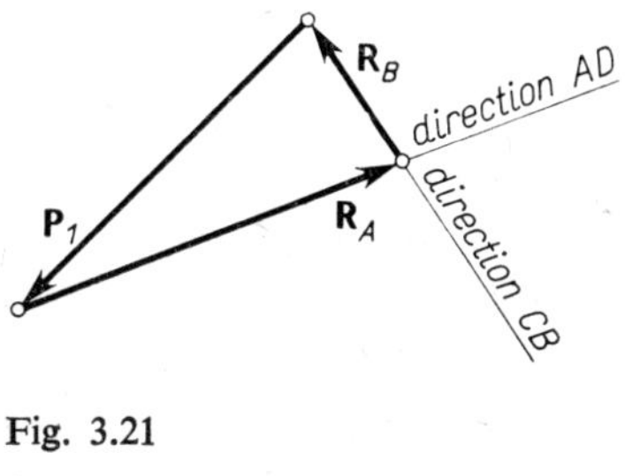

Fig. 3.21

We now come to the direct use of the equations of equilibrium. The member AC remains under the action of the sum of the forces

$$\mathbf{R}_B+\mathbf{R}_A+\mathbf{P}_1 = 0.$$

On the other hand,

$$\mathbf{R}_A = X_A\mathbf{i}+Y_A\mathbf{j}, \quad \mathbf{R}_B = X_B\mathbf{i}'.$$

We can therefore write:

$$\begin{aligned} X_A'-P\cos\alpha_1-X_B'\cos\beta &= 0,\\ Y_A'-P\sin\alpha_1+X_B'\sin\beta &= 0,\\ -a_1P_1\sin\alpha_1+l_1X_B'\sin\beta &= 0. \end{aligned} \tag{a}$$

Hence

$$\begin{aligned} X_B' &= P_1\frac{a_1\sin\alpha_1}{l_1\sin\beta},\\ X_A' &= P_1\left(\cos\alpha_1+\frac{a_1\sin\alpha_1}{l_1\sin\beta}\cos\beta\right),\\ Y_A' &= P_1\left(\sin\alpha_1-\frac{a_1\sin\alpha_1}{l_1}\right). \end{aligned} \tag{b}$$

Figure 3.22 shows a triple-jointed system loaded with forces $\mathbf{P}_1$ and $\mathbf{P}_2$. The problem which consists of finding the reaction at points A and B can be solved by seeking the reaction to the force $\mathbf{P}_1$ first and then to the force $\mathbf{P}_2$.

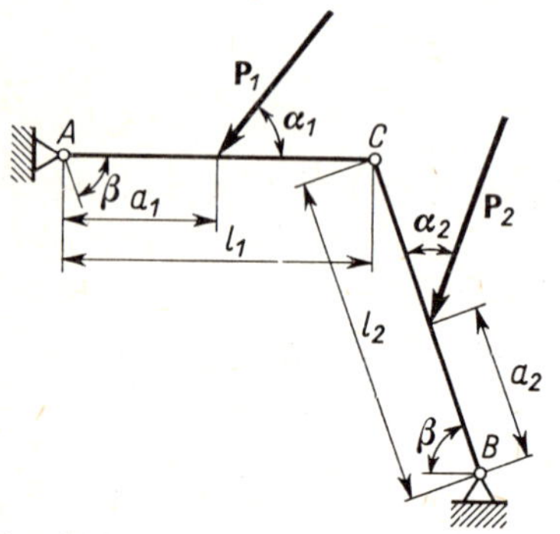

Fig. 3.22

The solution will be the sum of the reactions calculated in the first and in the second problem. The next step, as we have already anticipated, consists of discarding force $\mathbf{P}_1$ in order to examine the influence of force $\mathbf{P}_2$ on the reaction (Fig. 3.23). Comparing Fig. 3.20 with Fig. 3.23, it is easily seen that the solution to the posed problem is the system (b), with the difference, however, that a_1, l_1, $\mathbf{P}_1$, and α_1 are replaced by a_2, l_2, $\mathbf{P}_2$, and α_2. Besides, the points A and B change roles with each other. Therefore we can write

$$
\begin{aligned}
X_A'' &= P_2 \frac{a_2 \sin\alpha_2}{l_2 \sin\beta}, \\
X_B'' &= P_2 \left(\cos\alpha_2 + \frac{a_2 \sin\alpha_2}{l_2 \sin\beta} \cos\beta \right), \qquad \text{(c)} \\
Y_B'' &= P_2 \left(\sin\alpha_2 - \frac{a_2}{l_2} \sin\alpha_2 \right).
\end{aligned}
$$

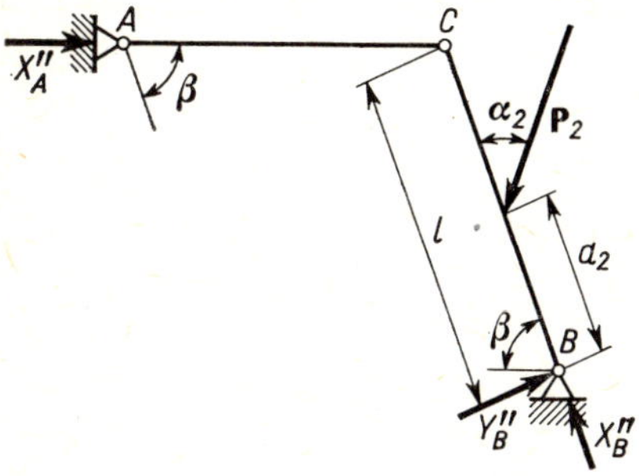

Fig. 3.23

In the case where member AC of the triple-jointed system is subjected to the simultaneous action of n forces $\mathbf{P}_{1i}$ $(i = 1, \ldots, n)$, localized like the force $\mathbf{P}_1$, and correspondingly the member CB is subjected to the action of m forces $\mathbf{P}_{2i}$ $(i = 1, \ldots, m)$, localized like the force $\mathbf{P}_2$ we should seek the solution to the problem on the basis of the superposition theorem applied to partial results. Thus, in place of the reactions described by equations (b) the following expressions come in:

$$
\begin{aligned}
X_B' &= \sum_{i=1}^{n} P_{1i} \frac{a_{1i} \sin\alpha_{1i}}{l_1 \sin\beta}, \\
X_A' &= \sum_{i=1}^{n} P_{1i} \left(\cos\alpha_{1i} + \frac{a_{1i}}{l_1} \sin\alpha_{1i} \cot\beta) \right), \qquad \text{(b1)} \\
Y_A' &= \sum_{i=1}^{n} P_{1i} \left(\sin\alpha_{1i} - \frac{a_{1i}}{l_1} \sin\alpha_{1i} \right).
\end{aligned}
$$

Analogously,

$$X''_A = \sum_{i=1}^{m} P_{2i}\frac{a_{2i}\sin\alpha_{2i}}{l_2\sin\beta},$$

$$X''_B = \sum_{i=1}^{m} P_{2i}\left(\cos\alpha_{2i}+\frac{a_{2i}}{l_2}\sin\alpha_{2i}\cot\beta\right), \tag{c1}$$

$$X''_B = \sum_{i=1}^{m} P_{2i}\left(1-\frac{a_{2i}}{l_2}\sin\alpha_{2i}\right).$$

This makes it demonstrable how the superposition theorem enables the generalization of results.

With the simultaneous action of the forces $\mathbf{P}_{1i}$ and $\mathbf{P}_{2i}$ the reactions should be added in the following manner (once more we make use of the superposition theorem):

$$X_A = X'_A+X''_A, \quad Y_A = Y''_A,$$
$$X_B = X'_B+X''_B, \quad Y_B = Y''_B. \tag{d}$$

Now we arrive at the same result by the graphical method. We build the scheme of solutions on the superposition theorem.

The solution of the problem (Fig. 3.24) gives a polygon of forces. Triangles *4-3-6* and *1-5-4* represent the equilibrium of the subsystems a) and b). The constructional principle of the triangles has been given in Fig. 3.20. Triangles *2-6-3* and *1-2-5* reflect the application of superposition. As we can see, sections *2-3* and *1-2* are proportional, in scale μ_p, to $\mathbf{R}_A = \mathbf{R}_{A1}+\mathbf{R}_{A2}$ and to $\mathbf{R}_B = \mathbf{R}_{B1}+\mathbf{R}_{B2}$. It should be appreciated that section *4-2* is proportional to the forces of interaction between rods *AC* and *CB* at point *C*.

In the case where a concentrated moment **M** acts on a particular part of the system, the problem can also be solved graphically. To do this we replace the moment **M** by a pair of forces, $\mathbf{P}_1$ and $\mathbf{P}_2$, which make a couple. The points of application of these forces are arbitrary as long as they remain in the section to which the moment

$$M = r_{12}P \quad (P_1 = P_2 = P). \tag{a}$$

has been applied.

We build the solution on the superposition theorem. It is evident that the reactions $\mathbf{R}_A$ and $\mathbf{R}_B$ constitute a couple. The successive stages of solving the problem are illustrated in Figs. 3.25, 3.26, and 3.27.

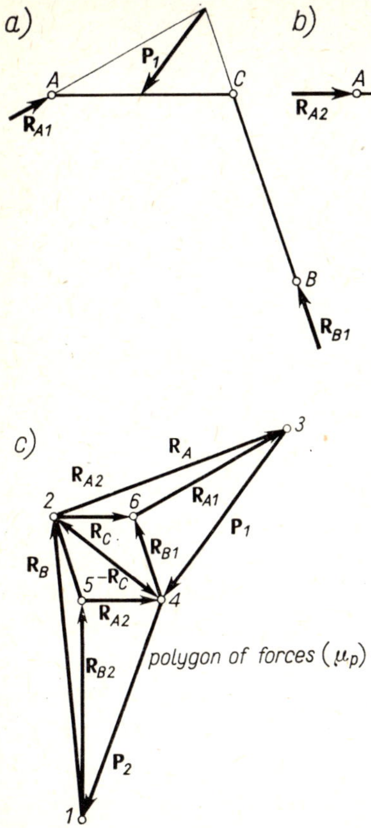

Fig. 3.24

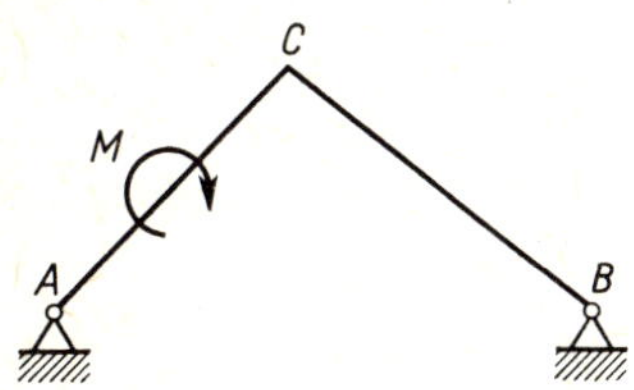

Fig. 3.25

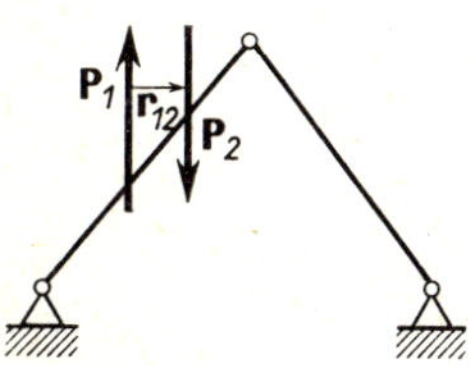

Fig. 3.26

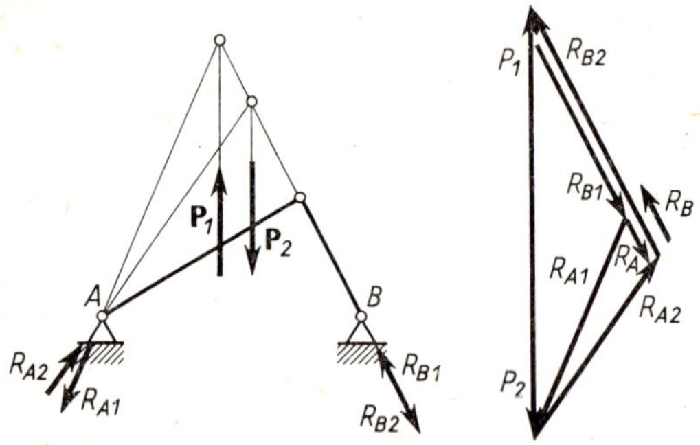

Fig. 3.27

Free-ends beam

A simple free-ends beam (Fig. 3.28) can be regarded as a degenerate triple-jointed system, in which one of the elements (CB) acts as the support. In this way, the direction of the reaction at point C is uniquely determined, it coincides, as we know, with the direction of CB. Such a fixing of the beam constitutes a support giving one unknown. Since in the articulation of A two support magnitudes, X_A and Y_A, are involved, we have altogether three unknowns, which correspond to the number of equations of equilibrium.

Fig. 3.28

It will be seen that all reactions on the beam at the support points can be solved on the basis of the equations of equilibrium. For this reason we can say that the considered beam constitutes an *externally determinate static system.*

Let us take a specific problem. A beam AB loaded with a force $\mathbf{P}$ acting at an angle α to the beam axis is given (Fig. 3.29). The dimensions a and l are known. The problem is to determine the reactions at points A and B of support. First, we introduce a frame of reference. The components of the force with senses conformant with the positive senses of the axes we shall regard as positive and the components with opposite senses as negative. The conditions of equilibrium for the considered case are as follows:

$$X_A + P\cos\alpha = 0, \tag{a}$$

$$Y_A + P\sin\alpha + R_B = 0, \tag{b}$$

$$aP\sin\alpha + lR_B = 0. \tag{c}$$

From the last of these equations it follows that

$$R_B = -P\frac{a}{l}\sin\alpha;$$

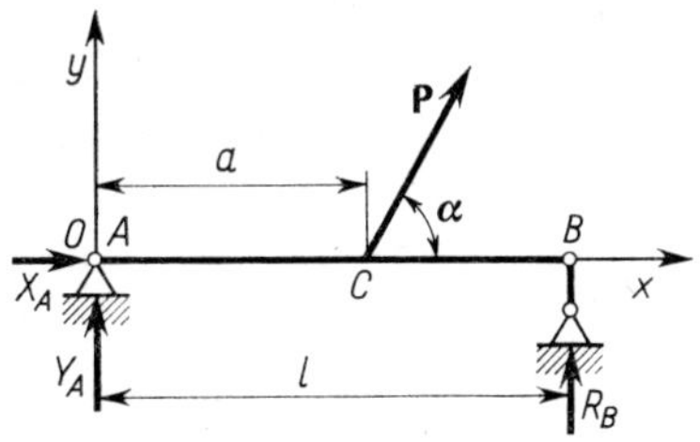

Fig. 3.29

From equation (a) we calculate

$$X_A = -P\cos\alpha$$

and from (a) and (b) we obtain

$$Y_A = -P\frac{l-a}{l}\sin\alpha,$$

The minus signs in the above equations indicate that the senses of forces $\mathbf{R}_B$, $\mathbf{X}_A$ and $\mathbf{Y}_A$ are opposite to the sense of the axes of reference frame Oxy. Force $\mathbf{P}$ can be resolved into $P\sin\alpha$, the normal to the beam axis, and $P\cos\alpha$, the tangential to the beam axis. The former causes bending and the latter, tension or compression in section AC (Fig. 3.29). As we can see, the reactions $\mathbf{R}_B$ and $\mathbf{Y}_A$ equilibrate the normal component and $\mathbf{X}_A$ equilibrates the tangential component.

The same problem can also be solved graphically (Fig. 3.30). We pass through point B a straight line having the given direction of the reaction R_B up to the point of intersection with the line of action of force $\mathbf{P}$. In this way, we find the point C at which all the forces on the beam are equilibrated. The points A and C give the direction of the reaction $\mathbf{R}_A$. In the polygon of forces we plot a section proportional by scale μ_p to the force $\mathbf{P}$, then we pass through the origin of the vector $\mathbf{P}$ a straight line having the direction of the reaction $\mathbf{R}_A$ and through its end a straight line having the direction of the reaction $\mathbf{R}_B$. The values of $\mathbf{R}_A$ and $\mathbf{R}_B$ we read off as the length products of the sides *3-2* and *1-3* multiplied by μ_p.

In the case of a concentrated moment we replace it with a couple. Making use of superposition, as in the case of the triple-jointed system, we find the unknown reactions of the supports. When the resultant of all forces

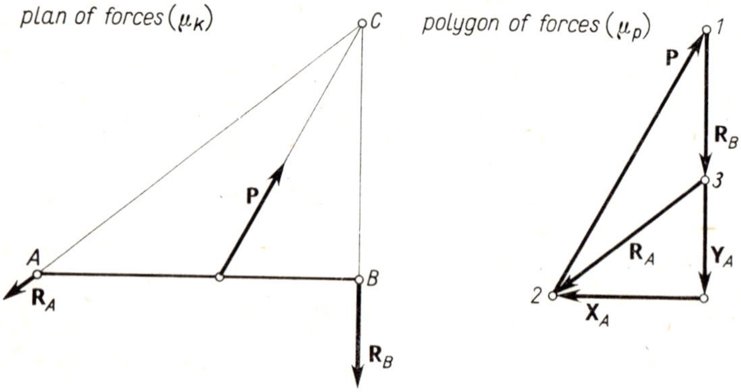

Fig. 3.30

on the beam is perpendicular to the beam axis (Fig. 3.31), we resolve it into two components $\mathbf{P}_1$ and $\mathbf{P}_2$ at point C. The straight lines a and b in which the reactions are lying are parallel to the direction of the resultant $\mathbf{P}$.

At point A, as we can see in Fig. 3.31, the forces $\mathbf{P}_1$, $\mathbf{Z}$, and the reaction $\mathbf{R}_A$ equilibrate. Similarly, at point B, the forces $\mathbf{P}_2$, $-\mathbf{Z}$, and $\mathbf{R}_B$ equilibrate; therefore

$$\mathbf{P}_1+\mathbf{Z}+\mathbf{R}_A = 0 \tag{a}$$

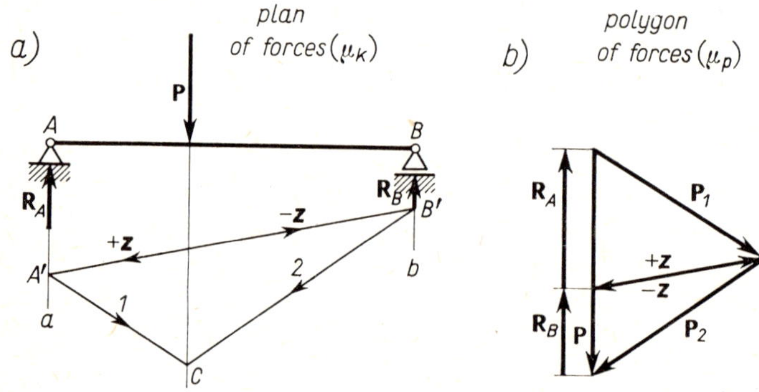

Fig. 3.31

and

$$\mathbf{P}_2 - \mathbf{Z} + \mathbf{R}_B = 0. \tag{b}$$

Adding the equations, we obtain

$$\mathbf{P}_1 + \mathbf{P}_2 + \mathbf{R}_A + \mathbf{R}_B = 0$$

and

$$\mathbf{P} + \mathbf{R}_A + \mathbf{R}_B = 0. \tag{c}$$

It will be seen from Fig. 3.31 that the set of forces **P**, $\mathbf{R}_A$, and $\mathbf{R}_B$ is not subject to reduction to a couple of forces, consequently the beam is in equilibrium.

Plane trusses

By the *truss* we understand a system of rods connected by articulations, constituting a rigid entity. We deal with a *plane truss* when all the rods are situated in one plane. Assume that we shall load the trusses at points of articulation with forces in the plane of the rods. In this connection, we can infer that the rods will be loaded along their axes.

We shall now consider externally and internally statically determinate trusses, i.e., such for which the number of unknowns is equal to the number of equations of statics.

If a truss remains in equilibrium, then all its nodes (articulations) also remain in equilibrium. For each node, two equilibrium conditions can be written. The third condition, that for moments, is identically satisfied. If the number of nodes is w, then the total number of equations of equilibrium will be $2w$.

Let us pursue the unknowns involved. These are the reactions of the supports (external magnitudes) and the tensions of the rods (internal magnitudes).

If we have r reactions and p rods, the sum total of unknowns will be $p+r$. Consequently, the plane truss is statically (externally and internally) determinate if the following condition is satisfied:

$$2w = p + r. \tag{a}$$

Trusses, such as in Fig. 3.32, will give three unknown external magnitudes, i.e., reactions. For this reason, for any such truss condition (a) will be

$$2w = p + 3.$$

In our case, $w = 10$, $p = 17$; thus condition (a) is satisfied, i.e., the truss is statically determinate.

The methods used in solving plane trusses can be classified into graphical and analytical. We shall now discuss the basic methods, most commonly used in engineering practice. We shall begin with the *node separation methods* and *Cremona's method.* The former will not be directly applicable to our considerations but we shall describe it for a better understanding of the construction of Cremona's polygon of forces.

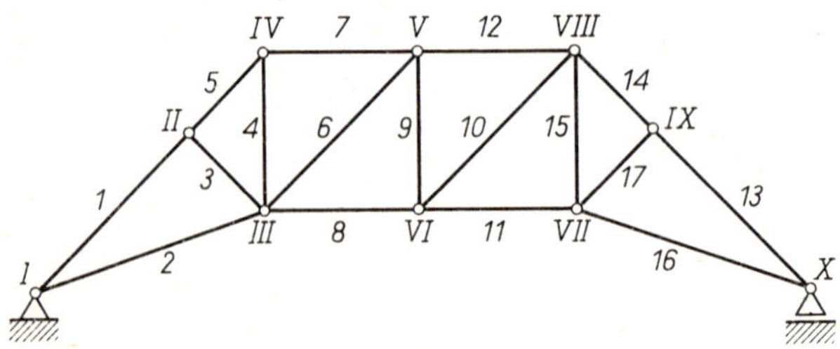

Fig. 3.32

We shall examine the plane truss given in Fig. 3.33a. It is statically determinate, $2w = 8$ and $p+3 = 5+3 = 8$.

Solving it by the node separation technique, we consider successively the state of equilibrium of the respective nodes, beginning with node *I*. Forces act on the nodes, and these in turn act on the rods by the law of action and reaction. We can infer therefrom, which of the rods meeting at the node considered are subject to tension and which to compression.

The forces specified in the plan of the construction are the forces acting on the nodes. If the triangles *I*, *II*, *III* are shifted so as make points A and A', D and D', and C, C' and C'' coincide, we shall obtain a polygon, known as *Cremona's polygon* (Fig. 3.33b). Note should be taken of the fact that the fields of the plan correspond to the points of Cremona's polygon, and vice versa. In Fig. 3.33b, the triangles ABC, ACD, and DCE correspond to nodes *I*, *II*, and *III*, and the polygon $EBAD$ corresponds to node *IV*, whereas the successive fields indicated in the plan of the construction correspond to the points of the polygon $ABCD$.

The sections of the polygon corresponding to the compressive forces are given the sign (−) placed before the rod number, and the sections corresponding to tension, the sign (+). Moreover, the sections corresponding to compressive forces are marked by heavy lines.

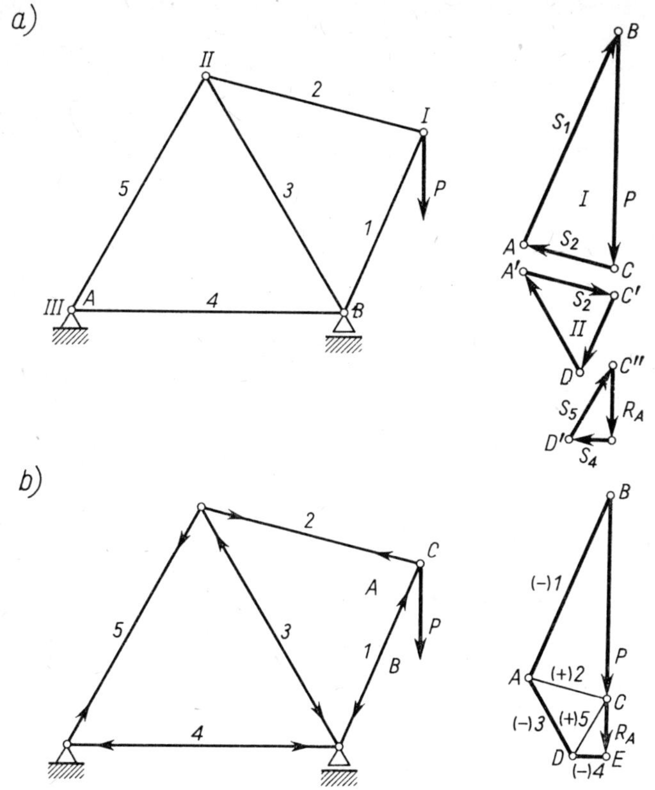

Fig. 3.33

We begin constructing Cremona's polygon from a node, at which two rods meet, e.g. *I* or *III*. We then go over to the next node at which again we have two unknowns. In the considered case having determined, for example, the tension in rod *2* by drawing the triangle *ABC* we treat this tension as the external load on node *II*.

In plotting polygons of forces we should observe the rule that in tracing out the perimeter of the polygon we should always go from one area to another in the plan of the construction in the same direction, either clockwise or anticlockwise. For example, plotting the triangle *ABC*, we move successively from area *C* to *A* and then to *B*, i.e., anticlockwise.

The same direction is always maintained for all the figures occurring in the polygon of forces. Reversal of the direction would turn a polygon as in Fig. 3.33 into its mirror reflection.

We shall now discuss *Culmann's method.* It is likewise a graphical method and is applicable whenever a truss problem cannot be solved by Cremona's method, e.g. in the case where at least three rods meet at all nodes or when we are concerned only with specified rods in a truss. To solve any such problem we make an imaginary division of the considered truss (Fig. 3.34) into two parts in cross-section, through three rods with axes which do not intersect at the same point. We discard one of the parts and consider the state of equilibrium of the other. The latter is under the action of external forces, in our case $\mathbf{P}_1$ and $\mathbf{R}_1$, and also of internal forces, that is, the reaction of the discarded parts through the cut rods $\mathbf{S}_4$, $\mathbf{S}_5$, $\mathbf{S}_6$. The resultant of the forces of the first group is equilibrated by the resultant of the second group.

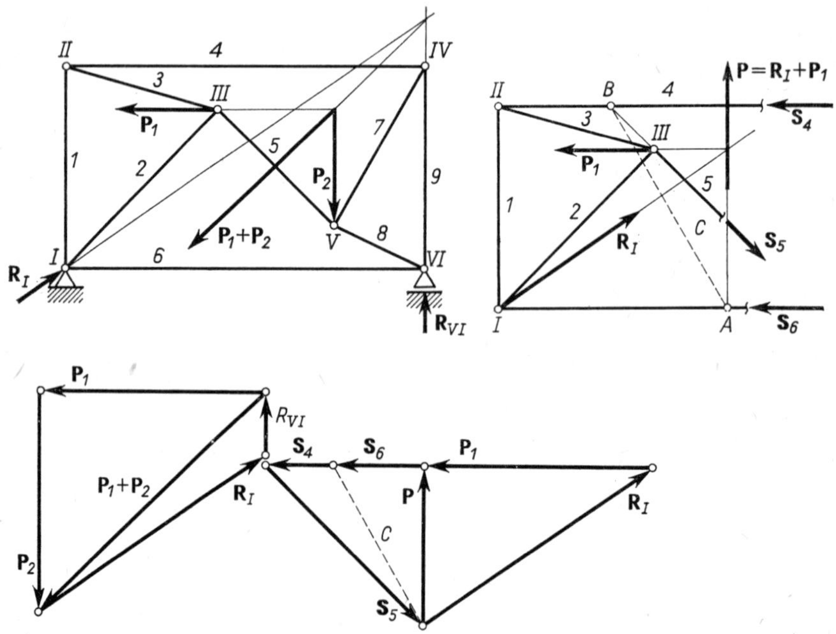

Fig. 3.34

As we can see, we are confronted here with a problem of equilibrium of four forces in a plane. The resultant of the external forces $\mathbf{P}$ as well as the directions of the rods, i.e., the directions of the internal forces, are known. If the system remains in equilibrium, then obviously the resultant of force $\mathbf{P}$ and one of the external forces equilibrates the resultant of the remaining two forces. This resultant therefore passes through the points at which the straight lines corresponding to the respective forces intersect.

For example, in the considered case the resultant of $\mathbf{P}$ and $\mathbf{S}_6$ must be in a straight line given by points A and B, since the resultant of $\mathbf{S}_4$ and $\mathbf{S}_5$ passes through point B and must equilibrate the resultant of $\mathbf{P}$ and $\mathbf{S}_6$ passing through A. The straight line given by points A and B is called *Culmann's line*.

The equivalent of Culmann's graphical method is the analytical method proposed by J. W. Ritter. Writing down the three conditions of equilibrium for the other part of the truss (Fig. 3.34) we find the three unknowns $\mathbf{S}_4$, $\mathbf{S}_5$ and $\mathbf{S}_6$.

3.6 The distributive character of transverse loads in simple rods

Deflection in the case of simple rods is induced by forces at right angles to the rod axis (transverse forces) or by concentrated moments. Let us consider a simple rod (Fig. 3.35) with a uniformly distributed load. The load distribution is characterized by the function $q(x)$. Its value at a given point defines the magnitude of the force per unit length of the rod axis:

$$q(x) \overset{\text{df}}{=} \frac{\mathrm{d}P}{\mathrm{d}x}. \tag{1}$$

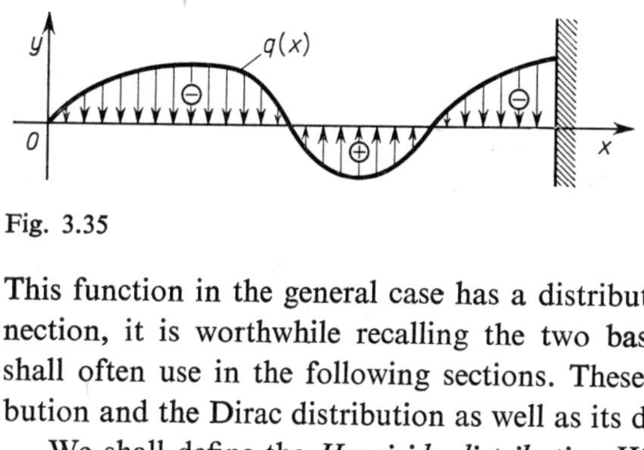

Fig. 3.35

This function in the general case has a distributive character. In this connection, it is worthwhile recalling the two basic distributions which we shall often use in the following sections. These are: the Heaviside distribution and the Dirac distribution as well as its derivatives.

We shall define the *Heaviside distribution* $H(x)$ as follows:

$$H(x) \overset{\text{df}}{=} \begin{cases} 0 & \text{for} \quad x \leqslant 0, \\ 1 & \text{for} \quad x > 0. \end{cases} \tag{2}$$

As we can see, it shows a unit step. This function can be treated as the limit of a certain class of series $\{f_n(x)\}$ such that

$$\lim_{n\to\infty} f_n(x) = H(x).$$

An example of such a series of functions might be

$$\{f_n(x)\} = \left\{\frac{1}{1+e^{-nx}}\right\}.$$

This idea of distributions [1] is both convenient and practical if only for that fact that the concept of differentiability ceases to be problem and that distributions can be operated as functions, as we shall find out further on, though they are not always functions in the ordinary sense of the word. It is the *Dirac delta* that belongs precisely to such distributions which we shall often call functions. We shall define it as being the derivative of $H(x)$:

$$\delta(x) \stackrel{\mathrm{df}}{=} \frac{\mathrm{d}H(x)}{\mathrm{d}x}. \tag{3}$$

Holding to our idea of distributions we treat the Dirac delta as the limit of a series $\{\varphi_n(x)\}$ of derivatives of $\{f_n(x)\}$. Thus

$$\delta(x) = \lim_{n\to\infty} \varphi_n(x),$$

where

$$\{\varphi_n(x)\} = \{f'_n(x)\}.$$

Following this line of reasoning we can easily see that

$$\delta(x) = \begin{cases} 0 & \text{for} \quad x < 0, \\ \infty & \text{for} \quad x = 0, \\ 0 & \text{for} \quad x > 0. \end{cases} \tag{4}$$

On the other hand, thinking in terms of approximations to the function $\delta(x)$, one can imagine its graph as a curve rapidly ascending in the vicinity of zero on the left and descending with equal rapidity on the right. We shall now prove an important property of the Dirac distribution.

$$\int_{-\infty}^{\infty} \delta(x) F(x) \mathrm{d}x = F(0). \tag{5}$$

This property may serve as the definition of the function $\delta(x)$ [36].

The proof is simple, since it will suffice to appreciate that for $a > 0$

$$\int_{-a}^{a} \delta(x) F(x) \mathrm{d}x = \int_{-a}^{a} \frac{\mathrm{d}H(x)}{\mathrm{d}x} F(x) \mathrm{d}x$$

$$= H(x)F(x)\Big|_{-a}^{a} - \int_{-a}^{a} H(x)\frac{dF(x)}{dx}dx$$

$$= F(a) - \int_{F(0)}^{F(a)} dF(x) = F(a) - F(a) + F(0).$$

Passing with the limits to infinity ($a \to \infty$) we prove our theorem.

An important part in our considerations will play a derivative of the Dirac delta function. This derivative has the following property:

$$\int_{-\infty}^{\infty} \delta'(x)F(x)dx = -F'(0). \tag{6}$$

The proof of this proposition is as follows: For $a > 0$

$$\int_{-a}^{a} \delta'(x)F(x)dx = \delta(x)F(x)\Big|_{-a}^{a} - \int_{-a}^{a} \delta(x)F'(x)dx = -F'(0).$$

The result will not change if $a \to \infty$. The result (6) can easily be generalized for the nth derivative of the function $\delta(x)$.

$$\int_{-\infty}^{\infty} \delta^{(n)}(x)F(x)dx = (-1)^n F^{(n)}(0). \tag{6a}$$

We shall now show how such concepts as continuous load of constant value, concentrated force, and concentrated moment bring forth the concept of the Heaviside distribution, the Dirac distribution, and the derivative of the Dirac delta in a natural way.

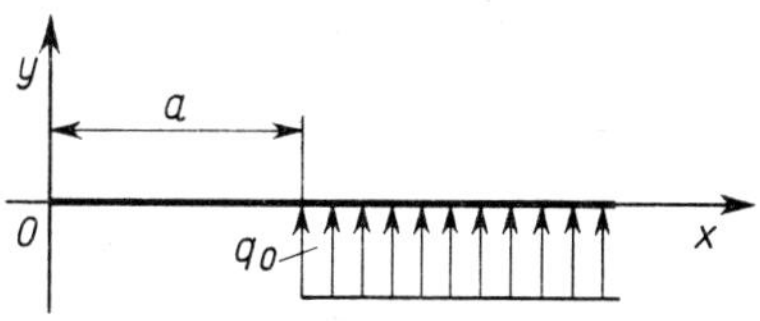

Fig. 3.36

Let a uniformly distributed load q_0 act on part of rod whose axis coincides with the x-axis (Fig. 3.36). We can see that the load is in the nature of a unit step, it can therefore be written, using the Heaviside's function:

$$q(x) = H(x-a)q_0. \tag{7}$$

Making use of superposition, the load applied to an arbitrary section, $b-a$ in length, can be expressed via (7) in the following manner:

$$q(x) = [H(x-a)-H(x-b)]q_0. \tag{7a}$$

Figure 3.37 gives graphical interpretation of equation (7a) in which the load summation method is demonstrated. One can see that a load other than zero occurs only in the section $b-a$. Naturally, with a larger number of loads of this type involved, it would be necessary to superpose expressions of the (7a) construction.

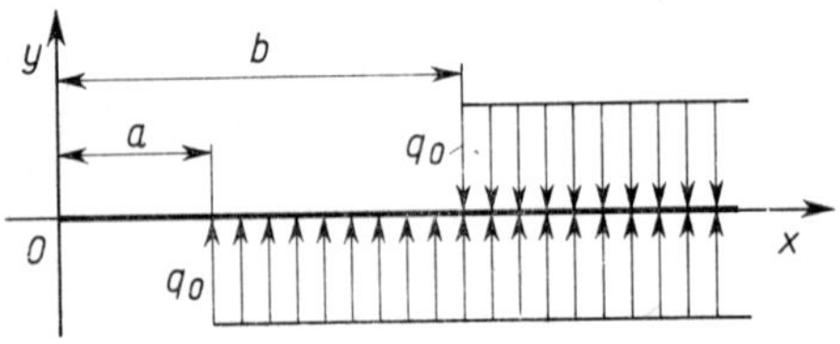

Fig. 3.37

Now let us consider the case of concentrated force **P**. Since

$$q_0 = \frac{P}{b-a},$$

we have

$$q(x) = P\frac{H(x-a)-H(x-b)}{b-a}. \tag{8}$$

Note that if $a \to b$ (the case of concentrated force), then the differential quotient

$$\frac{H(x-a)-H(x-b)}{b-a} \to \delta(x-b).$$

This can easily be checked, putting $x-b = \xi$, then $x-a = x-b+b-a = \xi+b-a$. Therefore,

$$\lim_{a\to b}\frac{H(x-a)-H(x-b)}{b-a} = \lim_{a\to b}\frac{H(\xi+b-a)-H(\xi)}{b-a} = \delta(x-b).$$

Thus

$$q(x) = P\delta(x-b). \tag{9}$$

In the case of several forces acting,

$$q(x) = \sum_j P_j\delta(x-b_j). \tag{9a}$$

Corresponding to plus values of P is the sense as specified in Fig. 3.38.

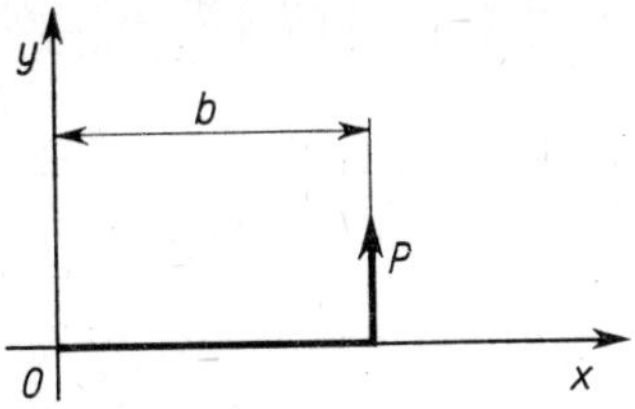

Fig. 3.38

We shall now consider a concentrated moment M (Fig. 3.39). We may imagine it as a couple (Fig. 3.39b):

$$M = P\Delta a. \tag{10}$$

We deal with a concentrated moment when $\Delta a \to 0$. The function $q(x)$ for this type of load may be given in the following form (we make use here of (9a) and (10)):

$$\begin{aligned} q(x) &= \lim_{\Delta a \to 0} [P\delta(x-a) - P\delta(x-a-\Delta a)] \\ &= M \lim_{\Delta a \to 0} \frac{\delta(x-a) - \delta(x-a-\Delta a)}{\Delta a} = M\delta'(x-a). \end{aligned}$$

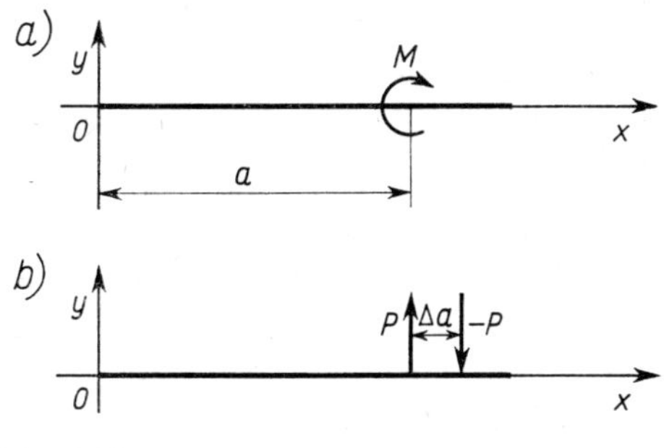

Fig. 3.39

In the case of several concentrated moments, we obtain an equation analogous to (9a):

$$q(x) = \sum_i M_i \delta'(x-a_i), \tag{11}$$

moreover the sense specified in Fig. 3.39 corresponds to a positive value of M.

In principle, all types of loads can be expressed with the use of the proposed distributions.

Example 1. Find the load function $q(x)$ for a rod subjected to the action of continuously distributed forces according to the function $f(x)$ in the section (a, b) (Fig. 3.40).

The problem consists of obtaining zero values for the function $q(x)$ outside the interval (a, b). It will easily be found that the following presentation satisfies this requirement:

$$q(x) = f(x)\,[H(x-a)-H(x-b)].$$

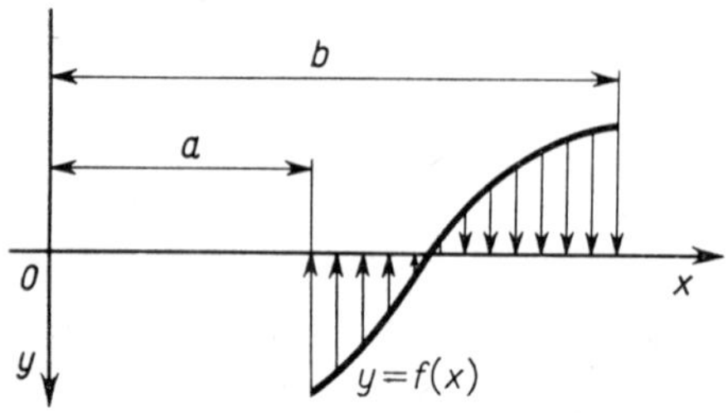

Fig. 3.40

By introducing the distribution concept, load can be treated uniformly, i.e., as being continuous. This is an easier line of approach to the problem, because the distribution language renders all the types of load encountered in a most natural way. Besides, the understanding of the sense of these few distributions should not present any particular difficulty provided, of course, that the reader is familiar with the concept of limits.

There is yet another reason why the departure from the traditional methods of handling the described problems is rational, namely it results in a marked simplification of the rod theory. One should add that the theory of distribution asserts itself very strongly in all fields of mechanics.

This was already apparent in Chapter 2 when we discussed the problems involved in the determination of the position of centre of mass in discrete systems.

Knowing the Dirac delta distribution, the definitions of the position vector $\mathbf{r}_s$ for discrete and continuous systems can be given a uniform wording. In the light of the present considerations the equation

$$\mathbf{r}_s = \frac{\int\limits_v \mu(\mathbf{r})\,\mathrm{d}v}{\int\limits_v \mu\,\mathrm{d}v} \tag{12}$$

becomes manifestly universal. Let us deal with the density function $\mu(\mathbf{r})$ in a more general way, i.e., for a system of n particles each with mass m_i:

$$\mu(\mathbf{r}) = \sum_{i=1}^{n} m_i\,\delta(x-x_i)\,\delta(y-y_i)\,\delta(z-z_i). \tag{13}$$

On substituting (13) into (12) and after integrating we obtain

$$\mathbf{r}_s = \frac{\sum_i m_i \mathbf{r}_i}{\sum_i m_i}. \tag{12a}$$

In this way the universality of equation (12) becomes apparent, provided that the density function $\mu(\mathbf{r})$ be treated more generally. Furthermore, equation (13) constitutes directly a precise definition of the discrete medium. One can therefore say that a discrete medium is such for which density $\mu(\mathbf{r})$ is defined by equation (13).

Yet another advantage of the distributive approach is that we can confine ourselves to the concept of integral in Riemann's sense.

3.7 The equilibrium of rods loaded with transverse forces

Suppose we have a rod of length l loaded only with transverse forces in one plane. Assuming the distributive character of load we deal with a problem of general nature. All the equilibrium conditions of the rod are satisfied by virtue of the mode of support. In the problem considered unidirectional forces are involved (Fig. 3.41), therefore only two equilibrium conditions come in here, namely the condition of projection of forces on the y-axis and the condition of moments

$$\int_0^l q(x)\,\mathrm{d}x = 0, \tag{1}$$

$$\int_0^l xq(x)\,\mathrm{d}x = 0. \tag{2}$$

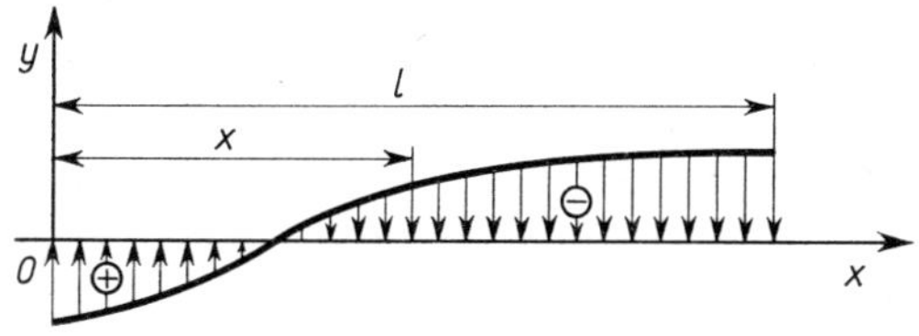

Fig. 3.41

Equations (1) and (2) are fundamental to all problems of the statics of a simple rod loaded with forces perpendicular to its axis.

Let us now examine several basic problems which will be of use to us later on.

Example 1. Consider a beam of length l supported at points A and B (Fig. 3.42). Let us isolate from the load $q(x)$ which we introduced previously an effective load $q_e(x)$. In this way, the total load is expressed by

$$q(x) = R_A\,\delta(x)+R_B\,\delta(x-l)+q_e(x). \tag{a}$$

Substituting (a) into (1) and (2) and integrating, we obtain the following equations:

$$R_A+R_B+\int_0^l q_e(x)\,\mathrm{d}x = 0, \tag{b}$$

$$R_B l+\int_0^l x q_e(x)\,\mathrm{d}x = 0. \tag{c}$$

Fig. 3.42

At this juncture the question may arise as to whether the results (b) and (c) are true in terms of the integration range. Another doubtful point might be whether $\int_0^l \delta(x)\mathrm{d}x$ equals to 1 or 1/2. These doubts will be dissipated by considering the fact that integration proceeds over the entire area in which $q(x) \neq 0$. Thus it becomes understandable that formally, instead of $\int_0^l (\cdot)\mathrm{d}x$, it would be justifiable to write $\int_{-\infty}^{\infty} (\cdot)\mathrm{d}x$, which is equivalent by virtue of the zero values which the function $q(x)$ assumes outside the range [0, 1]. To leave no room for doubt on this score, $\int_0^l (\cdot)\mathrm{d}x$ should be understood as $\int_{-\varepsilon}^{l+\varepsilon} (\cdot)\mathrm{d}x$.

Solving equations (a) and (b) with respect to R_A and R_B, we obtain

$$R_A = -\int_0^l \frac{l-x}{l} q_e(x)\,\mathrm{d}x, \tag{a1}$$

$$R_B = -\int_0^l \frac{x}{l} q_e(x)\,\mathrm{d}x. \tag{b1}$$

Example 2. A beam (Fig. 3.43) of length l with one end rigidly fixed and loaded with forces $q_e(x)$ is given. In this case

$$q(x) = M_u\,\delta'(x)+R\delta(x)+q_e(x). \tag{a}$$

Substituting (a) into (1) and (2), we have

$$R + \int_0^l q_e(x)\,dx = 0, \tag{b}$$

$$M_u + \int_0^l x q_e(x)\,dx = 0. \tag{c}$$

Equations (b) and (c) give the reactions R and M.

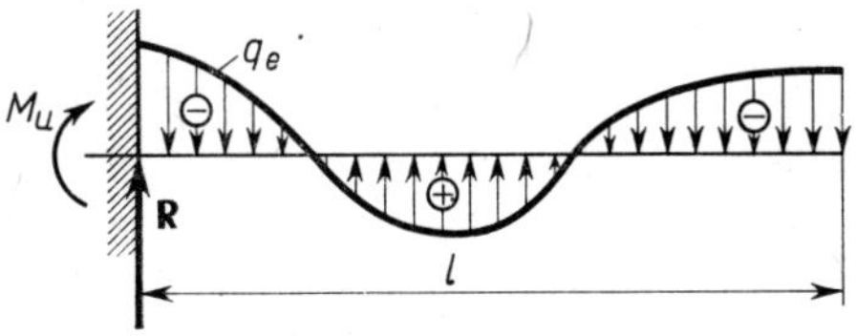

Fig. 3.43

We now come to the problem of determining the forces and moments in an arbitrary cross-section of simple rod. These forces and moments we shall call *internal magnitudes*. In order to find them in an arbitrary cross-section determined by a coordinate x, we shall adopt the following reasoning.

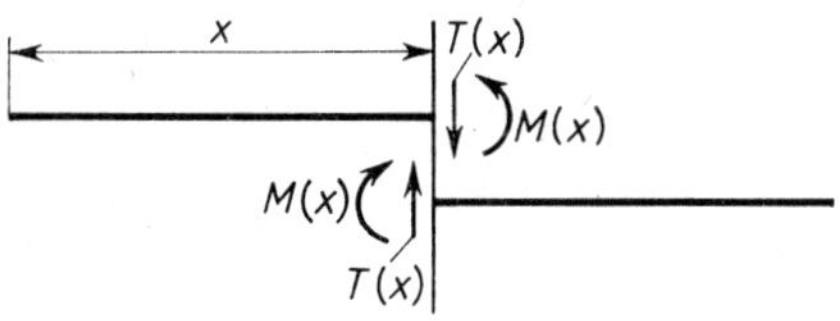

Fig. 3.44

The cross-section considered divides the rod into two parts; let them be called *left* and *right*. Each part remains in equilibrium, therefore all the forces acting on them satisfy the equilibrium conditions. For the left part (Fig. 3.44) we have

$$\int_0^x q_L(\xi)\,d\xi = 0, \tag{3}$$

$$\int^x \xi q_L(\xi)\,d\xi = 0. \tag{4}$$

For the right part only the integration limits change, i.e.,

$$\int_x^l q_R(\xi)\mathrm{d}\xi = 0, \tag{5}$$

$$\int_x^l \xi q_R(\xi)\mathrm{d}\xi = 0. \tag{6}$$

We have in equations (3)–(6) the functions $q_L(\xi)$ and $q_R(\xi)$ which define respectively the state of load of the left and right parts of the rod

$$q_L(\xi) = q(\xi) - M(x)\,\delta'(\xi - x) - T(x)\,\delta(\xi - x), \tag{7}$$

$$q_R(\xi) = q(\xi) + M(x)\,\delta'(\xi - x) + T(x)\,\delta(\xi - x). \tag{6}$$

To determine the external magnitudes, i.e., the bending moment $M(x)$ and the transverse force $T(x)$, we can make use of the equations of equilibrium for an arbitrarily chosen part. In the considered case, we have decided to consider the state of equilibrium of the left part.

Substituting (7) into (3) and (4), we obtain

$$\int_0^x q(\xi)\mathrm{d}\xi - T(x) = 0, \tag{3a}$$

$$\int_0^x \xi q(\xi)\mathrm{d}\xi + M(x) - T(x)x = 0. \tag{4a}$$

Solving the above system of equations with respect to $M(x)$ and $T(x)$, we obtain the following relations:

$$M(x) = \int_0^x (x - \xi) q(\xi)\mathrm{d}\xi, \tag{8}$$

$$T(x) = \int_0^x q(\xi)\mathrm{d}\xi. \tag{9}$$

Note that

$$\frac{\mathrm{d}M(x)}{\mathrm{d}x} = T(x). \tag{10}$$

Indeed, considering that

$$\frac{\mathrm{d}}{\mathrm{d}x}\left[\int_0^x f(x, \xi)\mathrm{d}\xi\right] = f(x, x) + \int_0^x \frac{\partial f(x, \xi)}{\partial x}\mathrm{d}\xi, \tag{11}$$

equation (10) becomes obvious. Differentiating (9) once more, we find that

$$\frac{dT(x)}{dx} = q(x) \tag{12}$$

and for this reason, too,

$$\frac{d^2M}{dx^2} = q(x). \tag{13}$$

Example 3. Let us consider a beam section loaded with concentrated moments (Fig. 3.45). The beam is in equilibrium $M_1 - M_2 + M_3 = 0$.

We have

$$q(\xi) = M_1\,\delta'(\xi-a_1) - M_2\,\delta'(\xi-a_2) + M_3\,\delta'(\xi-a_3). \tag{a}$$

To find the curve of moments $M(x)$ and of the transverse force $T(x)$, we shall make use of equations (8) and (9). After integrating as indicated there, we obtain

$$M(x) = M_1 H(x-a_1) - M_2 H(x-a_2) + M_3 H(x-a_3), \tag{b}$$

$$T(x) = 0. \tag{c}$$

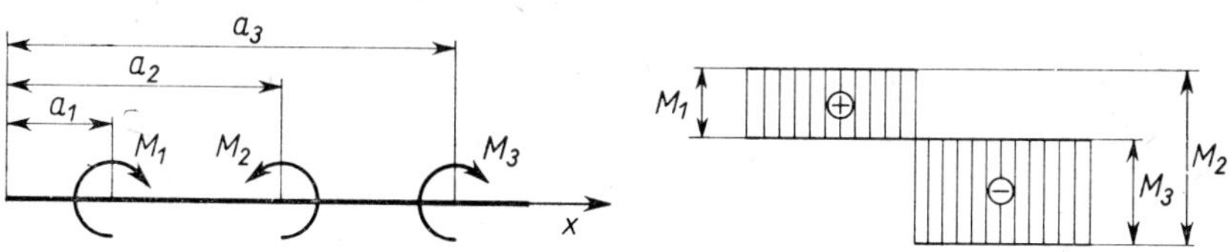

Fig. 3.45

Note that at points $x = a_1$, $x = a_2$, $x = a_3$ the graph is a discontinuous line. The ordinate at any point of the graph is proportional to the sum of the moments on one side of the considered cross-section.

Example 4. A rigidly built-in beam (Fig. 3.46) is loaded with concentrated forces such that

$$q(\xi) = P_1\,\delta(\xi-b_1) + P_2\,\delta(\xi-b_2) - P_3\,\delta(\xi-b_3) + P_4\,\delta(\xi-b_4).$$

Find the curve of bending moments and transverse forces.

Making use of equations (8) and (9), we obtain

$$M(x) = P_1(x-b_1)H(x-b_1) + P_2(x-b_2)H(x-b_2) - P_3(x-b_3)H(x-b_3) + P_4(x-b_4)H(x-b_4), \tag{a}$$

$$T(x) = P_1 H(x-b_1) + P_2 H(x-b_2) - P_3 H(x-b_3) + P_4 H(x-b_4). \tag{b}$$

The graphs of the functions (a) and (b) are given in Fig. 3.46. Note that the curve of the bending moments between change points is linear. The singular points (cusps) are at the same time the points of applicaton of the concentrated forces. The straight line rises when $T(x) > 0$ (an increasing function has a positive derivative) and falls when $T(x) < 0$.

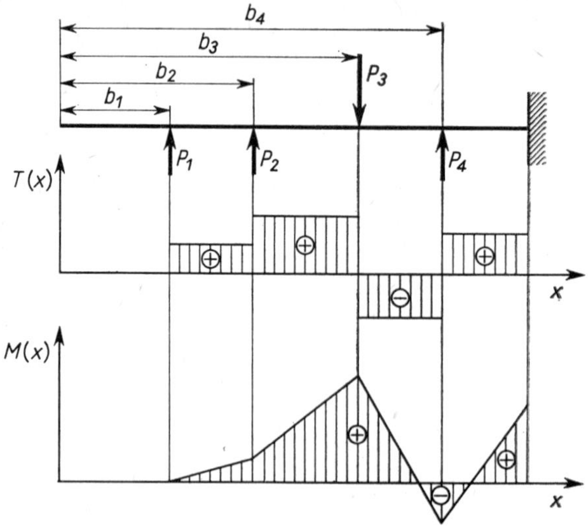

Fig. 3.46

Example 5. A beam with one end fixed is subjected to continuous loads of constant intensity q_1, q_2, q_3, originating at points with the coordinates c_1, c_2, c_3 (Fig. 3.47). Find the progress of forces $T(x)$ and moments $M(x)$.

The load can be represented as follows:

$$q(\xi) = -q_1 H(\xi-c_1)+q_2 H(\xi-c_2)-q_3 H(\xi-c_3). \tag{a}$$

The transverse forces and the moments we find, as in the foregoing examples, by making use of equations (8) and (9):

$$T(x) = -q_1(x-c_1)H(x-c_1)+q_2(x-c_2)H(x-c_2)-q_3(x-c_3)H(x-c_3),$$

$$M(x) = -q_1\frac{(x-c_1)^2}{2}H(x-c_1)+ \tag{b}$$
$$+q_2\frac{(x-c_2)^2}{2}H(x-c_2)-q_3\frac{(x-c_3)^2}{2}H(x-c_3).$$

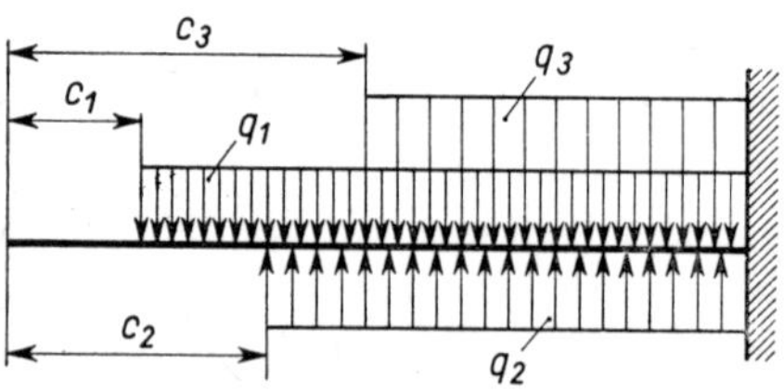

Fig. 3.47

The results obtained in the cited examples can easily be generalized for the case of a beam subjected to the simultaneous action of n concentrated moments M_j, s concentrated forces P_j, and r loads q_k. The load function for this more general case is of the following form:

$$q(x) = \sum_{i=1}^{n} M_i \delta'(x-a_i) + \sum_{j=1}^{s} P_j \delta(x-b_j) + \sum_{k=1}^{r} q_k H(x-c_k). \tag{14}$$

And in this connection,

$$T(x) = \sum_{j=1}^{s} P_j H(x-b_j) + \sum_{k=1}^{r} q_k (x-c_k) H(x-c_k), \tag{15}$$

$$M(x) = \sum_{i=1}^{n} M_i H(x-a_i) + \sum_{j=1}^{s} P_j (x-b_j) H(x-b_j) + \sum_{k=1}^{r} \frac{q_k (x-c_k)^2}{2} H(x-c_k). \tag{16}$$

Example 6. Calculate the bending moment and the transverse forces in a beam loaded with a constantly increasing continuous weight (Fig. 3.48), for which

$$q(x) = -\frac{q_0}{l} x.$$

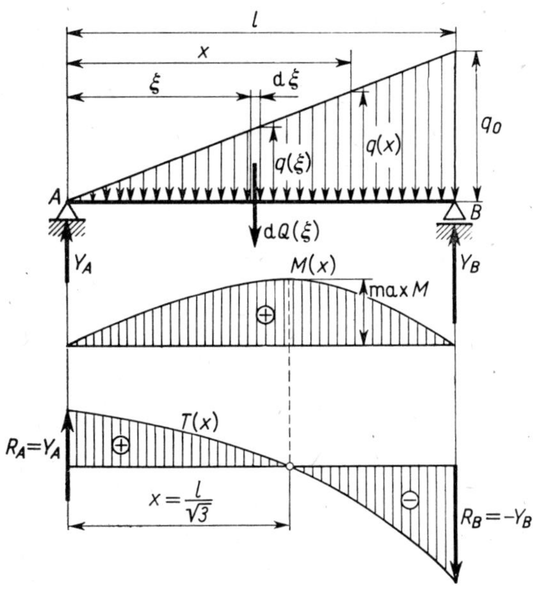

Fig. 3.48

We calculate the reaction at point A, making use of equations (a1) from Example 1:

$$Y_A = -\frac{1}{l}\int_0^l \left(-\frac{q_0}{l}\right)x(l-x)\,dx = \frac{q_0}{l^2}\int_0^l x(l-x)\,dx$$

$$= \frac{q_0}{l^2}\left(\frac{lx^2}{2}-\frac{x^2}{3}\right)\Big|_0^l = \frac{q_0}{l^2}\left(\frac{l^3}{2}-\frac{l^3}{3}\right) = \frac{1}{6}q_0 l.$$

The bending moment is expressed by

$$M(x) = Y_A x + \int_0^x \left(-\frac{q_0}{l}\right)\xi(x-\xi)\,d\xi = Y_A x - \frac{q_0}{l}\left(\frac{x\xi^2}{2}-\frac{\xi^3}{3}\right)\Big|_0^x$$

$$= Y_A x - \frac{1}{6l}q_0 x^3 = \frac{1}{6}q_0 lx - \frac{1}{6l}q_0 x^3.$$

Using equation (10), we find the magnitude of the transverse forces:

$$T(x) = \frac{q_0 l}{6} - \frac{q_0 3x^2}{6l} = \frac{q_0}{6l}(l^2-3x^2).$$

Hence we can also find the point of action of the maximum bending moment. Putting

$$\frac{dM}{dx} = T = 0,$$

we obtain $\frac{1}{6}q_0(l^3-3x^2) = 0$; hence $l^2-3x^2 = 0$, therefore $x = l/\sqrt{3}$. Figure 3.48 shows the curves of bending moments and transverse forces.

We shall now discuss graphical methods. The construction in Fig. 3.49 is a free-ends beam supported at points A and B. Consider the bending moment of the beam a cross-section taken at x_1. The moment is proportional, as we know, to the section contained between z and 1. This is the moment of the reaction force $\mathbf{R}_A$. Its magnitude will be according to equality (20) in Section 3.4

$$M_1 = \eta_1 h\mu_k\mu_p. \tag{17}$$

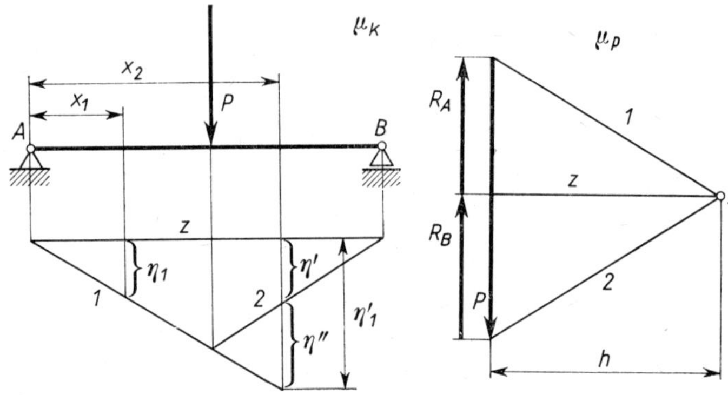

Fig. 3.49

But the moment in the cross-section at x_2 is the sum of the moments from forces on one side of the cross-section, i.e., from the reaction force $\mathbf{R}_A$ and the applied force $\mathbf{P}$ or, what amounts to the same, from the reaction force $\mathbf{R}_B$. The moment from $\mathbf{R}_A$ is

$$M = \eta'_1 h\mu_k\mu_p,$$

and from the force $\mathbf{P}$

$$M = \eta'' h\mu_k\mu_p.$$

Since $\mathbf{P}$ has the opposite sense to that of $\mathbf{R}_A$, the moment in the cross-sections on the right-hand side of $\mathbf{P}$ is described by

$$M = (\eta'_1 - \eta'') h\mu_k\mu_p.$$

In Fig. 3.50 we have a solution to the same problem in the case of simultaneous action of three forces at right angles to the beam axis. The diagram of the bending moments is a funicular polygon. The sections η are proportional in the scale $h\mu_k\mu_p$ to the bending moment.

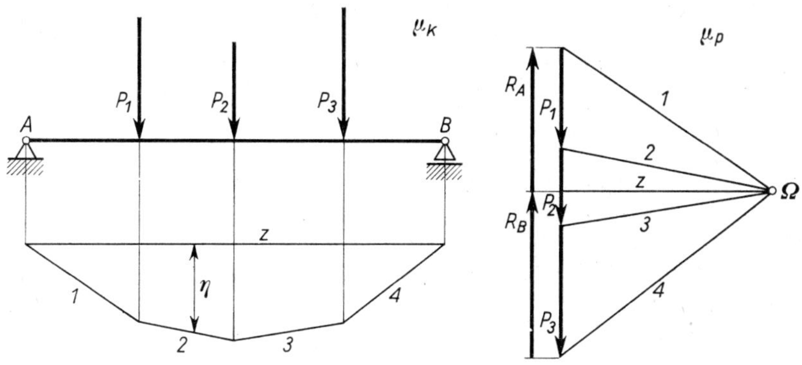

Fig. 3.50

When a beam is under the action of a continuous load, as shown in Fig. 3.51, the problem should be reduced to the case of concentrated forces in action. In this way, we reduce the problem to the example solved in Fig. 3.50. The accuracy of such an approximative solution increases with the density of the continuous load distribution, i.e., it is the greater the more forces $\mathbf{P}_i$ (Fig. 3.51) approximate a continuous load.

Figure 3.51 gives by the graphical method the curve of bending moments and transverse forces; the possibility of change of the pole Ω has also been demonstrated. It is worth noting that by moving the pole Ω to the point

Ω' we acquire a new closing side z' having the orientation of the horizontal line.

The possibility of bringing in such changes is useful on account of the better legibility of the diagram thus achieved. By discretization of the

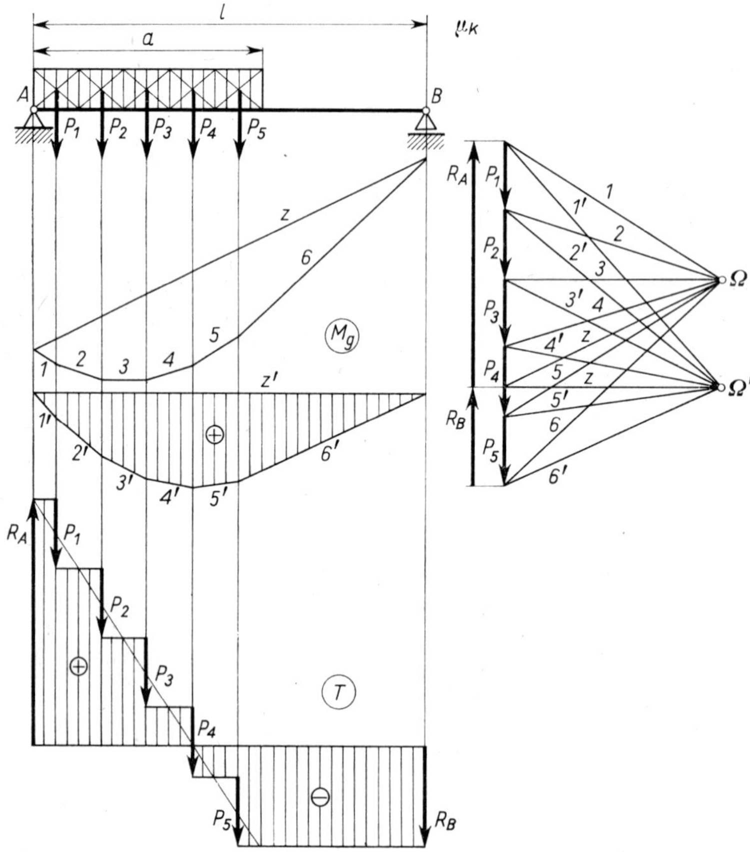

Fig. 3.51

continuous load we have obtained instead of a straight line a step curve. As the density of the load distribution increases, the curve tends to the straight line indicated in the drawing.

3.8 Friction

Friction is an effect in the course of which forces opposed to motion come in action. It occurs at the interface of two bodies moving against each other and is dependent on the pressure exerted by one body on the other. If, for example, we put a brick on the table (Fig. 3.52) and want to move it along the table's surface, we shall have to act on the brick with a force capable of overcoming friction.

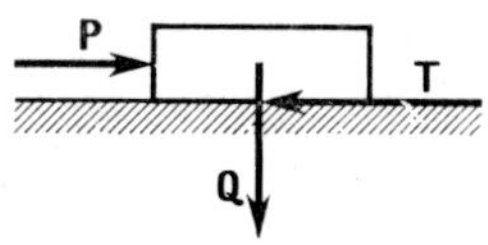

Fig. 3.52

The motion of the brick will be possible when $P \geqslant T$. Repeating the experiment in different conditions (different contact surfaces), we find that friction depends on the materials in contact. In the mathematical form, we shall write briefly

$$T \overset{\text{df}}{=} fN, \tag{1}$$

where T is the friction force, f the coefficient of friction dependent on materials subject to friction (the value of f will be found in physical tables), N thrust (in the considered case, equal to Q).

The value of the coefficient of friction is different when a body is at rest and from when it is in motion. Consequently, there are two values of f: the static value and the kinetic value.

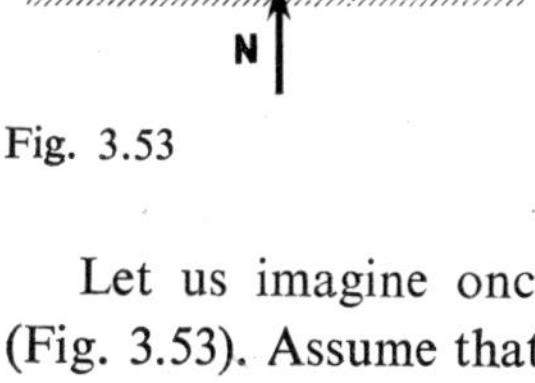

Fig. 3.53

Let us imagine once more the brick on a horizontal plane surface (Fig. 3.53). Assume that the brick is weightless. We apply a force P to the brick at an angle β to the normal. We shall consider for which values of β the brick will remain at rest. The conditions of equilibrium are

$$P\sin\beta \leqslant T, \quad P\cos\beta = N. \tag{2}$$

Dividing both sides we obtain

$$\frac{P\sin\beta}{P\cos\beta} \leqslant \frac{T}{N} = f,$$

or

$$\tan\beta \leqslant \frac{T}{N} = f. \tag{3}$$

This shows that the angle β can vary within certain limits. The equilibrium state will last for as long as

$$\beta \leqslant \tan^{-1}f. \tag{4}$$

The limit of angle $\varrho = \tan^{-1}f$ is called the *friction angle*. The force **P** in the limit position coincides with the generator of a cone with the apex angle 2ϱ. This cone is called the *cone of friction*. As we can see, the equilibrium state will be maintained for as long as the force remains within the cone of friction. Obviously, we may treat **P** as a resultant of several forces acting on the body.

We shall now discuss friction in different cases.

Friction on an inclined plane

For a body C to be in equilibrium or in uniform motion on an inclined plane (Fig. 3.54) it will suffice for the angle β to be not greater than the friction angle ϱ:

$$\beta \leqslant \varrho. \tag{5}$$

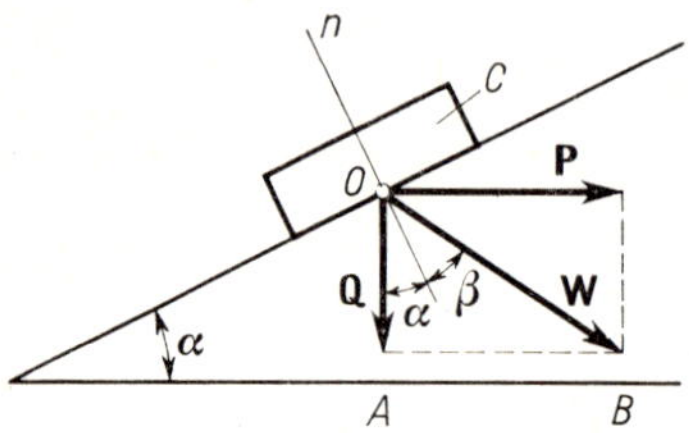

Fig. 3.54

The angle between the direction of the gravity force **Q** and the normal is α. Therefore, $\sphericalangle AOB = \alpha+\beta$. On the other hand,

$$\tan(\alpha+\beta) = \frac{P}{Q}. \tag{6}$$

In the limit case $\beta = \varrho$, then

$$P = Q\tan(\alpha+\varrho). \tag{7}$$

The other extreme case (Fig. 3.55) is when force P is so small that the resultant will coincide with the generator on the opposite side. A body can move down an inclined plane when $\beta = -\varrho$, and equation (6) takes the form

$$P = Q\tan(\alpha-\varrho). \tag{8}$$

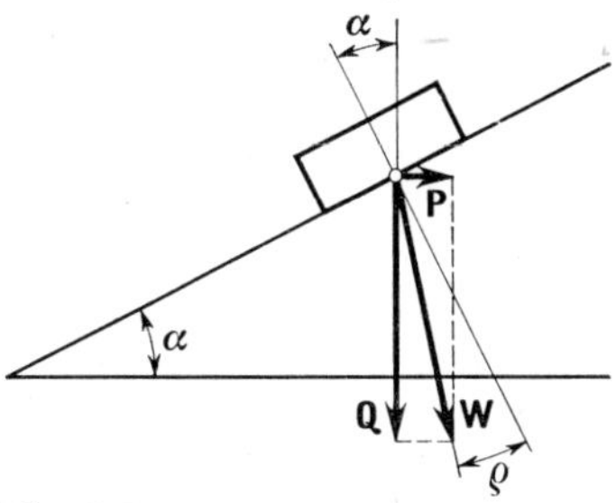

Fig. 3.55

If α is smaller than ϱ, the inclined plane is said to be *self-locking*. If we place some object on a movable board and then raise the board with one end so that the other end should become the axis of rotation, we shall obtain an inclined plane of variable angle α. In our experiment, by raising the end of the board, we increase the angle β. At a certain instant, $\alpha = \varrho$. On exceeding this value the object on the board begins to slide down, and so by measuring the angle α we can calculate approixmately the angle ϱ.

Equations (7) and (8) can be used without hindrance for calculating screw hoists with square thread (Fig. 3.56). The role of an inclined plane of angle $\tan\alpha = h/d_p\pi$ is that of the immovable nut and the role of a movable object, the screw. The moment needed for lifting or lowering a weight is

$$M = Q\frac{d_p}{2}\tan(\alpha\pm\varrho). \tag{9}$$

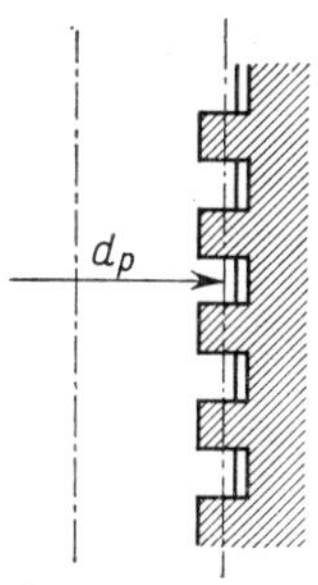

Fig. 3.56

Bearing friction

A different example is the friction of journals against bearing sleeves (Fig. 3.57). The journal is in contact with the sleeve along the generator passing through point O where sliding on the immovable cylindrical surface occurs. It is easily seen that

$$R = Q \quad \text{and} \quad T = R\sin\varrho,$$

therefore

$$T = Q\sin\varrho. \tag{10}$$

And the friction couple is

$$M = rT = Qr\sin\varrho. \tag{11}$$

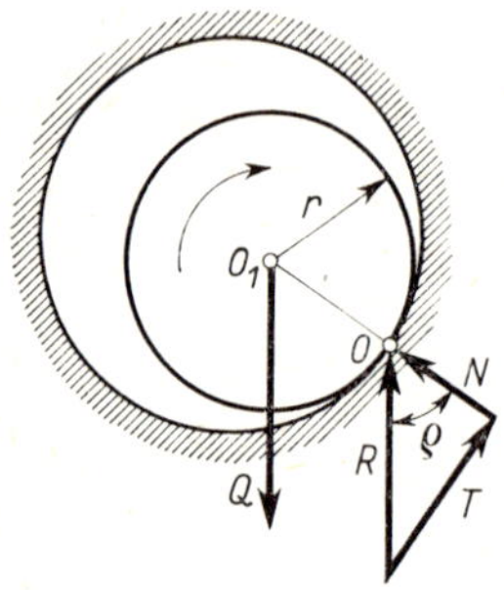

Fig. 3.57

Now we shall concern ourselves with the case of friction of a journal in an axial bearing (Fig. 3.57). Assume that the axial load Q is uniformly distributed over the surface of a circle with radius r. Consequently, the load per unit area is constant and equal to $q = Q/\pi r^2$.

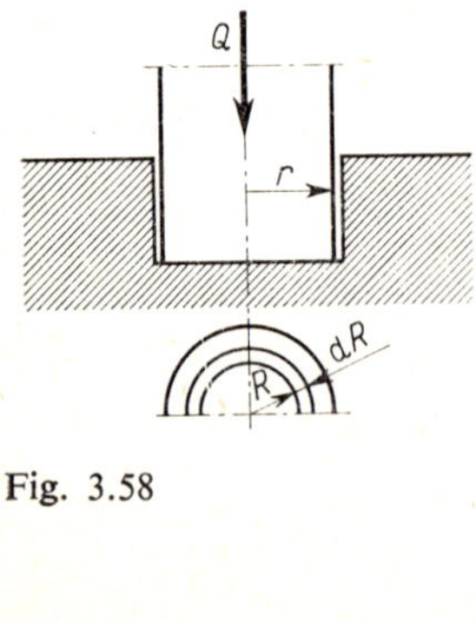

Fig. 3.58

We isolate in the area of the circle a strip of radius R and thickness $\mathrm{d}R$ (Fig. 3.58), area $\mathrm{d}A = 2\pi R\mathrm{d}R$. The friction couple occurring on the surface of the strip is

$$\mathrm{d}M = qfR\mathrm{d}A = 2\pi qfR^2\mathrm{d}R. \tag{12}$$

The total couple we obtain, integrating the expression (12),

$$M = 2\pi\frac{Qf}{\pi r^2}\int_0^r R^2\mathrm{d}R = \frac{2}{3}fQr. \tag{13}$$

Friction of rope against wheel

We shall consider the equilibrium of an elementary section of rope (Fig. 3.59):

$$\mathrm{d}N - 2S\frac{\mathrm{d}\alpha}{2} = 0, \quad \mathrm{d}S + \mathrm{d}T = 0, \tag{14}$$

but

$$\mathrm{d}T = f\,\mathrm{d}N, \tag{15}$$

therefore

$$\mathrm{d}S + f\mathrm{d}N = 0. \tag{16}$$

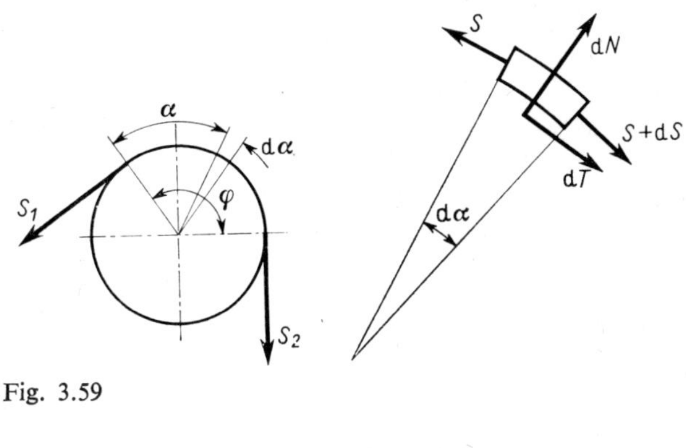

Fig. 3.59

From the first of the conditions of equilibrium (14) it follows that

$$\mathrm{d}N = S\mathrm{d}\alpha. \tag{17}$$

The friction of rope against wheel is utilized, for example, in the work of Koepe pulleys in hoisting machines. Here, friction performs a useful function as it enables the rope to be driven.

We shall examine how great a force S_1, applied to one end of a rope, is required to move it on a fixed wheel while overcoming the friction force as well as a force S_2 loading the other end of the rope (Fig. 3.59)

$$dS + fS\,d\alpha = 0; \tag{18}$$

hence

$$\frac{dS}{S} = -f\,d\alpha. \tag{19}$$

Integrating (19)

$$\int_{S_1}^{S_2} \frac{dS}{S} = -f \int_0^{\varphi} d\alpha,$$

we obtain

$$\ln \frac{S_2}{S_1} = -f\varphi,$$

or

$$S_2 = S_1 e^{-\varphi f}. \tag{20}$$

This is how the relationship between S_1 and S_2 stands when we want to force the motion of the rope in the direction of S_1.

Problem of friction in the case of a cylinder rolling on plane surface

A force P acts on a cylinder maintaining it in uniform motion (Fig. 3.60). From the equilibrium conditions it follows that

$$P = T, \quad Q = N, \quad Pr = Nk. \tag{21}$$

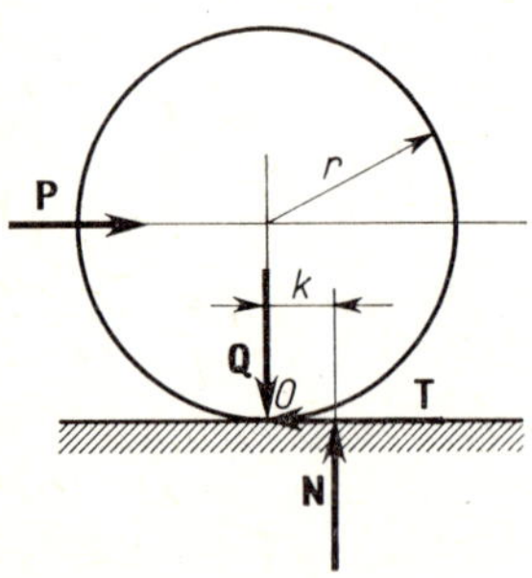

Fig. 3.60

From the last equation for the equilibrium of moments of forces about point O, it follows that

$$P = \frac{k}{r} N. \tag{22}$$

Taking into account the first two equalities, we write finally

$$T = \frac{k}{r} Q, \tag{23}$$

where k is the section by which the reaction force is moved as the result of material strain of the base.

It is evident from (23) that the coefficient of friction f in the case of rolling is the ratio k/r.

The cases of friction surveyed here are of course not confined to the cited examples. The main consideration in choosing these particular cases was that they are important from the engineering point of view.

CHAPTER 4

The Statics of Elastic Systems

4.1 Hooke's law

Up to now we have considered the equilibrium of systems consisting of perfectly rigid elements. Bodies having such an idealized internal structure are non-existent in reality, since the strain effect is always inseparable from any loading of a system. Generally, corresponding to greater strains are greater internal magnitudes which, it has to be remembered, cannot exceed a critical value which would cause the failure of a system. We should therefore work out methods enabling analysis of the state of distress of a system. The solution to such problems is furnished by the strength of materials.

The elementary system is a rod subjected to tension or compression. A simple experiment will answer what the relationship between its extension Δl and tensile force P is:

$$\Delta l = f(P). \tag{1}$$

Figure 4.1 shows the strain curve of a steel rod (low-carbon steel). Note that in the initial growth phase of the force, the curve is a straight line, i.e., the function $f(P)$ is linear (section OH)

$$\Delta l = kP. \tag{2}$$

For values higher than P_H (the limit of proportionality) the relationship (1) ceases to be linear. Above the limit of proportionality P_H is the elastic limit P_{el}. Any further growth of the tensile forces causes plastic strains (permanent deformation) and with force P_{pl}, the material becomes *fluid*; it is subject to strains without any marked increase of force.

Not for all materials, however, is the elastic limit and equally so the plastic limit clearly demarkated. For purposes of determination of the elastic or plastic limit conventional values have been adopted, specifying the stresses at which plastic strains reach a strictly defined value. The determination of this magnitude requires accurate measurement each time.

Thus the stress σ_{el} at which the permanent deformation $\varepsilon = (1 \div 5) \cdot 10^{-5}$ occurs, we shall regard as the *elastic limit*. The final choice of a value between 1 and 5 in the formula depends on the required accuracy.

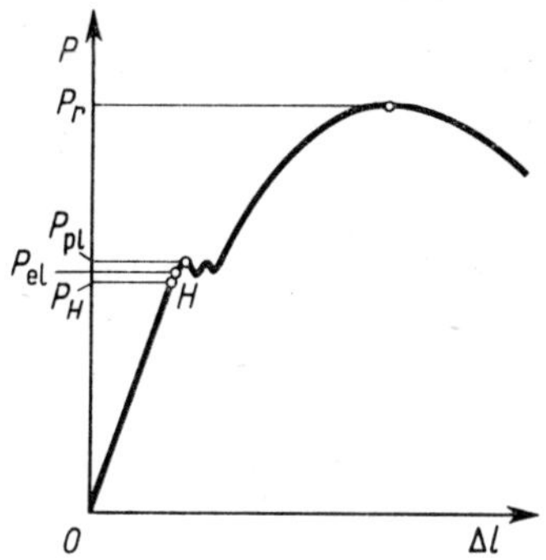

Fig. 4.1

Similarly, by the *plastic limit* σ_{pl} we shall understand such a stress value at which the permanent deformation $\varepsilon = (2 \div 5) \cdot 10^{-3}$. For soft steel $\sigma_{pl} \approx 25 \cdot 10^7$ N/m^2. The irregular section of the curve in Fig. 4.1 corresponds to material fluidity. With a stated strain magnitude, strengthening of the material takes place. The last stage culminates in the breakdown of the sample and its failure, which corresponds to the value of force P_r.

The above description refers to a soft steel sample. In our further considerations we shall be assuming that the elements of systems are subject to strain within the limits of the linear relationship (2) and therefore we shall confine ourselves to the range of small strains. In this connection, we shall assume that the strain of a system does not transgress the conditions for the equilibrium of forces. This last clause gives in essence the so-called *solidification hypothesis*.

Let us return for a moment to equation (2). It includes a constant k which we shall define as follows:

$$k = \frac{l}{AE}, \tag{2a}$$

where l length of rod, A cross-sectional area, E modulus of elasticity of material.

The mean values of the modulus E for some materials, given in N/m^2, are listed below:

steel	$(2.0\text{–}2.1)\ 10^{11}$,
copper	$(1.0\text{–}1.2)\ 10^{11}$,
pig iron	$(1.15\text{–}1.6)\ 10^{11}$,
aluminium and duralumin	$(0.7\text{–}0.8)\ 10^{11}$,
wood	$(0.08\text{–}0.12)\ 10^{11}$.

The product AE is called *tensile rigidity.*

We shall also introduce the concept of stress and strain. We shall call *stress* the ratio $\sigma = P/A$ in the case of uniform distribution of external forces on the cross-section of rod, or $\sigma = \mathrm{d}P/\mathrm{d}A$ when this distribution

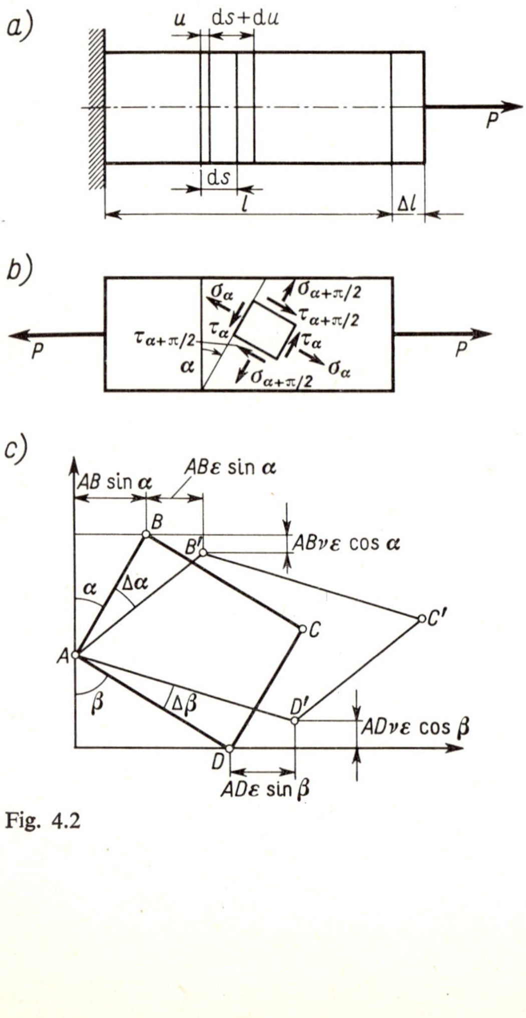

Fig. 4.2

is arbitrary. The strain of a rod we shall call the ratio $\varepsilon \doteq \Delta l/l$ when the rod is being uniformly extended over all its length. In the general case, $\varepsilon = du/ds$ which can easily be inferred from Fig. 4.2a.

After introducing these new concepts we can write

$$\frac{\Delta l}{l} = \frac{P}{A}\frac{1}{E} \tag{3}$$

or

$$\sigma = \varepsilon E. \tag{4}$$

Equation (4) gives *Hooke's law*.

Example 1. A steel rod of length $l = 1$ m and cross-section $A = 10^{-4}$ m^2 is being subjected to tension with force $P = 10{,}000$ N. Calculate the stresses and strains and the total extension of the rod.

$$\sigma = \frac{P}{A} = \frac{10{,}000\ \text{N}}{10^{-4}\ \text{m}^2} = 10^8 \frac{\text{N}}{\text{m}^2},$$

$$\varepsilon = \frac{\sigma}{E} = \frac{10^8}{2\cdot 10^{11}} = 0.5\cdot 10^{-3},$$

$$\Delta l = \varepsilon l = 0.5\cdot 10^{-3}\ \text{m} = 0.5\ \text{mm}.$$

Note how small the extension Δl is against the rod length. This is precisely what justifies the possibility of applying the solidification hypothesis.

Let us examine in greater detail the state of stresses and strains in a flat bar subjected to tensile forces P. Taking a section at an angle α (Fig. 4.2b), we find that

$$\sigma A = p \frac{A}{\cos\alpha}.$$

Hence

$$p = \sigma\cos\alpha. \tag{5}$$

Resolving p into the tangential and normal direction to the section, we have

$$\sigma_\alpha = \sigma\cos^2\alpha, \tag{6}$$

$$\tau_\alpha = \sigma\sin\alpha\cos\alpha = \tfrac{1}{2}\sigma\sin 2\alpha. \tag{7}$$

Therefore

$$\max\tau_\alpha = \tfrac{1}{2}\sigma, \tag{8}$$

which corresponds to the angle $\alpha = \frac{1}{4}\pi$.

Examining the equilibrium of the element *ABCD*, we find from the condition of moments about point C

$$\tau_\alpha \cdot AB \cdot BC = \tau_{\alpha+\frac{1}{2}\pi} \cdot BC \cdot AB,$$

i.e.,

$$\tau_\alpha = \tau_{\alpha+\frac{1}{2}\pi}. \tag{9}$$

It can be established experimentally that the axial strain of the flat bar is accompanied by a transverse strain ε' such that

$$\varepsilon' = -\nu\varepsilon, \tag{10}$$

where ν is the Poisson number. For metals this number is in the limit of

$$\nu = 0.25-0.35.$$

The transverse strain effect causes non-dilatational strain characterized by change of the angles of the rectangle *ABCD*. By *non-dilatational strain* we shall understand the angle

$$\gamma = \Delta\alpha+\Delta\beta.$$

From Fig. 4.2c we find

$$\Delta\alpha \approx \frac{AB \cdot \varepsilon \sin\alpha\cos\alpha + AB \cdot \nu\varepsilon\cos\alpha\sin\alpha}{AB} = \varepsilon(1+\nu)\sin\alpha\cos\alpha.$$

Similarly,

$$\Delta\beta = \varepsilon(1+\nu)\sin\beta\cos\beta.$$

Since

$$\beta = \tfrac{1}{2}\pi-\alpha,$$

we have

$$\Delta\beta = \varepsilon(1+\nu)\sin\alpha\cos\alpha,$$

whence

$$\gamma = 2\varepsilon(1+\nu)\sin\alpha\cos\alpha = 2\frac{\sigma}{E}(1+\nu)\sin\alpha\cos\alpha.$$

On the other hand,

$$\sigma\sin\alpha\cos\alpha = \tau_\alpha,$$

whence

$$\gamma = \frac{2(1+\nu)}{E}\tau_\alpha \tag{11}$$

or

$$\tau_\alpha = \frac{E}{2(1+\nu)}\gamma. \tag{12}$$

The derived equation is an analogue to $\sigma = E\varepsilon$, but only for non-dilatational strains γ. We shall denote the coefficient $E/2(1+\nu)$ by G and call it the *modulus of rigidity*:

$$\tau = G\gamma. \tag{13}$$

4.2 Safety factor

The simple rod tension test has furnished information about the relation $\varepsilon = f(\sigma)$, and what is more, it has made it possible to determine the characteristic points of the graph. To ensure safe operation of a rod, stresses must not be allowed to reach critical values (σ_{cr}). This we can achieve by a suitable design of the cross-section, so that the stresses occurring in real conditions should not exceed admissible values σ_{ad}.

We derive the admissible stress σ_{ad} by dividing the critical stress σ_{cr} by the so-called *safety factor* n:

$$\sigma_{ad} = \frac{\sigma_{cr}}{n}. \tag{1}$$

Number n should be chosen so that the stress magnitudes do not exceed the elastic limit (otherwise the rod will be subject to plastic strains).

In the case of variable loads, the admissible stresses should be reduced even more (as in the case of shock loads).

Thus, when designing the cross-section of a rod we must find beforehand the stress state recognized as dangerous (critical) and analyse the stress conditions of the element, and it is on this basis that we fix a safety factor n (and thereby the admissible stress σ_{ad}). Knowing the value of σ_{ad}, the requirement must be set for the stresses not to overstep the admissible level, i.e.,

$$\sigma_{ad} \geqslant \frac{P}{A}. \tag{2}$$

The enclosed two tables, quoted from Belayev [5], specify the approximate values of safety factors for various construction materials and for various types of load (Table 1), as well as the estimated magnitudes of basic admissible stresses (Table 2).

TABLE 1

Safety factors

Type of load	State of material	n
static load	plastic material	2.4–2.6
	brittle material	3.0–9.0
shock load	plastic material	2.8–5.0
variable load (tension-compression of identical magnitude)	plastic material	5.0–15.0

TABLE 2

Rough magnitudes of admissible stresses for basic construction materials

<table>
<tr><th rowspan="2">Material</th><th colspan="2">Admissible stresses $\times 10^5$ N/m^2</th></tr>
<tr><th>tension</th><th>compression</th></tr>
<tr><td>grey cast iron</td><td>280–800</td><td>1200–1500</td></tr>
<tr><td>steel</td><td colspan="2">1400–1600</td></tr>
<tr><td>constructional carbon steel</td><td colspan="2">600–2500</td></tr>
<tr><td>constructional alloy steel</td><td colspan="2">1000–4000 upward</td></tr>
<tr><td>copper</td><td colspan="2">300–1200</td></tr>
<tr><td>brass</td><td colspan="2">700–1400</td></tr>
<tr><td>bronze</td><td colspan="2">600–1200</td></tr>
<tr><td>aluminium</td><td colspan="2">300–800</td></tr>
<tr><td>aluminium bronze</td><td colspan="2">800–1500</td></tr>
<tr><td>duralumin</td><td colspan="2">800–1500</td></tr>
<tr><td></td><td colspan="2">300–400</td></tr>
<tr><td>bakelite</td><td colspan="2">400–500</td></tr>
<tr><td>pine along grain</td><td>70–100</td><td>100–120</td></tr>
<tr><td>pine across grain</td><td>–</td><td>15–20</td></tr>
<tr><td>oak along grain</td><td>90–130</td><td>110–150</td></tr>
<tr><td>oak across grain</td><td>–</td><td>20–35</td></tr>
<tr><td>stone wall</td><td>up to 3</td><td>4–40</td></tr>
<tr><td>brick wall</td><td>up to 2</td><td>6–25</td></tr>
<tr><td>concrete</td><td>1– 7</td><td>10–90</td></tr>
</table>

Example 1. A steel rod with plastic limit $\sigma_{pl} = 2500 \cdot 10^5$ N/m^2 was statically loaded with force $P = 10^4$ N. Find the cross-sectional area of the rod.

1. By virtue of the statical type of the load and the material involved we accept a safety factor $n = 2.5$.

2. We calculate the admissible stresses

$$\sigma_{ad} = \frac{\sigma_{pl}}{n} = \frac{2500 \cdot 10^5 \text{ N/m}^2}{2.5} = 10^8 \text{ N/m}^2.$$

3. We find the cross-section from the condition

$$\frac{P}{A} \leqslant \sigma_{ad};$$

hence

$$A \geqslant \frac{P}{\sigma_{ad}} = \frac{10^4 \text{ N}}{10^8 \text{ N/m}^2} = 10^{-4} \text{ m}^2.$$

4.3 Statically indeterminate systems

If the number of equations of equilibrium r is less than the number of unknowns n, such a system is called *statically indeterminate.* If the difference $n-r = k$, the system is said to be *k-fold statically indeterminate.*

Any k-fold statically indeterminate system can be solved by introducing in place of the missing equations the so-called *conditions of inseparability.* These conditions are consequent upon the geometric conformity of displacements. For any k-fold statically indeterminate system, k equations of inseparability can be written.

Example 1. A rigid beam was suspended from two identical rods of length l and cross-section A and a concentrated force P was applied. Calculate the forces in the rods.

The system is one-fold statically indeterminate (Fig. 4.3). It follows from the condition of moments about point A that

$$S_1 a + S_2 2a - P \cdot 3a = 0. \tag{a}$$

The condition of inseparability of strains is as follows:

$$\frac{\Delta l_1}{a} = \frac{\Delta l_2}{2a}, \tag{b}$$

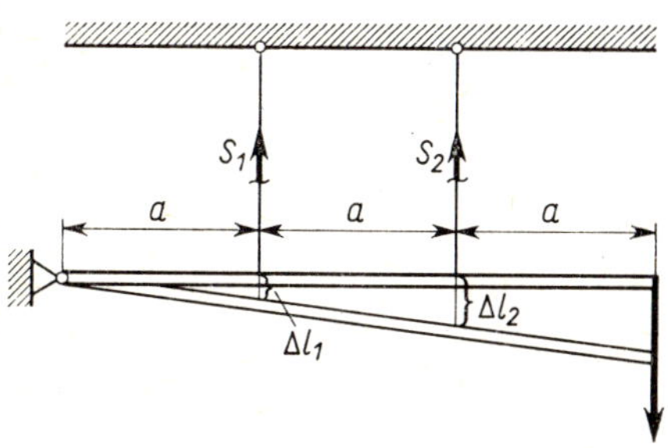

Fig. 4.3

but from Hooke's law it follows that

$$\Delta l_1 = \frac{S_1 l}{EA}, \tag{c}$$

$$\Delta l_2 = \frac{S_2 l}{EA}. \tag{d}$$

Substituting (c) and (d) into (b) we obtain

$$\frac{S_1 l/EA}{a} = \frac{S_2 l/EA}{2a}$$

or after simplification

$$2S_1 = S_2. \tag{e}$$

Solving equations (a) and (e) we obtain the result

$$S_1 = \tfrac{3}{5}P, \quad S_2 = \tfrac{6}{5}P. \tag{f}$$

Equations of inseparability of strains are linear functions of external and internal forces loading a system and, by virtue of this, hyperstatical systems like determinate statical systems obey the superpostion theorem.

Example 2. If a rod is fixed between two walls which are not subject to strain (Fig. 4.4), stresses due to temperature changes may develop. On raising the temperature of the rod from t_1 to t_2 reactions will occur. The system is statically indeterminate.

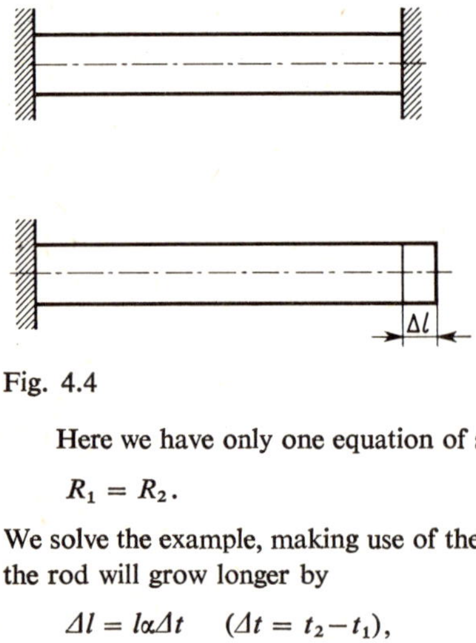

Fig. 4.4

Here we have only one equation of statics to proceed from:

$$R_1 = R_2. \tag{a}$$

We solve the example, making use of the superposition theorem. Allowing free dilatation the rod will grow longer by

$$\Delta l = l\alpha\Delta t \quad (\Delta t = t_2 - t_1), \tag{b}$$

where α is the coefficient of expansion of the material. The condition of inseparability is as follows (the rod length is not subject to change):

$$l\alpha\Delta t - \frac{R_2 l}{AE} = 0; \tag{c}$$

hence

$$R_2 = R_1 = AE\alpha\Delta t. \tag{d}$$

Evidently the stresses are a linear function of temperature changes.

Example 3. A rigid plate is supported by rods of different rigidity (Fig. 4.5). Moreover, the plate can move in slide bearings in translatory motion.

If we assume that the base on which the rods have been mounted is unstrainable, then all the rods are identically shortened by Δl under the force P. We can thus write

$$\frac{S_i}{A_i E_i} = \frac{S_j}{A_j E_j} \tag{a}$$

or

$$\frac{S_i}{S_j} = \frac{A_i E_i}{A_j E_j}, \tag{b}$$

which is the proof that the loads of the individual rods S_i are proportional to their rigidity $A_i E_i$.

Stresses in statically indeterminate systems may also develop by errors in construction.

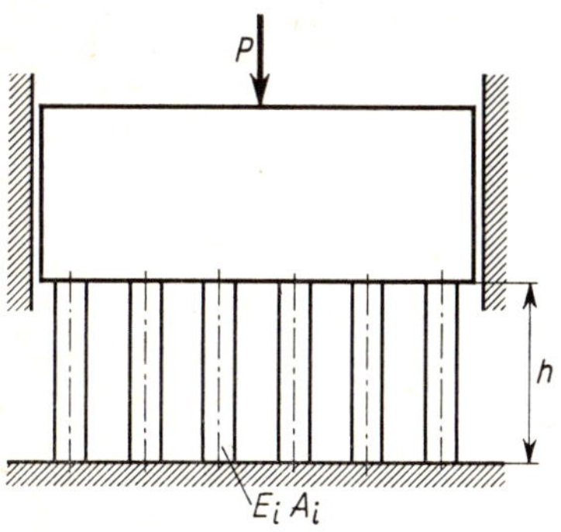

Fig. 4.5

Example 4. A truss consists of six rods (Fig. 4.6) but rod *5* is defective, being shorter than the others. What will be the stresses developing in the construction when put together?

To solve the problem we first replace the action of rod *5* with forces at points A and D. These forces will bring A and D closer to each other. The triangle ABC is subject to distortion (Fig. 4.6). We can easily see that

$$S_1 = S_2 = S_3 = S_4 = -\frac{P\sqrt{2}}{2}, \tag{a1}$$

and

$$S_6 = P. \tag{a2}$$

From the Pythagorean theorem applied to the distorted triangle *AOB* it follows that

$$a^2(1+\varepsilon_a)^2 = \tfrac{1}{4}c^2(1+\varepsilon_c)^2+\tfrac{1}{4}c^2(1+\varepsilon_x)^2.$$

Hence, since $c = a\sqrt{2}$, we obtain

$$1+2\varepsilon_a+\varepsilon_a^2 = \tfrac{1}{2}(1+2\varepsilon_c+\varepsilon_c^2)+\tfrac{1}{2}(1+2\varepsilon_x+\varepsilon_x^2).$$

Discarding the non-linear members we have

$$2\varepsilon_a = \varepsilon_c+\varepsilon_x. \tag{b}$$

Hence

$$\varepsilon_x = 2\varepsilon_a-\varepsilon_c. \tag{c}$$

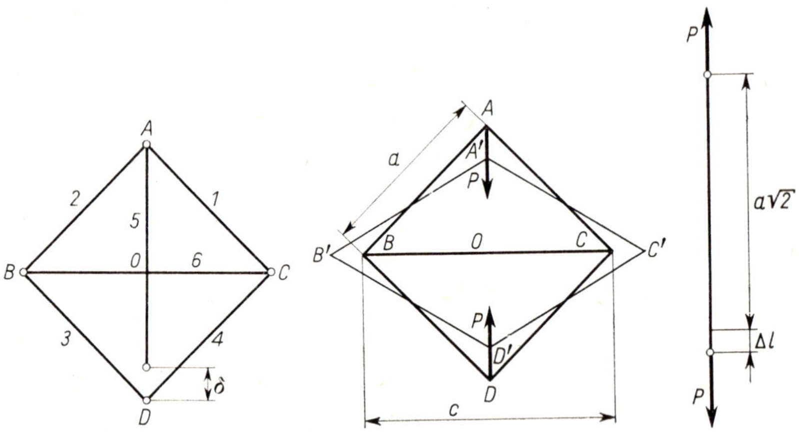

Fig. 4.6

It follows from Hooke's law that

$$\varepsilon_a = \frac{S_2}{AE} = -\frac{P\sqrt{2}}{2AE}, \tag{d}$$

$$\varepsilon_c = \frac{P}{AE}. \tag{e}$$

Substituting (d) and (e) into (c) we obtain

$$\varepsilon_x = -\frac{P}{AE}(\sqrt{2}+1). \tag{f}$$

The displacement of points *A* and *D* towards each other is

$$\Delta L = a\sqrt{2}\,|\varepsilon_x| = \frac{Pa}{AE}(\sqrt{2}+2). \tag{g}$$

The insufficient length of rod *5* must be compensated by its extension Δl and by the displacement of points *A* and *D*. Therefore

$$\delta = \Delta l+\Delta L = \frac{Pa}{AE}(2\sqrt{2}+2). \tag{h}$$

Hence

$$P = \frac{\delta AE}{2a(1+\sqrt{2})} = S_5. \tag{i}$$

The forces on the other rods are determined by substituting (i) into (a1):

$$S_1 = S_2 = S_3 = S_4 = -\frac{\delta AE\sqrt{2}}{4a(1+\sqrt{2})}. \tag{j}$$

Note that the forces on the rods of our system are in a linear dependence on the length deficiency δ of rod *5*.

The presented examples have been of the nature of an introduction into the problem of statically indeterminate systems. A more general aspect the problem will be considered in the following sections of this chapter.

4.4 Problems of simple beam-bending

Simple beam-bending is a problem which can be considered either in very precise terms, as is done in elasticity theory, or be viewed in a less accurate aspect, sufficient, however, to meet the needs of engineering practice. Strength of materials chooses the latter alternative.

The adopted method calls for a number of simplifying assumptions.

It is hypothesized in the theory of simple beam-bending (fairly well substantiated by experiment) that with pure bending, i.e., such, in which transverse forces do not come into action,

(1) a two-dimensional cross-section remains such after strain,

(2) longitudinal elements do not interact with each other but only transmit tension, or compression. within the bounds of Hooke's law.

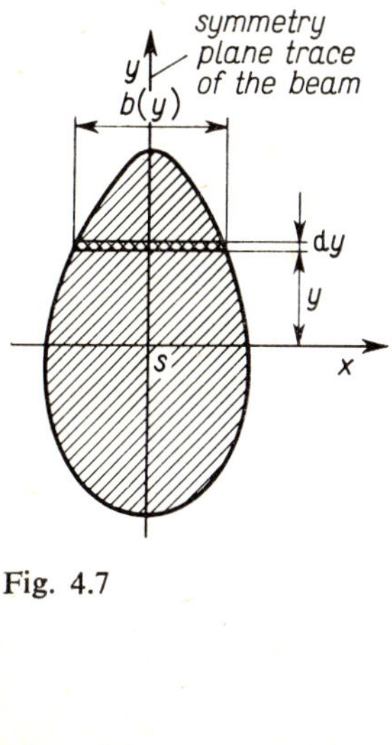

Fig. 4.7

We additionally assume that

(3) the beam cross-section has at least one plane of symmetry, which is at the same time the plane wherein the loading forces are situated (Fig. 4.7).

The state of the deflected beam element is shown in Fig. 4.8. If the strain of an element, running parallel to the rod axis OO_1 and separated from it by y is ε, then the stress in the fibre, according to the assumption (2), will reach a value

$$\sigma = E\varepsilon. \tag{1}$$

From Fig. 4.8 we infer that

$$\frac{\varrho+y}{AB} = \frac{\varrho}{OO_1}. \tag{2}$$

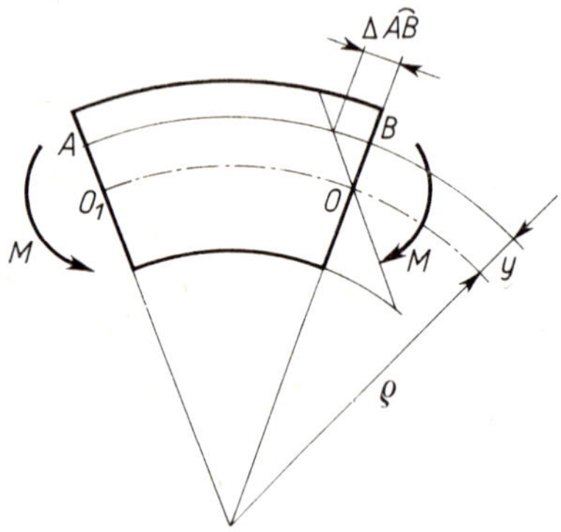

Fig. 4.8

This is the condition of the geometric conformity of strains. Since

$$AB = OO_1+\Delta(AB), \tag{3}$$

therefore, after considering (3) in (2), we have

$$\varrho\cdot OO_1+y\cdot OO_1 = \varrho\cdot OO_1+\varrho\Delta(AB). \tag{4}$$

From the last equation it follows that

$$\frac{\Delta(AB)}{OO_1} = \frac{y}{\varrho}. \tag{5}$$

Assume that no change in the length of the rod axis has taken place in the course of bending and the considered element has been cut out by two sections, at right angles to the beam axis; then $\Delta(AB)$ is the increase of the original length of the fibre or

$$\frac{\Delta(AB)}{OO_1} = \varepsilon \tag{5a}$$

is the corresponding extension. It follows from (5) and (5a) that

$$\varepsilon = \frac{y}{\varrho}. \tag{5b}$$

Finding ε from (1) and substituting into (5b) we obtain

$$\frac{\sigma}{E} = \frac{y}{\varrho}. \tag{6}$$

Hence

$$\sigma = E\frac{y}{\varrho}. \tag{6a}$$

With bending, the resultant of external forces is reduced to a moment, i.e. to a couple. The sum of projections of the couple is equal to zero, therefore

$$\int_A \sigma \,\mathrm{d}A = 0. \tag{7}$$

Putting in place of σ the value given by (6a), we write

$$\frac{E}{\varrho}\int_A y\,\mathrm{d}A = 0. \tag{8}$$

Since $E/\varrho \neq 0$,

$$\int_A y\,\mathrm{d}A = 0. \tag{9}$$

The expression (9) gives the position of the neutral axis, i.e. the straight line in which no stresses occur. The static moment about this axis is equal to zero, in other words, the axis passes through the centre of gravity of the beam's cross-section (Fig. 4.7, the axis $y = 0$).

The sum of moments of the external forces reckoned about the neutral axis corresponds to the sum of moments on one side of the cross-section, i.e.,

$$\int_A \sigma y\,\mathrm{d}A = M. \tag{10}$$

Considering expression (6a) in the integral (10) we write

$$\frac{E}{\varrho}\int y^2\,\mathrm{d}A = M. \tag{11}$$

The above equation contains the geometric moment of inertia of the cross-section as reckoned about the neutral axis of the cross-section:

$$I = \int_A y^2 \mathrm{d}A. \tag{12}$$

Consequently, relationship (11) can be shortened:

$$\frac{E}{\varrho} I = M;$$

hence

$$\frac{1}{\varrho} = \frac{M}{EI}. \tag{13}$$

Equation (13) shows the dependence of the curvature $1/\varrho$ on the bending moment M and beam rigidity EI. From (13) and (6a) we obtain

$$\sigma = \frac{My}{I}, \tag{14}$$

i.e., the equation giving the distribution of normal stresses in the cross-section of the beam (rod), implies a linear relation $\sigma = f(y)$ (Fig. 4.9). Consequently, maximum stresses occur in the extreme fibres.

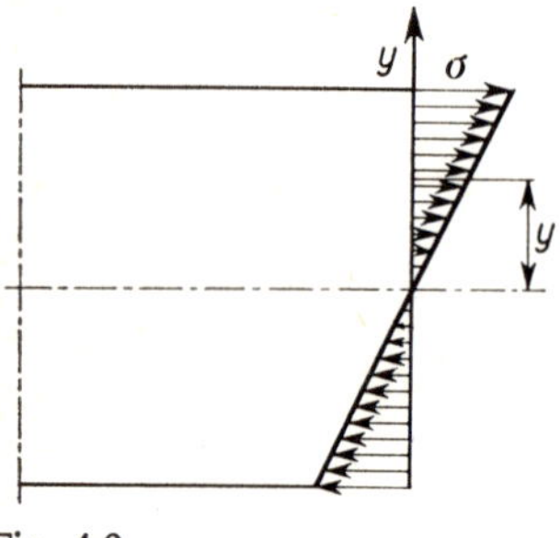

Fig. 4.9

The plane whose trace is the axis x (Fig. 4.7) is not a load plane, therefore we can write

$$\int_A \sigma x \mathrm{d}A = 0 \tag{15}$$

or after considering (6a)

$$\frac{E}{\varrho} \int_A yx \mathrm{d}A = 0;$$

hence, since $E/\varrho \neq 0$, we have

$$\int_A yx\mathrm{d}A = 0. \tag{16}$$

The vanishing of the integral (16) is mode passible by assumption (3), which will be explained further on. If it were not for the assumed symmetry of the cross-section, we would find that the changes in curvature were not taking place in the load plane.

The discussed case of bending we shall call *simple bending* as distinct from *oblique bending* which will be considered in one of the following sections.

4.5 Geometric moments of inertia

In the preceding section the following magnitudes appeared

$$I_x = \int_A y^2\mathrm{d}A, \tag{1}$$

called the moment of inertia of cross-section, and

$$I_{xy} = \int_A xy\mathrm{d}A \tag{2}$$

which we shall call the *moment of deviation* or *centrifugal moment.*

We shall begin with a few examples of determination of these magnitudes.

Example 1. The moment of inertia of a rectangle with base b and height h we calculate as follows:

$$I_x = \int_A y^2\mathrm{d}A = \iint_A y^2\mathrm{d}y\mathrm{d}x = \int_{-h/2}^{+h/2} y^2\Big(\int_{-b/2}^{+b/2} \mathrm{d}x\Big)\mathrm{d}y = b\int_{-h/2}^{h/2} y^2\mathrm{d}y = \frac{bh^3}{12}.$$

We can see, therefore, that the moment of the rectangle as reckoned about the axis parallel to the base and passing through the centre of gravity increases with the third power of its height h.

Example 2. The moment of inertia of a triangle about the axis passing through one of the bases we find similarly to the moment of a rectangle:

$$I_x = \int_0^h\int_{x_1}^{x_2} y^2\mathrm{d}x\mathrm{d}y = \int_0^h y^2\Big(\int_{x_1}^{x_2} \mathrm{d}x\Big)\mathrm{d}y = \int_0^h y^2(x_2-x_1)\mathrm{d}y = \int_0^h y^2 b(y)\mathrm{d}y.$$

The magitude $b(y)$ changes linearly since it is a section bounded by two straight lines (Fig. 4.10). We can therefore write

$$b(y) = C_1 y + C_2.$$

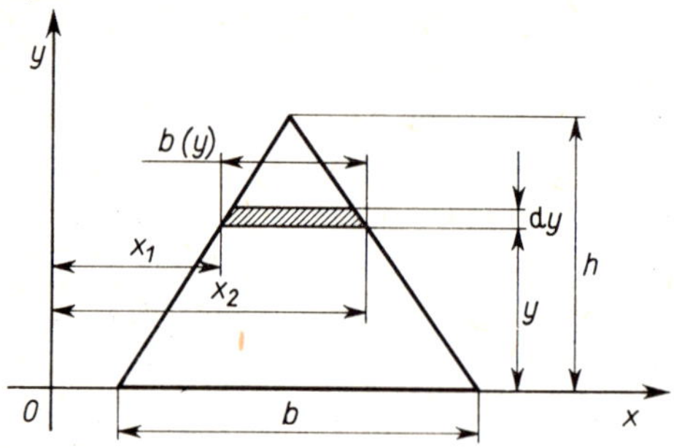

Fig. 4.10

For $y = 0$ we have $b(0) = b$ or $b = C_2$, whereas for $y = h$ we have $b(h) = 0$, therefore

$$0 = C_1 h + b.$$

Hence

$$C_1 = -\frac{b}{h}, \quad b(y) = -\frac{b}{h}y + b.$$

Considering the last equation in the integral giving the sought moment of inertia, we obtain

$$I_x = \int_0^h y^2 \left(b - \frac{b}{h}y\right) \mathrm{d}y = b\frac{h^3}{3} - b\frac{h^3}{4} = \frac{bh^3}{12}.$$

Example 3. We shall devote the present example to calculating the polar moment of a circle. We shall take the centre O as the pole (Fig. 4.11). The moment is expressed by the following equation:

$$I_O = \int_A r^2 \mathrm{d}A = \int_0^R \int_0^{2\pi} r^2 r \mathrm{d}\varphi \mathrm{d}r = 2\pi \int_0^R r^3 \mathrm{d}r = \frac{\pi R^4}{2}.$$

We shall also calculate the moments about the x and y axes. For a circle $I_x = I_y$, therefore

$$I_O = 2I_x = 2I_y,$$

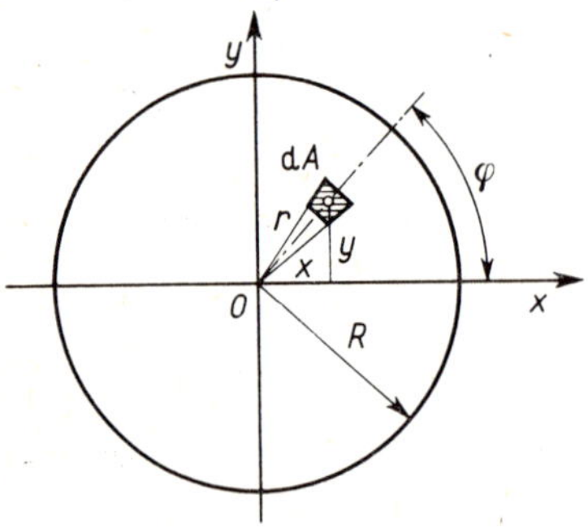

Fig. 4.11

whence

$$I_x = I_y = \frac{I_O}{2} = \frac{\pi R^4}{4}.$$

Example 4. Find the moment of inertia about one of the axes of symmetry of an ellipse.

$$I_x = \int_{A_e} y^2 \mathrm{d}A_e,$$

where $\mathrm{d}A_e = b_e \mathrm{d}y$ corresponds to the elemental field within the field of the ellipse (Fig. 4.12). Only a part of the field $\mathrm{d}A_e$ is contained within the circle of radius b:

$$\mathrm{d}A_k = b_k \mathrm{d}y, \quad \frac{\mathrm{d}A_e}{\mathrm{d}A_k} = \frac{b_e}{b_k} = \frac{a}{b},$$

whence

$$I = \frac{a}{b} \int_{A_k} y^2 \mathrm{d}A_k.$$

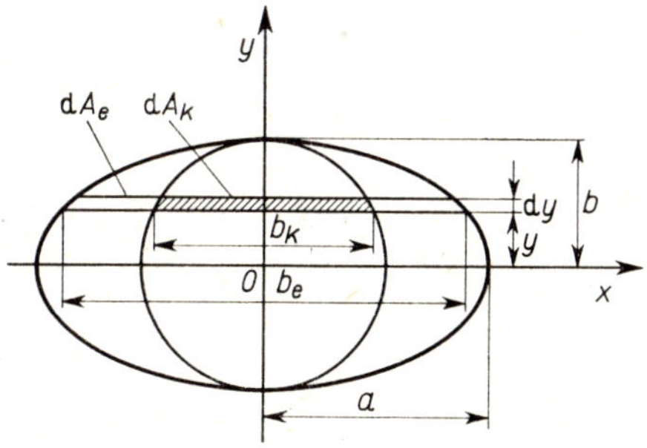

Fig. 4.12

The latter integral gives the moment of inertia of the circle of radius b. Consequently the moment of inertia for the ellipse is

$$I_x = \frac{\pi a b^3}{4}.$$

Having surveyed these few examples, we shall consider the problem of transformation of moments of inertia of plane figures with change of the frame of reference. We shall find first the relationship between moments of inertia as reckoned about the two parallel axes. From Fig. 4.13 we read

$$x = x_0 + x', \quad y = y_0 + y'.$$

Thus, for the moment of inertia, we have

$$I_x = \int_A (y_0^2 + 2y_0 y' + y'^2)\mathrm{d}A = y_0^2 \int_A \mathrm{d}A + 2y_0 \int_A y'\mathrm{d}A + \int_A y'^2 \mathrm{d}A.$$

If the point O' coincides with the centre of gravity, then $\int_A y'\,\mathrm{d}A = 0$. According to the theorem of static moments, we have

$$I_x = y_0^2 A + I_{x'}. \tag{3}$$

This is called *Steiner's equation*. We derive an analogous equation for moments of inertia about the y axis:

$$I_y = x_0^2 A + I_{y'}. \tag{4}$$

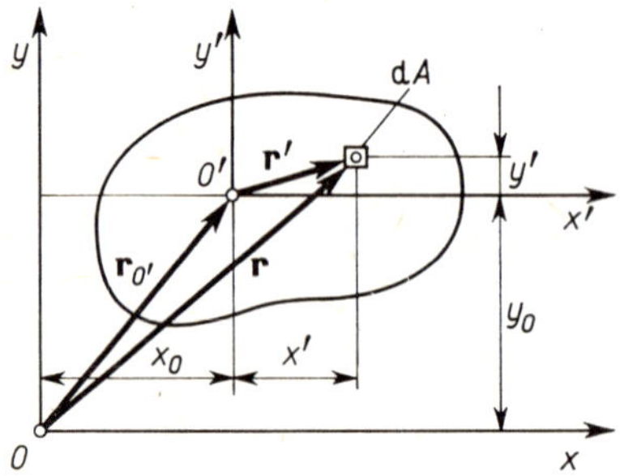

Fig. 4.13

Example 5. For a triangle, the moment of inertia about a straight line at right angles to one of its sides and passing through the centre of gravity we calculate from equation (3):

$$I_{x'} = I_x - y_0^2 A\,.$$

Since $I_{x'} = \dfrac{bh^3}{12}$, $A = \dfrac{bh}{2}$, $y_0 = \dfrac{h}{3}$, we have

$$I_{x'} = \frac{bh^3}{36}.$$

The moments of deviation with parallel translation of the axes x and y (Fig. 4.13) change according to the following relation:

$$I_{xy} = Ax_0 y_0 + I_{x'y'}, \tag{5}$$

where x_0 and y_0 are the coordinates of the centre of gravity of the figure. It is left to the reader to prove equation (5).

Example 6. Calculate the moment of deviation of a quadrant of radius R. We have

$$I_{xy} = \int_0^R x\Big(\int_0^{\sqrt{R^2-x^2}} y\,\mathrm{d}y\Big)\mathrm{d}x = \tfrac{1}{8}R^4.$$

Note that the moment of deviation, unlike the moment of inertia, may assume minus values.

Now we shall consider the transformation of a system by rotation about the reference origin. We shall answer the following question. What is the procedure to be followed to calculate moments of inertia and centrifugal moments in the new frame of reference?

$$I_\xi = \int_A \eta^2 \mathrm{d}A.$$

From Fig. 4.14

$$\eta = y\cos\varphi - x\sin\varphi, \qquad \xi = y\sin\varphi + x\cos\varphi,$$

therefore

$$I_\xi = \int_A (y^2\cos^2\varphi + x^2\sin^2\varphi - 2xy\sin\varphi\cos\varphi)\mathrm{d}A$$
$$= I_x\cos^2\varphi + I_y\sin^2\varphi - 2I_{xy}\sin\varphi\cos\varphi.$$

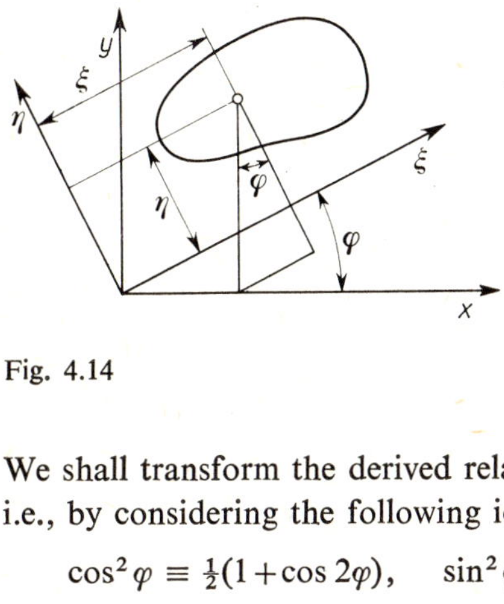

Fig. 4.14

We shall transform the derived relationship by introducing double angles, i.e., by considering the following identities

$$\cos^2\varphi \equiv \tfrac{1}{2}(1+\cos 2\varphi), \qquad \sin^2\varphi \equiv \tfrac{1}{2}(1-\cos 2\varphi),$$
$$2\sin\varphi\cos\varphi \equiv \sin 2\varphi.$$

Hence

$$I_\xi = \tfrac{1}{2}(I_x+I_y) + \tfrac{1}{2}(I_x-I_y)\cos 2\varphi - I_{xy}\sin 2\varphi. \tag{6}$$

The value of the moment of inertia about the axis η we shall calculate from (6), putting $\pi/2+\varphi$ for φ. After considering the reduction formulae we obtain

$$I_\eta = \tfrac{1}{2}(I_x+I_y) - \tfrac{1}{2}(I_x-I_y)\cos 2\varphi + I_{xy}\sin 2\varphi. \tag{7}$$

Adding (6) and (7), we find that

$$I_\xi + I_\eta = I_x + I_y = I_O,$$

for an arbitrary angle.

The sum of moments of inertia, as reckoned about two arbitrary axes at right angles to each other, is said to be an *invariant*.

The value of the centrifugal moment in the new frame of reference $\xi\eta$ can also be expressed by the moment of inertia and the centrifugal moment calculated in the original frame xy:

$$I_{\xi\eta} = \int_A \xi\eta \, dA = \int_A (y \sin\varphi + x \cos\varphi)(y \cos\varphi - x \sin\varphi) dA$$
$$= (I_x - I_y)\sin\varphi \cos\varphi + I_{xy}(\cos^2\varphi - \sin^2\varphi).$$

Making use of trigonometric identities, finally we write

$$I_{\xi\eta} = \tfrac{1}{2}(I_x - I_y)\sin 2\varphi + I_{xy}\cos 2\varphi. \tag{8}$$

We can now consider the problem of extreme values of moments of inertia. We find these values in the following condition:

$$\frac{dI_\xi}{d\varphi} = -(I_x - I_y)\sin 2\varphi - 2I_{xy}\cos 2\varphi = 0. \tag{9}$$

Hence

$$\tan 2\varphi = -\frac{2I_{xy}}{I_x - I_y}. \tag{10}$$

Note from (9) that

$$\frac{dI_\xi}{d\varphi} = -2I_{\xi\eta}.$$

Consequently, the disappearance of the centrifugal moment in a certain position of the axis $\xi\eta$ guarantees the existence of extreme moments of inertia. From the mathematical viewpoint, the requirement is that the second derivative should not be identically equal to zero, which actually is the case and which can be established without difficulty.

The axes for which the centrifugal moment disappears or, what amounts to the same, for which the moments of inertia assume extreme values, are called the *principal axes*. If in addition they pass through the centre of gravity, we shall call them the *principal* and *central axes of inertia*.

The values of principal moments of inertia are calculated, expressing $\cos 2\varphi$ and $\sin 2\varphi$ in equations (6) and (7) as $\tan 2\varphi$ from equation (10).

By simple transformations, these values are expressed by

$$I_{1,2} = \tfrac{1}{2}(I_x+I_y)\pm\tfrac{1}{2}\sqrt{(I_x-I_y)^2+4I_{xy}^2}. \tag{11}$$

Example 7. Find the principal and central moments of inertia for the plane figure given in Fig. 4.15.

We first calculate the moments of inertia about the axes of the reference frame:

$$I_x = \tfrac{1}{3}(b_1h_1^3+b_2h_2^3) = \tfrac{1}{3}(1\ \text{cm}\cdot 5^3\ \text{cm}^3+4\ \text{cm}\cdot 1^3\ \text{cm}^3) = 43\ \text{cm}^4,$$

$$I_y = \tfrac{1}{3}(1\ \text{cm}\cdot 4^3\ \text{cm}^3+5\ \text{cm}\cdot 1^3\ \text{cm}^3) = 23\ \text{cm}^4,$$

and next the moment of deviation I_{xy}:

$$I_{xy} = A_1x_{01}y_{01}+A_2x_{02}y_{02}$$

$$= 3\ \text{cm}^2\cdot 25\ \text{cm}\cdot 0.5\ \text{cm}+5\ \text{cm}^2\cdot 0.5\ \text{cm}\cdot 2.5\ \text{cm} = 10\ \text{cm}^4.$$

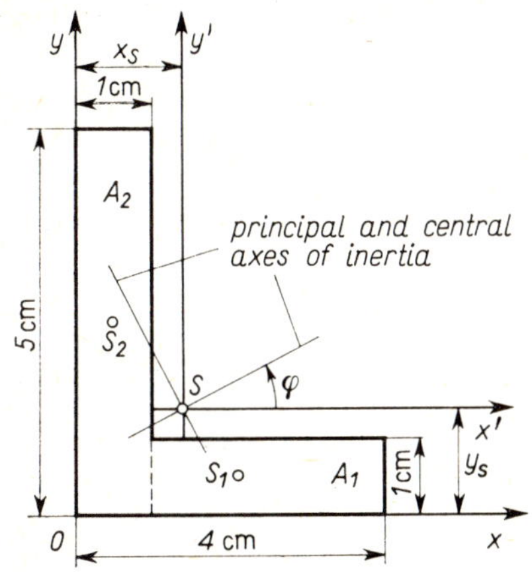

Fig. 4.15

We find the coordinates of the centre of gravity

$$y_s = \tfrac{1}{8}(3\cdot 0.5+5\cdot 2.5) = 1.75\ \text{cm}, \qquad x_s = \tfrac{1}{8}(3\cdot 2.5+5\cdot 0.5) = 1.25\ \text{cm}.$$

Now we calculate the moments of inertia and deviation about the translated *Oxy* frame of reference:

$$I_{x'} = I_x-Ay_s^2 = 43-8\cdot 3.05 = 18.5\ \text{cm}^4,$$

$$I_{y'} = I_y-Ax_s^2 = 23-8\cdot 1.56 = 10.5\ \text{cm}^4,$$

$$I_{x'y'} = I_{xy}-Ax_sy_s = 10-8\cdot 1.25\cdot 1.75 = -7.5\ \text{cm}^4.$$

For further calculation we can make use of equation (11):

$$I_{1,2} = \tfrac{1}{2}(I_x+I_y)\pm\tfrac{1}{2}\sqrt{(I_x-I_y)^2+4I_{xy}^2} = 14.5\pm 8.5\ \text{cm}^4,$$

$$I_1 = 23\ \text{cm}^4; \quad I_2 = 6\ \text{cm}^4.$$

The angle between the principal axes is calculated from

$$\tan 2\varphi = -\frac{2I_{x'y'}}{I_{x'}-I_{y'}} = \frac{15}{8} = 1.88; \qquad \varphi = 31°.$$

4.6 Strength calculations of beams

It follows from equation (14), Section 4.4, that the maximum bending stresses occur in extreme fibres, i.e., farthest away from the neutral axis:

$$\sigma_{\max} = \frac{My_{\max}}{I_x} = \frac{M}{I_x/y_{\max}}. \tag{1}$$

We shall introduce the magnitude $W_x = I_x/y_{\max}$ called the *modulus of section*. Hence

$$\sigma_{\max} = \frac{M}{W_x}. \tag{1a}$$

By the condition of strength the stresses in the beam must not be in excess of admissible stresses:

$$\sigma_{\max} \leqslant \sigma_{ad} \tag{2}$$

or

$$\frac{M}{W_x} \leqslant \sigma_{ad}, \tag{2a}$$

whence

$$W_x \geqslant \frac{M}{\sigma_{ad}}. \tag{3}$$

Taking the lowest of the admissible moduli of section, we replace the sign $\geqslant$ with the equality sign:

$$W_x = \frac{M}{\sigma_{ad}}. \tag{4}$$

Note that in the general case the moment is a function of the position of a cross-section and hence equality (4) gives the equistrength section of a beam. If the beam is assumed to be of rectangular cross-section with the height-breadth ratio $h/b = 2$, then

$$\frac{h^3}{12} = \frac{M}{\sigma_{ad}},$$

and so its sectional dimensions would be as follows:

$$h = \sqrt[3]{\frac{12M}{\sigma_{ad}}}, \tag{5}$$

$$b = \sqrt[3]{\frac{3M}{2\sigma_{ad}}}. \tag{6}$$

Only seldom do we have the opportunity for designing beams of variable cross-section, for commonly the available rods have a constant cross-section. We must then select such a cross-section as to make sure that in sections do stresses in excess of the admissible ones should occur. In practice, this amounts to working on sections in which maximum moments occur.

Example 1. Design the cross-section of a beam of length $l = 5$ m and of rectangular cross-section such that $h/b = 2$, where one end is rigidly fixed (Fig. 4.16) and a concentrated force $P = 10^5$ N is applied to the other end. Assume $\delta_{ad} = 10^8$ N/m².

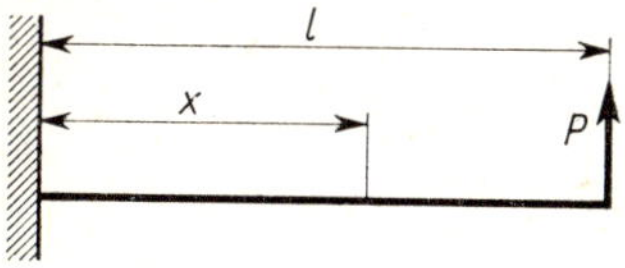

Fig. 4.16

In designing a beam of uniform strength, we make use of equations (5) and (6). Putting in these equations $M = P(l-x)$, as it is in reality, we obtain

$$h = \sqrt[3]{\frac{12P(l-x)}{\sigma_{ad}}}, \qquad b = \sqrt[3]{\frac{3P(l-x)}{2\sigma_{ad}}}.$$

If for some particular reason it is to be a beam of constant cross-section, it must be such that in sections do stresses exceed the admissible values. So we must take $\max\limits_{x} h$ and $\max\limits_{x} b$.

In the considered case we have

$$\max_{x} h = \sqrt[3]{\frac{12Pl}{\sigma_{ad}}} = \sqrt[3]{\frac{12 \cdot 10^5 \cdot 5}{10^8}} = 0.392 \text{ m} = 39.2 \text{ cm}.$$

Correspondingly $\max b = \frac{1}{2} \cdot 39.2 = 19.6$ cm. Rounding up, we take the dimensions

$$h = 40 \text{ cm} \quad \text{and} \quad b = 20 \text{ cm}.$$

Example 2. Design a *T-beam cross-section* (Fig. 4.17) loaded with a moment $M = 5.10^5$ Nm, if the admissible stress $\sigma_{ad} = 10^8$ N/m² and $h/a = 5$.

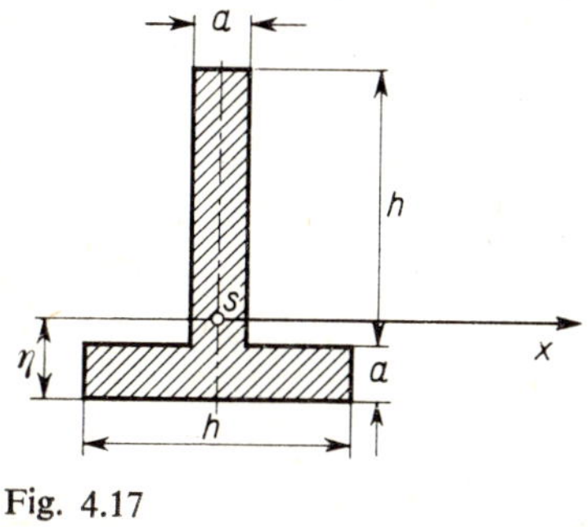

Fig. 4.17

First, we find the position of the neutral axis of the cross-section, i.e., the coordinate of the centre of gravity of the cross-section. The total cross-section A is

$$A = 2ha = 2 \cdot 5a \cdot a = 10a^2.$$

Using the theorem of static moments we write

$$\eta = \frac{1}{A}\left(5a^2 \cdot \frac{a}{2} + 5a^2 \cdot 3.5a\right) = \frac{1}{10a^2} 20a^2 = 2a.$$

Next, we calculate the moment of inertia of the cross-section I_x about the neutral axis:

$$I_x = 5a^2(1.5a)^2 + \tfrac{1}{12}5a \cdot a^3 + 5a^2(1.5a)^2 + \tfrac{1}{12}a(5a)^3 = 33.3a^4.$$

Consequently, the moduli of section corresponding to bottom and top elements are expressed by

$$W_{1x} = \frac{I_x}{2a} = 16.6a^3, \qquad W_{2x} = \frac{I_x}{4a} = 8.3a^3.$$

To find a safe cross-section we take the lower modulus. But it should be appreciated that when dealing with a material showing a lower tensile strength but a higher compressive strength, like let us say concrete, stresses in the extreme fibres on both sides of the neutral axis need to be considered.

In the present example we assume that the material exhibits symmetry in this respect, therefore it is justifiable to accept the lower of the two moduli.

Strength of materials requires the following relation to be hold

$$\sigma_{ad} \geqslant \frac{M}{W_{2x}};$$

hence

$$W_{2x} \geqslant \frac{M}{\sigma_{ad}} = \frac{5 \cdot 10^5 \text{ Nm}}{10^8 \text{N/m}^2} = 5 \cdot 10^{-3} \text{ m}^3,$$

therefore

$$8.30a^3 \geqslant 5 \cdot 10^{-3} \text{ m}^3,$$

and

$$a \geqslant \sqrt{0.6 \cdot 10^{-3}} \text{ m} = 0.00845 \text{ m}.$$

In some cases the rigidity of the system which must for constructional reasons be of sufficient magnitude, may be decisive in the choice of dimensions of the cross-section. If the construction is to satisfy not one but two conditions and two different cross-sections are obtained from calculations, then to have both conditions simultaneously satisfied ($\sigma \leqslant \sigma_{ad}$ and deflection $y \leqslant y_{ad}$), we take the larger of the two cross-sections. In conclusion, it should be noted that for standard section materials as available to the designer, it is sufficient to find the section modulus and on this basis choose the type of beam required.

4.7 The equation for the axis of a deflected beam

To derive the equation of axis of a deflected rod section subjected to the action of bending moments, we use the relationship (13), Section 4.4, giving the interdependence of curvature $1/\varrho$, bending moment M, and rigidity EI.

Assume that we deal with simple bending, i.e., that the axis of the deflected rod (beam) will remain in the load plane and its shape will be described by a plane curve

$$y = y(x). \tag{1}$$

The curvature of the curve (1) was determined in Example 8, Section 1.1 (equation (f)). Let us put the two results beside each other:

$$\frac{1}{\varrho} = \frac{y''}{[1+(y')^2]^{3/2}}, \qquad \frac{1}{\varrho} = \frac{M}{EI}.$$

It follows that

$$\frac{y''}{[1+(y')^2]^{3/2}} = \frac{M}{EI}. \tag{2}$$

We shall linearize the derived non-linear differential equation (2) of the axis for a deflected rod by assuming that only such cases are considered in which

$$(y')^2 \ll 1.$$

Consequently, omitting $(y')^2$ from equation (2) we obtain a linear differential equation of the deflected axis in the form

$$\frac{d^2y}{dx^2} = \frac{M}{EI}. \tag{3}$$

The right-hand side of (3) is a function of the coordinate x and it should be treated as a distribution. The distributive character of this function is consequential not only upon $M(x)$ being of the distributive type, but also on account of the possibility of discrete jumps in the bending rigidity on transition from one measure of the cross-section to another (Fig. 4.18). Making use of relationship (8), Section 3.7, we write

$$\frac{d^2y}{dx^2} = \frac{1}{EI}\int_0^x (x-\xi)q(\xi)d\xi. \tag{3a}$$

Integrating (3a) we obtain the slope $\theta = \mathrm{d}y/\mathrm{d}x$:

$$\theta(x) = C + \int_0^x \left[\frac{1}{EI}\int_0^\eta (\eta-\xi)q(\xi)\mathrm{d}\xi\right]\mathrm{d}\eta. \tag{4}$$

The integration constant C we find from the boundary condition: for $x = 0$, there is $\theta(0) = \theta_0 = C$. This constant is therefore the angular displacement of the rod (beam) end at the reference origin.

Fig. 4.18

Integrating (4) we obtain the sought function (1) in the form

$$y = D + \theta_0 x + \int_0^x\int_0^\zeta \left[\frac{1}{EI}\int_0^\eta (\eta-\xi)q(\xi)\mathrm{d}\xi\right]\mathrm{d}\eta\,\mathrm{d}\zeta. \tag{5}$$

The integration constant D, like the constant C, we find from the boundary condition: for $x = 0$, there is $y(0) = y_0 = D$. Using

$$\underbrace{\int_0^x \cdots \int_0^x}_{n} f(\xi)\mathrm{d}\xi^n = \int_0^x \frac{1}{(n-1)!}(x-\xi)^{n-1}f(\xi)\mathrm{d}\xi, \tag{6}$$

which is evident directly from (11), Section 3.7, we transform (5):

$$y = y_0 + \theta_0 x + \int_0^x \frac{x-\zeta}{EI}\left[\int_0^\zeta (\zeta-\xi)q(\xi)\mathrm{d}\xi\right]\mathrm{d}\zeta. \tag{7}$$

The solution (7) will be substantially simplified if EI be assumed constant, since then

$$\int_0^x \frac{x-\zeta}{EI}\left[\int_0^\zeta (\zeta-\xi)q(\xi)\mathrm{d}\xi\right]\mathrm{d}\zeta = \frac{1}{EI}\int_0^x\int_0^x\int_0^x\int_0^x q(\xi)\mathrm{d}\xi^4$$

$$= \frac{1}{EI}\int_0^x \frac{1}{3!}(x-\xi)^3 q(\xi)\mathrm{d}\xi, \tag{8}$$

which leads to the derivation of the equation of the deflected axis having the following form:

$$y = y_0+\theta_0 x+\frac{1}{EI}\int\limits_0^x \frac{(x-\xi)^3}{3!}\,q(\xi)\,\mathrm{d}\xi. \tag{9}$$

Differentiating (9) we obtain the expression for angular displacements

$$\theta = \theta_0+\frac{1}{EI}\int\limits_0^x \frac{(x+\xi)^2}{2!}\,q(\xi)\,\mathrm{d}\xi. \tag{10}$$

The equations derived for deflections y and slopes θ are structurally very simple and easy in application.

Example 1. Let a beam of constant rigidity EI, loaded simultaneously with concentrated moments, concentrated forces and continuous loads, be given (Fig. 4.19).

In this case, the continuous load takes the form

$$q(\xi) = \sum_{i=1}^{n} M_i\,\delta'(\xi-a_i)+\sum_{j=1}^{p} P_j\,\delta(\xi-b_j)+\sum_{k=1}^{m} q_k H(\xi-c_k). \tag{a}$$

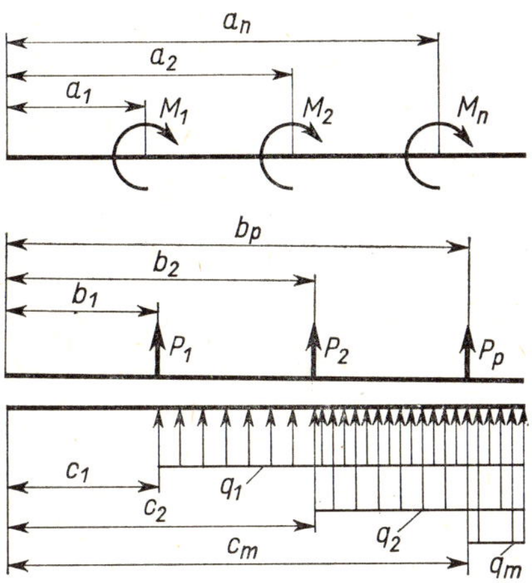

Fig. 4.19

Substituting (a) into (9) and (10) we obtain an expression giving the progress of deflections and slopes in the form

$$y = y_0+\theta_0 x+\frac{1}{EI}\left[\sum_{i=1}^{n} M_i\frac{(x-a_i)^2}{2!}H(x-a_i)+\right.$$
$$\left.+\sum_{j=1}^{p} P_j\frac{(x-b_j)^3}{3!}H(x-b_j)+\sum_{k=1}^{m} q_k\frac{(x-c_k)^4}{4!}H(x-c_k)\right], \tag{b}$$

$$\theta = \theta_0 + \frac{1}{EI}\left[\sum_{i=1}^{n} M_i \frac{(x-a_i)}{1!} H(x-a_i) + \right.$$

$$\left. + \sum_{j=1}^{p} P_j \frac{(x-b_j)^2}{2!} H(x-b_j) + \sum_{k=1}^{m} q_k \frac{(x-c_k)^3}{3!} H(x-c_k)\right]. \quad \text{(c)}$$

Example 2. Find the deflections of a cantilever of constant rigidity (Fig. 4.20) loaded with force P and concentrated moment M.

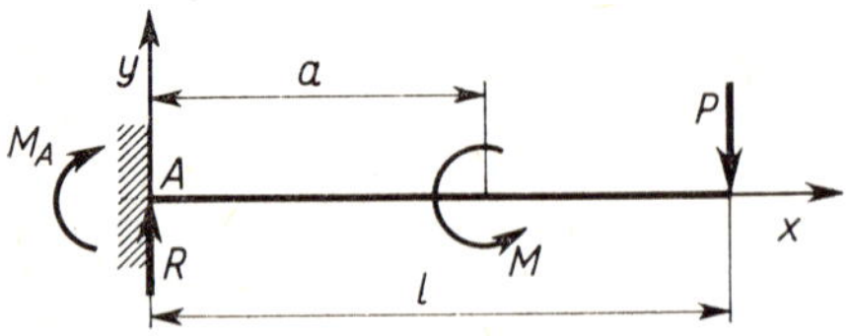

Fig. 4.20

We shall calculate first the reaction at support A. It gives two reactions: concentrated moment M_A and force R_A. Making use of the conditions of equilibrium we find

$$R_A = P, \qquad M_A = -Pl + M \quad \text{(a)}$$

Utilizing equation (b) from the preceding example,

$$y = \frac{1}{EI}\left[\frac{M_A x^2}{2!} + \frac{R_A x^3}{3!} - \frac{M(x-a)^2}{2!} H(x-a)\right], \quad \text{(b)}$$

since for $x = 0$ we have $y = 0$ and $y' = 0$ (rigid fixing). The same problem could be solved, taking the free end of the beam as the reference origin (Fig. 4.21). The deflection equation will take the following form here:

$$y = y_0 + \theta_0 x - \frac{1}{EI}\left[\frac{Px^3}{6} - M\frac{(x-(1-a))^2}{2} H(x-l+a)\right]. \quad \text{(c)}$$

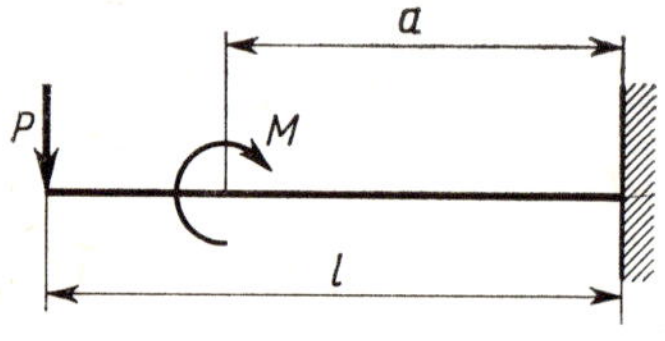

Fig. 4.21

The unknown deflection and angular displacement of the beam at the reference origin can be calculated from the conditions of the other boundary (where the beam is rigidly fixed), for it is there that we have

$$x = 1, \quad y' = 0, \quad y = 0.$$

Having these two conditions, it is possible to find y_0 and θ_0 and thus solve the problem.

Example 3. Find the progress of deflections for a free-ends beam to which a concentrated force P is applied at a distance a from the left support (Fig. 4.22). The beam is of length l and has constant rigidity EI over the entire span.

From the equilibrium conditions we find the reactions at the supports:

$$R_A = \frac{l-a}{l}P, \tag{a}$$

$$R_B = \frac{a}{l}P. \tag{b}$$

Making use of the results from Example 1, we get

$$y = \theta_0 x + \frac{1}{EI}\left(\frac{R_A x^3}{6} - \frac{P(x-a)^3}{6}H(x-a)\right). \tag{c}$$

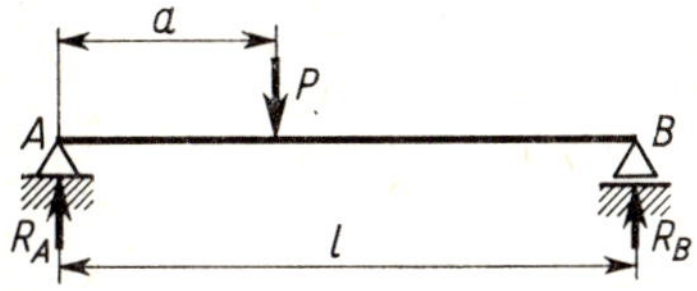

Fig. 4.22

From the condition that for $x = 1$, $y = 0$ we find θ_0:

$$\theta_0 = -\frac{1}{lEI}\left[\frac{R_A l^3}{6} - \frac{P(l-a)^3}{6}\right] = -\frac{P}{6lEI}[(l-a)l^2-(l-a)^3]. \tag{d}$$

Substituting the above result into (a) we obtain

$$y = -\frac{Px}{6lEI}[(l-a)l^2-(l-a)^3] + \frac{1}{EI}\left[\frac{l-a}{l}P\frac{x^3}{6} - \frac{P(x-a)^3}{6}H(x-a)\right]; \tag{e}$$

after transformation

$$y = \frac{Px(l-a)}{6lEI}[x^2-l^2+(l-a)^2] - \frac{P(x-a)^3}{6EI}H(x-a). \tag{f}$$

The deflection, for $x = l/2$ and $a = l/2$, is

$$y = -\frac{Pl^3}{48EI}. \tag{g}$$

4.8 Graphical methods of determining deflections of simple beams (Mohr's analogy)

It follows from the considerations on the equilibrium of beam elements that

$$q = \frac{d^2M}{dx^2}. \tag{1}$$

Comparing the above equation with the differential equation for deflections

$$\frac{M}{EI} = \frac{d^2y}{dx^2}, \qquad (2)$$

we find an analogy between them. From the mathematical point of view, these equations are identical.

Note that the integration of the first one, consisting of the determination of the bending moment, could be carried out by a graphical method (construction of a funicular polygon) whilst obeying the conditions of support.

In an analogous manner, equation (2) can be solved. In this equation, bending moments are replaced by deflections and the continuous loads by the quotient M/EI.

If EI = const, it will suffice to treat the moment as an imaginary load. In the case in which the bending rigidity EI varies continuously or discretely, the real beam can be replaced by one with a reduced rigidity E_rI_r loaded with a moment (Fig. 4.23)

$$M_r = M\frac{E_rI_r}{EI}.$$

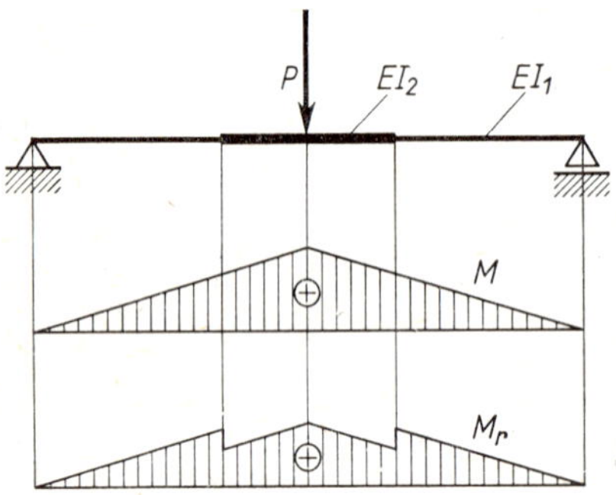

Fig. 4.23

The funicular polygon thus obtained gives a moment due to a moment treated as an imaginary load. If the load scale has been fixed by the coefficient $\mu_P^* = n$ Nm^2/cm, and the polar length of the imaginary forces by h^*, then the magnitude $M^* = \eta^*h^*\mu_k\mu_P^*$ is the imaginary moment.

The deflection will be equal to the ratio of M^* to bending rigidity E_rI_r, therefore

$$y = \frac{M^*}{E_rI_r} = \frac{\eta^*h^*\mu_k\mu_P^*}{E_rI_r}. \qquad (3)$$

Evidently, the ordinate η^* of the funicular polygon of imaginary forces is proportional in scale

$$\mu_y = \frac{h^* \mu_P^*}{E_r I_r} \mu_k \tag{4}$$

to the deflection of the beam. Equations (1) and (2) can be complemented with the relationships

$$q = \frac{\mathrm{d}T}{\mathrm{d}x}, \qquad \frac{M}{EI} = \frac{\mathrm{d}y'}{\mathrm{d}x}. \tag{5}$$

Evidently, in the cited analogy, the angle y' corresponds to the transverse force T.

Calculating the imaginary moment, the imaginary beam must be chosen so as to satisfy the conditions of support (the boundary conditions). We provide the considered beams either with rigid or hinged support. A rigid support admits neither slopes nor deflections, therefore in an imaginary system the free end of beam corresponds to the rigid support point. Conversely, to the free end corresponds the rigid fixing since it transmits transverse moments and forces other than zero, just as in the general case angular displacements and deflections of the free end of a beam are other than zero.

An articulation corresponds to a hinged support since the latter allows rotation but does not allow deflection, just as an articulation transmits a transverse force but does not transmit a bending moment. If the hinged support occurs at the beam end, then also to such support corresponds a hinged support. Following analogous reasoning, one will easily discover that a hinged support corresponds to an articulation.

An example of transition from a real to an imaginary beam is presented in Fig. 4.24. If the beam (*a*) is real, then the beam (*b*) is imaginary, and vice versa.

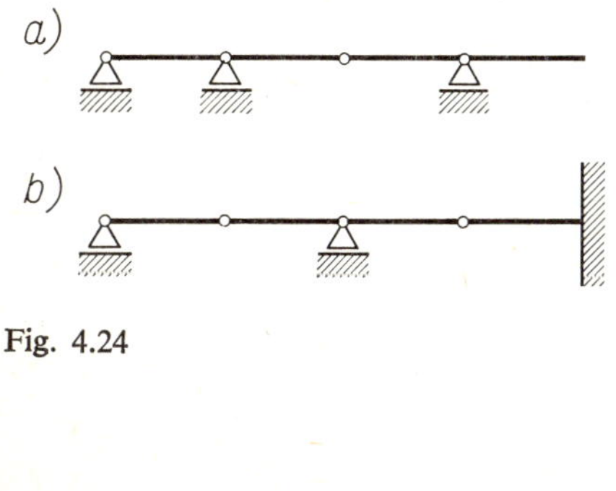

Fig. 4.24

4.9 Oblique bending

Oblique bending will occur in the case where the vector of the bending moment does not coincide with any of the principal and central axes of inertia (x and y in Fig. 4.25).

Considering the linear character of this system, it is possible to examine the influence of the components of moment **M** separately, i.e., M_x and M_y. The stresses due to moment M_x are

$$\sigma' = \frac{M_x}{I_y} y, \tag{1}$$

and due to moment M_y,

$$\sigma'' = \frac{M_y}{I_y} x. \tag{2}$$

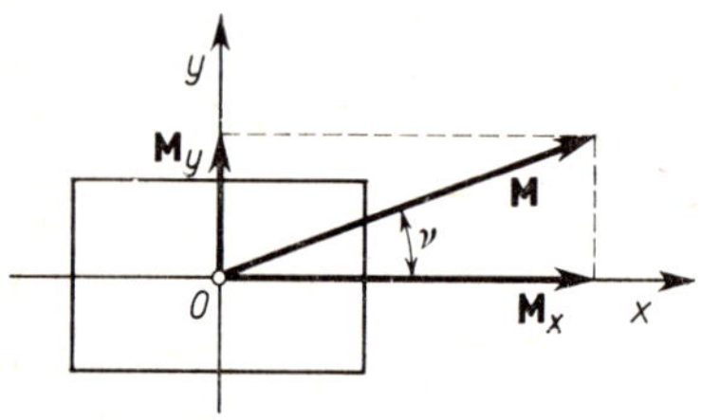

Fig. 4.25

Superposed stresses are represented by

$$\sigma = \frac{M_x}{I_x} y - \frac{M_y}{I_y} x. \tag{3}$$

From Fig. 4.25 it follows that

$$M_x = M\cos\nu, \quad M_y = M\sin\nu. \tag{4}$$

Substituting (4) into (3) we have

$$\sigma = M\left(\frac{y\cos\nu}{I_x} - \frac{x\sin\nu}{I_y}\right), \tag{5}$$

from which we can find the equation of neutral axis directly

$$\frac{y\cos\nu}{I_x} - \frac{x\sin\nu}{I_y} = 0 \tag{6}$$

or

$$y = x\frac{I_x}{I_y}\tan\nu. \tag{7}$$

This shows that the direction of the neutral axis coincides with the direction of the bending moment **M** only when $I_x = I_y$.

The deflected beam in the case of oblique bending is a space curve. Its projections on two perpendicular planes are respectively f_x and f_y. The resultant deflection f we shall define as the diagonal of a parallelogram built on the componential displacements; its value is

$$f = \sqrt{f_x^2+f_y^2}\,. \tag{8}$$

4.10 Some special problems of bending theory

In some cases we shall be assuming heterogeneity of the cross-section inasmuch as the modulus E will be a position function. A case like that will occur when a rod subject to bending consists of several laminae with different elastic properties.

Let us consider the bending of a bi-laminar rod. Assume that the cross-section has one axis of symmetry and the contact boundary between the two parts is a straight line parallel to the axis x (Fig. 4.26). We shall further assume that the vector of the bending moment has the direction of the axis x. In the considered problem only the bending rigidity EI, already considered in the foregoing sections, will be subject to change. One can speak here of a certain substitute value which is calculated as follows.

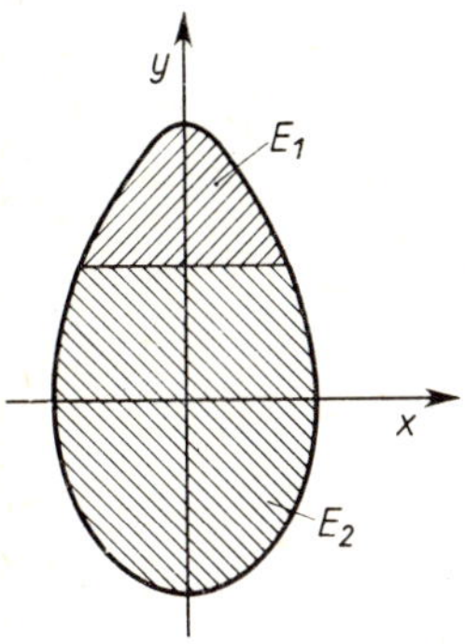

Fig. 4.26

First, we make use if condition (7), Section 4.4, from which it follows that

$$E_1 \int_{A_1} y\,\mathrm{d}A_1 + E_2 \int_{A_2} y\,\mathrm{d}A_2 = 0, \tag{1}$$

where E_1 and E_2 are the moduli of elasticity of the respective laminae. Bywirtue of the theorem of static moments, the integrals occurring in (1) can be exchanged for the products of the corresponding fields and the coordinates of centres of gravity, i.e.,

$$\int_{A_1} y\mathrm{d}A_1 = y_1 A_1$$

and

$$\int_{A_1} y\mathrm{d}A_2 = y_2 A_2.$$

Therefore

$$y_1 E_1 A_1 + y_2 E_2 A_2 = 0. \tag{2}$$

On the other hand, the distance between the centres of gravity of the two fields is given and is equal to η where

$$\eta = y_1 - y_2. \tag{3}$$

It follows from (2) that

$$y_2 = -\xi y_1, \tag{4}$$

where $\xi = \dfrac{E_1 A_1}{E_2 A_2}$. Hence

$$\eta = (1+\xi) y_1,$$

or

$$y_1 = \frac{\eta}{1+\xi} \tag{5a}$$

and

$$y_2 = -\frac{\xi\eta}{1+\xi}. \tag{5b}$$

Equations (5) give the position of the neutral axis of the cross-section. In the general case, this axis does not coincide with the central axis of inertia of the cross-section. Using next condition (10), we have

$$\frac{1}{\varrho}\left(E_1 \int_{A_1} y^2 \mathrm{d}A_1 + E_2 \int_{A_1} y^2 \mathrm{d}A_2\right) = M. \tag{6}$$

Introducing the notation

$$I_i = \int_{A_i} y^2 \mathrm{d}A_i, \tag{7}$$

we have

$$\frac{1}{\varrho} = \frac{M}{(EI)_{\text{sub}}} \frac{M}{\sum_{i=1}^{2} E_i I_i}; \tag{8}$$

hence

$$(EI)_{\text{sub}} = E_1 I_1 + E_2 I_2. \tag{9}$$

The derived results can easily be generalized for multilaminar rods. The position of the neutral axis is determined from the condition

$$\sum_i y_i E_i A_i = 0 \tag{2a}$$

and

$$\eta_{ik} = y_i - y_k. \tag{3a}$$

As for the substitute rigidity, we find it from

$$(EI)_{\text{sub}} = \sum_i E_i I_i, \tag{9a}$$

where E_i are the moduli of elasticity of individual laminae, and I_i moments of inertia of the cross-sections of individual laminae, as reckoned about the neutral axis. In order to maintain the validity of the results from the foregoing sections relating to bending, the rigidity EI should be replaced by the value given by equation (9a).

An example of a bi-laminar rod is the thermal bi-metal strip often used in control systems (Fig. 4.27). Owing to the different coefficients of thermal

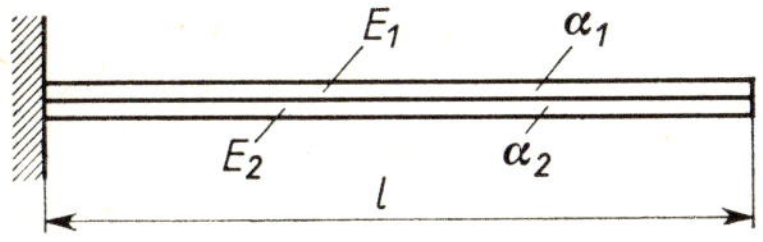

Fig. 4.27

expansion of the two laminae α_1 and α_2, the rod is subject to strains due to temperature change. Apart from strains due to stresses σ/E, there will also occur in this case stresses due to temperature increase Δt; as a result, equation (6a), Section 4.4, should be modified by putting $\varepsilon = \sigma/E + \alpha\Delta t$ in place of $\varepsilon = \sigma/E$ where α is the coefficient of thermal expansion. Since

$$\varepsilon = \frac{y}{\varrho},$$

we have

$$\sigma = E\left(\frac{y}{\varrho} - \alpha\Delta t\right), \tag{10}$$

where y denotes the coordinate defining the position of an element in relation to the neutral axis (which is at the same time the axis x). Translating parallelly the frame of reference so that $x \equiv x_1$ and $y = y_0 + y_1$, we obtain

$$\sigma = E\left(\frac{y_0}{\varrho} + \frac{y_1}{\varrho} - \alpha\Delta t\right). \tag{10a}$$

Assume that the axis x_1 is at the contact plane of the laminae in cross-section (Fig. 4.28); then making use of conditions (7) and (10), Section 4.4, we have

$$\int_{-g_2}^{0} E_2\left(\frac{y_0}{\varrho} + \frac{y_1}{\varrho} - \alpha_2\Delta t\right)dy_1 + \int_{0}^{g_1} E_1\left(\frac{y_0}{\varrho} + \frac{y_1}{\varrho} - \alpha_1\Delta t\right)dy_1 = 0,$$

$$\int_{-g_2}^{0} E_2\left(\frac{y_0}{\varrho} + \frac{y_1}{\varrho} - \alpha_2\Delta t\right)y_1\,dy_1 + $$
$$+ \int_{0}^{g_1} E_1\left(\frac{y_0}{\varrho} + \frac{y_1}{\varrho} - \alpha_1\Delta t\right)y_1\,dy_1 = 0.$$

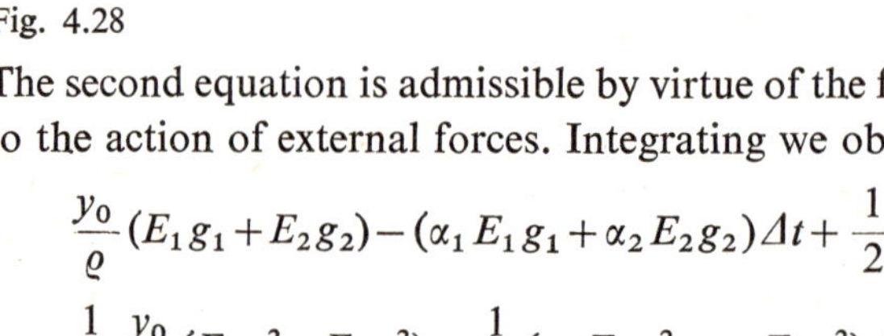

Fig. 4.28

The second equation is admissible by virtue of the fact that the rod is subject to the action of external forces. Integrating we obtain

$$\frac{y_0}{\varrho}(E_1 g_1 + E_2 g_2) - (\alpha_1 E_1 g_1 + \alpha_2 E_2 g_2)\Delta t + \frac{1}{2}\frac{1}{\varrho}(E_1 g_1^2 - E_2 g_2^2) = 0,$$

$$\frac{1}{2}\frac{y_0}{\varrho}(E_1 g_1^2 - E_2 g_2^2) - \frac{1}{2}(\alpha_1 E_1 g_1^2 - \alpha_2 E_2 g_2^2)\Delta t +$$
$$+ \frac{1}{3}\frac{1}{\varrho}(E_1 g_1^3 + E_2 g_2^3) = 0.$$

The derived equations have two unknowns: y_0/ϱ and $1/\varrho$. Solving this system with respect to the curvature $1/\varrho$, we write

$$\frac{1}{\varrho} =$$

$$\frac{[(E_1 g_1^2 - E_2 g_2^2)(\alpha_1 E_1 g_1 + \alpha_2 E_2 g_2) - (E_1 g_1 + E_2 g_2)(\alpha_1 E_1 g_1^2 - \alpha_2 E_2 g_2^2)]\Delta t}{\frac{1}{2}(E_1 g_1^2 - E_2 g_2^2)^2 - \frac{2}{3}(E_1 g_1 + E_2 g_2)(E_1 g_1^2 + E_2 g_2^3)}$$

$$= 6\Delta t \frac{E_1 E_2 g_1 g_2 (g_1 + g)(\alpha_1 - \alpha_2)}{4E_1 E_2 g_1 g_2 (g_1 + g_2)^2 + (E_1 g_1^2 - E_2 g_2^2)^2}. \tag{11}$$

To find the angles of rotation and deflection of the thermal bi-metal, it is enough to put $1/\varrho \approx y''$; then

$$y' = \varphi(x) = \frac{1}{\varrho} x + C \tag{12}$$

and

$$y(x) = \frac{1}{\varrho}\frac{x^2}{2} + Cx + D, \tag{13}$$

where C and D are integration constants and $1/\varrho$ is given by (11). If the conditions of rigid support be realized (Fig. 4.27), then $C = D = 0$, Therefore,

$$\varphi(x) = \frac{1}{\varrho} x, \quad y(x) = \frac{1}{\varrho}\frac{x^2}{2}.$$

The maximum values for the angle of rotation $\varphi(x)$ and for the angle of deflection $y(x)$ we obtain by putting $x = l'$. The ratio $k_t = \varphi(l)/l\Delta t$ is called the *sensitivity* of a thermal bi-metal. Note that

$$k_t = 6 \frac{E_1 E_2 g_1 g_2 (g_1 + g_2)(\alpha_1 - \alpha_2)}{4E_1 E_2 g_1 g_2 (g_1 + g_2)^2 + (E_1 g_1^2 - E_2 g_2^2)^2}. \tag{14}$$

The maximum sensitivity of a thermal bi-metal k_t follows from the form of equation (14) on satisfying the condition

$$E_1 g_1^2 - E_2 g_2^2 = 0. \tag{15}$$

Until now, we have been assuming that materials obey Hooke's law. This assumption can be abandoned for by putting

$$L_1(D)\sigma = L_2(D)\varepsilon, \tag{16}$$

were $L_1(D)$ and $L_2(D)$ are linear differential operators with respect to time ($D \equiv d/dt$), we shall be able to describe quasi-static processes of the changes of strains in time. Equation (6) describes the internal structure of linear visco-elastic bodies (*Boltzmann's bodies*) ([50], [56], [57]).

This model enables the description of transient states, such as creep and relaxation. *Creep* is a phenomenon of the changes in strains in time $\varepsilon(t)$ for of a steady load state (Fig. 4.29). It is very apparent in constructions

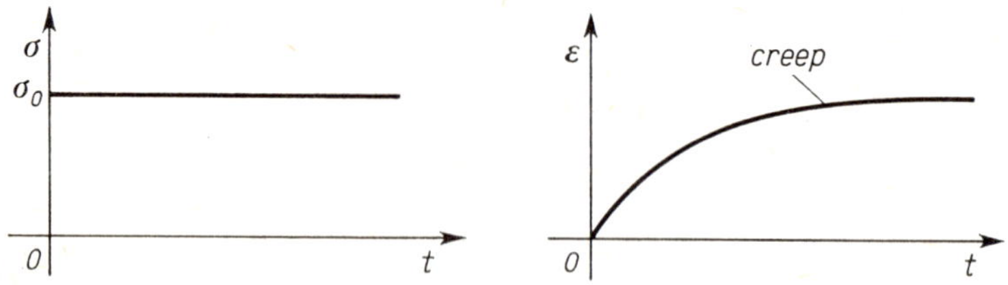

Fig. 4.29

working in high temperatures. The relaxation phenomenon is a process of changes of stresses $\sigma(t)$ in time, e.g. in a steady strain state (Fig. 4.30).

We shall analyse the bending problem referred to rods of visco-elastic materials as in Section 4.4. The only difference in the assumptions will be the replacement of Hooke's law by the equation of state (16).

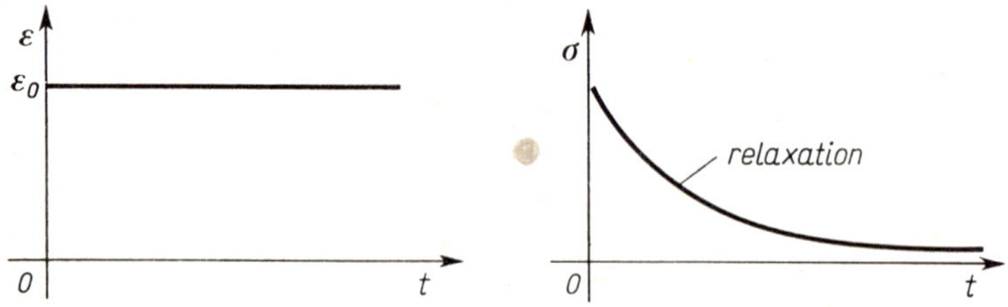

Fig. 4.30

Conditions (7) and (10), Section 4.4, remain valid. After considering the equation of state (16) we obtain

$$L_2(D) \int_A \varepsilon \mathrm{d}A = 0, \tag{17}$$

$$L_2(D) \int_A \varepsilon y \mathrm{d}A = L_1(D) M. \tag{18}$$

Putting $1/\varrho = \varkappa$, we may write

$$L_2(D) \varkappa \int_A y \mathrm{d}A = 0 \tag{17a}$$

and

$$L_2(D)\varkappa \int_A y^2 \mathrm{d}A = L_1(D)M. \tag{18a}$$

From (17a) it follows that

$$\int_A y \mathrm{d}A = 0;$$

consequently, rods of visco-elastic material have, like elastic rods, their neutral axis passing through the centre of gravity of the cross-section.

Equation (18a) is a differential relationship associating changes of curvature $\varkappa(t)$ with changes of a moment in each of the sections of a rod subject to bending.

The relationship (18a) is analogous to (13) in Section 4.4. To see this, let us divide both sides of equation (18a) by the moment of inertia of the cross-section $I = \int_A y^2 \mathrm{d}A$ and as the next step, perform Laplace's transformation. Then

$$L_2(s)\bar{\varkappa}(x, s) = L_1(s)\frac{\overline{M}(x, s)}{I}$$

or

$$\bar{\varkappa}(x, s) = \frac{\overline{M}(x, s)}{\overline{E}(s) I}, \tag{19}$$

where

$$\overline{E}(s) = \frac{L_2(s)}{L_1(s)}. \tag{19a}$$

Performing an integral transformation in equation (13), Section 4.4, we obtain

$$\bar{\varkappa}(x, s) = \frac{\overline{M}(x, s)}{EI}. \tag{20}$$

We find that for a visco-elastic rod (equation (19)) in place of the constant E (equation (20)) an operator function $\overline{E}(s)$ comes in. This fact can be utilized in solving the problem of visco-elastic rod bending.

Example 1. Find the deflection of a beam of visco-elastic material characterized by the equation of state

$$\sigma = E(1+TD)\varepsilon,$$

where E and T are constant.

The model thus described goes under the name of *Voight's model* ([50], [56], [57]). It represents a parallel connection of an elastic element with constant E and a viscous one (damper) with constant $ET = \eta$ (Fig. 4.31).

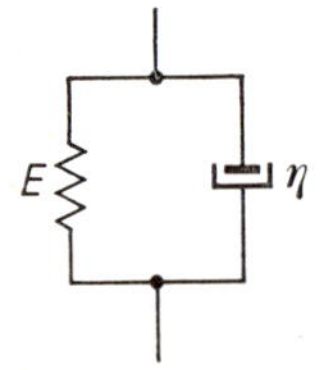

Fig. 4.31

Assume that our beam is rigidly fixed at one end and loaded at the other end (Fig. 4.32) with a force P such that

$$P = P(t).$$

We find first the solution for the beam as if it were only of elastic material. In the case considered

$$q(\xi) = Pl\delta'(\xi) - P\delta(\xi). \tag{a}$$

Making use of the solution (9), Section 4.7, we have

$$y = \frac{Plx^2}{2EI}\left[1 - \frac{1}{3}\left(\frac{x}{l}\right)^2\right], \tag{b}$$

since for the conditions of support $y_0 = 0$ and $\theta_0 = 0$. Performing Laplace's transformation on (b) we obtain

$$\bar{y}(x, s) = \frac{\bar{P}(s)lx^2}{2EI}\left[1 - \frac{1}{3}\left(\frac{x}{l}\right)^2\right]. \tag{c}$$

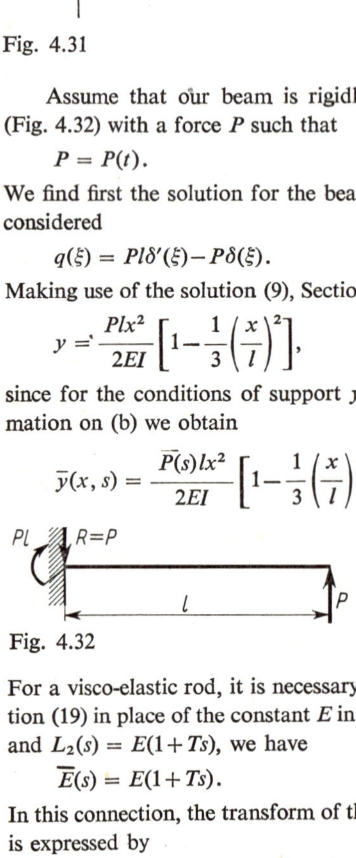

Fig. 4.32

For a visco-elastic rod, it is necessary to put the operator function $\bar{E}(s)$ defined by equation (19) in place of the constant E in equation (c). Since in the considered case $L_1(s) = 1$ and $L_2(s) = E(1+Ts)$, we have

$$\bar{E}(s) = E(1+Ts). \tag{d}$$

In this connection, the transform of the function $y(x, t)$ for a rod of visco-elastic material is expressed by

$$\bar{y}(x, s) = \frac{\bar{P}(s)}{1+Ts}\frac{l^2x^2}{2EI}\left[1 - \frac{1}{3}\left(\frac{x}{l}\right)^2\right]. \tag{e}$$

Transforming this result we obtain

$$y(x, t) = \frac{lx^2}{2EIT}\left[1 - \frac{1}{3}\left(\frac{x}{l}\right)^2\right]\int_0^t P(t-\tau)e^{-\tau/T}\,d\tau. \tag{f}$$

For $P(t) = P_0 H(t)$ where P_0 = const we have

$$y(x,t) = \frac{P_0 l x^2}{2EI}\left[1 - \frac{1}{3}\left(\frac{x}{l}\right)^2\right]\cdot(1 - e^{-t/T}). \tag{g}$$

Note that the deflections increase, tending to a limit value corresponding to the elastic model.

4.11 Clapeyron's systems

In our discussion of elastic systems so far, we have been assuming such a small deformation that its effects, consisting of a change in the distribution of both internal and external forces, could always be ignored. In other words, we could use the solidification hypothesis. These assumptions have led us to the conclusion that the equations occurring in individual problems were linear, and so the principle of superposition could be applied. In strength of materials problems, these findings have been generalized in the theory of Clapeyron's systems. This theory, starting out with the aforementioned premises, gives a decisive tool for finding displacements and forces.

Given an elastic system, we separate n points of application of forces and concentrated moments (generalized forces Q_i) and we will concern ourselves with their linear and angular displacements (generalized displacements q^j). The considered system is linear in the sense that the generalized displacements are linear function of generalized forces, which we write down as

$$q^j = c^{ji} Q_i, \tag{1}$$

where c^{ji} are elements of the so-called *influence matrix* (element c^{ji} gives the influence of force on displacement at point j). Equation (1) is written in the summation convention (we sum up according to indices i from 1 to n). Having assumed small deformations along with the solidification hypothesis and the linear physical law (Hooke's law), the potential energy accumulated in the system can be expressed by

$$V = \tfrac{1}{2} q^j Q_j. \tag{2}$$

The expressions (1) and (2) can be treated as a definition of *Clapeyron's system*.

We find that these systems obey the so-called *Betti's reciprocal theorem*. It states that if two groups of forces act on Clapeyron's system, then the work of group 1 on displacements of the forces of group 2 is equal to the work of the forces of group 2 on displacements of the forces of group 1.

The substantiation of the theorem is as follows. The work of the force in group 1, $Q_j^{(1)}$, on its displacements is

$$W_1^* = \tfrac{1}{2} \sum_{j=1}^{n_1} q^j Q_j^{(1)}. \tag{3}$$

To a system loaded with n_1 forces in group 1 we apply forces of group 2, $Q_k^{(2)}$. Since at the instant of application of group 2 forces, group 1 forces have already been in action, the work W^{**} is done by all the forces and its magnitude is given by

$$W_1^{**} = \tfrac{1}{2} \sum_{k=1}^{n_2} q^k Q_k^{(2)} + \sum_{j=1}^{n_1} q_{(2)}^j Q_j^{(1)}, \tag{4}$$

where $q_{(2)}^j$ are displacements at points of application of group 1 forces induced by group 2 forces.

The total work done by the forces in both groups is

$$W_1^* + W_1^{**} = W_1. \tag{5}$$

Let us change the load order, i.e., we first apply group 2 forces to the system and then group 1 forces. Repeating the above reasoning, we obtain

$$W_2 = W_2^* + W_2^{**}, \tag{6}$$

where

$$W_2^* = \tfrac{1}{2} \sum_{k=1}^{n_2} q^k Q_k^{(2)}, \tag{7}$$

and

$$W_2^{**} = \tfrac{1}{2} \sum_{j=1}^{n_1} q^j Q_j^{(1)} + \sum_{k=1}^{n_2} q_{(1)}^k Q_k^{(2)}. \tag{8}$$

Since $W_1 = W_2$,

$$\sum_{j=1}^{n_1} q_{(2)}^j Q_j^{(1)} = \sum_{k=1}^{n_2} q_{(1)}^k Q_k^{(2)}, \tag{9}$$

which gives the aforesaid Betti's reciprocal theorem.

It follows from (9) that the influence coefficients c^{ik} form a symmetrical matrix. Indeed, considering that

$$q_{(2)}^j = \sum_{k=1}^{n_2} c^{jk} Q_k^{(2)}$$

and

$$q^k_{(1)} = \sum_{j=1}^{n_1} c^{kj} Q_j^{(1)},$$

we have

$$\sum_{j=1}^{n_1} \sum_{k=1}^{n_2} c^{jk} Q_j^{(1)} Q_k^{(2)} = \sum_{j=1}^{n_1} \sum_{k=1}^{n_2} c^{kj} Q_j^{(1)} Q_k^{(2)}. \tag{10}$$

The relationship (10) occurs for an arbitrary division of forces into two groups, consequently $c^{jk} = c^{kj}$ for all j and k.

Example 1. Check the reciprocal theorem for the example of a beam fixed at one end (Fig. 4.33), with bending rigidity EI and length l, loaded by force P and moment M.

Let $Q^{(1)} \equiv P$, $Q^{(2)} \equiv M$, $q_{(1)} = \theta$ (θ is the angle of rotation at the point of application of moment M induced by force P) and $q_{(2)} = y$ (displacement due to moment M at the point of application of force P). According to the reciprocal theorem, the following should occur:

$$q_{(2)} Q^{(1)} = q_{(1)} Q^{(2)} \tag{a}$$

or

$$yP = \theta M, \tag{c}$$

where

$$y = \frac{Ml^2}{2EI}. \tag{d}$$

We should therefore have

$$\frac{Pl^2}{2EI} = \theta, \tag{e}$$

and this actually takes place (see equation (b) from Example 1, Section 4.10).

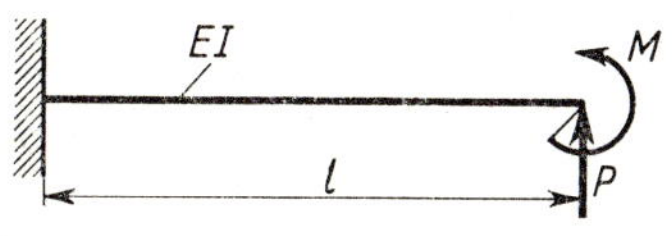

Fig. 4.33

Castigliano's theorem relating to Clapeyron's systems states that generalized displacements q^j (linear and angular) induced by generalized forces Q_j (force or moment) at any point of the system are derivatives of the elastic energy V of the system with respect to the generalized force causing the generalized displacement:

$$q^k = \frac{\partial V}{\partial Q_k}. \tag{11}$$

The proof of the theorem is given below. From (1) and (2) it follows that

$$V = \tfrac{1}{2}c^{ji}Q_jQ_i. \tag{12}$$

Differentiating (12) with respect to Q_k we obtain

$$\frac{\partial V}{\partial Q_k} = \frac{1}{2}c^{ji}\left(\frac{\partial Q_j}{\partial Q_k}Q_i + Q_j\frac{\partial Q_i}{\partial Q_k}\right).$$

Since $\partial Q_j/\partial Q_k = \delta_j^k$, we get

$$\frac{\partial V}{\partial Q_k} = \tfrac{1}{2}c^{ji}\delta_j^kQ_i + \tfrac{1}{2}c^{ji}Q_j\delta_j^k = \tfrac{1}{2}c^{ki}Q_i + \tfrac{1}{2}c^{jk}Q_j,$$

and since $c^{jk} = c^{kj}$, we get

$$\tfrac{1}{2}c^{ki}Q_i + \tfrac{1}{2}c^{jk}Q_j = q^k, \tag{13}$$

which concludes the proof.

Example 2. Find the displacement of the beam of length l and bending rigidity EI shown in Fig. 4.22 at the point of application of force P for the case $a = l/2$.

We shall first find the internal energy of the deflected beam. In the beam element dx, the energy is expressed by

$$dV = \tfrac{1}{2}M d\varphi,$$

where

$$d\varphi = \frac{dx}{\varrho} = \frac{M dx}{EI}.$$

Hence

$$dV = \frac{1}{2}\frac{M^2}{EI}dx.$$

Therefore, the elastic energy V accumulated in the beam subject to bending is expressed by

$$V = \frac{1}{2}\int_l \frac{M^2}{EI}dx, \tag{a}$$

where we integrate over the entire beam length. Displacement at the point of application of force P is

$$q_P = \frac{\partial V}{\partial P} = \int_l \frac{M}{EI}\frac{\partial M}{\partial P}dx. \tag{b}$$

In the considered case, the progress curve of moment M is given in Fig. 4.34. On account of symmetry, we can integrate on half the beam, i.e., in the limit of 0 to $l/2$. The curve of the moment in the interval $[0, l/2]$ is expressed by

$$M = \tfrac{1}{2}Px,$$

or

$$\frac{\partial M}{\partial P} = \tfrac{1}{2}x,$$

whence

$$q_P = 2\int_0^{l/2} \frac{Px^2}{4EI}\,dx = \frac{Pl^3}{48EI}. \tag{c}$$

We can easily check the conformity of the result (c) with the result (g) from Example 2, Section 4.7. The difference in signs should be attributed to the fact that when Castigliano's method is used, positive displacement is indicated by the sense of the applied force.

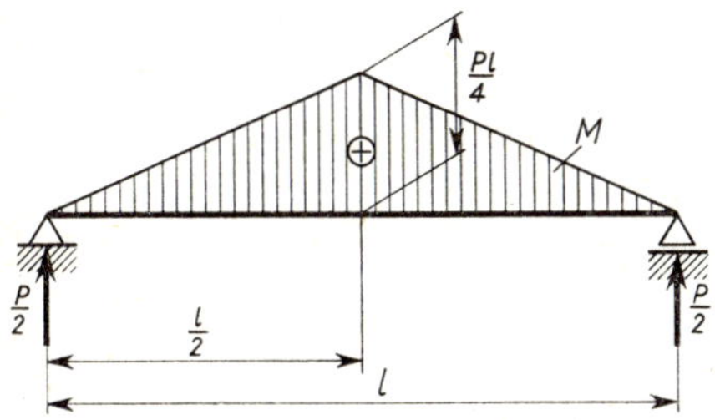

Fig. 4.34

Example 3. Calculate the displacement q of the beam end shown in Fig. 4.33 loaded with concentrated moment M at the end.

In order to solve the problem it is necessary to add a force P indicating the direction of displacement

$$q = \int_0^l \frac{M}{EI}\frac{\partial M}{\partial P}\,dx;$$

$$M = M_0 + Px, \qquad \frac{\partial M}{\partial P} = x,$$

therefore

$$q = \int_0^l \frac{M_0 + Px}{EI}\,x\,dx = \frac{M_0 l^2}{2EI} + \frac{Pl^3}{3EI}. \tag{a}$$

Bearing in mind that P is an imaginary force ($P = 0$), we have

$$q = \frac{M_0 l^2}{2EI}. \tag{b}$$

Example 4. Two rods making an angle α were provided with an articulated joint at point C and connected to a wall at points A and B (Fig. 4.35). Find the vertical component q_V and the horizontal component q_H of displacement at point C if it is known that the rods have identical tensile rigidity AE; the rod lengths are $AC = l_1$ and $BC = l_2$, and a force P acts on the articulation C.

To find the horizontal displacement q_H it is necessary to assume the existence of an imaginary force F acting in this direction. Determining the forces in rods *1* and *2* we obtain

$$S_1 = P\cot\alpha + F, \tag{a}$$

$$S_2 = \frac{P}{\sin\alpha}. \tag{b}$$

The potential energy is expressed by

$$V = \frac{1}{2}\sum_{i=1}^{2}\frac{S_i^2 l_i}{AE};$$

hence the displacements: horizontal

$$q_H = \frac{\partial V}{\partial F} = \sum_{i=1}^{2}\frac{S_i l_i}{AE}\frac{\partial S_i}{\partial F} = \frac{Pl_1}{AE}\cot\alpha,$$

and vertical

$$q_V = \frac{\partial V}{\partial P} = \frac{Pl_1}{AE}\cot^2\alpha + \frac{Pl_2}{\sin^2\alpha AE}.$$

In the case of k-fold statically indeterminate (hyperstatic) Clapeyron's systems, there are always k conditions of inseparability of displacements. They can be represented as follows:

$$\frac{\partial V}{\partial X_j} = 0 \quad (j = 1, \ldots, k), \tag{14}$$

where X_j is a hyperstatic magnitude.

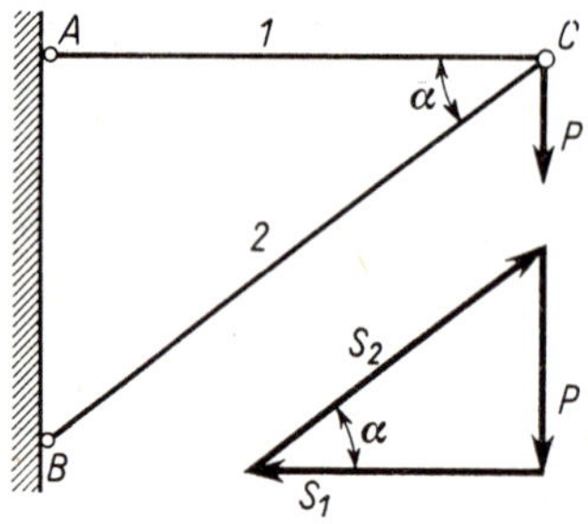

Fig. 4.35

Menabrea's theorem (14) states that the derivatives of the complete energy of a system with respect to hyperstatic magnitudes are equal to zero.

We shall substantiate (14) in two stages: for external forces (reactions) and internal statically indeterminate forces. Note that the expression (14) gives the displacement in the direction of force X_j. If this force is the reaction of the support (external magnitude) and the support is fixed, no displacement takes place ((14) occurs). In the case of an internally hyperstatic magnitude, we divide the system into two subsystems by a section through an element in which X_j occurs.

Let V_1 and V_2 denote the respective potential energies accumulated in members 1 and 2:

$$V = V_1 + V_2. \tag{15}$$

Substituting (15) into (14) gives

$$\frac{\partial V_1}{\partial X_j} = -\frac{\partial V_2}{\partial X_j},$$

which is the condition of inseparability following from the geometric conformity of the cross-section considered. This is apparent from Fig. 4.36.

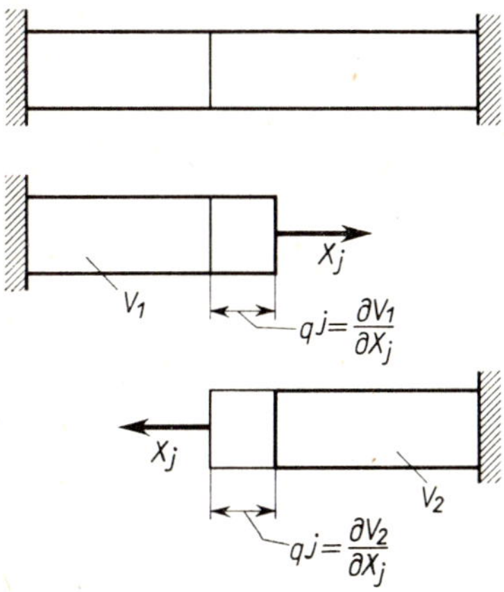

Fig. 4.36

In order to solve a statically indeterminate system we proceed in the following manner. We modify the system by discarding the supports (in the case of an internally statically indeterminate system) or else we make sections in which internal hyperstatic magnitudes occur, so that the system thus obtained should become statically determinate (the basic system). Next, we load the basic system with effective external forces acting and then successively with forces *1* at the points of application of the hyperstatic magnitudes. In each case, we calculate the internal forces and then make use of the superposition theorem in Clapeyron's system. This procedure leads to a linear system of equations with respect to the magnitudes X_j, constituting the conditions of inseparability (14). Indeed,

$$V = \frac{1}{2} \sum_{i=1+k}^{n+k} q^i Q_j + \frac{1}{2} \sum_{m=1}^{k} q^m X_m,$$

therefore

$$\frac{\partial V}{\partial X_j} = \frac{1}{2} \sum_{i=1+k}^{n+k} \frac{\partial q^i}{\partial X_j} Q_i + \frac{1}{2} \sum_{m=1}^{k} \left(\frac{\partial q^m}{\partial X_j} X_m + q^m \frac{\partial X_m}{\partial X_j} \right) = 0.$$

But

$$q^i = \sum_{l=1+k}^{n+k} c^{il} Q_l + \sum_{r=1}^{k} c^{ir} X_r,$$

therefore

$$\frac{1}{2} \sum_{i=1+k}^{n+k} \sum_{r=1}^{k} c^{ir} \frac{\partial X_r}{\partial X_j} Q_i + \frac{1}{2} \sum_{m=1}^{k} \sum_{r=1}^{k} c^{mr} \frac{\partial X_r}{\partial X_j} X_m +$$

$$+ \frac{1}{2} \sum_{m=1}^{k} \sum_{l=1+k}^{n+k} c^{ml} Q_l \frac{\partial X_m}{\partial X_j} + \frac{1}{2} \sum_{m=1}^{k} \sum_{r=1}^{k} c^{mr} X_r \frac{\partial X_m}{\partial X_j} = 0;$$

hence

$$\sum_{i=1+k}^{n+k} c^{ji} Q_i + \sum_{r=1}^{k} c^{jr} X_r = 0. \tag{14a}$$

The first term in (14a) gives the displacement induced by the active forces Q_i at the points of action of the magnitude X_j. Denoting this member by $q^j_{(Q)}$, we obtain what is known as the *Mohr–Maxwell canonical system of equations*:

$$q^j_{(Q)} + \sum_{r=1}^{k} c^{jr} X_r = 0 \quad (j = 1, \dots, k). \tag{14b}$$

Let us discuss the possibilities of applying the derived Mohr–Maxwell system of equations to different cases. Consider the problem of trusses internally statically indeterminate. For such trusses, the doubled number of nodes $2w$ is less than the number of rods p and of the reactions of supports:

$$r + p - 2w = k > 0. \tag{15a}$$

Let $S_i^{(0)}$ be the tension of the ith rod of the basic system. The hyperstatic forces X_r induce in the ith rod the force

$$S_i^{(x)} = \sum_{r=1}^{k} \alpha_i^r X_r.$$

Making use of the superposition theorem, we find the real tension S_i in the ith rod:

$$S_i = S_i^{(0)} + \sum_{r=1}^{k} \alpha_i^r X_r. \tag{16}$$

The potential energy of the truss is expressed by

$$V = \frac{1}{2} \sum_{i=1}^{p} \frac{S_i^2 l_i}{A_i E_i}, \tag{17}$$

where l_i is the length of the ith rod and $A_i E_i$ its tensile rigidity. Using (14) in this particular case, we obtain

$$\sum_{i=1}^{p} \frac{S_i l_i}{A_i E_i} \frac{\partial S_i}{\partial X_j} = 0 \tag{18}$$

or after considering (16) and (17),

$$\sum_{i=1}^{p} \frac{S_i^{(0)} \alpha_i^j l_i}{A_i E_i} + \sum_{r=1}^{k} \left(\sum_{i=1}^{p} \frac{\alpha_i^j \alpha_i^r l_i}{A_i E_i} \right) X_r = 0 \quad (j = 1, \ldots, k). \tag{19}$$

On comparing the result (19), Section 4.6, with (14b) we note that in the case of trusses

$$c^{jr} = \sum_{i=1}^{p} \frac{\alpha_i^j \alpha_i^r l_i}{A_i E_i} \tag{20}$$

and

$$q_{(Q)}^j = \sum_{i=1}^{p} \frac{S_i^{(0)} \alpha_i^j l_i}{A_i E_i}. \tag{21}$$

Example 5. A truss (Fig. 4.37) was loaded with forces P. Calculate the values of forces in the individual rods if it is known that they have the same tensile rigidity AE.

The truss is one-fold statically indeterminate, since $p = 6$, $w = 4$, $r = 3$. From the condition $r+p-2w = k$ we obtain $k = 1$.

Cutting rod 6 we obtain the basic system (Fig. 4.37b). The canonical equation is as follows:

$$q_{(Q)}+c^{11}X = 0; \tag{a}$$

hence

$$X = -\frac{q_{(Q)}}{c^{11}}. \tag{b}$$

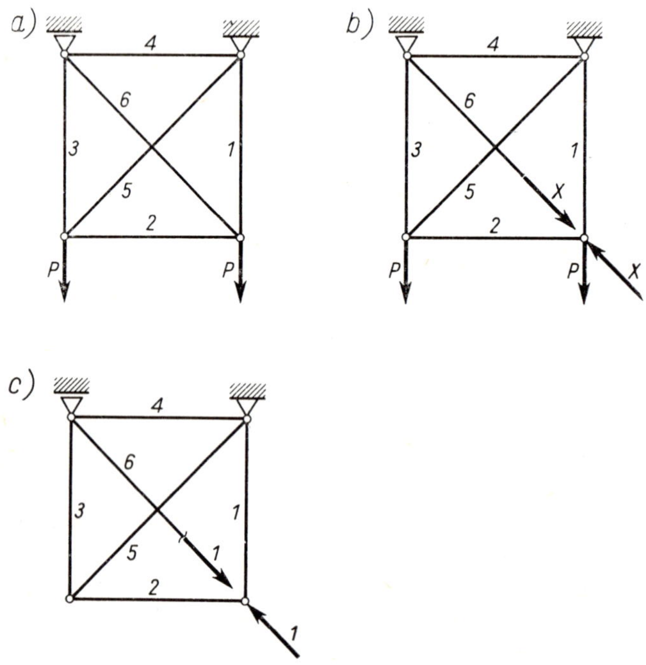

Fig. 4.37

Using equations (20) and (21), we can write

$$X = -\frac{\frac{1}{AE}\sum_{i=1}^{6} S_i^{(0)}\alpha_i^1 l_i}{\frac{1}{AE}\sum_{i=1}^{6}(\alpha_i^1)^2 l_i} = -\frac{\sum_{i=1}^{6} S_i^{(0)}\alpha_i^1 l_i}{\sum_{i=1}^{6}(\alpha_i^1)^2 l_i}. \tag{c}$$

To calculate the hyperstatic magnitude, we make the following table:

Rod No	l_i	$S_i^{(0)}$	α_i^1	$S^{(0)}\alpha_i^1 l_i$	$(\alpha_i^1)^2 l_i$	S_i
1	l	P	$-\frac{\sqrt{2}}{2}$	$-\frac{Pl\sqrt{2}}{2}$	$\frac{l}{2}$	$\frac{P}{2}\frac{1+2\sqrt{2}}{1+\sqrt{2}}$
2	l	0	$-\frac{\sqrt{2}}{2}$	0	$\frac{l}{2}$	$\frac{-P}{2(1+\sqrt{2})}$
3	l	P	$-\frac{\sqrt{2}}{2}$	$-\frac{Pl\sqrt{2}}{2}$	$\frac{l}{2}$	$\frac{P}{2}\frac{1+2\sqrt{2}}{1+\sqrt{2}}$
4	l	0	$-\frac{\sqrt{2}}{2}$	0	$\frac{l}{2}$	$\frac{-P}{2(1+\sqrt{2})}$
5	$l\sqrt{2}$	0	1	0	$l\sqrt{2}$	$\frac{P\sqrt{2}}{2(1+\sqrt{2})}$
6	$l\sqrt{2}$	0	1	0	$l\sqrt{2}$	$\frac{P\sqrt{2}}{2(1+\sqrt{2})}$
				$\Sigma = -Pl\sqrt{2}$	$2l(1+\sqrt{2})$	

Therefore

$$X = \frac{P\sqrt{2}}{2(1+\sqrt{2})}.$$

Thus the last column of the table can be completed.

Now let us consider the case of a statically indeterminate construction, in which bending plays the dominant part. We shall consider beams and systems of rigidly connected rods (so-called *frameworks*) foremost among such constructions. Here we shall look only at bending energy (see equation (a) Example 2):

$$V = \int_l \frac{M^2}{2EI}\mathrm{d}x. \tag{22}$$

Since the moment M is a linear function of loads Q_i

$$M = \sum_i m^i Q_i, \tag{23}$$

therefore

$$q^j = \frac{\partial V}{\partial Q_j} = \int_l \frac{M}{EI}\frac{\partial M}{\partial Q_j}\mathrm{d}x = \sum_i \left(\int_l \frac{m^i m^j}{EI}\mathrm{d}x\right) Q_i = \sum_i c^{ij} Q_j; \tag{24}$$

hence

$$c^{ij} = \int_l \frac{m^i m^j}{EI}\mathrm{d}x. \tag{25}$$

The displacement $q^j_{(Q)}$, which in the basic system comes from the active force Q_i, is given by the relation (24)

$$q^j_{(Q)} = \int_l \frac{\left(\sum_i m^i Q_i\right) m^j}{EI}\mathrm{d}x.$$

Introducing the notation

$$M^{(0)} = \sum_i m^i Q_i,$$

we obtain

$$q^j_{(Q)} = \int_l \frac{M^{(0)} m^j}{EI}\mathrm{d}x. \tag{26}$$

Example 6. Two identical rods of length l and bending rigidity EI were rigidly connected at point C and then secured to a fixed base x (rigidly at point A and pivotally at point B). The member CB of the system was loaded with uniformly distributed weight (Fig. 4.38) of constant intensity Q/l. Find the reaction at point B.

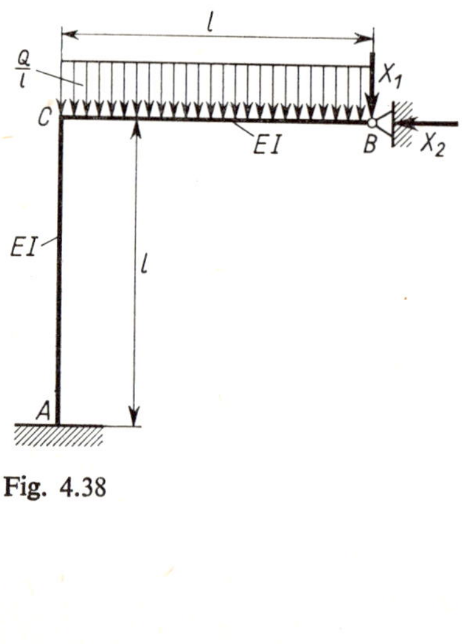

Fig. 4.38

Discarding the support B, we obtain the basic system (Fig. 4.39). It is twofold statically indeterminate, therefore the number of Mohr–Maxwell equations is reduced to two:

$$q^1_{(Q)}+c^{11}X_1+c^{12}X_2 = 0,$$
$$q^2_{(Q)}+c^{21}X_1+c^{22}X_2 = 0. \tag{a}$$

Making use of equations (25) and (26), we may write

$$q^1_{(Q)} = \frac{-1}{EI}\int_0^l \frac{qx_1^2}{2}(-x_1)\mathrm{d}x_1+\int_0^l \frac{ql^3}{2EI}\mathrm{d}x_2 = \frac{5}{8}\frac{Ql^3}{EI}, \tag{b}$$

$$q^2_{(Q)} = \frac{-1}{EI}\int_0^l \frac{qx_1^2}{2}(0)\mathrm{d}x_1-\int_0^l \frac{ql^2}{2EI}x_2\mathrm{d}x_2 = -\frac{Ql^3}{4EI}, \tag{c}$$

$$c^{11} = \int_0^l \frac{(-x_1)(-x_1)}{EI}\mathrm{d}x_1+\int_0^l \frac{(-l)(-l)}{EI}\mathrm{d}x_2 = \frac{4}{3}\frac{l^3}{EI}, \tag{d}$$

$$c^{12} = c^{21} = \int_0^l \frac{(-x_1)(0)}{EI}\mathrm{d}x_1+\int_0^l \frac{(-l)x_2}{EI}\mathrm{d}x_2 = -\frac{l^3}{2EI}, \tag{e}$$

$$c^{22} = \int_0^l \frac{(0)^2}{EI}\mathrm{d}x_1+\int_0^l \frac{x_2^2}{EI}\mathrm{d}x_2 = \frac{l^3}{3EI}. \tag{f}$$

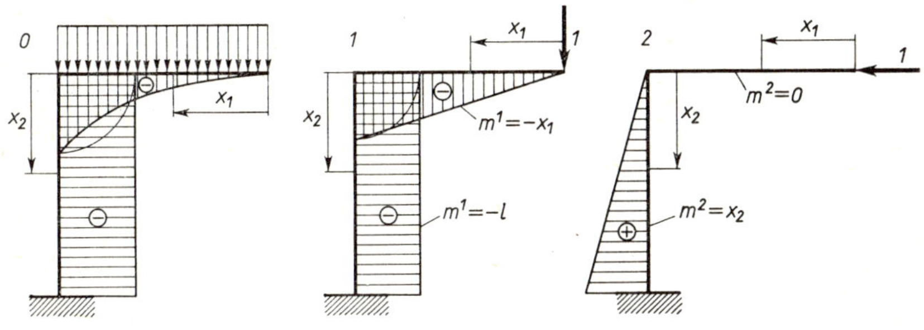

Fig. 4.39

Substituting the coefficients given by equations from (b) to (f) into equations (a) and then solving them, we find

$$X_1 = -\tfrac{3}{7}Q, \tag{g}$$

$$X_2 = \tfrac{3}{28}Q. \tag{h}$$

The displacements $q^1_{(Q)}$ and $q^2_{(Q)}$ were induced by the load Q. The problem remains analogous when $q^1_{(Q)}$ and $q^2_{(Q)}$ are the effect of temperature.

Example 7. Find the hyperstatic magnitudes X_1 and X_2 from the previous example if the temperature of the construction when not loaded increases by Δt and the coefficient of expansion of the material of which the rods are made is α.

The rods are subject to extension under the influence of temperature by a magnitude $\Delta l = \alpha l \Delta t$. Displacements in the direction of X_1 and X_2 are

$$q^1_{(t)} = -\alpha l \Delta t = q^2_{(t)}. \tag{a}$$

We therefore write the canonical equations in the following form:

$$\frac{4}{3}\frac{l^3}{EI}X_1 - \frac{l^3}{2EI}X_2 = \alpha l \Delta t, \qquad -\frac{l^3}{2EI}X_1 + \frac{l^3}{3EI}X_2 = \alpha l \Delta t. \tag{b}$$

Solving equations (b) we obtain

$$X_1 = \frac{30}{7}\frac{EI}{l^2}\alpha \Delta t, \qquad X_2 = \frac{66}{7}\frac{EI}{l^2}\alpha \Delta t. \tag{c}$$

Example 8. A rectangular framework with sides l and l_1 and corresponding rigidity EI and $E_1 I_1$ were loaded with forces N (Fig. 4.40) acting in an axis of symmetry. Find the mutual displacements of the points of application of the forces loading the system.

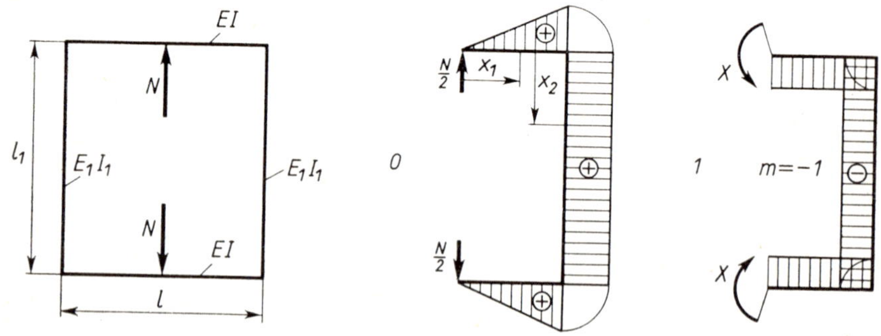

Fig. 4.40

The system is one-fold statically indeterminate,

$$q^1_{(Q)} + c^{11}X = 0; \tag{a}$$

hence

$$X = -\frac{q^1_{(Q)}}{c^{11}}, \tag{a1}$$

where

$$q^1_{(Q)} = -2\int_0^{l/2} \frac{1}{EI}\frac{N}{2}x_1 \, dx_1 - \int_0^{l_1} \frac{1}{E_1 I_1}\frac{Nl}{4} dx_2 = -\frac{Nl}{4}\left(\frac{l}{2EI} + \frac{l_1}{E_1 I_1}\right), \tag{b}$$

$$c^{11} = 2\int_0^{l/2} \frac{1}{EI} dx_1 + \int_0^{l_1} \frac{1}{E_1 I_1} dx_2 = 2\frac{l}{2EI} + \frac{l_1}{E_1 I_1}. \tag{c}$$

Substituting (b) and (c) into (a1), we obtain

$$X = \frac{Nl}{8}\,\frac{l+2l_1\dfrac{EI}{E_1I_1}}{l+l_1\dfrac{EI}{E_1I_1}}. \tag{d}$$

The deflection is

$$\Delta = \frac{\partial V}{\partial N} = 2\left[2\int_0^{l/2}\frac{\frac{1}{2}N(x_1-\varkappa)\frac{1}{2}(x_1-\varkappa)}{EI}\,dx_1 + \right.$$

$$\left. + \int_0^{l_1}\frac{\frac{1}{2}N(l/2-\varkappa)\frac{1}{2}(l/2-\varkappa)}{E_1I_1}\,dx_2\right]$$

$$= \frac{Nl}{2}\left[\frac{1}{EI}\left(\varkappa^2-\frac{\varkappa}{2}+\frac{l^2}{12}\right)+\frac{1}{E_1I_1}\left(\varkappa^2+\frac{l^2}{4}-\varkappa l\right)\right], \tag{e}$$

where

$$\varkappa = \frac{l}{4}\,\frac{l+2l_1\,EI/E_1I_1}{l+l_1\,EI/E_1I_1}. \tag{f}$$

4.12 Buckling problems of axially compressed rods

We shall consider the loss of stability of axially compressed rod (Fig. 4.41) in two cases:

(1) when stresses have not exceeded the limit of proportionality (*elastic buckling*);

(2) when stresses induced by a compressive force are contained between the limit of proportionality and the plastic limit (*postelastic buckling*).

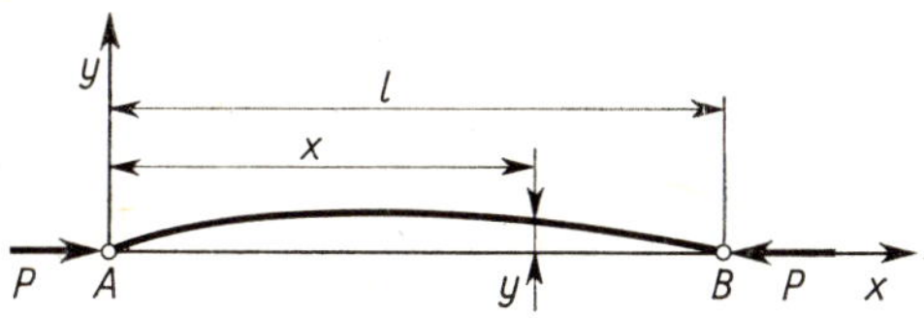

Fig. 4.41

Force P induces in the considered cross-section a moment of value $M = -Py$. For the elastic state the equation of the deflected axis takes the form

$$EI\frac{d^2y}{dx^2} = -Py. \tag{1}$$

Equation (1) is solved as follows:

$$y = C_1 \cos \sqrt{\frac{P}{EI}}x + C_2 \sin \sqrt{\frac{P}{EI}}x, \tag{2}$$

where C_1 and C_2 are integration constants. At point A we have $y = 0$. Hence

$$y = C_2 \sin \sqrt{\frac{P}{EI}}x.$$

The deflected axis is sinusoidal (Fig. 4.42).

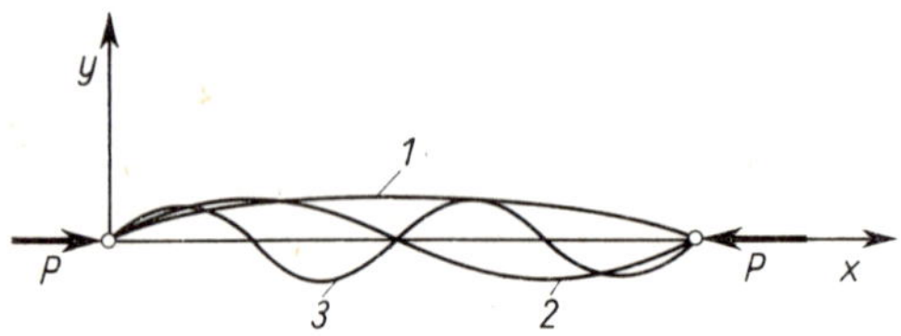

Fig. 4.42

The successive values of forces, for which states *1*, *2*, and *3* occur, are determined from the condition that, for $x = 1$, $y = 0$. Since we are interested in the state of equilibrium with a deflected axis of the rod, $C_2 \neq 0$. Therefore

$$\sin \sqrt{\frac{P}{EI}} l = 0,$$

consequently

$$\sqrt{\frac{P}{EI}} l = n\pi,$$

where $n = 0, \pm 1, \pm 2, \ldots$; hence

$$P_n = \frac{n^2 \pi^2 EI}{l^2}. \tag{3}$$

The buckling plane is at the same time the plane of reduced bending rigidity of the rod, and for this reason I is the lower value of the moment of inertia of the cross-section. For the critical value of the axial force P we shall take the lowest of the terms of the series given by (3):

$$P_{cr} = \frac{\pi^2 EI}{l^2}. \tag{4}$$

Repeating the reasoning for different methods of fixing (Fig. 4.43), equations can be derived
for case a)

$$P_{cr} = \frac{\pi^2 EI}{4l^2}, \tag{5}$$

for case b)

$$P_{cr} = \frac{\pi^2 EI}{l^2}, \tag{6}$$

for case c)

$$P_{cr} = \frac{4\pi^2 EI}{l^2}. \tag{7}$$

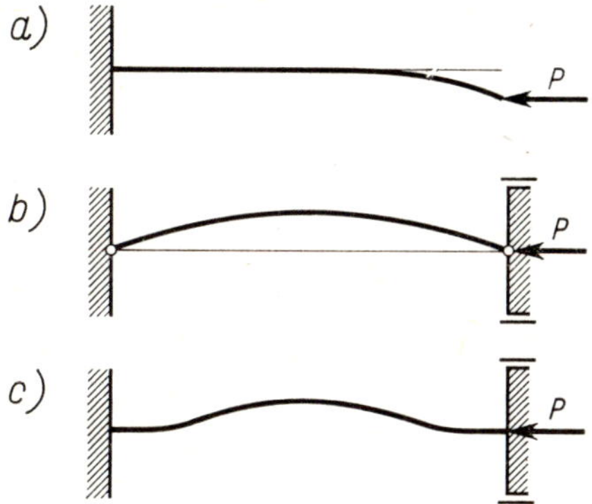

Fig. 4.43

More generally,

$$P_{cr} = \frac{\pi^2 EI}{(\alpha l)^2}. \tag{8}$$

Dividing P_{cr} by the cross-sectional area of the rod, we obtain

$$\sigma_{cr} = \frac{\pi^2 E}{(\alpha l)^2}\frac{I}{A} = \frac{\pi^2 E}{(\alpha\lambda)^2}. \tag{9}$$

We shall call the magnitude $\lambda = l\sqrt{A/I}$ the *slenderness ratio* of rod. The relationship $\sigma_{cr} = f(\lambda)$, shown in Fig. 4.44, is important for such λ for which $\sigma_{cr} \leqslant \sigma_H$, where σ_H denotes the stress at the limit of Hooke's law. The stress $\sigma_{cr} = \sigma_H$ corresponds to the limit value of the slenderness ratio

$$\lambda_{lim} = \frac{\pi}{\alpha}\sqrt{\frac{E}{\sigma_H}}. \tag{10}$$

For steel $\lambda_{lim} \approx 100$.

When the slenderness ratio λ is less than λ_{lim}, we deal with the so-called *postelastic buckling*. In this case, the following linear relationship is accepted:

$$\sigma_{\text{cr}} = \lambda a + b, \tag{11}$$

To determine the magnitude of coefficients a and b, we must have a straight line, called *Tetmajer's line*, passing through two points, one with the coordinates $(\lambda_{\text{lim}}, \sigma_H)$ and the other $(0, \sigma_{\text{pl}})$. Hence

$$\sigma_H = \lambda_{\text{lim}} a + b, \qquad \sigma_{\text{pl}} = 0 \cdot a = b. \tag{12}$$

Solving the equations (12) with respect to a and b, we obtain

$$a = -\frac{\sigma_{\text{pl}} - \sigma_H}{\lambda_{\text{lim}}}, \qquad b = \sigma_{\text{pl}}. \tag{13}$$

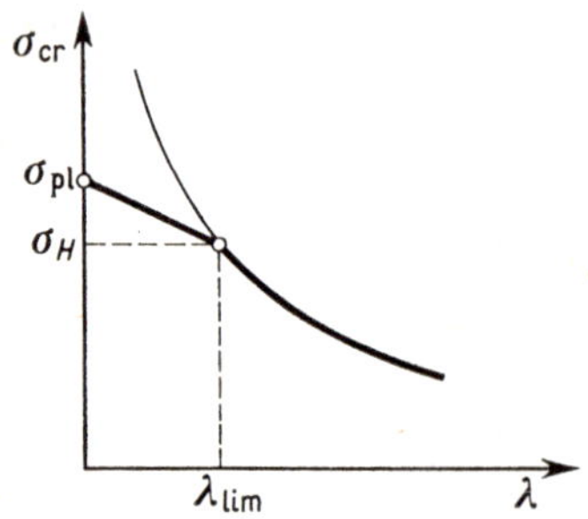

Fig. 4.44

Equations (9) and (11) can be written as one equation by introducing the Heaviside distribution:

$$\sigma_{\text{cr}} = \frac{\pi^2 E}{(\alpha\lambda)^2} H(\lambda - \lambda_{\text{lim}}) + (\lambda a + b) H(\lambda_{\text{lim}} - \lambda). \tag{14}$$

Example 1. Calculate the admissible axial force compressing a steel rod ($E = 2.1 \cdot 10^{11}$ N/m^2) (Fig. 4.41) with rectangular cross-section $b \times h$ ($b = 40$ mm, $h = 5$ mm) and length $l = 0.2$ m, assuming a factor of safety $n = 2.5$.

We shall first determine the smaller of the principal central moments of inertia:

$$I = \frac{4 \cdot (0.5)^3}{12} = 0.04166 \text{ cm}^4;$$

hence

$$\lambda = 20 \frac{20}{0.04166} = 440,$$

therefore $\lambda > \lambda_{lim}$. So we can use equation (4):

$$P_{cr} = \frac{\pi EI}{l^2} = 21{,}149 \text{ N}.$$

The admissible force is

$$P_{ad} = \frac{P_{cr}}{n} = \frac{21{,}149}{2.5} \text{ N} = 8{,}400 \text{ N}.$$

4.13 Highly curved rods

The assumptions applicable to the theory of highly curved rods are exactly the same as for simple beams. Acting with moment M, we induce the displacement of points, according to the hypothesis of plane sections, in such a way that the extension of an element placed at a distance y from the neutral axis is

$$\Delta \mathrm{d}s = y\Delta \mathrm{d}\varphi. \tag{1}$$

On the other hand, ds also depends on y and considering the original high curvature of the elements, it will be as in (Fig. 4.45)

$$\mathrm{d}s = (\varrho+y)\mathrm{d}\varphi. \tag{2}$$

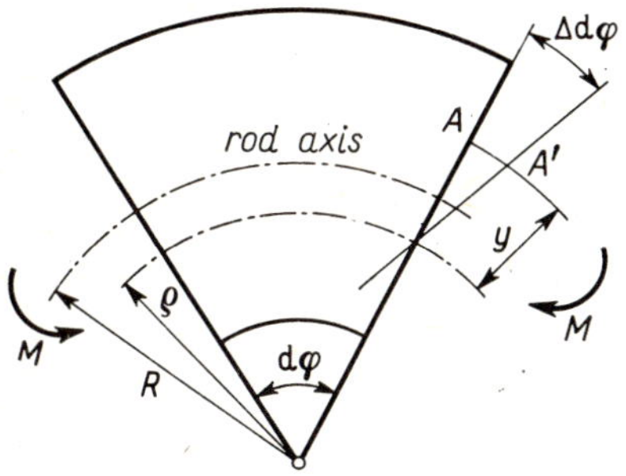

Fig. 4.45

From (1) and (2) we find the equation giving the deflection of elements:

$$\varepsilon = \frac{\Delta \mathrm{d}s}{\mathrm{d}s} = \frac{\Delta \mathrm{d}\varphi}{\mathrm{d}\varphi}\frac{y}{\varrho+y}. \tag{3}$$

According to Hooke's law,

$$\sigma = E\frac{\Delta \mathrm{d}\varphi}{\mathrm{d}\varphi}\frac{y}{\varrho+y}. \tag{4}$$

Since the load is reduced to a couple,

$$\int_A \sigma \mathrm{d}A = E\frac{\Delta \mathrm{d}\varphi}{\mathrm{d}\varphi}\int_A \frac{y}{\varrho+y}\mathrm{d}A = 0.$$

But

$$E\frac{\Delta \mathrm{d}\varphi}{\mathrm{d}\varphi} \neq 0, \quad \text{thus} \quad \int_A \frac{y}{\varrho+y}\mathrm{d}A = 0. \tag{5}$$

Adding and subtracting ϱ in the numerator, we obtain

$$\int_A \mathrm{d}A - \int_A \frac{\varrho}{\varrho+y}\mathrm{d}A = 0. \tag{6}$$

Denoting the sum $\varrho+y$ by u, we write

$$A - \varrho\int_A \frac{\mathrm{d}A}{u} = 0,$$

whence

$$\varrho = A/\int_A \frac{\mathrm{d}A}{u}. \tag{7}$$

The sum of the moments of internal forces is reduced to a couple of the value M or

$$M = \int_A \sigma y \mathrm{d}A = E\frac{\Delta \mathrm{d}\varphi}{\mathrm{d}\varphi}\int_A \frac{y^2}{\varrho+y}\mathrm{d}A = E\frac{\Delta \mathrm{d}\varphi}{\mathrm{d}\varphi}\int_A \frac{y(y+\varrho)-\varrho y}{\varrho+y}\mathrm{d}A$$

$$= E\frac{\Delta \mathrm{d}\varphi}{\mathrm{d}\varphi}\int_A y\mathrm{d}A - E\frac{\Delta \mathrm{d}\varphi}{\mathrm{d}\varphi}\int_A \frac{\varrho y}{\varrho+y}\mathrm{d}A.$$

By reason of (5),

$$M = E\frac{\Delta \mathrm{d}\varphi}{\mathrm{d}\varphi}\int_A y\mathrm{d}A. \tag{8}$$

Dividing both sides of (4) by (8), we have

$$\frac{\sigma}{M} = \frac{y}{(\varrho+y)\int_A y\mathrm{d}A}; \tag{9}$$

then, denoting the static moment $\int_A y\mathrm{d}A$ by S, we write

$$\sigma = \frac{My}{(\varrho+y)S}. \tag{10}$$

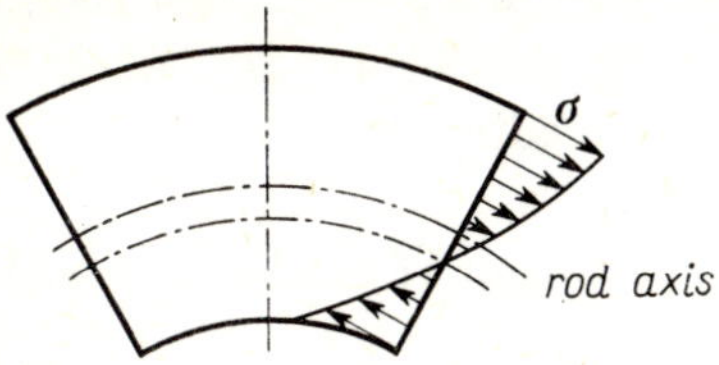

Fig. 4.46

The stress distribution is non-linear, as it has been shown in Fig. 4.46. Considering the stresses from the axial force N, we can write

$$\sigma = \frac{N}{A} + \frac{My}{(\varrho+y)S}. \tag{11}$$

Example 1. Find the position of the neutral axis for a rectangular and circular cross-section (Fig. 4.47).

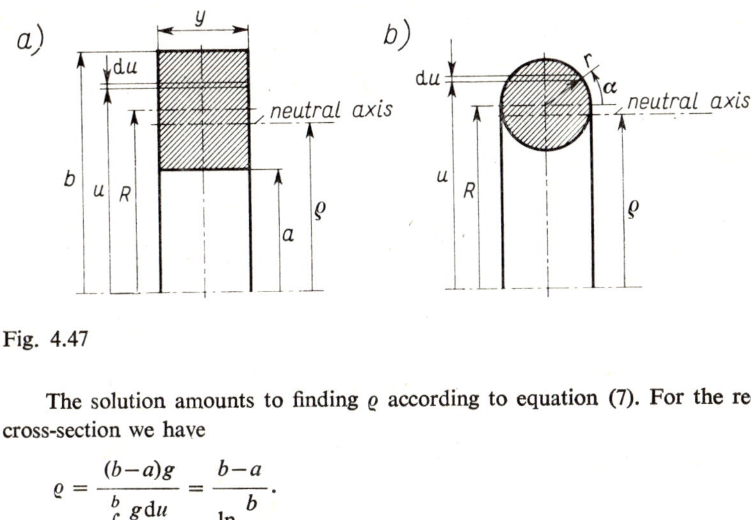

Fig. 4.47

The solution amounts to finding ϱ according to equation (7). For the rectangular cross-section we have

$$\varrho = \frac{(b-a)g}{\int_a^b \frac{g\,du}{u}} = \frac{b-a}{\ln\frac{b}{a}}. \tag{a}$$

For the circular cross-section

$$u = R+r\sin\alpha, \quad dA = 2r\cos\alpha\, du = 2r^2\cos^2\alpha\, d\alpha,$$

$$\varrho = \frac{r^2}{\int_{-\pi/2}^{\pi/2} \frac{2r^2\cos^2\alpha\, d\alpha}{R+r\sin\alpha}} = \frac{r^2}{2(R+\sqrt{R^2-r^2})}. \tag{b}$$

We shall now discuss deflections of highly curved rods. The solution to the problem we shall seek by reference to the energy method on the basis of Castigliano's theorem. We begin by finding the potential energy accumulated in a rod element of length ds. Assume that a moment M and an axial force N act in the cross-section. The moment M induces an angular displacement of the cross-section by a magnitude

$$\Delta \mathrm{d}\varphi_1 = \frac{M\mathrm{d}\varphi}{ES} = \frac{M\mathrm{d}s}{ESR}. \tag{12}$$

The axial force N will induce an additional angular displacement of the cross-section

$$\Delta \mathrm{d}\varphi_2 = \frac{N}{EA}\frac{\mathrm{d}s}{R}. \tag{13}$$

The total is therefore

$$\Delta \mathrm{d}\varphi = \frac{M}{ESR}\mathrm{d}s + \frac{N}{EAR}\mathrm{d}s. \tag{14}$$

The work of the moment M on angular displacement $\Delta \mathrm{d}\varphi$ is equal to

$$\mathrm{d}W_M = \frac{1}{2}\frac{M^2}{ESR}\mathrm{d}s + \frac{1}{2}\frac{NM}{EAR}\mathrm{d}s.$$

Considering Betti's reciprocal theorem, the work of moment M on the angular displacement induced by force N is equal to the work of force N on the displacement induced by moment M, which is

$$\mathrm{d}W_{MN} = \frac{1}{2}\frac{MN}{EAR}\mathrm{d}s. \tag{15}$$

Thus the work of force N on the linear displacement $\Delta \mathrm{d}s$ is

$$\mathrm{d}W_N = \frac{1}{2}\frac{N^2\mathrm{d}s}{EA} + \frac{1}{2}\frac{MN}{EAR}\mathrm{d}s. \tag{16}$$

A transverse force T will also do work on the displacement resulting from the non-dilatational strain of the rod section ds. From (13), Section 4.1, we assume that the non-dilatational strain in expressed by

$$y = k\frac{\tau_m}{G} = k\frac{T}{AG}, \tag{17}$$

where τ_m is a mean shear stress, and k is a coefficient related to the shape of the rod cross-section ($k = 1.2$ for rectangular cross-section, and

$k = 1.185$ for circular cross-section). The displacement of force T is $\gamma \mathrm{d}s$ and the work it does equals to

$$\mathrm{d}W_T = \frac{1}{2} T\gamma \mathrm{d}s = \frac{1}{2} k \frac{T^2}{AG} \mathrm{d}s. \tag{18}$$

The potential energy is an integral from the sum $\mathrm{d}W_M + \mathrm{d}W_N + \mathrm{d}W_T$ and is expressed by

$$V = \frac{1}{2} \int_l \frac{M^2 \mathrm{d}s}{ESR} + \int_l \frac{MN \mathrm{d}s}{EAR} + \frac{1}{2} \int_l \frac{N^2 \mathrm{d}s}{EA} + \frac{1}{2} k \int_l \frac{T^2 \mathrm{d}s}{GA}. \tag{19}$$

We find the generalized displacement q^j from Castigliano's theorem:

$$q^j = \int_l \frac{M}{ESR} \frac{\partial M}{\partial Q_j} \mathrm{d}s + \int_l \frac{M}{EAR} \frac{\partial N}{\partial Q_j} \mathrm{d}s + \int_l \frac{N}{EAR} \frac{\partial M}{\partial Q_j} \mathrm{d}s +$$

$$+ \int_l \frac{N}{EA} \frac{\partial N}{\partial Q_j} \mathrm{d}s + k \int_l \frac{T}{GA} \frac{\partial T}{\partial Q_j} \mathrm{d}s. \tag{20}$$

Example 2. A highly curved circular rod of rectangular cross-section (Fig. 4.48a) is loaded with a force P. Find the angular displacement of the planes forming the slit and the relative displacement of points O and O'.

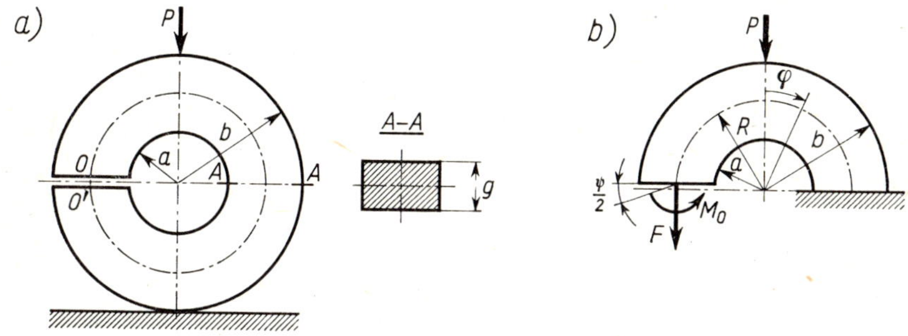

Fig. 4.48

In order to solve the problem we first apply imaginary magnitudes to the cross-section of the slit: a force F and a moment M_o. The system is statically determinate. The internal forces of moment M, the longitudinal force N, and the transverse force T are expressed respectively by

$$M = M_o + FR(1+\sin\varphi) + PRH(\varphi)\sin\varphi, \tag{a}$$

$$N = -[F + PH(\varphi)]\sin\varphi, \tag{b}$$

$$T = [F + PH(\varphi)]\cos\varphi, \tag{c}$$

where $H(\varphi)$ is Heaviside's function, $R = \frac{1}{2}(a-b)$.

We find from (20) the angular displacement ψ, making use simultaneously of the conditions of symmetry (for this reason we consider only the top part of the rod as shown in Fig. 4.48b):

$$\psi = 2\left[\int_{-\pi/2}^{\pi/2} \frac{M}{ES}\frac{\partial M}{\partial M_o}\,d\varphi + \int_{-\pi/2}^{\pi/2} \frac{M}{EA}\frac{\partial N}{\partial M_o}\,d\varphi + \int_{-\pi/2}^{\pi/2} \frac{N}{EA}\frac{\partial M}{\partial M_o}\,d\varphi + \right.$$

$$\left. + \int_{-\pi/2}^{\pi/2} \frac{N}{EA}\frac{\partial N}{\partial M_o}\,R\,d\varphi + k\int_{-\pi/2}^{\pi/2} \frac{T}{GA}\frac{\partial T}{\partial M_o}\,R\,d\varphi\right], \tag{d}$$

$$\frac{\partial M}{\partial M_o} = 1, \qquad \frac{\partial N}{\partial M_o} = 0, \qquad \frac{\partial T}{\partial M_o} = 0.$$

Therefore

$$\psi = 2\left(\frac{1}{ES}\int_0^{\pi/2} PR\sin\varphi\,d\varphi - \frac{1}{EA}\int_0^{\pi/2} P\sin\varphi\,d\varphi\right) = \frac{2P}{E}\left(\frac{R}{S} - \frac{1}{A}\right). \tag{e}$$

The static moment S is expressed by the following equation (see Example 1):

$$S = A(R-\varrho) = g(b-a)\left(\frac{a+b}{2} - \frac{b-a}{\ln(b/a)}\right). \tag{f}$$

We find the mutual displacement of the points in a similar way:

$$\Delta = 2\left(\int_{-\pi/2}^{\pi/2} \frac{M}{ES}\frac{\partial M}{\partial F}\,d\varphi + \int_{-\pi/2}^{\pi/2} \frac{M}{EA}\frac{\partial N}{\partial F}\,d\varphi + \int_{-\pi/2}^{\pi/2} \frac{N}{EA}\frac{\partial M}{\partial F}\,d\varphi + \right.$$

$$\left. + \int_{-\pi/2}^{\pi/2} \frac{N}{EA}\frac{\partial N}{\partial F}\,R\,d\varphi + k\int_{-\pi/2}^{\pi/2} \frac{T}{GA}\frac{\partial T}{\partial F}\,R\,d\varphi\right), \tag{g}$$

$$\frac{\partial M}{\partial F} = R(1+\sin\varphi), \qquad \frac{\partial N}{\partial F} = -\sin\varphi, \qquad \frac{\partial T}{\partial F} = \cos\varphi.$$

Hence

$$\Delta = 2\left(\frac{1}{ES}\int_0^{\pi/2} PR^2\sin\varphi(1+\sin\varphi)\,d\varphi - \frac{1}{EA}\int_0^{\pi/2} PR\sin^2\varphi\,d\varphi + \right.$$

$$\left. + \int_0^{\pi/2} PR\sin\varphi(1+\sin\varphi)\,d\varphi - \int_0^{\pi/2} PR\sin^2\varphi\,d\varphi + \frac{k}{GA}\int_0^{\pi/2} PR\cos^2\varphi\,d\varphi\right)$$

$$= \frac{2PR}{E}\left(\frac{R}{S} - \frac{1}{A}\right)\left(1 + \frac{\pi}{4}\right) + k\,\frac{\pi PR}{2GA}, \tag{h}$$

where $k = 1.2$ (rectangular cross-section).

4.14 Torsion of rods with circular cross-section

Torsion of the components of a construction is rather a common occurrence in our experience. All type of shafts, connectors or springs and even an ordinary door handle or key are examples of constructions subject to torsion. Unlike the general case, the torsion of rods with circular cross-section belongs to a comparatively simple group of problems, since in this case no deformation of the cross-section occurs.

If a part of such a rod (Fig. 4.49) is loaded with moments M and $-M$ on surfaces $\mathrm{d}x$ apart from each other, the effect will be expressed by their

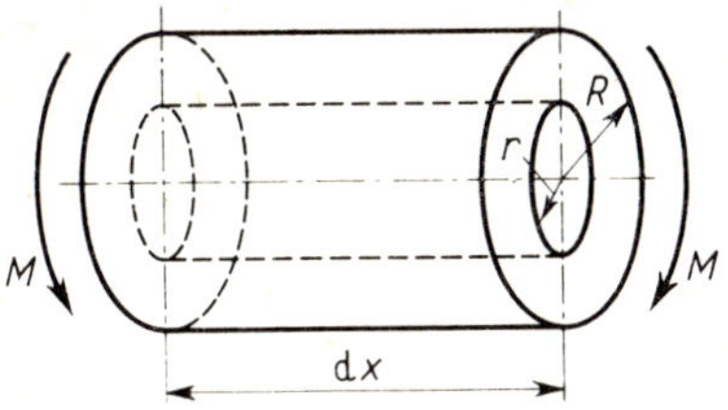

Fig. 4.49

angular displacement $\Delta\varphi$. In this connection, the generators of a cylinder will turn by an angle γ (Fig. 4.50), being the measure of non-dilatational strain (see Section 4.1). Since according to equation (13), Section 4.1,

$$\gamma = \tau/G$$

and

$$AA' = r\mathrm{d}\varphi = \gamma\mathrm{d}x,$$

we have

$$\frac{\tau}{G}\mathrm{d}x = r\mathrm{d}\varphi; \quad (1)$$

hence

$$\tau = \frac{\mathrm{d}\varphi}{\mathrm{d}x}Gr. \quad (2)$$

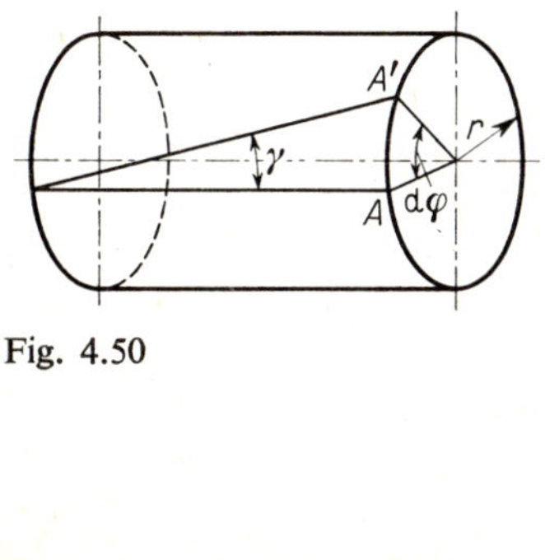

Fig. 4.50

Note that shear stresses grow proportionally to r (Fig. 4.51), and the distribution is radially symmetrical. In order to find the stress relative to torsional moment M, we shall utilize the fact that the sum of the moments of tangential forces is equal to M:

$$M = \int_A \tau r \mathrm{d}A. \tag{3}$$

Considering (2) in (3), we have

$$M = \frac{\mathrm{d}\varphi}{\mathrm{d}x} G \int_A r^2 \mathrm{d}A = \frac{\mathrm{d}\varphi}{\mathrm{d}x} GI_0, \tag{3a}$$

where $I_0 = \int_A r^2 \mathrm{d}A$ is, as we know, the polar geometric moment of inertia of circular cross-section

$$I_0 = \frac{\pi R^4}{2}. \tag{4}$$

The product GI_0 we shall call *torsional rigidity.*

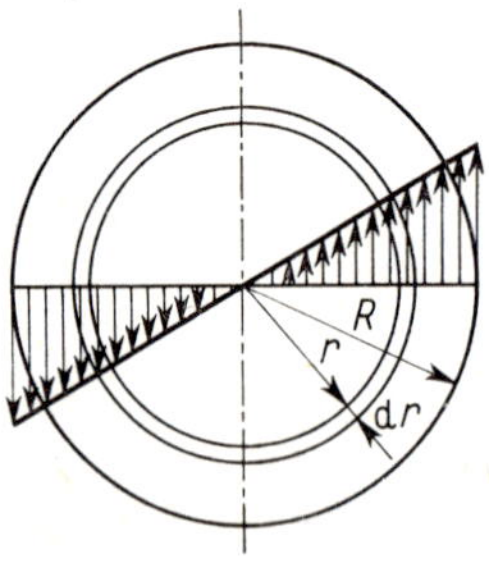

Fig. 4.51

From (2) and (3a) we find the equation giving the shear stresses τ relative to the torsional moment M

$$\tau = \frac{Mr}{I_0}. \tag{5}$$

If the constructional detail under design is to safely transmit the moment M, the following strength condition must be satisfied:

$$\max \tau \leqslant \tau_{\mathrm{ad}}, \tag{6}$$

i.e., maximum tangential stresses τ must not exceed the admissible values τ_{ad}

$$\tau_{\mathrm{ad}} = \frac{\tau_{\mathrm{cr}}}{n}, \tag{7}$$

where n is the safety factor.

Maximum stresses occur at the extreme points of the cross-section and they are equal to

$$\max \tau = \frac{M}{I_0/r_{max}}. \tag{8}$$

As in the case of bending we also introduce here the concept of *section modulus for torsion*

$$W_0 = \frac{I_0}{r_{max}} = \frac{\pi r_{max}^3}{2}. \tag{9}$$

Example 1. Find the diameter of a shaft spinning at 300 r.p.m. and transmitting a power $N = 65$ hp, assuming $\sigma_{ad} = 45 \cdot 10^6$ N/m².

Note that power, according to equation (9), Section 2.1, can be defined by

$$N = Pr\omega = M\omega; \tag{a}$$

hence

$$M = \frac{N}{\omega}. \tag{b}$$

Since the shaft spins at 300 r.p.m.,

$$\omega = \frac{300 \cdot 2\pi}{60 \text{ s}} = 10\pi \text{ s}^{-1}, \tag{c}$$

or

$$M = \frac{65 \cdot 750 \text{ Nm}}{10\pi} = 1{,}550 \text{ Nm}.$$

Making use of strength condition (6), we write

$$\frac{M}{\tau_{ad}} \leqslant W_0 \quad \text{or} \quad r_{max} = \frac{d}{2} = \sqrt[3]{\frac{2M}{\pi \tau_{ad}}} = 2.8 \cdot 10^{-2} \text{ m}.$$

We shall now determine the energy of a rod subject to torsion. This energy, accumulated in a section of length dx, is equal to half the product of the torsional moment and the angular displacement dφ:

$$dV = \tfrac{1}{2} M d\varphi. \tag{10}$$

Calculating the angle dφ from the relationship (3a) and then substituting into (10), we obtain

$$dV = \frac{1}{2} \frac{M^2}{GI_o} dx. \tag{11}$$

Hence the total energy is expressed by

$$V = \frac{1}{2} \int_l \frac{M^2}{GI_o} dx. \tag{12}$$

Example 2. A rod *ABCD* with circular cross-section and torsional rigidity GI_0 was rigidly fixed between two walls (Fig. 4.52). The system was stiffened at point *C* with a spring of a constant *c* Nm/rad and loaded at point *B* with a torsional moment *M*. Calculate the reactions of the walls, knowing that the respective rod sections have the following length: $AB = a$, $BC = b$, $CD = l$.

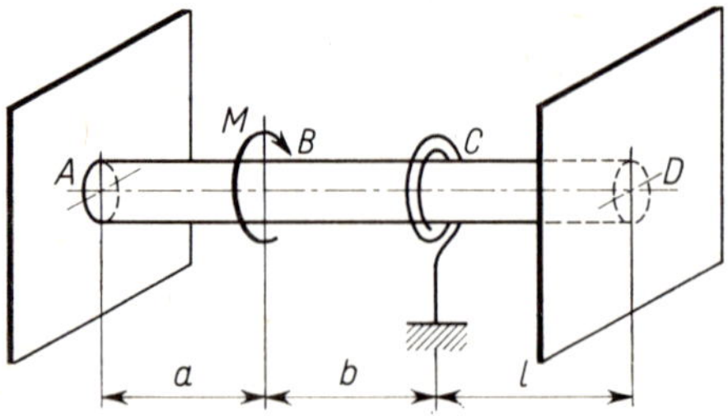

Fig. 4.52

First we isolate the basic system. In the considered case it is a rod relieved of the fixings at points *A* and *D*. Assuming hyperstatic moments X_A and X_D at these points, we set up an equation expressing the conditions of inseparability (*Mohr–Maxwell equations*):

$$-M\left(\frac{1}{c}+\frac{b}{GI_0}\right)+\left(\frac{1}{c}+\frac{a+b}{GI_0}\right)X_1+\frac{1}{c}X_2 = 0, \tag{a}$$

$$-\frac{M}{c}+\frac{1}{c}X_1+\left(\frac{1}{c}+\frac{1}{GI_0}\right)X_2 = 0. \tag{b}$$

Equations (a) and (b) express the absence of angular displacement of cross-sections *A* and *D*. Solving equations (a) and (b) relative to the hyperstatic magnitudes, we obtain

$$X_1 = \frac{l+b\left(1+\dfrac{cl}{GI_0}\right)}{l+(a+b)\left(1+\dfrac{cl}{GI_0}\right)}M, \tag{c}$$

$$X_2 = \frac{a}{l+(a+b)\left(1+\dfrac{cl}{GI_0}\right)}M. \tag{d}$$

4.15 Springs

Compression or tension springs withstanding considerable strains are commonly used in servomotors, springing of all types, and in other parts of mechanisms. The most common type of spring is the spiral (coil) spring of steel wire with circular cross-section.

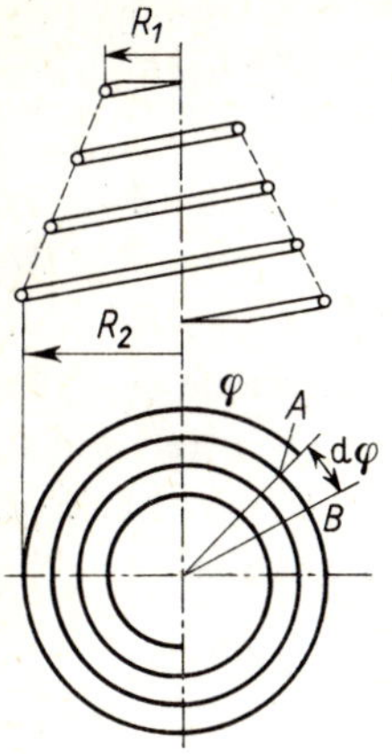

Fig. 4.53

We shall discuss here the case of a tapering spring. The vertical projection of such a spring is the spiral of Archimedes (Fig. 4.53) expressed by

$$R = C_1\varphi + C_2. \tag{1}$$

For $\varphi = 0$ we have $R = R_1$, whereas for $\varphi = 2\pi n$ (n is the number of coils) $R = R_2$. Hence

$$C_1 = \frac{R_2 - R_1}{2\pi n}, \qquad C_1 = R_1.$$

Therefore

$$R = \frac{R_2 - R_1}{2\pi n}\varphi + R_1. \tag{2}$$

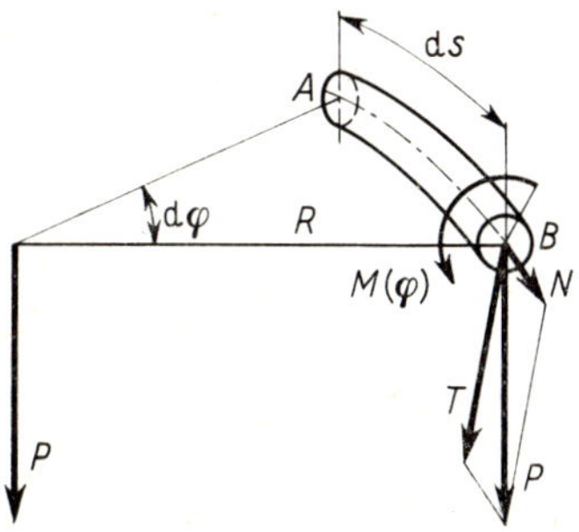

Fig. 4.54

Let us make sections at A and at B to isolate a piece of wire of length ds, contained in the angle dφ (Fig. 4.54). Translating the force P we find that the vector of the moment $M(\varphi) = PR(\varphi)$ in section B is not in a plane perpendicular to the rod axis because of the inclination of the coils. Hence the moment $M(\varphi)$ will have two components: the bending component M_g and the torsional component M_t. Similarly, the force P has an axial component N and the transverse component T. With a small angle ψ of the helix, both bending and tension are assumed to be so small in relation to torsion and shear that they can be neglected (they are of the order of $\sin\psi$). For $\psi \leqslant 20°$ the error does not excess 5 per cent. Hence the stresses can be calculated with the use of an approximate equation

$$\tau = \frac{P}{\pi r^2} + \frac{PR}{W_0}, \tag{3}$$

where $r = d/2$ is half the wire diameter. Analysing the deflection of the piece ds, we find that section B will turn about section A by an angle

$$\mathrm{d}\alpha = \frac{M(\varphi)\,\mathrm{d}s}{GI_o} = \frac{M(\varphi)\,R\,\mathrm{d}\varphi}{GI_o}.$$

The potential energy accumulated in the considered piece of the spring is

$$\mathrm{d}V = \tfrac{1}{2}M(\varphi)\mathrm{d}\alpha = \tfrac{1}{2}\frac{[M(\varphi)]^2 R\,\mathrm{d}\varphi}{GI_o}. \tag{4}$$

We find the total potential energy of the spring by integrating the expression (4):

$$V = \tfrac{1}{2}\int\limits_0^{2\pi n} \frac{[M(\varphi)]^2 R\,\mathrm{d}\varphi}{GI_o}. \tag{5}$$

Deflection, on the other hand, is expressed according to Castigliano's theorem by

$$f = \frac{\partial V}{\partial P} = \int\limits_0^{2\pi n} \frac{M}{GI_o}\frac{\partial M(\varphi)}{\partial P} R\,\mathrm{d}\varphi \tag{6}$$

or

$$f = \int\limits_0^{2\pi n} \frac{PR^3}{GI_o}\mathrm{d}\varphi. \tag{7}$$

Changing the variables under the integral, we obtain

$$f = \int_{R_1}^{R_2} \frac{2\pi n}{R_2 - R_1} PR^3 \, \mathrm{d}R = \frac{\pi n P}{2GI_o} (R_1^2 + R_2^2)(R_1 + R_2). \tag{8}$$

When dealing with a cylindrical spring, we put $R_1 = R_2 = R$ in equation (8). Then

$$f = 2\frac{\pi n P}{GI_o} R^3. \tag{9}$$

It should be appreciated that in deriving equation (8) we have ignored the influence of transverse forces on the deflection of the spring, as we did for the simple beam theory. The influence of these forces on deflections is slight.

We introduce the so-called *constant of spring* described by

$$c = \frac{P}{f} = \frac{GI_o}{2\pi n R^3}. \tag{10}$$

It characterizes the rigidity of spring and corresponds, let us say, to the value AE/l in the case of simple beam tension.

Springs, like electrical resistances or condensers, are connected either in parallel or in series. In the case of parallel connection of two springs (Fig. 4.55). The substitute constant c_s is calculated according to

$$c_s = c_1 + c_2, \tag{11}$$

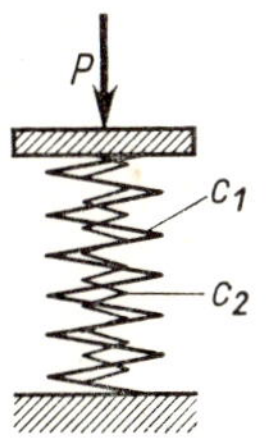

Fig. 4.55

since under the force P both springs are subjected to the same deflection Δ, and the sum of forces $P_1 = c_1\Delta$ and $P_2 = c_2\Delta$ is equal to force $P = c_s\Delta$. For a larger number of springs joined parallel, we generalize equation (11)

$$c_s = \sum_{i=1}^{n} c_i. \tag{11a}$$

When springs are joined in series (Fig. 4.56) the substitute constant is expressed by

$$\frac{1}{c_s} = \frac{1}{c_1} + \frac{1}{c_2}. \tag{12}$$

Fig. 4.56

Equation (12) is substantiated as follows. Total extension is $\Delta = P/c_s$. On the other hand, Δ is the sum of $\Delta_1 = P/c_1$ and $\Delta_2 = P/c_2$. With a larger number of springs involved we have

$$\frac{1}{c_s} = \sum_{i=1}^{n} \frac{1}{c_i}. \tag{12a}$$

Example 1. Calculate the constant c_s for a system that consists of a fixed beam AB of bending rigidity EI and length l and a spring with a constant c (Fig. 4.57).

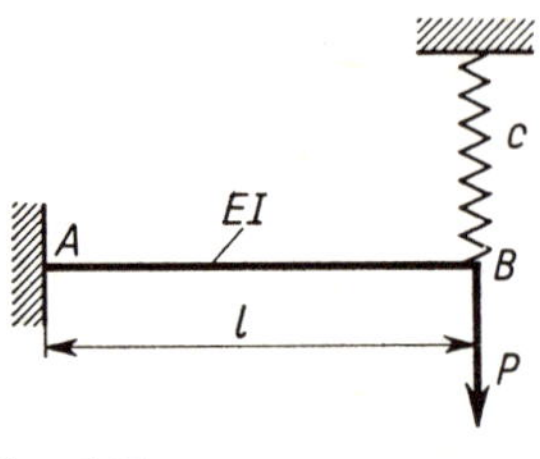

Fig. 4.57

In solving the problem, it should be noted that the beam AB can be treated as a spring with a constant $c = 3EI/l^3$ ($f = Pl^3/3EI$). We deal therefore with the case of parallel connection of two springs. Thus the substitute constant is expressed by

$$c_s = c + \frac{3EI}{l^3}. \tag{a}$$

Example 2. Calculate the substitute constant of the system given in Fig. 4.58, consisting of three springs with a constant c and a beam of bending rigidity EI. By the substitute constant we shall understand here the ratio of force P to displacement of point II.

We form a subsystem by removing the spring between points II and IV. We obtain in this way a system with a constant (parallel connection with two springs connected in series):

$$c_{s1} = \frac{1}{l^3/48EI+1/2c}. \tag{a}$$

Reconnecting the removed spring (parallel connection), we obtain the required substitute constant:

$$c_s = c_{s1}+c. \tag{b}$$

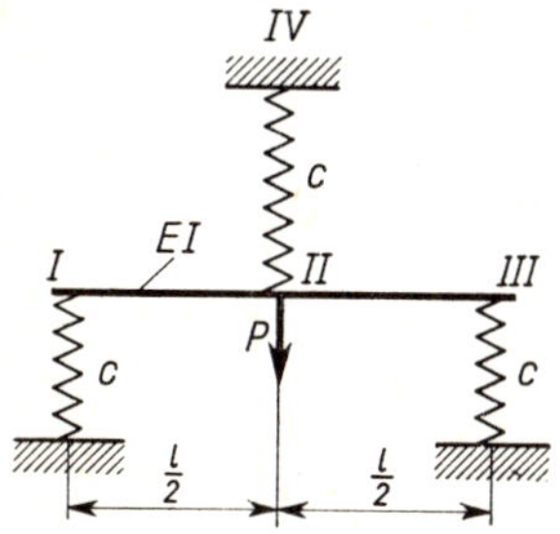

Fig. 4.58

4.16 Simultaneous bending and torsion of rods with circular cross-section

The case of simultaneous torsion and bending of rods of circular cross-section belongs to compound states, therefore the following question arises: How should the strength of the system be estimated? (Experimental data comes from simple bending tests). One could make an experiment corresponding to what actually happens to the rod but such a test would be troublesome: This can only be avoided when we know the criteria of comparison between a compound state of stresses and the best investigated of the simplest states, which is simple tension. Such criteria exist in the form of the so-called *effort hypotheses*. These can be divided into three groups:

(a) strain
(b) stress
(c) energy

In group (a) it is considered that on comparing the tension state with a compound state, account should be taken of strain: in group (b) account should be taken of stress, and in group (c), of non-dilatational strain.

Without going into details, we shall make use of the energy hypothesis called the *Huber–Mises–Hencky hypothesis*. After comparing non-dilatational strain energy for simple tension with non-dilatational strain energy for simultaneous bending and torsion we obtain the following relationship:

$$\sigma_t = \sqrt{\sigma_b^2+3\tau^2}, \tag{1}$$

where σ_t—stress of sample subject to tension (substitute stress), σ_b—stress due to bending, and τ—stress due to torsion.

Note that

$$\sigma_b = \frac{M_t}{W_x}$$

and that

$$\tau = \frac{M_t}{W_0} = \frac{M_t}{2W_x};$$

hence

$$\sigma_t = \sqrt{M_b^2+\tfrac{3}{4}M_t^2}/W_x. \tag{2}$$

The value of the square root on the right-hand side of (2) is called the *reduced moment*, i.e., such a bending moment that leads to material breakdown as in the case of the simultaneous action of M_b and M_t.

Example 1. A system consists of beam AB of length l_1 and bending rigidity E_1I_1, and a lever BCD with dimensions a and l and bending rigidity EI. The part in torsion, CD, has a circular cross-section with diameter d, torsional rigidity equals GI_o (Fig. 4.59). Find the substitute stresses at point D of the rigid fixing of the lever when a force P is applied to the articulation B connecting the two subsystems.

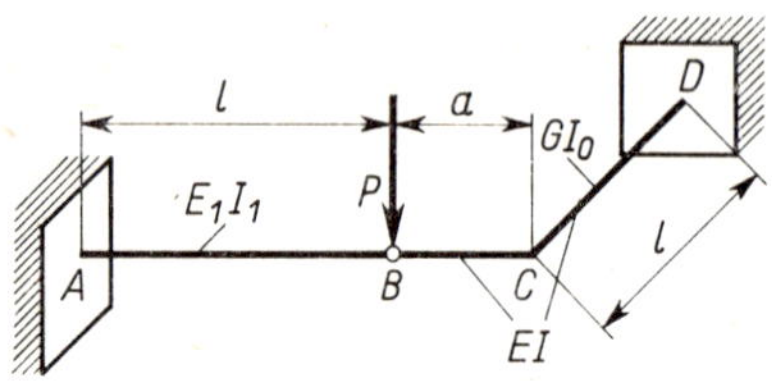

Fig. 4.59

It is convenient to take note of the fact that the two subsystems can be treated as parallel springs. The constant of spring for part AB is

$$c_1 = \frac{3E_1I_1}{l_1^3},$$

and for the lever

$$c = \frac{1}{\dfrac{a^2l}{GI_o}+\dfrac{l^3+a^3}{3EI}}.$$

Hence

$$c_s = \frac{3E_1 I_1}{l_1^3} + \frac{1}{\dfrac{a^2 l}{GI_o} + \dfrac{l^3+a^3}{3EI}}. \tag{a}$$

The displacement of point B is

$$\Delta = \frac{P}{c_s}.$$

Therefore the force transmitted to the lever BCD at point B is

$$P_B = c\Delta = \frac{Pc}{c_s} = \frac{P}{3\dfrac{E_1 I_1}{l_1^3}\left(\dfrac{a^2 l}{GI_o} + \dfrac{l^3+a^3}{3EI}\right)+1}. \tag{b}$$

On this basis we find the bending moment M_b and the torsional moment M_t in cross-section D

$$M_b = P_B l, \tag{c}$$

$$M_t = P_B a. \tag{d}$$

Considering relationship (b) in (c) and (d) and then substituting the derived expressions for M_b and M_t into equation (2), we obtain the sought equation for reduced stresses in cross-section D of the lever

$$\sigma_t = \frac{1}{W_x} \frac{P\sqrt{l^2+\frac{3}{4}a^2}}{1+3\dfrac{E_1 I_1}{l_1^3}\left(\dfrac{a^2 l}{GI_o} + \dfrac{l^3+a^2}{3EI}\right)}, \tag{e}$$

where by virtue of the circular cross-section

$$W_x = \frac{d^3}{32}\pi. \tag{f}$$

CHAPTER 5

The Dynamics of Rigid Bodies

5.1 Moments of inertia of rigid bodies

Let a rigid body filling an area V, which is bounded by a surface S, be given. We introduce an orthogonal frame of reference $x^1x^2x^3$ fixed against the body considered. Nine magnitudes defined as

$$J^{ik} \overset{\text{df}}{=} \int_V (\delta_{mn}x^m x^n \delta^{ik} - x^i x^k)\mu\,\mathrm{d}V, \tag{1}$$

where δ_{mn} and δ^{ik} are Kronecker's symbols, μ density of the body, form the inertia matrix characterizing the dynamical structure of the body. By virtue of the adopted frame of reference, $J^{ik} = J_{ik}$.

The magnitudes on the diagonal of the matrix J are moments of inertia about the axes x^1, x^2, x^3. The remaining ones with the opposite sign are all moments of deviation.

We shall now present a few examples to illustrate the methods of calculating moments of inertia of bodies with circular symmetry. In these examples we shall use the symbols x, y, and z to denote the coordinates x^1, x^2, x^3, which will simplify the notation to some extent. For example, the moment of inertia about the axis x^1 should be denoted according to definition (1) by J^{11} or J_{11}. In our examples they will be symbolized briefly as J_x.

Example 1. We shall calculate here the moment of inertia of a homogeneous body ($\mu(x, y, z) = \text{const}$) axisymmetrical to its axis of symmetry.

Let the axis of symmetry be the axis x (Fig. 5.1)

$$J_x = \iiint_V (y^2+z^2)\mu\,\mathrm{d}x\,\mathrm{d}y\,\mathrm{d}z\,. \tag{a}$$

Changing the variables of the integral (a) by transition to cylindrical coordinates

$$y = R\cos\varphi, \qquad z = R\sin\varphi, \qquad x = x, \tag{b}$$

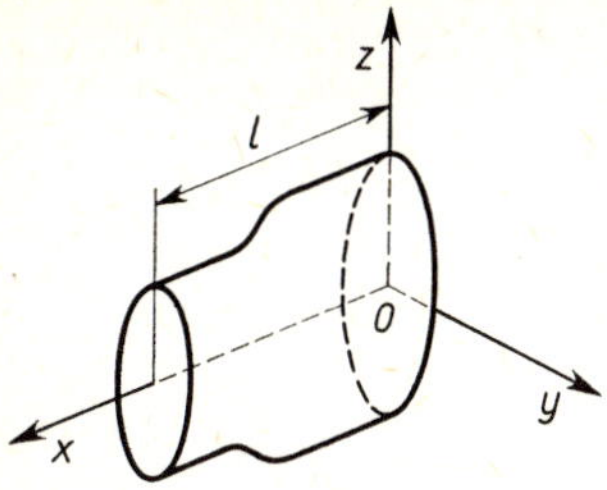

Fig. 5.1

we obtain

$$J_x = \iiint\limits_V R^2\mu \mathrm{d}V = \iiint\limits_{V'} R^3\mu\, \mathrm{d}R\mathrm{d}\varphi \mathrm{d}x. \tag{c}$$

We integrate over the area

$$0 \leqslant R \leqslant f(x), \quad 0 \leqslant \varphi \leqslant 2\pi, \quad 0 \leqslant x \leqslant l.$$

Iterating integral (c) and integrating, we write

$$J_x = \tfrac{1}{2}\pi\mu \int_0^l [f(x)]^4 \mathrm{d}x. \tag{d}$$

In the case of a circular cylinder with radius R and mass m (Fig. 5.2)

$$J_x = \tfrac{1}{2}\pi\mu R^4 \int_0^l \mathrm{d}x = \tfrac{1}{2}\pi\mu R^4 l.$$

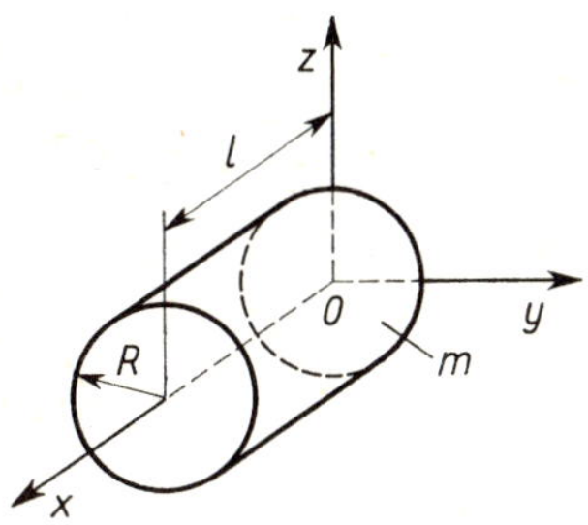

Fig. 5.2

Since $\pi R^2 l\mu = m$,

$$J_x = \tfrac{1}{2}mR^2, \tag{e}$$

For a cone with base radius R and mass m (Fig. 5.3) the moment of inertia is

$$J_x = \frac{\pi\mu}{2}\left(\frac{R}{l}\right)^4 \int_0^l x^4 \mathrm{d}x = \tfrac{3}{10}\cdot\tfrac{1}{3}\pi\mu R^2 l R^2.$$

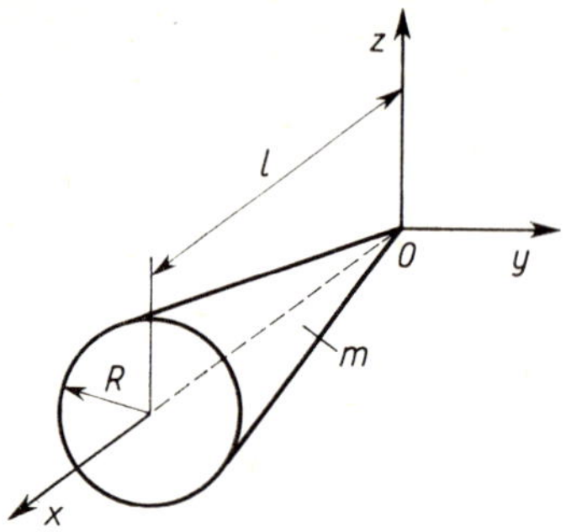

Fig. 5.3

But $\frac{1}{3}\pi\mu R^2 l = m$. Therefore

$$J_x = \tfrac{3}{10} m R^2. \tag{f}$$

In the case of a sphere with radius R and mass m (Fig. 5.4), we have

$$f(x) = \sqrt{R^2 - x^2},$$

whence

$$J_x = \tfrac{1}{2}\pi\mu \int_{-R}^{+R} (R^2 - x^2)^2 \mathrm{d}x = \tfrac{2}{5}\tfrac{4}{3}\pi\mu R^3 R^2.$$

But

$$\tfrac{4}{3}\pi\mu R^3 = m.$$

Therefore

$$J_x = \tfrac{2}{5} m R^2.$$

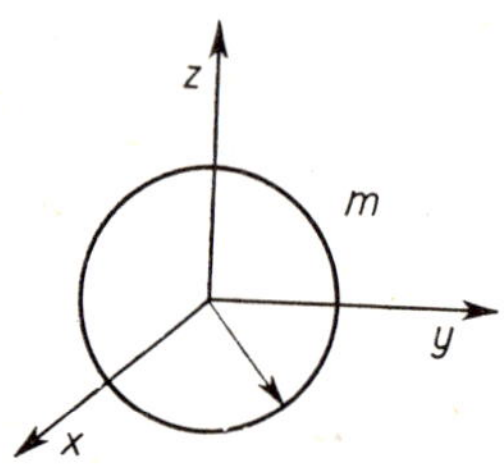

Fig. 5.4

We shall now consider the change of elements of the inertia matrix on transition from one frame of reference to another. Let the transformation rule be

$$x^i = x_0^i + A^i_{j'} \xi^{j'},^* \tag{2}$$

* We are using the same notation as in Section 1.2 (see equations (29)).

where $A^i_{j'}$ form the matrix of cosines of angles between the coordinate axes x^1, x^2, x^3, and ξ^1, ξ^2, ξ^3. Substituting (2) into (1), we obtain

$$\begin{aligned} J^{ik} &= (x_0^m x_0^n \delta_{mn}\delta^{ik} - x_0^i x_0^k)m + \\ &+ \int_V (x_0^n A^m_{j'}\xi^j \delta_{mn}\delta^{ik} - x_0^i A^k_{j'}\xi^j)\mu\,\mathrm{d}V + \\ &+ \int_V (x_0^m A^n_{j'}\xi^j \delta_{mn}\delta^{ik} - x_0^k A^i_{j'}\xi^j)\mu\,\mathrm{d}V + \\ &+ \int_V (A^m_{r'} A^n_{j'}\xi^r\xi^j \delta_{mn}\delta^{ik} - A^i_{j'}A^k_{r'}\xi^j\xi^r)\mu\,\mathrm{d}V, \end{aligned} \tag{3}$$

where $m = \int_V \mu\,\mathrm{d}V$ is the mass of the body.

Let us now consider two cases, namely translation which must be such that the reference origin for $\xi^1\xi^2\xi^3$ coincides with the centre of mass of the body, and angular displacement without translation, $x_0^i = 0$.

In the first case

$$A^i_{j'} = \delta^i_j, \tag{4}$$

where δ^i_j is Kronecker's delta. Considering that the reference origin for $\xi^1\xi^2\xi^3$ coincides with the centre of mass of the body, all static moments are identically equal to zero $\left(\bigwedge_i \int_V \mu\xi^i \mathrm{d}V \equiv 0\right)$, consequently we have

$$J^{ik} = (x_0^m x_0^n \delta_{mn}\delta^{ik} - x_0^i x_0^k)m + \int_V (\xi^m\xi^n \delta_{mn}\delta^{ik} - \xi^i\xi^k)\mu\,\mathrm{d}V. \tag{5}$$

The integral expresses the components of the inertia matrix $J^{ik}_{(\xi)}$ in the frame of reference $\xi^1\xi^2\xi^3$. Thus we can write (5) in the form

$$J^{ik} = (x_0^m x_0^n \delta_{mn}\delta^{ik} - x_0^i x_0^k)m + J^{ik}_{(\xi)}. \tag{5a}$$

It this way, we have obtained what is known as *Steiner's equation* which corresponds to equations (3) and (5), Section 4.5, which have been derived for a two-dimensional problem.

The second case amounts to finding transformation formulae corresponding to the case of angular displacement of the $\xi^1\xi^2\xi^3$ reference frame, without translation $\left(\bigwedge_i x_0^i \equiv 0\right)$. Here equation (3) takes the form

$$J^{ik} = \int_V (A^m_{r'} A^n_{j'}\xi^r\xi^j \delta_{mn}\delta^{ik} - A^i_{j'}A^k_{r'}\xi^j\xi^r)\mu\,\mathrm{d}V. \tag{6}$$

Let us calculate the moment of inertia about the axis $x^1 \equiv x$:

$$J^{11} = J_x = \int_V (A^1_{r'} A^1_{j'} \xi^r \xi^j + A^2_{r'} A^2_{j'} \xi^r \xi^j + A^3_{r'} A^3_{j'} \xi^r \xi^j - A^1_{j'} A^1_{r'} \xi^j \xi^r) \mu \, dV$$

$$= \int_V (A^2_{r'} A^2_{j'} + A^3_{r'} A^3_{j'}) \xi^r \xi^j \mu \, dV. \tag{7}$$

Considering that orthogonal frames of reference are involved, we have

$$A^1_{r'} A^1_{j'} + A^2_{r'} A^2_{j'} + A^3_{r'} A^3_{j'} = \delta_{rj}. \tag{8}$$

therefore

$$J_x = \int_V (\delta_{rj} - A^1_{r'} A^1_{j'}) \xi^r \xi^j \mu \, dV \tag{9}$$

or after expanding

$$J_x = \int_V \{[1-(A^1_{1'})^2](\xi^1)^2 + [1-(A^1_{2'})^2](\xi^2)^2 + [1-(A^1_{3'})^2](\xi^3)^2 -$$

$$-2A^1_{1'} A^1_{2'} \xi^1 \xi^2 - 2A^1_{2'} A^1_{3'} \xi^2 \xi^3 - 2A^1_{3'} A^1_{1'} \xi^3 \xi^1\} \mu \, dV. \tag{10}$$

By virtue of the relationship $(A^1_{1'})^2 + (A^1_{2'})^2 + (A^1_{3'})^2 = 1$ we have

$$\begin{aligned} 1-(A^1_{1'})^2 &= (A^1_{2'})^2 + (A^1_{3'})^2, \\ 1-(A^1_{2'})^2 &= (A^1_{1'})^2 + (A^1_{3'})^2, \\ 1-(A^1_{3'})^2 &= (A^1_{1'})^2 + (A^1_{2'})^2. \end{aligned} \tag{11}$$

Substituting (11) into (10) we obtain

$$J_x = (A^1_{1'})^2 \int_V [(\xi^2)^2 + (\xi^3)^2] \mu \, dV + (A^1_{2'})^2 \int_V [(\xi^1)^2 + (\xi^3)^2] \mu \, dV +$$

$$+ (A^1_{3'})^2 \int_V [(\xi^1)^2 + (\xi^2)^2] \mu \, dV - 2A^1_{1'} A^1_{2'} \int_V \xi^1 \xi^2 \mu \, dV -$$

$$-2A^1_{2'} A^1_{3'} \int_V \xi^2 \xi^3 \mu \, dV - 2A^1_{3'} A^1_{1'} \int_V \xi^3 \xi^1 \mu \, dV. \tag{12}$$

Integrals of the type $\int_V [(\xi^i)^2 + (\xi^k)^2] \mu \, dV$ $(i \neq k)$ occurring in equation (12) are moments of inertia about the axes ξ^1, ξ^2, ξ^3 (J_{ξ^1}, J_{ξ^2}, J_{ξ^3}). The remaining integrals we identify as being moments of deviation ($J_{\xi^i \xi^k} = \int_V \xi^i \xi^k \mu \, dV$).

It will be more convenient at the present stage of transformation to introduce the notations

$$\xi^1 \equiv \xi, \quad \xi^2 \equiv \eta, \quad \xi^3 \equiv \zeta.$$

We rewrite equation (12) with the new notations

$$J_x = J_\xi(A^1_{1'})^2+J_\eta(A^1_{2'})^2+J_\zeta(A^1_{3'})^2-$$
$$-2J_{\xi\eta}A^1_{1'}A^1_{2'}-2J_{\eta\zeta}A^1_{2'}A^1_{3'}-2J_{\zeta\xi}A^1_{3'}A^1_{1'}. \tag{13}$$

From the derived equation we can find the moment of inertia of a body about a straight line passing through the reference origin for $\xi\eta\zeta$ (in the considered case, that straight line is the axis x), if we know the values of moments of inertia J_ξ, J_η, J_ζ, moments of deviation and angles α, β, γ ($A^1_{1'} = \cos\alpha$, $A^1_{2'} = \cos\beta$, $A^1_{3'} = \cos\gamma$), which the x-axis makes with axes ξ, η, and ζ (Fig. 5.5).

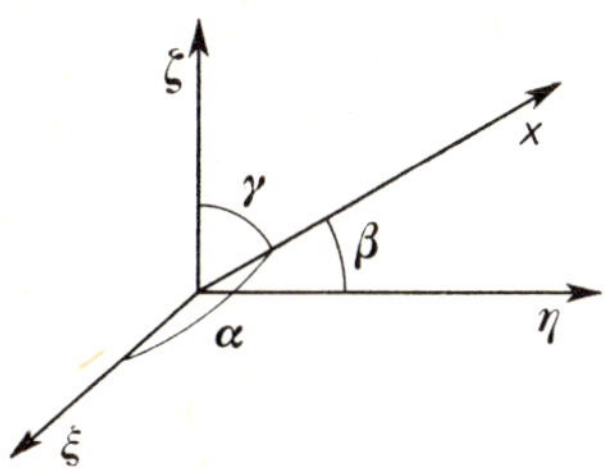

Fig. 5.5

We shall also introduce the concept of radius of gyration b_l:

$$b_x \stackrel{\text{df}}{=} \sqrt{\frac{J_x}{m}}. \tag{14}$$

Returning to the derived transformation formula (13), we shall assume for the purposes of further considerations that the body in question is not a set of particles in a straight line passing through the reference origin. Excluding this case, we assume at the same time that $J_x \neq 0$. Next, construct in the x-axis a section OA having a length inversely proportional to the radius of gyration

$$OA = \frac{k}{b_x} = k\sqrt{\frac{m}{J_x}}. \tag{15}$$

Let point A have the following coordinates:

$$\xi = OA\cdot\cos\alpha, \quad \eta = OA\cdot\cos\beta, \quad \zeta = OA\cdot\cos\gamma. \tag{16}$$

Now note that if we derive the cosines of angles from (16) and then substitute them into (13), we obtain

$$J_x = J_\xi \frac{\xi^2}{(OA)^2} + J_\eta \frac{\eta^2}{(OA)^2} + J_\zeta \frac{\zeta^2}{(OA)^2} -$$

$$-2J_{\xi\eta}\frac{\xi\eta}{(OA)^2} - 2J_{\eta\zeta}\frac{\eta\zeta}{(OA)^2} - 2J_{\zeta\xi}\frac{\zeta\xi}{(OA)^2}. \tag{17}$$

Considering (15) and denoting the product mk^2 by C^2, we write

$$J_\xi \xi^2 + J_\eta \eta^2 + J_\zeta \zeta^2 - 2J_{\xi\eta}\xi\eta - 2J_{\eta\zeta}\eta\zeta - 2J_{\zeta\xi}\zeta\xi = C^2. \tag{18}$$

Equation (18) describes the area of an ellipsoid; it is the so-called *inertia ellipsoid* (the section OA has a finite maximum, therefore the area (18) is limited).

We can see from the aforesaid that there are three values of moments about three mutually perpendicular axes with two of them assuming the maximum and the minimum value, respectively. These are called the *principal central axes of inertia* (Fig. 5.6).

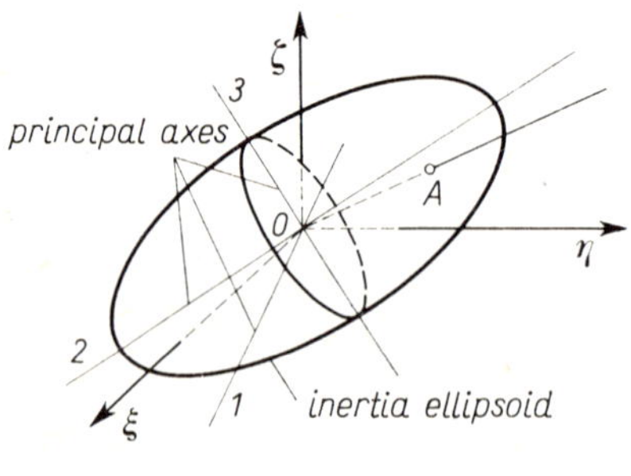

Fig. 5.6

5.2 The angular momentum of a rigid body in general motion

Let us take a certain point O in space and make it the $Oxyz$ reference origin (Fig. 5.7). According to definition the angular momentum of a body is described by the integral

$$\mathbf{K} = \int_V \mathbf{r} \times \mathbf{v}\mu \, dV. \tag{1}$$

As shown in Fig. 5.7,

$$\mathbf{r} = \mathbf{r}_\Omega + \boldsymbol{\rho}. \tag{2}$$

Therefore

$$\mathbf{K} = \mathbf{r}_\Omega \times \int_V \mathbf{v}\mu \mathrm{d}V + \int_V \boldsymbol{\rho} \times \mathbf{v}\mu \mathrm{d}V. \tag{3}$$

Velocity $\mathbf{v}$ in the general motion of an arbitrary point is described by equation (41), Section 1.2, whence

$$\mathbf{K} = \mathbf{r}_\Omega \times m\mathbf{v}_\Omega + \mathbf{r}_\Omega \times \left(\boldsymbol{\omega} \times \int_V \boldsymbol{\rho}\mu \mathrm{d}V\right) -$$

$$- \mathbf{v}_\Omega \times \int_V \boldsymbol{\rho}\mu \mathrm{d}V + \int_V \boldsymbol{\rho} \times (\boldsymbol{\omega} \times \boldsymbol{\rho})\mu \mathrm{d}V. \tag{4}$$

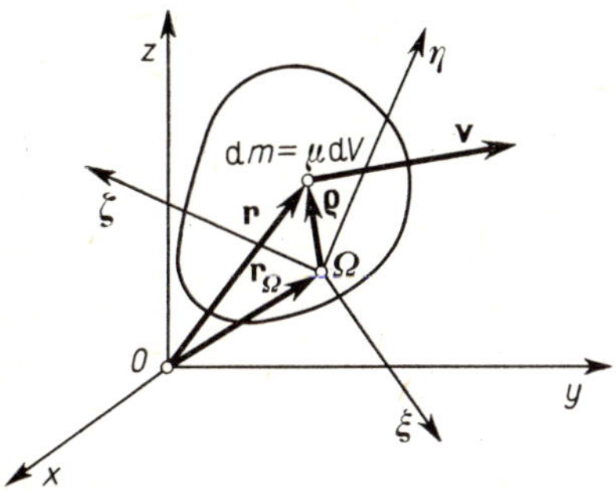

Fig. 5.7

If the reference origin $\Omega\xi\eta\zeta$, rigidly bound with a moving body, coincides with the centre of mass of the body, s, then the second and third terms on the right-hand side of equation (4) vanish. For this case we have

$$\mathbf{K} = \mathbf{r}_s \times m\mathbf{v}_s + \int_V \boldsymbol{\rho} \times (\boldsymbol{\omega} \times \boldsymbol{\rho})\mu \mathrm{d}V. \tag{4a}$$

The angular momentum in general motion is the sum of the angular momentum of a body in translatory motion with the velocity of the centre of mass and the angular momentum in circular motion.

According to equation (25) from the Introduction

$$\boldsymbol{\rho} \times (\boldsymbol{\omega} \times \boldsymbol{\rho}) = \boldsymbol{\omega}\varrho^2 - \boldsymbol{\rho}(\boldsymbol{\omega} \cdot \boldsymbol{\rho}),$$

therefore expression (4a) can be presented in a somewhat different form, namely

$$\mathbf{K} = \mathbf{r}_s \times m\mathbf{v}_s + \boldsymbol{\omega} \int_V \varrho^2 \mu \mathrm{d}V - \int_V \boldsymbol{\rho}(\boldsymbol{\omega} \cdot \boldsymbol{\rho})\mu \mathrm{d}V. \tag{4b}$$

5.3 Angular momentum in circular motion

Going over circular motion as a special case of general motion means assuming that $\mathbf{v}_\Omega = 0$ and $\mathbf{r}_\Omega = 0$. Making such a substitution in equation (4), Section 5.2, we obtain

$$\mathbf{K} = \int_V \boldsymbol{\rho}\times(\boldsymbol{\omega}\times\boldsymbol{\rho})\mu\,\mathrm{d}V, \tag{1}$$

or

$$\mathbf{K} = \boldsymbol{\omega}\int_V \varrho^2\mu\,\mathrm{d}V - \int_V \boldsymbol{\rho}(\boldsymbol{\omega}\cdot\boldsymbol{\rho})\mu\,\mathrm{d}V. \tag{1a}$$

The integral constituting the first term on the right-hand side is the moment of inertia of the body, reckoned about the reference origin. Now

$$\boldsymbol{\omega} = \omega_\xi\mathbf{e}_\xi+\omega_\eta\mathbf{e}_\eta+\omega_\zeta\mathbf{e}_\zeta, \qquad \boldsymbol{\rho} = \xi\mathbf{e}_\xi+\eta\mathbf{e}_\eta+\zeta\mathbf{e}_\zeta, \tag{2}$$

therefore

$$\begin{aligned}
\mathbf{K} = {} & \mathbf{e}_\xi\Big(\omega_\xi\int_V \xi^2\mu\,\mathrm{d}V+\omega_\xi\int_V \eta^2\mu\,\mathrm{d}V+\omega_\xi\int_V \zeta^2\mu\,\mathrm{d}V-\\
& -\omega_\xi\int_V \xi^2\mu\,\mathrm{d}V-\omega_\eta\int_V \xi\eta\mu\,\mathrm{d}V-\omega_\zeta\int_V \xi\zeta\mu\,\mathrm{d}V\Big)+\\
& +\mathbf{e}_\eta\Big(\omega_\eta\int_V \xi^2\mu\,\mathrm{d}V+\omega_\eta\int_V \eta^2\mu\,\mathrm{d}V+\omega_\eta\int_V \zeta^2\mu\,\mathrm{d}V-\\
& -\omega_\xi\int_V \xi\eta\mu\,\mathrm{d}V-\omega_\eta\int_V \eta^2\mu\,\mathrm{d}V-\omega_\zeta\int_V \eta\zeta\mu\,\mathrm{d}V\Big)+\\
& +\mathbf{e}_\zeta\Big(\omega_\zeta\int_V \xi^2\mu\,\mathrm{d}V+\omega_\zeta\int_V \eta^2\mu\,\mathrm{d}V+\omega_\zeta\int_V \zeta^2\mu\,\mathrm{d}V-\\
& -\omega_\xi\int_V \xi\zeta\mu\,\mathrm{d}V-\omega_\eta\int_V \eta\zeta\mu\,\mathrm{d}V-\omega_\zeta\int_V \zeta^2\mu\,\mathrm{d}V\Big).
\end{aligned} \tag{3}$$

The individual components of the angular momentum contain moments of inertia and moments of deviation of the body in the frame of reference $\Omega\xi\eta\zeta$:

$$\begin{aligned}
K_\xi &= J_\xi\,\omega_\xi-J_{\xi\eta}\,\omega_\eta-J_{\xi\zeta}\,\omega_\zeta,\\
K_\eta &= -J_{\eta\xi}\,\omega_\xi+J_\eta\,\omega_\eta-J_{\eta\zeta}\,\omega_\zeta,\\
K_\zeta &= -J_{\zeta\xi}\,\omega_\xi-J_{\zeta\eta}\,\omega_\eta+J_\zeta\,\omega_\zeta.
\end{aligned} \tag{4}$$

Note that in the summation convention these equations are written as follows:

$$K^i = \mathbf{J}^{ij}\omega_j \tag{4a}$$

or in the matrix notation

$$\mathbf{K} = \mathbf{J}\cdot\boldsymbol{\omega}. \tag{4b}$$

Problem. Prove that if the angular momentum **K** or the angular velocity vector has the direction of one of the principal axes of inertia, then the angular momentum and the angular velocity have an identical direction.

Example 1. When for any reason an artificial satellite begins spinning, the unwanted motion should be eliminated. To this end, it is necessary to release two masses attached at two points, which unwind two threads of the same length l. As soon as the threads have completely unwound and acquired radial directions, they are free from the points of attachment. Given the masses m, the moment of inertia of the satellite **J** and the radius R of the core on which threads have been wound (Fig. 5.8). How long should be the threads to stop after the unwinding the rotation of the satellite?

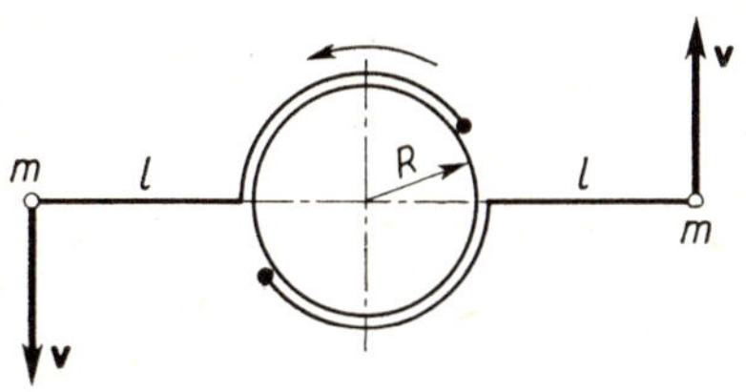

Fig. 5.8

Suppose the satellite was spinning with angular velocity $\boldsymbol{\omega}$; then its angular momentum is

$$K_1 = (J+2mR^2)\boldsymbol{\omega}.$$

After the masses have unwound the requisite amount of thread the satellite should stop spinning, therefore the angular momentum of the system is

$$K_2 = 2m(R+l)\mathbf{v}$$

where $\mathbf{v}$ is the velocity of the masses. Making use of the principle of conservation of angular momentum and the principle of conservation of energy, we write

$$(J+2mR^2)\boldsymbol{\omega} = 2m\cdot(R+l)v, \tag{a}$$

$$\tfrac{1}{2}(J+2mR^2)\cdot\omega^2 = mv^2.$$

Solving the system for l we obtain

$$l = \sqrt{\frac{J+2mR^2}{2m}} - R. \tag{b}$$

5.4 Euler's equations

We shall begin this section by taking interest in the derivative of the angular momentum of a rigid body. The problem can be approached just formally by differentiating expression (4), Section 5.2, but it might be worthwhile seeing whether or not we have already considered a similar problem, in other words, if we could make us of some results discussed earlier.

Introducing the frame of reference $\Omega\xi\eta\zeta$, intimately associated with the moving body (Fig. 5.9), with angular velocity $\boldsymbol{\omega}$, we discover that, by making use of equation (6), Section 1.3, which transforms the vector derivative from a moving to a fixed frame of reference, the procedure of finding this derivative will be greatly facilitated, and what is more, one specific moving frame will be suggested to us, thus substantially simplifying the results.

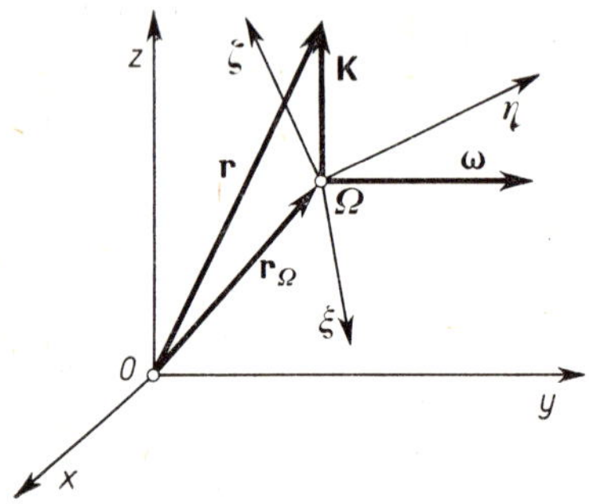

Fig. 5.9

Assume first that the origin of the moving frame of reference coincides with the centre of mass of the body considered. In this way, as we have noted earlier, the angular momentum will be the sum of the angular momentum in translatory motion with the velocity of the centre of mass and the angular momentum in circular motion. In the case of circular motion, this assumption is altogether unnecessary, since then $r_\Omega = 0$.

As we can readily see, the derivative of the first term on the right-hand side of (4), Section 5.2, is the moment of forces acting on the system after their parallel translation to the $\Omega\xi\eta\zeta$ reference origin. This term is of litlle interest to us, therefore we shall concern ourselves with the derivative of angular momentum in circular motion.

The components of this vector are given by equations (4), Section 5.3. Therefore

$$\mathbf{K} = K_\xi \mathbf{e}_\xi + K_\eta \mathbf{e}_\eta + K_\zeta \mathbf{e}_\zeta, \tag{1}$$

where $\mathbf{e}_i$ are unit vectors in the $\Omega\xi\eta\zeta$ frame of reference. Making use of the transformation formula (6), Section 1.3, we write

$$\frac{\mathrm{d}\mathbf{K}}{\mathrm{d}t} = \frac{\check{\mathrm{d}}\mathbf{K}}{\mathrm{d}t} + \boldsymbol{\omega}\times\mathbf{K}. \tag{2}$$

Considering the first of equations (2), Section 5.3, and equation (1) we have

$$\frac{\mathrm{d}\mathbf{K}}{\mathrm{d}t} = \mathbf{e}_\xi\left(\frac{\mathrm{d}K_\xi}{\mathrm{d}t} + K_\zeta\omega_\eta - K_\eta\omega_\zeta\right) + \\ + \mathbf{e}_\eta\left(\frac{\mathrm{d}K_\eta}{\mathrm{d}t} + K_\xi\omega_\zeta - K_\zeta\omega_\xi\right) + \mathbf{e}_\zeta\left(\frac{\mathrm{d}K_\zeta}{\mathrm{d}t} + K_\eta\omega_\xi - K_\xi\omega_\eta\right). \tag{3}$$

If the moment of the forces acting on the system about the moving reference origin be M, then according to the principle of angular momentum, it is equal to the derivative of angular momentum

$$\mathbf{M} = M_\xi\mathbf{e}_\xi + M_\eta\mathbf{e}_\eta + M_\zeta\mathbf{e}_\zeta. \tag{4}$$

Comparing (3) with (4) we can write

$$\begin{aligned} M_\xi &= \frac{\mathrm{d}K_\xi}{\mathrm{d}t} + K_\zeta\omega_\eta - K_\eta\omega_\zeta, \\ M_\eta &= \frac{\mathrm{d}K_\eta}{\mathrm{d}t} + K_\xi\omega_\zeta - K_\zeta\omega_\xi, \\ M_\zeta &= \frac{\mathrm{d}K_\zeta}{\mathrm{d}t} + K_\eta\omega_\xi - K_\xi\omega_\eta. \end{aligned} \tag{5}$$

Equations (5) are equations of circular motion.

If we take a moving frame of reference, so that the axes ξ, η, ζ be the principles axes of inertia, then

$$K_\xi = J_\xi\omega_\xi, \quad K_\eta = J_\eta\omega_\eta, \quad K_\zeta = J_\zeta\omega_\zeta, \tag{6}$$

and the equations of motion will be

$$\begin{aligned} M_\xi &= J_\xi\dot{\omega}_\xi + (J_\zeta - J_\eta)\omega_\zeta\omega_\eta, \\ M_\eta &= J_\eta\dot{\omega}_\eta + (J_\xi - J_\zeta)\omega_\xi\omega_\zeta, \\ M_\zeta &= J_\zeta\dot{\omega}_\zeta + (J_\eta - J_\xi)\omega_\eta\omega_\xi. \end{aligned} \tag{7}$$

Equations (7) are called *Euler's equations.*

Example 1. A homogeneous circular disc has been mounted on a shaft (Fig. 5.10) in such a way that its plane makes an angle α with a plane perpendicular to the shaft axis. Calculate the dynamic reactions at bearings A arn B, knowing that the mass of the disc is m and its radius R, and the angular velocity ω = const is given.

Adopting the frame of reference $\Omega\xi\eta\zeta$ in such a way that its axes are the principal (as well as the central in this case) axes of inertia we find the components of angular velocity ω_ξ, ω_η and ω_ζ.

$$\omega_\xi = \omega\cos\alpha, \quad \omega_\eta = \omega\sin\alpha, \quad \omega_\zeta = 0.$$

But

$$J_\xi = \frac{mR^2}{2}, \quad J_\eta = \frac{mR^2}{4}, \quad J_\zeta = \frac{mR^2}{4};$$

therefore, making use of Euler's equations, we have

$$M_\xi = 0, \quad M_\eta = 0, \quad M_\zeta = -\frac{mR^2}{4}\omega^2\sin\alpha\cos\alpha.$$

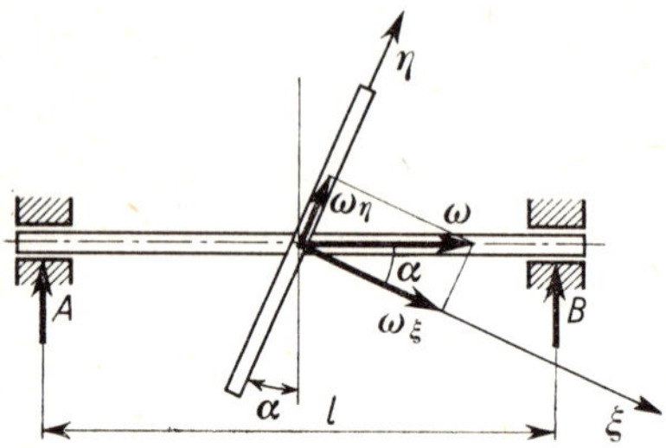

Fig. 5.10

The dynamic reactions, equal in absolute value but having opposite signs, are

$$A = -B = \frac{-M_\zeta}{l}.$$

Problem. Prove from (7) that the necessary and sufficient condition for dynamic reactions to vanish is to superimpose one of the axes of the $\Omega\xi\eta\zeta$ frame of reference on the axis of rotation.

If the axis of rotation is one of the principal axes of inertia, a rotating body is said to be *dynamically balanced.* If that axis is at the same time the central axis, then the body is also statically balanced, i.e., there will be no additional reactions resulting from the centre of mass being in the field of centrifugal forces. The proof of this statement is immediate from equations (8), Section 2.4.

Presented in Fig. 5.11 is the successive objective picture of (a) a dynamically, (b) statically unbalanced system, and (c) simultaneously statically and dynamically unbalanced system.

Example 2. To conclude this section we shall discuss the problem of dynamic reactions of bearings in the case of impact on a body in angular motion. Figure 5.12 shows the

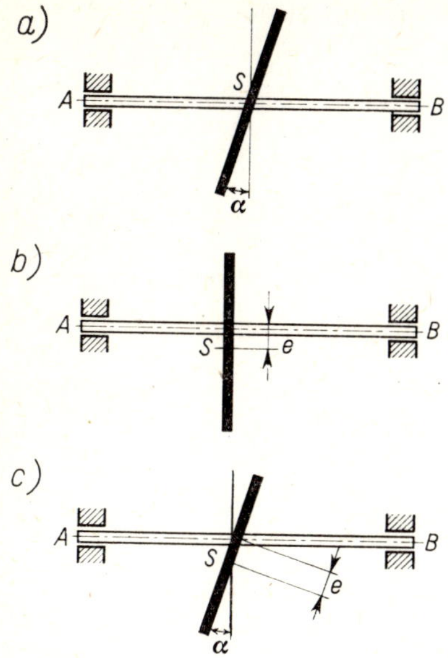

Fig. 5.11

body at the instant of collision. The body can rotate about point O. The distance of the centre of mass from the axis of rotation is l_S. The point at which impact occurs is l_u distant from point O. Assume that the axis of rotation is the principal axis of inertia and that the impact conditions are such that the impact force P has the direction of the velocity of the striking point (the horizontal line in our drawing).

Making use of one of equations (7) we write

$$J\ddot{\varphi} = l_u P, \tag{a}$$

where J is the moment of the body about the axis of rotation. Denoting the reaction at the bearings by the letter R and considering the principle of momentum, we obtain

$$P - R = ml_S\ddot{\varphi}, \tag{b}$$

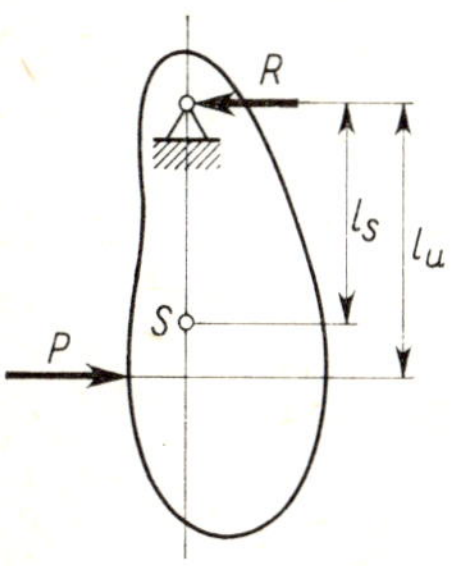

Fig. 5.12

where m is the mass of the body. From (a) and (b) it follows that

$$P-R = ml_S\frac{Pl_u}{J}. \tag{c}$$

Finding from (c) the reaction R, we write

$$R = P\left(1-m\frac{l_S l_u}{J}\right). \tag{d}$$

Note that we can choose l_u so that the dynamic reaction R equals zero:

$$R = 0, \quad \text{when} \quad 1-m\frac{l_S l_u}{J} = 0. \tag{e}$$

Equation (e) will be satisfied when

$$l_u = \frac{J}{ml_S}.$$

The point of the body thus determined is called the *centre of impact*.

Note further that examining the motion of the body in the vicinity of the position of equilibrium (Fig. 5.13), the following equation of motion is valid:

$$J\ddot{\varphi} = -l_S Q\sin\varphi. \tag{g}$$

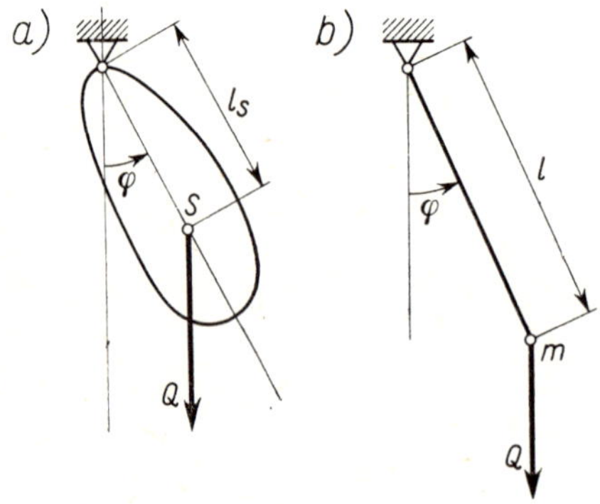

Fig. 5.13

For small angles φ, $\sin\varphi \approx \varphi$ can be assumed, therefore

$$\ddot{\varphi}+\frac{l_S Q}{J}\varphi = 0. \tag{h}$$

The derived equation is the equation for small vibrations of physical pendulum.

The solution of equation (h) for initial conditions $\varphi(0)$ and $\dot{\varphi}(0)$ is as follows:

$$\varphi(t) = \varphi(0)\cos\sqrt{\frac{l_S Q}{J}}\,t+\frac{\dot{\varphi}(0)}{\sqrt{\frac{l_S Q}{J}}}\sin\sqrt{\frac{l_S Q}{J}}\,t. \tag{i}$$

Obviously, the frequency of oscillation for such a pendulum is

$$k = \sqrt{\frac{l_S Q}{J}}, \tag{j}$$

and the period of oscillation

$$T = 2\pi\frac{1}{k} = 2\pi\sqrt{\frac{J}{l_S Q}}. \tag{k}$$

On the other hand, examining the vibrations of a mathematical pendulum, that is, one in which the entire mass is concentrated at the end (Fig. 5.13b), we write

$$l^2 m\ddot{\varphi} = -lQ\sin\varphi. \tag{l}$$

For small angles φ, the equation of motion is

$$\ddot{\varphi} + \frac{Q}{lm}\varphi = 0. \tag{m}$$

Let us assume that a physical pendulum has the same mass as the mathematical pendulum just considered and that the periods of motion of the two systems are equal. Then

$$\frac{J}{l_S Q} = \frac{lm}{Q}. \tag{n}$$

If the masses are equal, then the weights Q are also equal, therefore

$$l = \frac{J}{ml_S}. \tag{o}$$

The assumption of equal masses and periods of oscillation for the two systems leads to the conclusion that the length of the mathematical pendulum is equal to the distance between the centre of impact and the axis of rotation. The above observation can therefore be readily utilized in experimental determinations of the position of the centre of impact. It will suffice for this purpose to measure the period of oscillation of a physical pendulum and find the length of a mathematical pendulum giving the same period of oscillation

$$T = 2\pi\sqrt{\frac{l_u}{g}}; \tag{p}$$

hence

$$l_u = \frac{T^2 g}{4\pi^2}. \tag{r}$$

5.5 The kinetic energy of rigid bodies in general motion

As we know the kinetic energy is half the sum of the products of masses and their square velocities. In the case of a continuous medium, these sums change to integrals and therefore we can write

$$E = \tfrac{1}{2}\int_V \mathbf{v}^2 \mu \mathrm{d}V. \tag{1}$$

Velocity in general motion is expressed by

$$\mathbf{v} = \mathbf{v}_\Omega + \boldsymbol{\omega} \times \boldsymbol{\rho}. \tag{2}$$

Therefore

$$E = \frac{1}{2} \int_V \mathbf{v}_\Omega^2 \mu \mathrm{d}V + \int_V \mathbf{v}_\Omega \cdot (\boldsymbol{\omega} \times \boldsymbol{\rho}) \mu \mathrm{d}V + \frac{1}{2} \int_V (\boldsymbol{\omega} \times \boldsymbol{\rho})^2 \mu \mathrm{d}V. \tag{3}$$

But $\mathbf{v}_\Omega(\boldsymbol{\omega} \times \boldsymbol{\rho}) = (\mathbf{v}_\Omega \times \boldsymbol{\omega}) \cdot \boldsymbol{\rho}$, therefore

$$\int_V \mathbf{v}_\Omega \cdot (\boldsymbol{\omega} \times \boldsymbol{\rho}) \mu \mathrm{d}V = (\mathbf{v}_\Omega \times \boldsymbol{\omega}) \cdot \int_V \boldsymbol{\rho} \mu \mathrm{d}V = (\mathbf{v}_\Omega \times \boldsymbol{\omega}) \cdot \boldsymbol{\rho}_S m,$$

where $\boldsymbol{\rho}_S$ is a position vector of the centre of mass of a body relative to a moving reference frame.

If the reference origin $(\Omega\xi\eta\zeta)$ coincides with the centre of mass S, then the second term of the sum on the right-hand side of (3) is equal to zero, since $\boldsymbol{\rho}_S = 0$. In this case, kinetic energy is the sum of energies in translatory and circular motion

$$E = \frac{1}{2} m \mathbf{v}_S^2 + \frac{1}{2} \int_V (\boldsymbol{\omega} \times \boldsymbol{\rho})^2 \mu \mathrm{d}V. \tag{4}$$

Expressing the vectors $\boldsymbol{\omega}$ and $\boldsymbol{\rho}$ by components in the $\Omega\xi\eta\zeta$ frame of reference (equations (2), Section 5.3) and putting next the vectors thus determined under the integral in equation (4), we write

$$\begin{aligned} \frac{1}{2} \int_V (\boldsymbol{\omega} \times \boldsymbol{\rho})^2 \mu \mathrm{d}V = {} & \frac{1}{2}\omega_\xi^2 \int_V (\eta^2+\zeta^2) \mu \mathrm{d}V + \frac{1}{2}\omega_\eta^2 \int_V (\xi^2+\zeta^2) \mu \mathrm{d}V + \\ & + \frac{1}{2}\omega_\zeta^2 \int_V (\xi^2+\eta^2) \mu \mathrm{d}V - \omega_\xi \omega_\eta \int_V \xi\eta\mu \mathrm{d}V - \\ & - \omega_\zeta \omega_\xi \int_V \zeta\xi\mu \mathrm{d}V - \omega_\eta \omega_\zeta \int_V \eta\zeta\mu \mathrm{d}V. \end{aligned}$$

Considering that the integrals connotate successively moments of inertia and deviation about a moving reference frame, we obtain

$$\begin{aligned} E = {} & \frac{1}{2} m v_S^2 + \frac{1}{2} J_\xi \omega_\xi^2 + \frac{1}{2} J_\eta \omega_\eta^2 + \frac{1}{2} J_\zeta \omega_\zeta^2 - \\ & - J_{\xi\eta} \omega_\xi \omega_\eta - J_{\eta\zeta} \omega_\eta \omega_\zeta - J_{\zeta\xi} \omega_\zeta \omega_\xi. \end{aligned} \tag{5}$$

In index notation

$$E = \frac{1}{2} m v_S^2 + \frac{1}{2} J^{ij} \omega_i \omega_j \tag{5a}$$

or in matrix notation

$$E = \frac{1}{2} m v_S^2 + \frac{1}{2} \boldsymbol{\omega}^T \cdot \mathbf{J} \cdot \boldsymbol{\omega}. \tag{5b}$$

If we assume that the axes ξ, η, ζ are not only central but also principal axes, then the expression for the kinetic energy of a rigid body in general motion will be considerably simplified since the last three terms on the right-hand side of (5) are all equal to zero by virtue of the moments of deviation $J_{\xi\eta}$, $J_{\eta\zeta}$, and $J_{\zeta\xi}$ being identically equal to zero.

Example 1. Find the point at which a spherical nodule detaches itself from stationary sphere after rolling on it without sliding (Fig. 5.14).

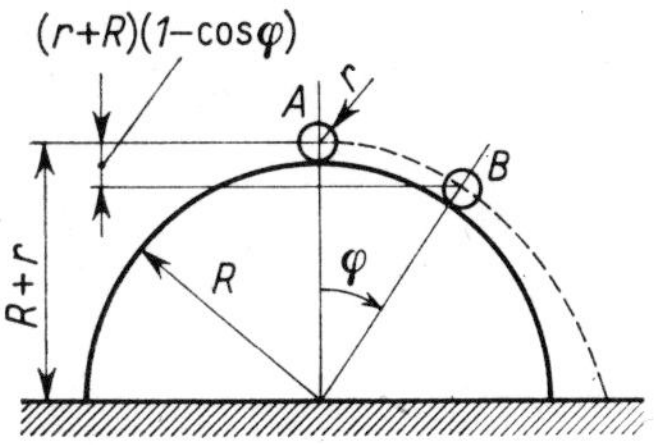

Fig. 5.14

In the initial instant, the nodule is at point A. Assume that at this point both the kinetic and the potential energy are equal to zero. At a certain instant the nodule begins rolling. Its angular velocity vector $\boldsymbol{\omega}$ is directed at right angles to the plane of the drawing:

$$\omega = \frac{v}{r}.$$

At point B it achieves a certain kinetic energy E_B and a certain potential energy V_B. If at that particular point the nodule separates from the base, this means that the component of the gravity force $mg\cos\varphi$ is equilibrated by the normal force of inertia $mv^2/(R+r)$. Therefore,

$$mg\cos\varphi = m\frac{v^2}{R+r}. \tag{a}$$

From the law of conservation of mechanical energy we know that

$$E_A + V_A = E_B + V_B. \tag{b}$$

Since

$$V_B = -mg(r+R)(1-\cos\varphi), \tag{c}$$

$$E_B = \tfrac{1}{2}mv^2 + \tfrac{1}{2}J\omega^2 = \tfrac{1}{2}mv^2 + \tfrac{1}{5}mv^2 = \tfrac{7}{10}mv^2, \tag{d}$$

by virtue of (b), (c) and (d) we can write

$$\tfrac{7}{10}v^2 - g(R+r)(1-\cos\varphi) = 0. \tag{e}$$

It follows from (a) and (e) that

$$\tfrac{7}{10}\cos\varphi - (1-\cos\varphi) = 0; \tag{f}$$

hence

$$\cos\varphi = \tfrac{10}{17}. \tag{g}$$

In this way we find the angle φ. Note that it does not depend on the radius and mass of the nodule nor on the size of the sphere on which the nodule is rolling.

Example 2. A sphere of radius r is rolling without sliding on a horizontal plane surface with a velocity v_0 and runs into a sill of heigh h ($h < r$) (Fig. 5.15).

Calculate the velocity of the sphere when it reaches the top of the sill, assuming that there is no sliding when touching the sill. Discuss the possibilities of solving the problem.

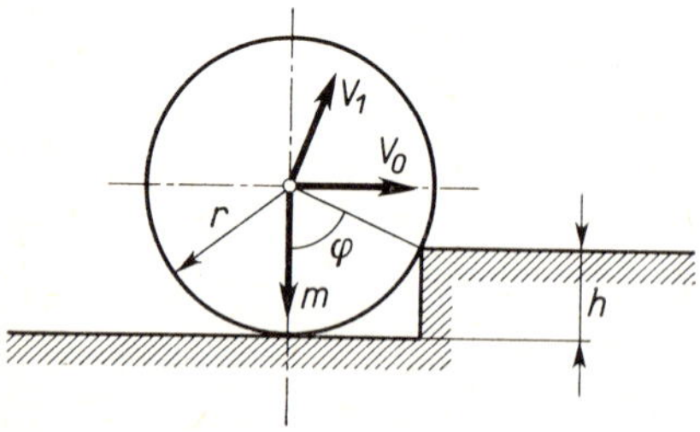

Fig. 5.15

From the principle of angular momentum we have

$$mv_0(r-h)+\tfrac{2}{5}mr^2\frac{v_0}{r} = mv_1r+\tfrac{2}{5}mr^2\frac{v_1}{r}; \qquad \text{(a)}$$

hence

$$v_1 = v_0\left(1-\frac{5}{7}\frac{h}{r}\right), \qquad \text{(b)}$$

which is the velocity with which the sphere travelled on running into the sill but before it detached itself from the surface on which it rolled.

The velocity of the centre of the sphere after getting across the sill can be calculated from the law of conservation of energy

$$\tfrac{7}{10}mv_1^2 = mgh+\tfrac{7}{10}mv_2^2; \qquad \text{(c)}$$

hence

$$v_2 = \sqrt{v_1^2-\tfrac{10}{7}gh}. \qquad \text{(d)}$$

The value under the root should be positive, therefore the limitation on v_0 is such that

$$v_0 > \frac{\sqrt{\tfrac{10}{7}hg}}{1-\dfrac{5}{7}\dfrac{h}{r}}. \qquad \text{(e)}$$

Another limitation is the condition that while the sphere is rolling onto the sill, the two should be in firm contact. This will be satisfied when in the lowest position of the sphere, the component of the gravity force in the direction of the point of contact $mg\cos\varphi$ is greater than that of the normal force of inertia mv_1^2/r. Since $\cos\varphi = (r-h)/r$, the condition of inseparability is expressed by

$$(r-h)g > v_0^2\left(1-\frac{5}{7}\frac{h}{r}\right)^2;$$

$$v_0 < \frac{\sqrt{g(r-h)}}{1-\frac{5}{7}\frac{h}{r}}. \tag{f}$$

From conditions (e) and (f) the following should occur:

$$h < \tfrac{7}{17}r.$$

Example 3. The sphere from the previous example was set in motion on a horizontal plane surface, however, in this example, it moved in such a way that it slides in the initial instant with velocity v, with no rotation. Friction existed between the sphere and the horizontal surface is defined by the coefficient of friction f. Find the distance s covered by the sphere before it started rolling without sliding. What was its final velocity? (friction T is given by the equation $T \overset{\mathrm{df}}{=} fN$, where N is thrust on bodies subject to friction; f depends on the type of surface of bodies subject to friction).

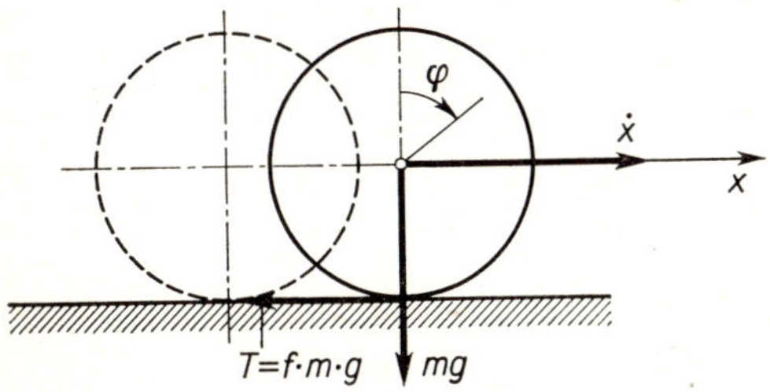

Fig. 5.16

By the law of variation of linear and angular momentum (Fig. 5.16) referred to the considered case we have

$$\tfrac{2}{5}mr\ddot{\varphi} = rfmg, \tag{a}$$

$$m\ddot{x} = -fmg. \tag{b}$$

The integration of equations (a) and (b) gives

$$\dot{\varphi} = \tfrac{5}{2}f\frac{g}{r}t, \tag{c}$$

$$\dot{x} = -fgt+v. \tag{d}$$

From the condition of contact,

$$\dot{x} = r\dot{\varphi}$$

which gives

$$-fgt+v = \tfrac{5}{2}fgt;$$

hence

$$t = \frac{2v}{7fg} \tag{e}$$

where t denotes the instant at which the sphere will start rolling without sliding. Consequently, the velocity with which the sphere will be rolling without sliding is

$$v_0 = -fg\frac{2v}{7fg}+v = \tfrac{5}{7}v. \tag{f}$$

Integrating (d) we obtain

$$x = -\frac{fgt^2}{2}+vt. \tag{g}$$

Substituting (e) into (g) we find the looked for value of distance s which the sphere will cover before it begins rolling without sliding:

$$s = \frac{12v^2}{49fg}. \tag{h}$$

The considered problem may also be solved on the basis of the law of conservation of energy. The reader is suggested here to solve the problem himself along this line but he is warned at the same time that the calculation of the work of friction on the traverse s involves a certain "danger"—so it is an interesting case.

CHAPTER 6

Dynamics in relative motion

6.1 Differential equation of the motion of a particle in a non-inertial system

The equations of motion hitherto considered referred to the case of motion in inertial systems. We shall now explore how Newton's basic equation of motion changes on transition to a non-inertial system:

$$m\mathbf{a} = \mathbf{P}. \tag{1}$$

We remember that acceleration in relative motion is the sum of three terms, the first being acceleration of transportation, the second, Coriolis acceleration, and the third, relative acceleration:

$$\mathbf{a} = \mathbf{a}_{tr} + \mathbf{a}_{Cor} + \mathbf{a}_r. \tag{2}$$

Let us substitude (2) into Newton's equation of motion (1):

$$m\mathbf{a}_{tr} + m\mathbf{a}_{Cor} + m\mathbf{a}_r = \mathbf{P}. \tag{3}$$

Our interest is focussed on defining the vector $\boldsymbol{\rho}(t)$. It will give the path along which a particle moves in a moving frame of reference. Using the relations

$$\mathbf{a}_{Cor} = 2\boldsymbol{\omega} \times \frac{\check{d}\boldsymbol{\rho}}{dt}, \quad \mathbf{a}_r = \frac{\check{d}^2\boldsymbol{\rho}}{dt^2}, \tag{4}$$

we rewrite equation (3)

$$m\left(\frac{\check{d}^2\boldsymbol{\rho}}{dt^2} + 2\times\frac{\check{d}\boldsymbol{\rho}}{dt} + a_{tr}\right) = \mathbf{P}. \tag{5}$$

Equation (5) is the differential equation of the motion of a particle in a non-inertial system. As we can see, besides force $\mathbf{P}$ there also occur in this case the force of transportation $m\mathbf{a}_{tr}$ and Coriolis force $2m\boldsymbol{\omega} \times \check{d}\boldsymbol{\rho}/dt$.

The motion of the Earth serves as an example of the motion of a body in a non-inertial system.

Example 1. A mathematical pendulum of mass m and length ϱ was attached at the end of a rod of length r (Fig. 6.1). Find the equation of motion of the pendulum in a moving frame of reference $\Omega\xi\eta$ if it is known that the rod $O\Omega$ moves such that

$$\alpha = \alpha(t).$$

Starting from the basic equation we have

$$m\mathbf{a}_r = \mathbf{P} - m\mathbf{a}_{tr} - m\mathbf{a}_{Cor}. \tag{a}$$

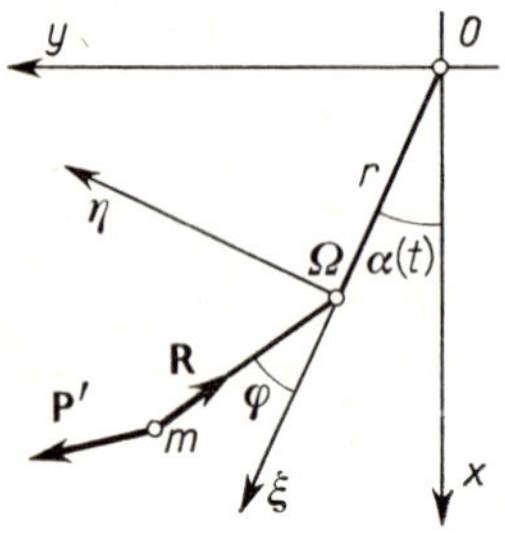

Fig. 6.1

We proceed to find its individual terms. Let us begin with the acceleration of transportation $\mathbf{a}_{tr}$

$$\mathbf{a}_{tr} = \mathbf{a}_\Omega + \boldsymbol{\epsilon}\times\boldsymbol{\rho} + \boldsymbol{\omega}\times(\boldsymbol{\omega}\times\boldsymbol{\rho}), \tag{b}$$

$\mathbf{a}_\Omega$ is the acceleration of point Ω and it is

$$\mathbf{a}_\Omega = -r\dot{\alpha}^2\mathbf{e}_\xi + r\ddot{\alpha}\mathbf{e}_\eta, \tag{c}$$

since in the considered case

$$\omega = \dot{\alpha} \quad \text{and} \quad \varepsilon = \ddot{\alpha}.$$

Going further,

$$\boldsymbol{\epsilon}\times\boldsymbol{\rho} = \begin{vmatrix} \mathbf{e}_\xi & \mathbf{e}_\eta & \mathbf{e}_\zeta \\ 0 & 0 & \ddot{\alpha} \\ \varrho\cos\varphi & \varrho\sin\varphi & 0 \end{vmatrix} = -\mathbf{e}_\xi\ddot{\alpha}\varrho\sin\varphi + \mathbf{e}_\eta\ddot{\alpha}\varrho\cos\varphi.$$

Finally,

$$\boldsymbol{\omega}\times(\boldsymbol{\omega}\times\boldsymbol{\rho}) = \boldsymbol{\omega}\times\begin{vmatrix} \mathbf{e}_\xi & \mathbf{e}_\eta & \mathbf{e}_\zeta \\ 0 & 0 & \dot{\alpha} \\ \varrho\cos\varphi & \varrho\sin\varphi & 0 \end{vmatrix} = -\mathbf{e}_\xi\dot{\alpha}^2\varrho\cos\varphi - \mathbf{e}_\eta\dot{\alpha}^2\varrho\sin\varphi.$$

Therefore,

$$\mathbf{a}_{tr} = \mathbf{e}_\xi(-r\dot{\alpha}^2 - \ddot{\alpha}\varrho\sin\varphi - \dot{\alpha}^2\varrho\cos\varphi) + \mathbf{e}_\eta(r\ddot{\alpha} + \ddot{\alpha}\varrho\cos\varphi - \dot{\alpha}^2\varrho\sin\varphi). \tag{d}$$

We now determine Coriolis acceleration derived from

$$\mathbf{a}_{\mathrm{Cor}} = 2\begin{vmatrix} \mathbf{e}_\xi & \mathbf{e}_\eta & \mathbf{e}_\zeta \\ 0 & 0 & \dot{\alpha} \\ v_\xi & v_\eta & 0 \end{vmatrix} = -2\mathbf{e}_\xi \dot{\alpha} v_\eta + 2\mathbf{e}_\eta \dot{\alpha} v_\xi = -2\dot{\alpha}\dot{\varphi}\varrho(\mathbf{e}_\xi \cos\varphi + \mathbf{e}_\eta \sin\varphi). \tag{e}$$

Relative acceleration

$$\mathbf{a}_r = a_{r\xi}\mathbf{e}_\xi + a_{r\eta}\mathbf{e}_\eta,$$

where

$$a_{r\xi} = \frac{\mathrm{d}^2\xi}{\mathrm{d}t^2} = \frac{\mathrm{d}^2}{\mathrm{d}t^2}(\varrho\cos\varphi) = -\varrho\dot{\varphi}^2\cos\varphi - \varrho\ddot{\varphi}\sin\varphi,$$

$$a_{r\eta} = \frac{\mathrm{d}^2\eta}{\mathrm{d}t^2} = \frac{\mathrm{d}^2}{\mathrm{d}t^2}(\varrho\sin\varphi) = -\varrho\dot{\varphi}^2\sin\varphi + \varrho\ddot{\varphi}\cos\varphi.$$

Now we find the forces acting on the considered mass: We should appreciate here besides an external force, the reaction of constraints having the direction of the pendulum axis R. Therefore,

$$\mathbf{P} = \mathbf{e}_\xi(P'_\xi - R\cos\varphi) + \mathbf{e}_\eta(P'_\eta - R\sin\varphi). \tag{f}$$

Considering in (a) the above results, we obtain a vector equation, in which two components occur: one with the direction of the ξ-axis and the other with the direction of the η-axis. Comparing the components on each side of the equation, we have

$$-m\varrho\dot{\varphi}^2\cos\varphi - m\varrho\ddot{\varphi}\sin\varphi$$
$$= P'_\xi - R\cos\varphi + m(r\dot{\alpha}^2 + \ddot{\alpha}\varrho\sin\varphi + \dot{\alpha}^2\varrho\cos\varphi) + 2m\dot{\alpha}\dot{\varphi}\varrho\cos\varphi, \tag{g}$$

$$-m\varrho\dot{\varphi}^2\sin\varphi + m\varrho\ddot{\varphi}\cos\varphi$$
$$= P'_\eta - R\sin\varphi - m(r\ddot{\alpha} + \ddot{\alpha}\varrho\cos\varphi - \dot{\alpha}^2\varrho\sin\varphi) + 2m\dot{\alpha}\dot{\varphi}\varrho\sin\varphi. \tag{h}$$

Multiplying equation (g) by $\sin\varphi$ and (h) by $\cos\varphi$ and then subtracting, we obtain

$$-m\varrho\ddot{\varphi} = P'_\xi\sin\varphi - P'_\eta\cos\varphi + m(r\dot{\alpha}^2\sin\varphi + r\ddot{\alpha}\cos\varphi + \ddot{\alpha}\varrho). \tag{i}$$

In this way we have eliminated the unknown reaction R from the system. Equation (i) is the sought equation of motion of a mathematical pendulum in a moving frame of reference.

6.2 The dynamics of rigid bodies in relative motion

We shall refer in this section to the results obtained in Section 1.3. Figure 1.24 shows a rigid body associated with a frame of reference $\Omega_1\xi_1\eta_1\zeta_1$ in relative motion to another background frame of reference $\Omega_2\xi_2\eta_2\zeta_2$ with angular velocity $\boldsymbol{\omega}_r$. At the same time, the frame $\Omega_2\xi_2\eta_2\zeta_2$ is subject to transportation with angular velocity $\boldsymbol{\omega}_{tr}$ against a fixed frame of reference $Oxyz$.

The momentum problem is reduced to the problem of a point (centre of mass of a body), discussed in the foregoing section, and therefore we shall now consider the angular momentum problem.

The angular momentum of a rigid body in general motion is described by equation (4a), Section 5.2, while assuming that the centre of the body coincides with the point (Fig. 5.7):

$$\mathbf{K} = \mathbf{e}_i \varepsilon_{ikj} x_k v_j m + J_{ik} \omega_k \mathbf{e}'_i ,^* \tag{1}$$

where

ε_{ikj} permutation symbol (see the Introduction, equation (19)),
x_k coordinate of point Ω_1 in the frame of reference $Oxyz$,
v_j components of absolute velocity vector of point Ω_1 in the frame of reference $Oxyz$,
m mass of the body,
J_{ik} elements of the matrix of the moment of inertia of the body in the $\Omega_1 \xi_1 \eta_1 \zeta_1$ system.
ω_k components of absolute angular velocity in $\Omega_1 \xi_1 \eta_1 \zeta_1$ system,
$\mathbf{e}_i$ and $\mathbf{e}'_i$ are unit vectors in $Oxyz$ and $\Omega_1 \xi_1 \eta_1 \zeta_1$, respectively.

We find the nth component in the moving frame of reference $\Omega_1 \xi_1 \eta_1 \zeta_1$ or in the fixed frame $Oxyz$ by scalar multiplication of equation (1) by corresponding unit vectors. Multiplying (1) by $\mathbf{e}'_n$ we obtain the component K_n of angular momentum in the frame of reference $\Omega_1 \xi_1 \eta_1 \zeta_1$

$$K_n = \mathbf{e}'_n \mathbf{e}_i \varepsilon_{ikj} x_k v_j m + J_{nk} \omega_k . \tag{1a}$$

Since

$$\omega_k = \omega_{(\mathrm{tr})k} + \omega_{(r)k}$$

(the parenthetised subscripts (tr) and (r) denote respectively transportation and relative magnitudes), we have

$$K_n = \mathbf{e}'_n \mathbf{e}_i \varepsilon_{ikj} x_k v_j m + J_{nk}(\omega_{(\mathrm{tr})k} + \omega_{(r)k}) . \tag{2}$$

The differential equation of motion is obtained by differentiating equation (1) or (1a) and by making use of the law of variation of angular momentum.

For this purpose, equation (6), Section 1.3, is used twice, once for transition from the reference frame $\Omega_1 \xi_1 \eta_1 \zeta_1$ to $\Omega_2 \xi_2 \eta_2 \zeta_2$ and for the second time for transition from the frame $\Omega_2 \xi_2 \eta_2 \zeta_2$ to $Oxyz$.

We resolve the vector $\mathbf{K}$ into two components $\mathbf{K}_{(t)}$ and $\mathbf{K}_{(c)}$

$$\mathbf{K} = \mathbf{K}_{(t)} + \mathbf{K}_{(c)} , \tag{3}$$

---

* In this section summation will be made with respect to indices written in one level.

where $K_{(t)}$ is an angular momentum in translatory motion and $K_{(c)}$ in circular motion;

$$\begin{aligned} \mathbf{K}_{(t)} &= \mathbf{e}_i \varepsilon_{ikj} x_k v_j m, \\ \mathbf{K}_{(c)} &= J_{ik}\omega_k \mathbf{e}'_i. \end{aligned} \tag{3a}$$

We bear in mind that the point Ω_1 is the centre of mass of the body.

In the first step we go over from the frame of reference $\Omega_1 \xi_1 \eta_1 \zeta_1$ to the frame $\Omega_2 \xi_2 \eta_2 \zeta_2$:

$$\frac{\check{\mathrm{d}}\mathbf{K}_{(c)}}{\mathrm{d}t} = \boldsymbol{\omega}_{(r)} \times \mathbf{K}_{(c)} + \frac{\check{\check{\mathrm{d}}}\mathbf{K}_{(c)}}{\mathrm{d}t} \tag{4}$$

where $\check{\mathrm{d}}\mathbf{K}_{(c)}/\mathrm{d}t$ is the derivative of angular momentum $\mathbf{K}_{(c)}$ in the frame of reference $\Omega_2 \xi_2 \eta_2 \zeta_2$, and $\check{\check{\mathrm{d}}}\mathbf{K}_{(c)}/\mathrm{d}t$ is the derivative in the frame $\Omega_1 \xi_1 \eta_1 \zeta_1$. In the next step, we go over from the frame of reference $\Omega_2 \xi_2 \eta_2 \zeta_2$ to the frame $Oxyz$

$$\frac{\mathrm{d}\mathbf{K}_{(c)}}{\mathrm{d}t} = \boldsymbol{\omega}_{\mathrm{tr}} \times \mathbf{K}_{(c)} + \frac{\check{\mathrm{d}}\mathbf{K}_{(c)}}{\mathrm{d}t}. \tag{5}$$

Substituting (4) into (5) we have

$$\frac{\mathrm{d}\mathbf{K}_{(c)}}{\mathrm{d}t} = \boldsymbol{\omega}_{\mathrm{tr}} \times \mathbf{K}_{(c)} + \boldsymbol{\omega}_r \times \mathbf{K}_{(c)} + \frac{\check{\check{\mathrm{d}}}\mathbf{K}_{(c)}}{\mathrm{d}t},$$

where $\mathrm{d}\mathbf{K}_{(c)}/\mathrm{d}t$ is the derivative of angular momentum in the $Oxyz$ frame of reference.

Since $\mathrm{d}\mathbf{K}/\mathrm{d}t = \mathbf{M}$, we have

$$\frac{\mathrm{d}\mathbf{K}_{(t)}}{\mathrm{d}t} + \boldsymbol{\omega}_{\mathrm{tr}} \times \mathbf{K}_{(c)} + \boldsymbol{\omega}_{(r)} \times \mathbf{K}_{(c)} + \frac{\check{\check{\mathrm{d}}}\mathbf{K}_{(c)}}{\mathrm{d}t} = \mathbf{M}, \tag{6}$$

where $\mathbf{K}_{(t)}$ and $\mathbf{K}_{(c)}$ are described by equations (3a).

The absolute angular velocity $\boldsymbol{\omega}$ is the sum

$$\boldsymbol{\omega} = \mathbf{e}'_i \omega_i = \boldsymbol{\omega}_{\mathrm{tr}} + \boldsymbol{\omega}_r = \mathbf{e}'_i(\omega_{(\mathrm{tr})i} + \omega_{(r)i});$$

hence

$$\mathbf{K}_{(c)} = \mathbf{e}'_i J_{im}(\omega_{(\mathrm{tr})m} + \omega_{(r)m}). \tag{7}$$

On the other hand,

$$\frac{\check{\check{\mathrm{d}}}\mathbf{K}_{(c)}}{\mathrm{d}t} = \mathbf{e}'_i J_{im} \varepsilon_m, \tag{8}$$

but according to equation (22), Section 1.3,

$$\boldsymbol{\epsilon} = \mathbf{e}'_m \varepsilon_m = \boldsymbol{\epsilon}_{\mathrm{tr}} + \boldsymbol{\epsilon}_{(r)} + \boldsymbol{\omega}_{\mathrm{tr}} \times \boldsymbol{\omega}_r = \mathbf{e}'_m(\varepsilon_{(\mathrm{tr})m} + \varepsilon_{(r)m} + \varepsilon_{mjk}\omega_{(\mathrm{tr})j}\omega_{(r)k}).$$

Therefore

$$\frac{\check{\mathrm{d}}\mathbf{K}_{(c)}}{\mathrm{d}t} = \mathbf{e}'_i J_{im}(\varepsilon_{(\mathrm{tr})m} + \varepsilon_{(r)m} + \varepsilon_{mjk}\omega_{(\mathrm{tr})j} \cdot \omega_{(r)k}). \tag{9}$$

Considering the results (3a), (7), (8) and (9) in equation (6) and also that $\mathbf{M} = \mathbf{e}'_i M_i$, we obtain

$$\varepsilon_{ijk} x_j a_{km} + \varepsilon_{ijk}(\omega_{(\mathrm{tr})j} + \omega_{(r)j}) J_{km}(\omega_{(\mathrm{tr})m} + \omega_{(r)m}) + \\ + J_{im}(\varepsilon_{(\mathrm{tr})m} + \varepsilon_{(r)m} + \varepsilon_{mjk}\, \omega_{(\mathrm{tr})j}\omega_{(r)k}) = M_i, \tag{10}$$

where a_k is the acceleration of point Ω and at the same time of the centre of mass of the system.

Now we shall consider circular motion. It will be convenient to assume here all three frames of reference: $Oxyz$, $\Omega_1 \xi_1 \eta_1 \zeta_1$, and $\Omega_2 \xi_2 \eta_2 \zeta_2$ (Fig. 6.2), in such a way that their origins should be at a common point O.

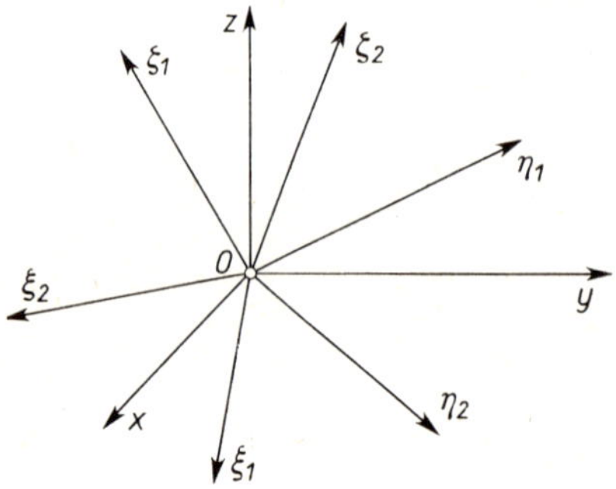

Fig. 6.2

The point O is assumed to be fixed $r_{\Omega_1} = r_{\Omega_2} = O$. Here the point O need not necessarily coincide with the centre of mass of the body (which follows from the analysis of equation (4), Section 5.2). The equation of motion can be written in the following form:

$$\varepsilon_{ijk}(\omega_{(\mathrm{tr})j} + \omega_{(r)j}) J_{km}(\omega_{(\mathrm{tr})m} + \omega_{(r)m}) + \\ + J_{im}(\varepsilon_{(\mathrm{tr})m} + \varepsilon_{(r)m} + \varepsilon_{mjk}\omega_{(\mathrm{tr})j}\omega_{(r)k}) = M_i. \tag{11}$$

Note that if the frame of reference $\Omega_2\xi_2\eta_2\zeta_2$ be made fixed, i.e., if it be assumed that $\omega_{(tr)} = 0$, the system of equations (11) passes into equations (5), Section 5.4.

Example 1. A homogeneous circular disc was mounted on a shaft (Fig. 6.3) in such a way that its surface makes an angle α with the plane perpendicular to the shaft axis. The shaft with a horizontal axis rotates with angular velocity ω_w; it is housed in bearings in a casing which is in angular motion about the vertical axis with angular velocity ω_u. Find the moment acting on the disc if it is known that both ω_w and ω_u are constant. The present problem is a generalization of Example 1, Section 5.4.

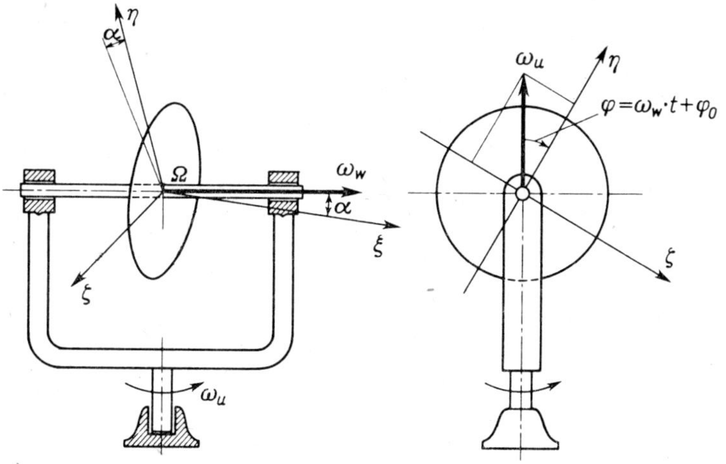

Fig. 6.3

The angular velocity ω_w of the disc about a fixed casing we regard as relative velocity whereas we regard the angular velocity of the casing ω_u as velocity of transportation.

We shall first find the components of vectors ω_w and ω_u in the frame of reference $\Omega\xi\eta\zeta$ rigidly bound to the disc (Fig. 6.3). Looking at the drawing we write down:

$$\begin{aligned}
\omega_{(w)1} &\equiv \omega_{(w)\xi} = \omega_{(w)}\cos\alpha, \\
\omega_{(w)2} &\equiv \omega_{(w)\eta} = \omega_{(w)}\sin\alpha, \\
\omega_{(w)3} &\equiv \omega_{(w)\zeta} = 0;
\end{aligned} \tag{a}$$

$$\begin{aligned}
\omega_{(u)1} &\equiv \omega_{(u)\xi} = -\omega_{(u)}\sin\alpha\cos(\omega_{(w)}t+\varphi_0), \\
\omega_{(u)2} &\equiv \omega_{(u)\eta} = \omega_{(u)}\cos\alpha\cos(\omega_{(w)}t+\varphi_0), \\
\omega_{(u)3} &\equiv \omega_{(u)\zeta} = -\omega_{(u)}\sin(\omega_{(u)}t+\varphi_0),
\end{aligned} \tag{b}$$

where φ_0 is the angle giving the initial position of the shaft. The matrix of the moment of inertia can be written as follows:

$$[J_{ik}] = \begin{bmatrix} \frac{1}{2}mR^2 & 0 & 0 \\ 0 & \frac{1}{4}mR^2 & 0 \\ 0 & 0 & \frac{1}{4}mR^2 \end{bmatrix}. \tag{c}$$

Substituting (a), (b), and (c) into equations (11), we obtain

$$M_1 = M_\xi = \frac{1}{2} mR^2 \omega_u \omega_w \sin(\omega_w t + \varphi_0) \sin\alpha,$$

$$M_2 \equiv M_\eta = \frac{1}{4} mR^2 (\omega_u^2 \sin\alpha \cos(\omega_w t + \varphi_0) - 2\omega_u \omega_w \cos\alpha) \times \sin(\omega_w t + \varphi_0).$$

$$M_3 = M_\zeta = -\frac{1}{4} mR^2 (2\omega_u \omega_w \cos^2\alpha \cos(\omega_w t + \varphi_0) +$$

$$+ \omega_u^2 \sin\alpha \cos\alpha - \omega_u^2 \sin\alpha \cos\alpha \cos^2(\omega_w t + \varphi_0)).$$

Note that the moments do not vanish when $\alpha = 0$ (the disc plane at right angles to the shaft axis), then we have

$$\begin{aligned} M_1 &= 0, \\ M_2 &= -\tfrac{1}{2} mR^2 \omega_u \omega_w \sin(\omega_w t + \varphi_0), \\ M_3 &= -\tfrac{1}{2} mR^2 \omega_u \omega_w \cos(\omega_w t + \varphi_0). \end{aligned} \tag{e}$$

Therefore the modulus of the moment is

$$M = \tfrac{1}{2} mR^2 \omega_u \cdot \omega_w. \tag{f}$$

When $\omega_u = 0$ (fixed casing) we obtain the result from Example 1, Section 5.4.

CHAPTER 7

Fundamentals of Analytical Mechanics

7.1 Generalized coordinates and degrees of freedom of a mechanical system

To determine the configuration (position) of a mechanical system it is enough to specify its coordinates. If we want to find the position of, let us say, a crank system (Fig. 7.1), the angle between the x-axis and the crank radius r should be specified.

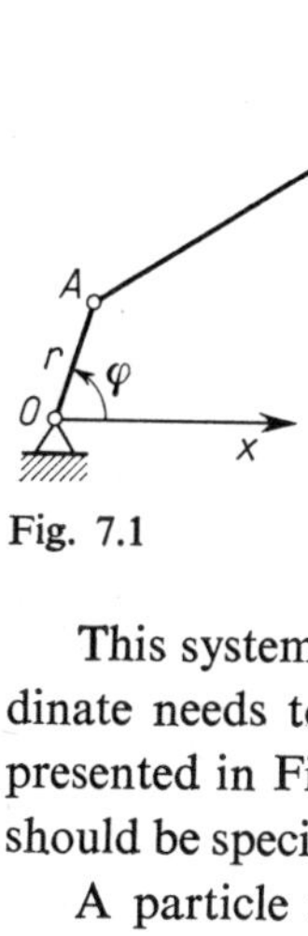

Fig. 7.1

This system is said to have *one degree of freedom* because only one coordinate needs to be known to describe its position. Examining the system presented in Fig. 7.2 we note that to find its configuration two coordinates should be specified, namely x and φ. It has, therefore, *two degrees of freedom.*

A particle in a three-dimensional space has *three degrees of freedom,* since three independent values are necessary and sufficient to fix its position uniquely. Obviously, the number of degrees of freedom corresponds in this case to the number of dimensions of a given space. With a mechanical system consisting of n particles we find that to determine uniquely its configuration $3n$ coordinates need to be specified. It has therefore $3n$ degrees of freedom (bearing in mind that we are in a three-dimensional space).

If now we restrict the motion of some particles to a motion on surfaces or interrelate in any way the positions and velocities of these particles (while excluding the equations of motion), the system is then said to be

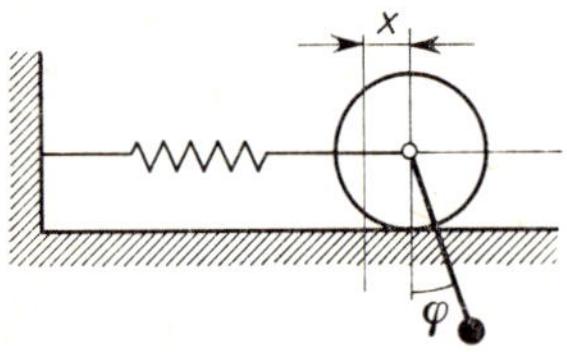

Fig. 7.2

constrained. In this case, the number of degrees of freedom decreases by the number of conditions imposed on the system. More generally: if we have n points in a three-dimensional space, then the number of degrees of freedom is

$$s = 3n - l \tag{1}$$

where l is the number of independent equations of constraints determinable for the system. To explain what we understand by dependence or independence of equations we shall cite an example.

Consider a rigid body. To find the position of such a body in a three-dimensional space, it is sufficient to know the position of three points which are not in the same straight line. Since the distance between these points remains unchanged, we can write three equations of constraint:

$$l_{ik} = \sqrt{\sum_{j=1}^{3} [x_{j(i)} - x_{j(k)}]^2}. \tag{2}$$

Consequently, the system has $s = 3 \cdot 3 - 3 = 6$ degrees of freedom. Of course, the position of the rigid body is determined by six coordinates: three linear displacements and three angular displacements. Isolating n points of a rigid body, we find that the numbers l_{ik} forming a quadratic matrix possess $(n^2 - n)/2$ independent components ($l_{ik} = 0$ when $i = k$ and $l_{ik} = l_{ki}$). Since however the body in question has $s = 6$ degrees of freedom, we have

$$6 = 3n - \frac{n(n-1)}{2} + m, \tag{3}$$

where m is the number of dependent equations of type (2). One will easily notice that only in the cases where $n = 4$ and $n = 3$ we have $m = 0$.

If the same rigid body is considered in a two-dimensional Euclidean space, the number of degrees of freedom with n points of the body separated is

$$s = 2n - \frac{n(n-1)}{2} + m. \tag{4}$$

For two points, $m = 0$, whence

$$s = 2\cdot 2 - \frac{2(2-1)}{2} = 3.$$

In this particular case two points uniquely determine the position of the body, therefore $s = 3$ is the number of degrees of freedom of the body (two linear and one angular displacements). It follows that

$$m = \frac{n^2-5n+6}{2}$$

or only in the case of $n = 2$ or $n = 3$ all equations of constraint of type (2) are independent.

Note now that if we abandon the rigid body model, our system will have an infinite number of degrees of freedom. An infinitely large number of points can be considered on such a body, and the distance between each pair of points is liable to change. It is seen that the model adopted determines the number of degrees of freedom.

But let us return to a rigid body. If a system consists of c bodies in a three-dimensional space and if l independent equations of constraints can be derived for it, then

$$s = \sum_{i=1}^{c} s_i - l \tag{5}$$

is the number of degrees of freedom (s_i is the number of degrees of freedom of each element).

Taking, for example, a quadruple-jointed (crank-rocker) system (Fig. 7.3), we find that it has one degree of freedom. Note that the rods AB and CD, each having one degree of freedom (enabling them to turn about A and D), cannot move independently of each other because of the rigid rod BC. Therefore

$$s = 2-1 = 1. \tag{5a}$$

Examining the possible position of the crank-rocker system, we come to the conclusion that the length of the rod BC has a direct influence on the range of changes of position. It tends to zero with the decrease of the length of BC. In the case where B superimposes on C we have already a triple-jointed system with zero degrees of freedom. This is due to the fact that the common point B cannot simultaneously move in two circles: it is therefore arrested at the point of their intersection. Such a point may be determined, for example in a three-dimensional space, by three surfaces:

$$F_i(x, y, z) = 0 \qquad (i = 1, 2, 3). \tag{6}$$

If we limit the number of equations (6) to two ($i = 1, 2$), then the point will be able to move along a curve and it will have therefore one degree of freedom.

Fig. 7.3

The constraints described here are geometric constraints and the system on which these constraints are imposed are called *holonomic systems.* Any mechanical system can be viewed as a set of points which we can move in a certain limited way. These geometrical restrictions may vary in time; therefore for the general case we shall write the equations of constraints as follows:

$$F_\mu(x_1, y_1, z_1, x_2, y_2, z_2, \ldots, x_n, y_n, z_n, t) = 0 \qquad (\mu = 1, 2, \ldots, l). \tag{7}$$

The time t in equations (7) may occur either in explicit form or not. In the first instance, the constraints are called *rheonomic* and in the second, *scleronomic.*

We can also consider here constraints written as inequalities [4]

$$F_{\mu'}(x_1, y_1, z_1, x_2, y_2, z_2, \ldots, x_n, y_n, z_n, t) \leqslant 0 \qquad (\mu' = 1, 2, \ldots, l'). \tag{8}$$

The former are called *bilateral* (equations (7)) and the latter *unilateral* (equations (8)). An example of unilateral constraints may be the motion in a space limited by a surface.

Besides geometric constraints differential constraints may occur

$$R_\nu(x_1, y_1, z_1, x_2, y_2, z_2, \ldots, x_n, y_n, z_n, \dot{x}_1, \dot{y}_1, \dot{z}_1, \dot{x}_2, \dot{y}_2, \dot{z}_2, \ldots$$
$$\ldots, \dot{x}_n, \dot{y}_n, \dot{z}_n, t) = 0 \qquad (\nu = 1, 2, \ldots, l_r). \tag{9}$$

We assume the latter to be linear forms with respect to velocity, i.e., they have the following construction:

$$\sum_{k=1}^{n} \mathbf{A}_{\nu k}^{(2)} \cdot \mathbf{v}_k + B_\nu^{(2)} = 0. \tag{9a}$$

If the differential equations (9) are integrable, i.e., if system (9) can be reduced by integration to the form (7), then we deal with a holonomic system. Otherwise the system is *non-holonomic*. An example of a system with unintegrable (non-holonomic) constraints is a sphere which can move on a rough plane surface [37]. Relationships between virtual displacements $\delta \mathbf{r}_k$ for holonomic constrains are characterized by variation, without time variation, of expression (7):

$$\sum_{k=1}^{n} \mathbf{A}_{\mu k}^{(1)} \cdot \delta \mathbf{r}_k = 0, \tag{10a}$$

where

$$\mathbf{A}_{\mu k}^{(1)} = \frac{\partial F_\mu}{\partial x_k} \mathbf{i} + \frac{\partial F_\mu}{\partial y_k} \mathbf{j} + \frac{\partial F_\mu}{\partial z_k} \mathbf{k}$$

and for non-holonomic by equations of the following form:

$$\sum_{k=1}^{n} \mathbf{A}_{\nu k}^{(2)} \cdot \delta \mathbf{r}_k = 0. \tag{10b}$$

We shall now focus our attention on the question of choice of coordinates. The best thing to do is to choose a frame of reference so that it defines in a most natural way the position of a system. Equations (7) and (9) can be substantially simplified if we introduce generalized coordinates. These may be angular displacements or distances between points under examination, and the like. These new generalized coordinates q^i are associated with cartesian coordinates by the following relationships:

$$\mathbf{r}_k = \mathbf{r}_k(q^1, q^2, \ldots, q^r, t) \qquad (k = 1, \ldots, n), \tag{11}$$

where $\mathbf{r}_k$ is the position vector of a point. It is more convenient to select generalized coordinates so that their number should correspond to the number of degrees of freedom. This can always be achieved in holonomic systems.

Where the number of coordinates is higher than the number of degrees of freedom, constraints exist, which may be written as follows:

$$\Phi_i(q^1, q^2, \ldots, q^r, t) = 0 \qquad (i = 1, \ldots, l_g). \tag{12a}$$

For a non-holonomic system we should consider additionally constraints described by unintegrable differential equations of the form

$$C_{\nu j}^{(2)} \dot{q}^j + D_\nu^{(2)} = 0 \qquad (\nu = l, \ldots, l_r), \tag{12b}$$

where $C_{\nu j}^2$ and D_ν^2 are functions of coordinates and time. The number of degrees of freedom is

$$s = r - l_g - l_r.$$

From the geometrical point of view, equations (7) represent hypersurfaces of an r-dimensional space.

A set of generalized coordinates gives the position of a point moving in a certain subspace built in the original r-dimensional space and determined by equations of constraint.

Virtual displacements $\delta \mathbf{r}_k$ we find from equations (10), as variables of the functions occurring there, without variation in time:

$$\delta \mathbf{r}_k = \frac{\partial \mathbf{r}_k}{\partial q^i} \delta q^i \qquad (k = 1, 2, \ldots, n). \tag{13}$$

We calculate the possible velocities from

$$\frac{\mathrm{d}\mathbf{r}_k}{\mathrm{d}t} = \frac{\partial \mathbf{r}_k}{\partial q^j} \frac{\mathrm{d}q^j}{\mathrm{d}t} + \frac{\partial \mathbf{r}_k}{\partial t}. \tag{14}$$

Vectors $\delta \mathbf{r}_k$ and $\mathrm{d}\mathbf{r}_k/\mathrm{d}t$ have the same direction in scleronomic systems only since in such instances

$$\frac{\partial \mathbf{r}_k}{\partial t} = 0.$$

If the number of degrees of freedom corresponds to the number of generalized coordinates, then the (virtual) displacements δq^j are independent, otherwise the following relationships occur:

$$\frac{\partial \Phi_i}{\partial q^j} \delta q^j = 0 \qquad (i = 1, \ldots, l_g). \tag{15}$$

The relationships between velocities are represented by

$$\frac{\partial \Phi_i}{\partial t} + \frac{\partial \Phi_i}{\partial q^k} \dot{q}^k = 0 \qquad (i = 1, \ldots, l_g) \tag{16}$$

and (12b).

One will easily see that virtual displacements δq_j are characterized by a vector tangential to a hypersurface for a certain specified time t = const (Fig. 7.4). Vectors with the coordinates $\dot{q}_j$ and δq_j superimpose in the case of scleronomic systems, for then

$$\frac{\partial \Phi_i}{\partial t} = 0 \quad \text{and} \quad D_\nu^{(2)} = 0.$$

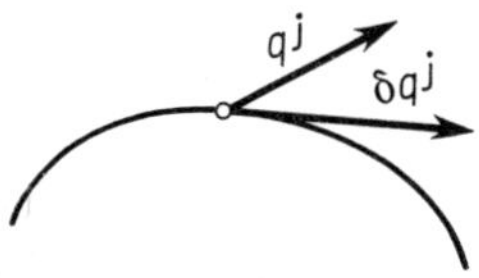

Fig. 7.4

7.2 D'Alembert's principle

According to Suslov [70] we differentiate between the *concept of principle*, and the *concept of law*. It seems that we have two terms in use which are not synonymous, thus both their content and purview should be differentiated.

The concept of principle is more general than that of law which, being a theorem, cannot fully characterize the motion of a system. For example, the law of variation of momentum does not wholly characterize the motion but constitutes only a component part necessary for its description in the general case.

We shall now introduce an important principle known as *d'Alembert's principle*. If we consider the motion of a single point with mass m then, where constraints are present, motion is influenced by the reaction forces of restraints $\mathbf{R}$ beside external forces $\mathbf{P}$. According to Newton's law we can write

$$m\mathbf{a} = \mathbf{P} + \mathbf{R}. \tag{1}$$

If there is more than one such point and $n > 1$, we can make use of n equations. To these we add $l + l_r$ equations of constraints. It is evident, however, that the number of unknowns $2n$ ($\mathbf{a}_k, \mathbf{R}_k$) is greater than the number of equations $(n+l+l_r)$. In this way the problem thus formulated is indeterminate. As a result, it is necessary to introduce an additional assumption concerning constraints on the system. This assumption refers to the work of the reaction of constraints on virtual displacements; it should further be assumed that this work is equal to zero, which we shall write down as

$$\delta W_r = \sum_{k=1}^{n} \mathbf{R}_k \cdot \delta \mathbf{r}_k = 0. \tag{2}$$

Let us now put the question: Which forces in mechanical systems do no work? The answer is as follows. Considering the definition of work it is evident that the only forces that do no work are those in a direction at right angles to the displacements of points.

We reckon forces on smooth supports to be such as do no work since a smooth support involves no friction. Nor do the forces of interaction of particles of rigid bodies do any work. This follows directly from the fact that all displacements in the direction of a straight line connecting two particles of such a body are equal to each other and so, in keeping with the law of action and reaction, are due forces of interaction acting in opposite senses.

We have a similar situation with rough supports, e.g., in the case of bodies rolling without sliding (no resistance to rolling). By accepting (2) the general problem of mechanics is made specific. If however friction did exist, the friction force involved would have to be treated as external, and in this way the general character of the solution to our problem would not be disturbed. The constraints realized by (2) are called *ideal.* It is for such systems that d'Alembert's principle is derivable.

For every point in a mechanical system on which beside an effective force $\mathbf{P}_k$, there acts a reaction force $\mathbf{R}_k$, the following equation is valid:

$$m_k \mathbf{a}_k = \mathbf{P}_k + \mathbf{R}_k. \tag{3}$$

Multiplying (3) by the vector $\delta \mathbf{r}_k$ and summing all points of the system, we obtain

$$\sum_{k=1}^{n} m_k \mathbf{a}_k \cdot \delta \mathbf{r}_k = \sum_{k=1}^{n} \mathbf{P}_k \cdot \delta \mathbf{r}_k + \sum_{k=1}^{n} \mathbf{R}_k \cdot \delta \mathbf{r}_k. \tag{4}$$

Considering (2) in (4) and transferring the residual sum of the right-hand side to the left, we finally obtain

$$\sum_{k=1}^{n} (m_k \mathbf{a}_k - \mathbf{P}_k) \cdot \delta \mathbf{r}_k = 0. \tag{5}$$

Equation (5) gives *d'Alembert's principle.*

An important theorem concerning statical problems follows from d'Alembert's principle. What we have in mind is that, in the case where a system remains at rest $\left(\bigwedge_k a_k = 0\right)$, the sum of the work of external forces $\mathbf{P}_k$ on virtual displacements is equal to zero:

$$\sum_{k=1}^{n} \mathbf{P}_k \cdot \delta \mathbf{r}_k = 0. \tag{5a}$$

Example 1. Let a crank system (Fig. 7.5) be given. First we find the relationship of force $\mathbf{P}_A$ (applied at point A) to force $\mathbf{P}_B$ (applied at point B) where the system remains in equilibrium:

$$-\mathbf{P}_A \cdot \delta \mathbf{r}_A + \mathbf{P}_B \cdot \delta \mathbf{r}_B = 0. \tag{a}$$

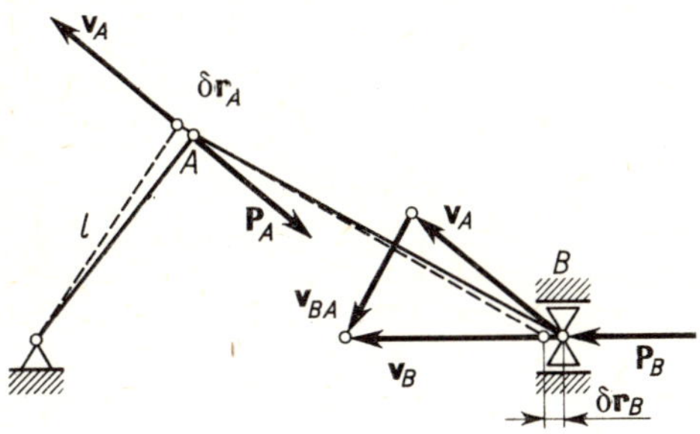

Fig. 7.5

Note that the considered system has scleronomic constraints. Consequently the virtual displacements are proportional to their velocities. We can therefore write

$$-\mathbf{P}_A \cdot \mathbf{v}_A + \mathbf{P}_B \cdot \mathbf{v}_B = 0. \tag{b}$$

Since the directions of forces and velocities accord with each other (Fig. 7.5), we can rewrite equation (b) in the following form:

$$-P_A v_A + P_B v_B = 0. \tag{b1}$$

We should further note that the determination of the velocity ratio v_A/v_B gives the relationship between the forces $\mathbf{P}_A$ and $\mathbf{P}_B$. This can be achieved, for example, by a graphical method (see Chapter 1). Equation (b) can be generalized:

$$\sum_{k=1}^{n} \mathbf{P}_k \cdot \mathbf{v}_k = 0. \tag{6}$$

The relation (6) is very convenient for applications.

7.3 Hamilton's principle

Every theory embodies theorems, some of which are axioms and others are corollaries from those axioms. For a theory itself, it is of no consequence which of the theorems are regarded as axioms and which as corollaries. However, for a better understanding of a theory it is better to accept axiomatically that system of theorems which is more obvious.

Newton's laws seem to be intuitively perceptible without difficulty and their acceptance without proof does not raise any major objections, but there are other such examples in mechanics, e.g., a theorem known under the name of *Hamilton's principle* which can be accepted without proof in order to be able to prove other theorems of mechanics, including Newton's laws or d'Alembert's principle.

We shall now discuss Hamilton's principle. According to this principle, every mechanical system is characterized by a specified function dependent on the position of the system, generalized velocities and time. We shall write this function as follows:

$$L = L(q^1, q^2, \dots, q^r, \dot{q}^1, \dot{q}^2, \dots, \dot{q}^r, t). \tag{1}$$

If the system now moves from position A to position B, the function L exhibits the property that $\int_{t_A}^{t_B} L\mathrm{d}t$ assumes a stationary value.

According to the notations of variation calculus we write Hamilton's principle in the form

$$\delta \int_{t_A}^{t_B} L\mathrm{d}t = 0 \tag{2}$$

or

$$\int_{t_A}^{t_B} \delta L\mathrm{d}t = 0. \tag{3}$$

The symbol δ denotes the variation sign. We shall demonstrate that the principle is legitimate. We define δL as the sum of the variation of kinetic energy and of work [4]:

$$\delta L = \delta E + \delta' W, \tag{4}$$

where

$$\delta E = \sum_{i=1}^{n} m_i \mathbf{v}_i \cdot \delta \mathbf{r}_i, \tag{5}$$

$$\delta' W = \sum_{i=1}^{n} \mathbf{P}_i \cdot \delta \mathbf{r}_i. \tag{6}$$

But

$$\mathbf{v}_i \cdot \delta \dot{\mathbf{r}}_i = \frac{\mathrm{d}}{\mathrm{d}t}(\mathbf{v}_i \cdot \delta \mathbf{r}_i) - \dot{\mathbf{v}}_i \cdot \delta \mathbf{r}_i,$$

therefore

$$\begin{aligned}\delta L &= \frac{\mathrm{d}}{\mathrm{d}t}\Big(\sum_{i=1}^{n} m_i \mathbf{v}_i \cdot \delta \mathbf{r}_i\Big) - \sum_{i=1}^{n} m_i \cdot \dot{\mathbf{v}}_i \delta \mathbf{r}_i + \sum_{i=1}^{n} \mathbf{P}_i \cdot \delta \mathbf{r}_i \\ &= \frac{\mathrm{d}}{\mathrm{d}t}\Big(\sum_{i=1}^{n} m_i \mathbf{v}_i \cdot \delta \mathbf{r}_i\Big) - \sum_{i=1}^{n} (m_i \dot{\mathbf{v}}_i - \mathbf{P}_i) \cdot \delta \mathbf{r}_i.\end{aligned} \tag{7}$$

On the strength of d'Alembert's principle,

$$\sum_{i=1}^{n} (m_i \dot{\mathbf{v}}_i - \mathbf{P}_i) \cdot \delta \mathbf{r}_i = 0,$$

therefore

$$\delta L = \frac{\mathrm{d}}{\mathrm{d}t}\Big(\sum_{i=1}^{n} m_i \mathbf{v}_i \cdot \delta \mathbf{r}_i\Big). \tag{8}$$

Integrating the above equation we obtain

$$\delta \int_{t_A}^{t_B} L \mathrm{d}t = \sum_{i=1}^{n} m_i (\underset{B}{\mathbf{v}_i} \cdot \underset{B}{\delta \mathbf{r}_i} - \underset{A}{\mathbf{v}} \cdot \underset{A}{\delta \mathbf{r}_i}). \tag{9}$$

Since the variations $\delta \mathbf{r}_i$ at points A and B are equal to zero, the entire expression vanishes, and this substantiates our proposition.

Hamilton's principle is of major importance to the dynamics of mechanical systems, and it can also be used by analogy, for example, in the theory of electromagnetic systems. If forces have a potential V,* then

$$\delta W = -\delta V, \tag{10}$$

whence

$$\delta L = \delta(E-V). \tag{11}$$

The difference between the kinetic and the potential energy of a system is called the *Lagrange kinetic potential.* If V does not contain derivative $\dot{q}^i$, the quadratic form

$$A = \sum_{i,j} \frac{\partial^2 L}{\partial q^i \partial q^j} \eta^i \eta^j.$$

is positively determined. Therefore, in keeping with Legendre's condition, the integral $\int_{t_A}^{t_B} L\,dt$ assumes a minimum value for a space problem.

Example 1. Find the differential equation for a mathematical pendulum of variable length $l = l(t)$ (Fig. 7.6).

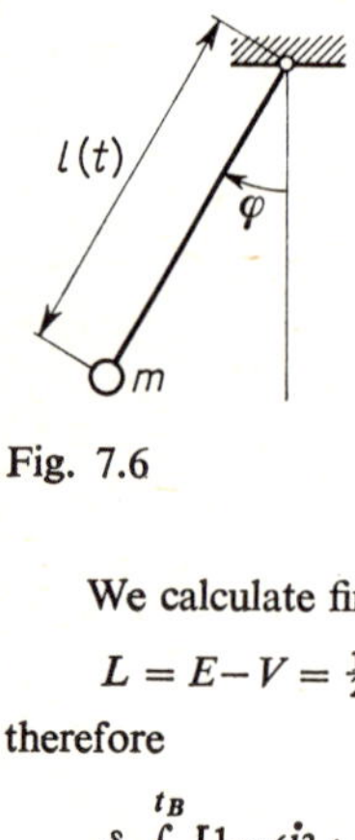

Fig. 7.6

We calculate first the Lagrange kinetic potential. In the considered case

$$L = E-V = \tfrac{1}{2} m(\dot{l}^2 + l^2\dot{\varphi}^2 + 2gl\cos\varphi), \tag{a}$$

therefore

$$\delta \int_{t_A}^{t_B} [\tfrac{1}{2}m(\dot{l}^2 + l^2\dot{\varphi}^2 + 2gl\cos\varphi)]\,dt = 0. \tag{b}$$

Euler's equation for a functional determinant thus constructed is

$$l^2\ddot{\varphi} + 2l\dot{l}\dot{\varphi} + lg\sin\varphi = 0 \tag{c}$$

which brings the discussion of the problem to an end.

---

* In our considerations the terms *potential* and *potential energy* are synonymous.

7.4 Lagrange equations of the first order

We shall now derive differential equations of motion while assuming the presence of geometric constraints, as well as differential constraints of the following form:

$$F_\mu(x_1, y_1, z_1, \ldots, x_n, y_n, z_n, t) = 0 \quad (\mu = 1, 2, \ldots, l) \tag{1}$$

and

$$\sum_{k=1}^{n} \mathbf{A}_{\nu k}^{(2)} \cdot \mathbf{v}_k + B_\nu^{(2)} = 0 \quad (\nu = 1, 2, \ldots, l_r), \tag{2}$$

where $\mathbf{A}_{\nu k}^{(2)}$ and $B_{\nu k}^{(2)}$ are functions of coordinates and time in the case of a rheonomic system and $\mathbf{v}_k$ are point velocities of the system.

Note that we limit ourselves to the case where the differential constraints are linear forms in relation to velocities. Equation (2) is, at the same time, an equation of possible velocities, without exhausting however, all the equations of this type. It should be complemented with relationships arising from constraints of the geometric type (1). Differentiating (1), we obtain

$$\frac{\mathrm{d}F_\mu}{\mathrm{d}t} = \sum_{k=1}^{n} \left(\frac{\partial F_\mu}{\partial x_k} v_{xk} + \frac{\partial F_\mu}{\partial y_k} v_{yk} + \frac{\partial F_\mu}{\partial z_k} v_{zk} \right) + \frac{\partial F_\mu}{\partial t} = 0 \tag{3}$$
$$(\mu = 1, 2, \ldots, l)$$

or written shortly

$$\sum_{k=1}^{n} \mathbf{A}_{\mu k}^{(1)} \cdot \mathbf{v}_k + B_\mu^{(1)} = 0 \quad (\mu = 1, 2, \ldots, l), \tag{3a}$$

where

$$\mathbf{A}_{\mu k}^{(1)} = \frac{\partial F_\mu}{\partial x_k}\mathbf{i} + \frac{\partial F_\mu}{\partial y_k}\mathbf{j} + \frac{\partial F_\mu}{\partial z_k}\mathbf{k}, \tag{4}$$

$$\mathbf{v}_k = v_{xk}\mathbf{i} + v_{yk}\mathbf{j} + v_{zk}\mathbf{k}, \tag{5}$$

$$B_\mu^{(1)} = \frac{\partial F_\mu}{\partial t}. \tag{6}$$

Going over to equations of virtual displacement we write

$$\sum_{k=1}^{n} \mathbf{A}_{\mu k}^{(1)} \cdot \delta\mathbf{r}_k = 0, \tag{7}$$

$$\sum_{k=1}^{n} \mathbf{A}_{\nu k}^{(2)} \cdot \delta\mathbf{r}_k = 0. \tag{8}$$

Multiplying the sums by the multipliers $\lambda_\mu^{(1)}$ and $\lambda_\nu^{(2)}$, temporarily undetermined, we form the expression

$$\sum_{\mu=1}^{l} \lambda_\mu^{(1)} \sum_{k=1}^{n} \mathbf{A}_{\mu k}^{(1)} \cdot \delta \mathbf{r}_k + \sum_{\nu=1}^{l_r} \lambda_\nu^{(2)} \sum_{k=1}^{n} \mathbf{A}_{\nu k}^{(2)} \cdot \delta \mathbf{r}_k = 0$$

or

$$\sum_{k=1}^{n} \left(\sum_{\mu=1}^{l} \lambda_\mu^{(1)} \mathbf{A}_{\mu k}^{(1)} + \sum_{\nu=1}^{l_r} \lambda_\nu^{(2)} \mathbf{A}_{\nu k}^{(2)} \right) \cdot \delta \mathbf{r}_k = 0. \tag{9}$$

Note further that the work of the reactions of ideal constraints on virtual displacements is equal to zero

$$\sum_{k=1}^{n} \mathbf{R}_k \cdot \delta \mathbf{r}_k = 0. \tag{10}$$

Equations (9) and (10) have an identical construction, and consequently the reactions of ideal constraints can be expressed as follows:

$$\mathbf{R}_k = \sum_{\mu=1}^{l} \lambda_\mu^{(1)} \mathbf{A}_{\mu k}^{(1)} + \sum_{\nu=1}^{l_r} \lambda_\nu^{(2)} \cdot \mathbf{A}_{\nu k}^{(2)}. \tag{11}$$

These reactions are determined with an accuracy to the coefficients $\lambda_\mu^{(1)}$ and $\lambda_\nu^{(2)}$ which are called *Lagrange multipliers*. Since, besides the external force $\mathbf{P}_k$, the reaction of constraints $\mathbf{R}_k$ acts on the kth point, the Newtonian equation of motion assumes the following form:

$$m_k \mathbf{a}_k = \mathbf{P}_k + \sum_{\mu=1}^{l} \lambda_\mu^{(1)} \mathbf{A}_{\mu k}^{(1)} + \sum_{\nu=1}^{l_r} \lambda_\nu^{(2)} \mathbf{A}_{\nu k}^{(2)}. \tag{12}$$

Equation (12) is called a *Lagrange equation of the first order*.

The system (12) contains n unknown accelerations $\mathbf{a}_k$ and $l + l_r$ Lagrange multipliers. To find all the unknowns, it is necessary to add to n first order Lagrange equations the equations of possible accelerations derived by differentiating the equations of possible velocities, i.e. equations (2) and (3a):

$$\sum_{k=1}^{n} (\dot{\mathbf{A}}_{\mu k}^{(1)} \cdot \mathbf{v}_k + \mathbf{A}_{\mu k}^{(1)} \cdot \mathbf{a}_k) + \dot{B}_\mu^{(1)} = 0 \quad (\mu = 1, 2, \ldots, l) \tag{13}$$

and

$$\sum_{k=1}^{n} (\dot{\mathbf{A}}_{\nu k}^{(2)} \cdot \mathbf{v}_k + \mathbf{A}_{\nu k}^{(2)} \cdot \mathbf{a}_k) + \dot{B}_\nu^{(2)} = 0 \quad (\nu = 1, 2, \ldots, l_r). \tag{14}$$

Thus, the system of equations (12), (13), and (14) constitutes the basis for the solution to the problem of constrained motion of a system. This system of equations is convenient for finding the reactions of constraints. Note that the multipliers $\lambda_{\mu}^{(1)}$, $\lambda_{\nu}^{(2)}$ can easily be expressed by velocities. All that needs to be done is to eliminate accelerations $\mathbf{a}_k$ from the system of equations. In this way we obtain $l+l_r$ equations from which $l+l_r$ Lagrange multipliers can be determined.

Example 1. A point is moving on a surface $F(x, y, z, t) = 0$. Find the equations of possible velocities, possible accelerations, and virtual displacements.

The solution consists of making use of equations (3a), (7), and (13):

$$\mathbf{A}^{(1)}\cdot\mathbf{v}+B^{(1)} = 0, \tag{a}$$

where

$$\mathbf{A}^{(1)} = \frac{\partial F}{\partial x}\mathbf{i}+\frac{\partial F}{\partial y}\mathbf{j}+\frac{\partial F}{\partial z}\mathbf{v}, \quad B^{(1)} = \frac{\partial F}{\partial t}, \tag{a1}$$

$$\mathbf{A}^{(1)}\cdot\delta\mathbf{r} = 0, \tag{b}$$

$$\dot{\mathbf{A}}^{(1)}\cdot\mathbf{v}+\mathbf{A}^{(1)}\cdot\mathbf{a}+\dot{B}^{(1)} = 0. \tag{c}$$

Example 2. Consider the motion of a point with mass m on the surface of a sphere satisfying the following equation:

$$x^2+y^2+z^2-(r_0+ct)^2 = 0. \tag{a}$$

In the considered case we have only constraints of the geometric type. From the equation of constraint (a) we find the vector $\mathbf{A}_{11}^{(1)}$:

$$\mathbf{A}_{11}^{(1)} = 2x\mathbf{i}+2y\mathbf{j}+2z\mathbf{k} \tag{b}$$

and

$$B_1^{(1)} = -2c(r_0+ct), \tag{c}$$

where c = const. From (13) we have

$$v^2+x\ddot{x}+y\ddot{y}+z\ddot{z}-c^2 = 0. \tag{d}$$

Using the Langrange equation (12) and going over to scalar notation, we obtain

$$m\ddot{x} = P_x+2\lambda^{(1)}x, \quad m\ddot{y} = P_y+2\lambda^{(1)}y, \quad m\ddot{z} = P_z+2\lambda^{(1)}z. \tag{e}$$

Eliminating the components of acceleration from equation (d) using equations (e), we have

$$mv^2+P_yx+P_yy+P_zz+2\lambda_1^{(1)}(x^2+y^2+z^2)-mc^2 = 0; \tag{f}$$

hence

$$\lambda_1^{(1)} = \frac{mc^2-(P_xx+P_yy+P_zz)-mv^2}{2(x^2+y^2+z^2)}. \tag{g}$$

As a special case, let us consider the problem of the motion of this particle under gravitational force $\mathbf{P} = -mg\mathbf{k}$ (Fig. 7.7). Assume further that the radius of our sphere is constant, i.e., $c = 0$. Then

$$\dot{\lambda}_1^{1)} = \frac{mgz - mv^2}{2r_0^2}. \tag{h}$$

Making use of equations (11) we find the reactions of constraints:

$$R_x = mg\frac{xz}{r_0^2} - \frac{mv^2}{r_0}\frac{x}{r_0}, \quad R_y = mg\frac{yz}{r_0^2} - \frac{mv^2}{r_0}\frac{y}{r_0}, \quad R_z = mg\frac{z^2}{r_0^2} - \frac{mv^2}{r_0}\frac{z}{r_0}. \tag{i}$$

The interpretation of the results is evident.

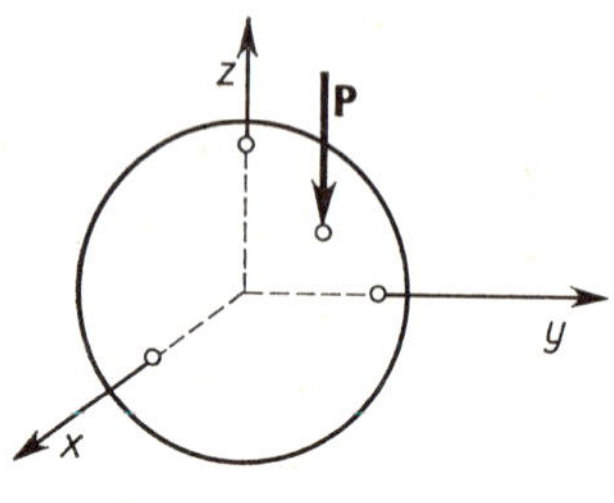

Fig. 7.7

Example 3. Calculate the reactions of a surface $F(x, y, z, t) = 0$ during the internal motion (without effective external forces acting) of a point with mass m on this surface.

We shall solve the problem along the same lines as in Example 2. Since only the reaction of the surface acts on the mass, the equation of motion is

$$m\mathbf{a} = \lambda^{(1)}\mathbf{A}^{(1)}, \tag{a}$$

where $\mathbf{A}^{(1)}$ is given by equation (a1) from Example 1. The equation of possible accelerations (equation (c) from Example 1) makes it possible to find the acceleration $\mathbf{a}$ which we subsequently eliminate from equation (a). We have

$$m\mathbf{A}^{(1)}\cdot\mathbf{a} = \lambda^{(1)}(\mathbf{A}^{(1)})^2,$$

$$m\frac{\mathrm{d}\mathbf{A}^{(1)}}{\mathrm{d}t}\cdot\mathbf{v} + m\mathbf{A}^{(1)}\cdot\mathbf{a} + m\frac{\mathrm{d}B^{(1)}}{\mathrm{d}t} = 0.$$

Eliminating the middle term of the second equation, we obtain

$$m\dot{\mathbf{A}}^{(1)}\cdot\mathbf{v} + \lambda^{(1)}\cdot(\mathbf{A}^{(1)})^2 + m\dot{B}^{(1)} = 0.$$

Hence

$$\lambda^{(1)} = \frac{-m}{(\mathbf{A}^{(1)})^2}(\dot{\mathbf{A}}^{(1)}\cdot\mathbf{v} + \dot{B}^{(1)}).$$

Therefore the reaction of the surface is expressed by

$$\mathbf{R} = -\frac{m\dot{\mathbf{A}}^{(1)}}{(\mathbf{A}^{(1)})^2}(\dot{\mathbf{A}}^{(1)}\cdot\mathbf{v} + \dot{B}^{(1)}).$$

7.5 Lagrange equations of the second order

Now we shall transform the equations of motion to generalized Lagrange coordinates. The cartesian coordinates can be expressed by means of the latter, as we have previously noted, by the following relationships:

$$\mathbf{r}_k = \mathbf{r}_k(q^1, q^2, \ldots, q^r, t) \qquad (k = 1, 2, \ldots, n). \tag{1}$$

Considering relationship (1) in equation (1), Section 7.4, we obtain

$$\Phi_\mu(q^1, q^2, \ldots, q^r, t) = 0 \qquad (\mu = 1, 2, \ldots, l_q). \tag{2}$$

By a suitable choice of generalized coordinates the number of equations (2) can even be reduced to zero; we deal then with the case of independent coordinates. Obviously, this is possible only when the considered system is holonomic. In the case of unintegrable constraints of the differentiable type it is possible, if necessary, to take curvilinear coordinates q^i so that the number of equations of type (2) is reduced to zero but, owing to the existence of the constraints (2), Section 7.4, the coordinates will not be independent. It follows that only in the absence of unintegrable differential constraints can as many coordinates q^i be chosen as there are degrees of freedom of a system. In the case of a nonholonomic system the number of coordinates is always greater than the number of degrees of freedom. The difference between the number of coordinates and the number of degrees of freedom can be reduced to the number of unintegrable differential equations of constraints. Substituting (1) into (2), Section 7.4, we obtain

$$C_{\nu j}^{(2)} \dot{q}^j + D_\nu^{(2)} = 0 \qquad (\nu = 1, 2, \ldots, l_r). \tag{3}$$

We shall now proceed to express first order Lagrange equations in generalized coordinates. For this purpose, we multiply (12), Section 7.4, by the derivatives $\partial \mathbf{r}_k / \partial q^j$ and perform a summation:

$$\sum_{k=1}^{n} m_k \mathbf{a}_k \cdot \frac{\partial \mathbf{r}_k}{\partial q^j} = \sum_{k=1}^{n} \mathbf{P}_k \cdot \frac{\partial \mathbf{r}_k}{\partial q^j} + \\ + \sum_{k=1}^{n} \sum_{\mu=1}^{l} \lambda_\mu^{(1)} \mathbf{A}_{\mu k}^{(1)} \cdot \frac{\partial \mathbf{r}_k}{\partial q^j} + \sum_{k=1}^{n} \sum_{\nu=1}^{r} \lambda_\nu^{(2)} \mathbf{A}_{\nu k}^{(2)} \cdot \frac{\partial \mathbf{r}_k}{\partial q^j}. \tag{4}$$

The sum on the left-hand side of equation (4) we transform in the following manner:

$$\sum_{k=1}^{n} m_k \mathbf{a}_k \cdot \frac{\partial \mathbf{r}_k}{\partial q^j} = \sum_{k=1}^{n} m_k \frac{\mathrm{d}\dot{\mathbf{r}}_k}{\mathrm{d}t} \frac{\partial \mathbf{r}_k}{\partial q^j}$$

$$= \sum_{k=1}^{n} \frac{\mathrm{d}}{\mathrm{d}t}\left(m_k \dot{\mathbf{r}}_k \cdot \frac{\partial \mathbf{r}_k}{\partial q^j}\right) - \sum_{k=1}^{n} m_k \dot{\mathbf{r}}_k \cdot \frac{\mathrm{d}}{\mathrm{d}t}\left(\frac{\partial \mathbf{r}_k}{\partial q^j}\right). \tag{5}$$

Note that

$$\frac{\partial \mathbf{r}_k}{\partial q^j} = \frac{\partial \dot{\mathbf{r}}_k}{\partial \dot{q}^j}. \tag{6}$$

The validity of relationship (6) can easily be checked, considering that

$$\dot{\mathbf{r}}_k = \frac{\partial \mathbf{r}_k}{\partial q^j} \dot{q}^j + \frac{\partial \mathbf{r}_k}{\partial t}. \tag{7}$$

Differentiating (7) with respect to $\dot{q}^j$, we find (6) to hold. Going further

$$\frac{\mathrm{d}}{\mathrm{d}t}\left(\frac{\partial \mathbf{r}_k}{\partial q^j}\right) = \frac{\partial \dot{\mathbf{r}}_k}{\partial q^j}, \tag{8}$$

since

$$\frac{\mathrm{d}}{\mathrm{d}t}\left(\frac{\partial \mathbf{r}_k}{\partial q^j}\right) = \frac{\partial^2 \mathbf{r}_k}{\partial q^j \partial q^i} \dot{q}^i + \frac{\partial^2 \mathbf{r}_k}{\partial q^j \partial t} \tag{9}$$

and

$$\frac{\partial \dot{\mathbf{r}}_k}{\partial q^j} = \frac{\partial^2 \mathbf{r}_k}{\partial q^j \partial q^i} \dot{q}^i + \frac{\partial^2 \mathbf{r}_k}{\partial q^j \partial t}. \tag{10}$$

Evidently, the right-hand sides of (9) and (10) are equal, which proves the validity of (8). Equations (6) and (8) enable us to make further transformations on (5):

$$\sum_{k=1}^{n} m_k \mathbf{a}_k \cdot \frac{\partial \mathbf{r}_k}{\partial q^j} = \sum_{k=1}^{n} \frac{\mathrm{d}}{\mathrm{d}t}\left(m_k \dot{\mathbf{r}}_k \cdot \frac{\partial \dot{\mathbf{r}}_k}{\partial \dot{q}^j}\right) - \sum_{k=1}^{n} m_k \dot{\mathbf{r}}_k \cdot \frac{\partial \dot{\mathbf{r}}_k}{\partial q^j}$$

$$= \frac{\mathrm{d}}{\mathrm{d}t}\left[\frac{\partial}{\partial \dot{q}^j}\left(\frac{1}{2}\sum_{k=1}^{n} m_k \dot{\mathbf{r}}_k \cdot \dot{\mathbf{r}}_k\right)\right] - \frac{\partial}{\partial q^j}\left(\frac{1}{2}\sum_{k=1}^{n} m_k \dot{\mathbf{r}}_k \cdot \dot{\mathbf{r}}_k\right). \tag{11}$$

Note that

$$\frac{1}{2}\sum_{k=1}^{n} m_k \dot{\mathbf{r}}_k \cdot \dot{\mathbf{r}}_k = E$$

is the kinetic energy of the system. We can therefore write

$$\sum_{k=1}^{n} m_k \mathbf{a}_k \cdot \frac{\partial \mathbf{r}_k}{\partial q^j} = \frac{\mathrm{d}}{\mathrm{d}t}\left(\frac{\partial E}{\partial \dot{q}^j}\right) - \frac{\partial E}{\partial q^j}. \tag{12}$$

What matters here is that the structure of the right-hand side of (12) is independent of the choice of the frame of reference. Examining the first term on the right-hand side of equation (4), we find that it expresses the law of transformation of the sum of vectors $\mathbf{P}_k$ from an orthogonal frame of reference to a generalized frame. This member represents a generalized force. We shall denote it by the symbol Q_j:

$$Q_j = \sum_{k=1}^{n} \mathbf{P}_k \cdot \frac{\partial \mathbf{r}_k}{\partial q^j}. \tag{13}$$

Continuing,

$$\sum_{k=1}^{n} \sum_{\mu=1}^{l} \lambda_\mu^{(1)} \mathbf{A}_{\mu k}^{(1)} \cdot \frac{\partial \mathbf{r}_k}{\partial q^j} = \sum_{\mu=1}^{l} \left(\lambda_\mu^{(1)} \sum_{k=1}^{n} \mathbf{A}_{\mu k}^{(1)} \cdot \frac{\partial \mathbf{r}_k}{\partial q^j} \right)$$

$$= \sum_{\mu=1}^{l} \lambda_\mu^{(1)} \sum_{k=1}^{n} \left(\frac{\partial F_\mu}{\partial x_k} \frac{\partial x_k}{\partial q^j} + \frac{\partial F_\mu}{\partial y_k} \frac{\partial y_k}{\partial q^j} + \frac{\partial F_\mu}{\partial z_k} \frac{\partial z_k}{\partial q^j} \right)$$

$$= \sum_{\mu=1}^{l} \lambda_\mu^{(1)} \frac{\partial \Phi_\mu}{\partial q^j}. \tag{14}$$

Naturally,

$$\begin{aligned} &\Phi_\mu(q^1, q^2, \ldots, q^r, t) \\ &\quad = F_\mu[x_1(q,t), y_1(q,t), z_1(q,t), \ldots, x_n(q,t), y_n(q,t), z_n(q,t), t] \\ &\quad = 0 \quad (\mu = 1, 2, \ldots, l). \end{aligned} \tag{15}$$

It should be appreciated that out of l equations of constraints (15), $l - l_q$ are identically satisfied, which obviously requires an appropriate choice of coordinates q. In this way, the number of equations of constraints is reduced to l_q which has been indicated in equation (2). Consequently,

$$\sum_{k=1}^{n} \sum_{\mu=1}^{l} \lambda_\mu^{(1)} \mathbf{A}_{\mu k}^{(1)} \cdot \frac{\partial \mathbf{r}_k}{\partial q^j} = \sum_{\mu=1}^{l_q} \lambda_\mu^{(1)} \frac{\partial \Phi_\mu}{\partial q^j}. \tag{16}$$

It still remains to analyse the change of the last term on the right-hand side of equation (4) on transition to generalized coordinates:

$$\sum_{k=1}^{n}\sum_{\nu=1}^{l_r}\lambda_\nu^{(2)}\mathbf{A}_{\nu k}^{(2)}\cdot\frac{\partial\mathbf{r}_k}{\partial q^j}=\sum_{\nu=1}^{l_r}\lambda_\nu^{(2)}\sum_{k=1}^{n}\mathbf{A}_{\nu k}^{(2)}\cdot\frac{\partial\mathbf{r}_k}{\partial q^j}. \tag{17}$$

On the other hand, we notice that with a change of variables the equations of nonholonomic constraints (2), Section 7.4, transform to (3); for we have

$$\sum_{k=1}^{n}\mathbf{A}_{\nu k}^{(2)}\cdot\mathbf{v}_k+B_\nu^{(2)}=\sum_{k=1}^{n}\mathbf{A}_{\nu k}^{(2)}\cdot\frac{\partial\mathbf{r}_k}{\partial q^j}\dot{q}^j+\sum_{k=1}^{n}\mathbf{A}_{\nu k}^{(2)}\cdot\frac{\partial\mathbf{r}_k}{\partial t}+B_\nu^{(2)}. \tag{18}$$

Reversing the order of summation of the first of the sums on the right-hand side of (18) and introducing the notations

$$C_{\nu j}^{(2)}=\sum_{k=1}^{n}\mathbf{A}_{\nu k}^{(2)}\cdot\frac{\partial\mathbf{r}_k}{\partial q^j}, \tag{19}$$

$$D_\nu^{(2)}=\sum_{k=1}^{n}\mathbf{A}_{\nu k}^{(2)}\cdot\frac{\partial\mathbf{r}_k}{\partial t}+B_\nu^{(2)}, \tag{20}$$

we obtain equation (3). Considering (19) in (17) we have

$$\sum_{k=1}^{n}\sum_{\nu=1}^{l_r}\lambda_\nu^{(2)}\mathbf{A}_{\nu k}^{(2)}\cdot\frac{\partial\mathbf{r}_k}{\partial q^j}=\sum_{\nu=1}^{l_r}\lambda_\nu^{(2)}C_{\nu j}^{(2)}. \tag{21}$$

We can now find the final form which the Lagrange equations of the first order take on transition from orthogonal to generalized coordinates. For this purpose we consider equalities (12), (13), (14), and (21) in the equation of motion (4):

$$\frac{\mathrm{d}}{\mathrm{d}t}\left(\frac{\partial E}{\partial\dot{q}^j}\right)-\frac{\partial E}{\partial q^j}=Q_j+\sum_{\mu=1}^{l_q}\lambda_\mu^{(1)}\frac{\partial\Phi_\mu}{\partial q^j}+\sum_{\nu=1}^{l_r}\lambda_\nu^{(2)}C_{\nu j}^{(2)}$$
$$(j=1,2,\ldots,r). \tag{22}$$

The derived equations are *Lagrange equations of the second order*. They contain $r+l_q+l_r$ unknowns. We should add the equations of possible accelerations to equation (22). In this way, the number of unknowns corresponds to the number of equations:

$$\frac{\mathrm{d}}{\mathrm{d}t}\left(\frac{\partial\Phi_\mu}{\partial q^j}\right)\dot{q}^j+\frac{\partial\Phi_\mu}{\partial q^j}\ddot{q}^j+\frac{\mathrm{d}}{\mathrm{d}t}\left(\frac{\partial\Phi_\mu}{\partial t}\right)=0, \tag{23}$$

$$\dot{C}_{\nu j}^{(2)}\dot{q}^j+C_{\nu j}^{(2)}\ddot{q}^j+\dot{D}_\nu^{(2)}=0. \tag{24}$$

The generalized forces Q_j occurring in second order Lagrange equations may be potential in character. In this case, there exists such a function V that

$$\mathbf{P}_k = -\left(\frac{\partial V}{\partial x_k}\mathbf{i}+\frac{\partial V}{\partial y_k}\mathbf{j}+\frac{\partial V}{\partial z_k}\mathbf{k}\right). \tag{25}$$

Whence

$$Q_j = -\sum_{k=1}^{n}\left(\frac{\partial V}{\partial x_k}\frac{\partial x_k}{\partial q^j}+\frac{\partial V}{\partial y_k}\frac{\partial y_k}{\partial q^j}+\frac{\partial V}{\partial z_k}\frac{\partial z_k}{\partial q^j}\right) = -\frac{\partial V}{\partial q^j}. \tag{26}$$

Forces other than potential may occur and therefore we write

$$\frac{\mathrm{d}}{\mathrm{d}t}\left(\frac{\partial E}{\partial \dot{q}^j}\right)-\frac{\partial E}{\partial q^j} = -\frac{\partial V}{\partial q^j}+Q_j+\sum_{\mu=1}^{l_q}\lambda_\mu^{(1)}\cdot\frac{\partial \Phi_\mu}{\partial q^j}+\sum_{\nu=1}^{l_r}\lambda_\nu^{(2)}C_{\nu j}^{(2)}$$
$$(j = 1, 2, \ldots, r). \tag{27}$$

Assuming that the potential is independent of velocity, we form a function $L = E-V$ (*Lagrange's kinetic potential function*). Introducing L into Lagrange equations of the second order, we obtain

$$\frac{\mathrm{d}}{\mathrm{d}t}\left(\frac{\partial L}{\partial \dot{q}^j}\right)-\frac{\partial L}{\partial q^j} = Q_j+\sum_{\mu=1}^{l_q}\lambda_\mu^{(1)}\frac{\partial \Phi_\mu}{\partial q^j}+\sum_{\nu=1}^{l_r}\lambda_\nu^{(2)}C_{\nu j}^{(2)}$$
$$(j = 1, 2, \ldots, r). \tag{28}$$

In a more general case, where V is a function of generalized velocities, equation (28) continues to be valid provided that the following equality is true:

$$Q_i = \frac{\mathrm{d}}{\mathrm{d}t}\left(\frac{\partial V}{\partial \dot{q}^i}\right)-\frac{\partial V}{\partial q^i}. \tag{29}$$

Generalized coordinates on which the Lagrange kinetic potential does not depend explicitly are called *cyclic coordinates*. Obviously, for such coordinates the following occurs:

$$\frac{\partial L}{\partial q^j} = 0 \quad (j = m+1, m+2, \ldots, r). \tag{30}$$

As we have already ascertained, for holonomic systems it is always possible to choose such a number of coordinates as corresponds to the number of degrees of freedom of a system. The virtual displacements and the possible velocities and displacements are then independent. In the considered case the Lagrange equations have a much simpler form:

$$\frac{\mathrm{d}}{\mathrm{d}t}\left(\frac{\partial L}{\partial \dot{q}^j}\right)-\frac{\partial L}{\partial q^j}=Q_j \quad (j=1,2,\ldots,r), \tag{31}$$

where r is simultaneously the number of degrees of freedom of the system.

Example 1. A gearing consists of two toothed wheels. Moments M_1 and M_2 act on these wheels (Fig. 7.8). Find the tangential force of action between the wheels and the acceleration of the first wheel.

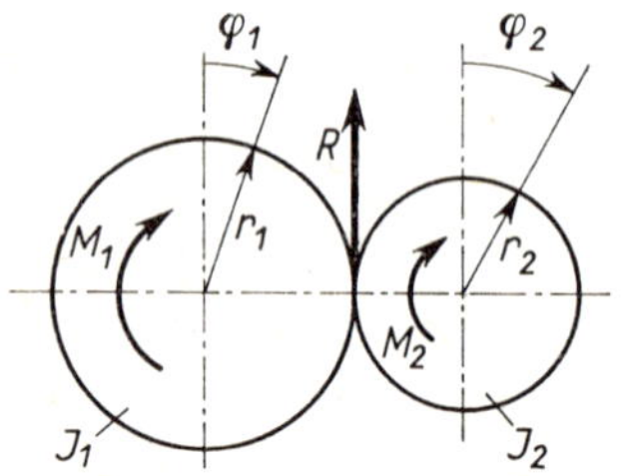

Fig. 7.8

We choose generalized coordinates $q^1 \equiv \varphi_1$ and $q^2 \equiv \varphi_2$. The kinetic energy of the system is

$$E = \tfrac{1}{2}J_1\dot{\varphi}_1^2+\tfrac{1}{2}J_2\dot{\varphi}_2^2. \tag{a}$$

The equation of constraints is as follows:

$$\Phi = r_1\varphi_1+r_2\varphi_2 = 0. \tag{b}$$

The Lagrange equations (22) for the considered case takes the form

$$\frac{\mathrm{d}}{\mathrm{d}t}\left(\frac{\partial E}{\partial \dot{\varphi}_1}\right)-\frac{\partial E}{\partial \varphi_1}=M_1+\lambda\frac{\partial \Phi}{\partial \varphi_1}, \qquad \frac{\mathrm{d}}{\mathrm{d}t}\left(\frac{\partial E}{\partial \dot{\varphi}_2}\right)-\frac{\partial E}{\partial \varphi_2}=M_{12}+\lambda\frac{\partial \Phi}{\partial \varphi_2}$$

or after considering (a) and (b)

$$J_1\ddot{\varphi}_1 = M_1+\lambda r_1, \quad J_2\ddot{\varphi}_2 = M_2+\lambda r_2. \tag{c}$$

We now make use of the equation of possible accelerations:

$$\frac{\ddot{\varphi}_1}{\ddot{\varphi}_2} = -\frac{r_2}{r_1}.$$

Therefore

$$-\frac{J_1}{J_2}\frac{r_2}{r_1} = \frac{M_1+\lambda r_1}{M_2+\lambda r_2},$$

and hence

$$\lambda = -\frac{J_1 r_2 M_2 + M_1 J_2 r_1}{J_1 r_2^2 + r_1^2 J_2}. \tag{d}$$

We have thus obtained the answer to the first part of the question.

To calculate the acceleration, we shall return to equations (c):

$$\ddot{\varphi} = \frac{M_1}{J_1} - \frac{r_1}{J_1}\frac{J_1 r_2 M_2 + M_1 J_2 r_1}{J_1 r_2^2 + r_1^2 J_2} = \frac{M_1}{J_1} - \frac{M_2 + M_1 \dfrac{J_2}{J_1}\dfrac{r_1}{r_2}}{J_1 \dfrac{r_2}{r_1} + J_2 \dfrac{r_1}{r_2}}. \tag{e}$$

We can arrive at this result proceeding in a somewhat different way. Due to equation (b) we can express the energy E with the use of one coordinate:

$$E = \frac{1}{2} J_1 \dot{\varphi}_1^2 + \frac{1}{2} J_2 \left(\frac{r_1}{r_2}\right)^2 \dot{\varphi}_1^2. \tag{f}$$

Now a second order Lagrange equation will be valid:

$$\frac{\mathrm{d}}{\mathrm{d}t}\left(\frac{\partial E}{\partial \dot{\varphi}_1}\right) - \frac{\partial E}{\partial \varphi_1} = Q_{\varphi_1}. \tag{g}$$

We find the generalized force Q_{φ_1} in the following manner:

$$Q_{\varphi_1}\delta\varphi_1 = M_1 \delta\varphi_1 + M_2 \delta\varphi_2.$$

Since

$$\delta\varphi_2 = -\frac{r_1}{r_2}\delta\varphi_1,$$

we have

$$Q_{\varphi_1}\delta\varphi_1 = \left(M_1 - \frac{r_1}{r_2} M_2\right)\delta\varphi_1.$$

Hence

$$Q_{\varphi_1} = M_1 - \frac{r_1}{r_2} M_2. \tag{h}$$

Substituting (f) and (h) into (g), we obtain

$$\left[J_1 + J_2\left(\frac{r_1}{r_2}\right)^2\right]\ddot{\varphi}_1 = M_1 - \frac{r_1}{r_2} M_2,$$

whence

$$\ddot{\varphi}_1 = \frac{M_1 - \dfrac{r_1}{r_2} M_2}{J_1 + J_2 \left(\dfrac{r_1}{r_2}\right)^2}. \tag{i}$$

The result (i) coincides with the result (e), which can easily be checked.

7.6 Kinetic energy of a system

We have already determined the kinetic energy of a system with the use of equation (7a), Section 2.1. Now we shall express it using generalized velocities and coordinates:

$$E = \tfrac{1}{2}\sum_{k=1}^{n} m_k \dot{\mathbf{r}}_k \cdot \dot{\mathbf{r}}_k, \tag{1}$$

but

$$\begin{aligned}\dot{\mathbf{r}}_k \dot{\mathbf{r}}_k &= \left(\frac{\partial \mathbf{r}_k}{\partial q^j}\dot{q}^j + \frac{\partial \mathbf{r}_k}{\partial t}\right)\cdot\left(\frac{\partial \mathbf{r}_k}{\partial q^i}\dot{q}^i + \frac{\partial \mathbf{r}_k}{\partial t}\right) \\ &= \frac{\partial \mathbf{r}_k}{\partial q^i}\cdot\frac{\partial \mathbf{r}_k}{\partial q^j}\dot{q}^i\dot{q}^j + 2\frac{\partial \mathbf{r}_k}{\partial t}\cdot\frac{\partial \mathbf{r}_k}{\partial q^i}\dot{q}^i + \left(\frac{\partial \mathbf{r}_k}{\partial t}\right)^2.\end{aligned} \tag{2}$$

Evidently, in generalized coordinates, the kinetic energy is expressed by

$$E = \tfrac{1}{2}a_{ij}\dot{q}^i\dot{q}^j + a_i\dot{q}^i + a_0, \tag{3}$$

where

$$a_{ij} = \sum_{k=1}^{n} m_k \frac{\partial \mathbf{r}_k}{\partial q^i}\cdot\frac{\partial \mathbf{r}_k}{\partial q^j}, \qquad a_i = \sum_{k=1}^{n} m_k \frac{\partial \mathbf{r}_k}{\partial q^i}\cdot\frac{\partial \mathbf{r}_k}{\partial t},$$
$$a_0 = \frac{1}{2}\sum_{k=1}^{n} m_k\left(\frac{\partial \mathbf{r}_k}{\partial t}\right)^2. \tag{4}$$

Note that for scleronomic systems $a_0 = 0$ and for any i there is $a_i = 0$, since in this case $\partial \mathbf{r}_k/\partial t = 0$ for any k. Thus we can see that in the case where the constraints are independent of time, the kinetic energy of a system is a homogeneous quadratic form of generalized velocities.

Returning to a more general case, we write equation (3) as follows:

$$E = E_2 + E_1 + E_0, \tag{5}$$

where

$$E_2 = \frac{1}{2}a_{ij}\dot{q}^i\dot{q}^j, \qquad E_1 = a_i\dot{q}^i, \qquad E_0 = a_0. \tag{5a}$$

Notation (5) will be used further on.

7.7 Impulsive motion

In Section 7.6 we discussed the application of the law of variation of momentum in the case of collision of two masses. This was an example of impulsive motion. In the present section we shall use the methods of analytical mechanics to formalize the problem. From equation (7), Section 2.4, we have:

$$\Delta\mathbf{p} = \int_{t_0}^{t_0+\Delta t} \mathbf{P}\,dt = \mathbf{P}_{\text{mean}}\cdot\Delta t, \tag{1}$$

which means that the change in momentum of a mass $\Delta\mathbf{p}$ is equal to the impulse of a force. The force $\mathbf{P}_{\text{mean}}$ is the mean value of the force acting on the mass in the course of collission (the force inducing the impulse). In the instant after impulse $\Delta\mathbf{p}_k$ has been given to mass m_k, its velocity will change from $\mathbf{v}_{0k}$ to $\mathbf{v}_k$, which we write

$$m_k(\mathbf{v}_k - \mathbf{v}_{0k}) = \Delta\mathbf{p}_k. \tag{2}$$

The considered system consists of n particles. Examining the change of the momentum of mass m_k due to impulse $\Delta\mathbf{p}_k$, we notice that it is a resultant of the impulses of external forces $\mathbf{P}_{k(1)}$ and internal forces $\mathbf{P}_{k(2)}$. Hence

$$\Delta\mathbf{p}_k = \Delta\mathbf{p}_{k(1)} + \Delta\mathbf{p}_{k(2)}, \tag{3}$$

therefore

$$m_k(\mathbf{v}_k - \mathbf{v}_{0k}) = \Delta\mathbf{p}_{k(1)} + \Delta\mathbf{p}_{k(2)}, \tag{4}$$

where $\Delta\mathbf{p}_{k(1)}$ is the resultant impulse of external forces and $\Delta\mathbf{p}_{k(2)}$ the resultant impulse of internal force.

To derive the equations of impulsive motion in curvilinear coordinates we shall perform analogous operations to those in the derivation of the second order Lagrange equations. Thus we construct from (4) a system of equations analogous to (4), Section 7.5:

$$\sum_{k=1}^{n} m_k(\mathbf{v}_k - \mathbf{v}_{0k})\frac{\partial\mathbf{r}_k}{\partial q^j} = \sum_{k=1}^{n} \Delta\mathbf{p}_{k(1)}\frac{\partial\mathbf{r}_k}{\partial q^j} + \sum_{k=1}^{n} \Delta\mathbf{p}_{k(2)}\frac{\partial\mathbf{r}_k}{\partial q^j}. \tag{4a}$$

Note first that the last term on the right-hand side of (4a) is equal to zero. Indeed, making use of equality (1) we have

$$\sum_{k=1}^{n} \Delta\mathbf{p}_{k(2)}\cdot\frac{\partial\mathbf{r}_k}{\partial q^j} = \sum_{k=1}^{n} \mathbf{P}_{k(\text{mean})}\cdot\Delta t\frac{\partial\mathbf{r}_k}{\partial q^j}.$$

But

$$\mathbf{P}_{k(\text{mean})} = \sum_{i=1}^{n} \mathbf{P}_{ik(\text{mean})}$$

where $\mathbf{P}_{ik(\text{mean})}$ are mean values of action and reaction forces between points of the system

$$\sum_{k=1}^{n} \Delta \mathbf{p}_{k(2)} \cdot \frac{\partial \mathbf{r}_k}{\partial q^j} = \sum_{k=1}^{n} \sum_{i=1}^{n} \mathbf{P}_{ik(\text{mean})} \cdot \Delta t \frac{\partial \mathbf{r}_k}{\partial q^j}$$

$$= \Delta t \frac{1}{2} \sum_{k=1}^{n} \sum_{i=1}^{n} (\mathbf{P}_{ik(\text{mean})} + \mathbf{P}_{ki(\text{mean})}) \frac{\partial \mathbf{r}_k}{\partial q^j}.$$

Now

$$\bigwedge_{i,k} \mathbf{P}_{ik(\text{mean})} + \mathbf{P}_{ki(\text{mean})} = 0 \qquad (\textit{law of action and reaction});$$

therefore

$$\sum_{k=4}^{n} \Delta \mathbf{p}_{k(2)} \frac{\partial \mathbf{r}_k}{\partial q^j} = 0.$$

In this way the number of terms on the right-hand side of equation (4a) is decreased.

We shall now concern ourselves with a transformation of the left-hand side of (4a). We have

$$\sum_{k=1}^{n} m_k (\mathbf{v}_k - \mathbf{v}_{0k}) \frac{\partial \mathbf{r}_k}{\partial q^j} = \sum_{k=1}^{n} m_k \dot{\mathbf{r}}_k \frac{\partial \mathbf{r}_k}{\partial q^j} - \sum_{k=1}^{n} m_k \dot{\mathbf{v}}_{0k} \frac{\partial \mathbf{r}_k}{\partial q^j}. \tag{5}$$

By virtue of (6), Section 7.5,

$$\sum_{k=1}^{n} m_k \dot{\mathbf{r}}_k \frac{\partial \mathbf{r}_k}{\partial q^j} = \sum_{k=1}^{n} m_k \dot{\mathbf{r}}_k \frac{\partial \dot{\mathbf{r}}_k}{\partial \dot{q}^j} = \frac{\partial}{\partial \dot{q}^j} \left(\frac{1}{2} \sum_{k=1}^{n} m_k \dot{\mathbf{r}}_k \cdot \dot{\mathbf{r}}_k \right) = \frac{\partial E}{\partial \dot{q}^j}. \tag{6}$$

Analogously,

$$\sum_{k=1}^{n} m_k \dot{\mathbf{r}}_{0k} \frac{\partial \mathbf{r}_k}{\partial q^j} = \left(\frac{\partial E}{\partial \dot{q}^j} \right)_0, \tag{7}$$

which means that the derivative $(\partial E / \partial \dot{q}^j)_0$ corresponds to the derivative $\partial E / \partial \dot{q}^j$ but refers to the instant preceding the impact.

The right-hand side of equation (4a) is a transformation of the impulse vector to curvilinear coordinates

$$\sum_{k=1}^{n} \Delta \mathbf{P}_{k(1)} \frac{\partial \mathbf{r}_k}{\partial q^j} = I_j \tag{8}$$

where I_j is a generalized impulse.

Note that the transformations had the character of tensor transformations.

Considering (6), (7) and (8) in equation (4a), we have

$$\frac{\partial E}{\partial \dot{q}^j} - \left(\frac{\partial E}{\partial \dot{q}^j}\right)_0 = I_j. \tag{9}$$

Equations of type (9) are known under the name of *Lagrange equations of impulsive motion.*

Example 1. Find the equation for the path $s = s(t)$ of a homogeneous ingot sel in motion without sliding on a horizontal conveyor table with initial velocity v_0. The conveyor table (Fig. 7.9) consists of horizontally mounted rollers each having equat diameters, $2r$, and moments of inertia, J. Assume that the rollers rotate without friction

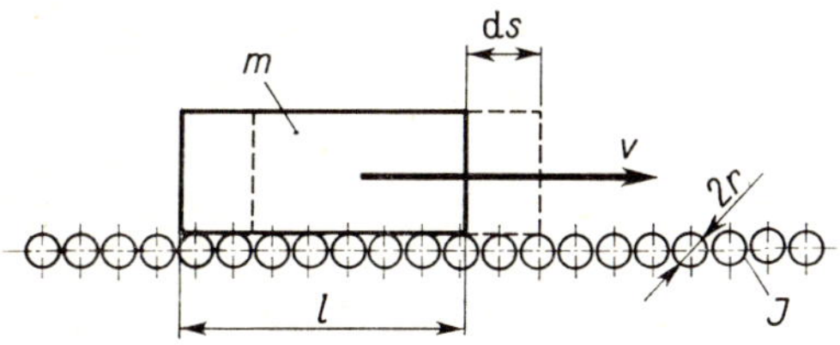

Fig. 7.9

and are uniformly distributed at the rate of n rollers per 1 m of table length. The mass of the ingot is m, and length l. During the impact of the ingot against the roller an internal impulse occurs, therefore the equation of impulsive motion is

$$\frac{\partial E}{\partial V} - \left(\frac{\partial E}{\partial V}\right)_0 = 0. \tag{a}$$

Prior to the impact the kinetic energy of the system is

$$E = \frac{1}{2}\left[m + \frac{nlJ}{r^2}\right]v^2, \tag{b}$$

and after the impact

$$E = \frac{1}{2}\left[m+\frac{nlJ}{r^2}\right](v+\mathrm{d}v)^2+\frac{1}{2}\frac{nJ}{r^2}v^2\,\mathrm{d}s. \tag{c}$$

Substituting (b) and (c) into (a), we have

$$\left[m+\frac{nlJ}{r^2}\right]v-\left[m+\frac{nlJ}{r^2}\right](v+\mathrm{d}v)+\frac{nJ}{r^2}v\,\mathrm{d}s = 0. \tag{d}$$

Substituting

$$\varkappa = \frac{nJ}{[mr+nlJ/r]r}, \tag{e}$$

after simplification we obtain

$$\frac{\mathrm{d}v}{v} = -\varkappa\,\mathrm{d}s; \tag{f}$$

integrating

$$\ln\frac{v}{v_0} = -\varkappa s; \tag{g}$$

hence

$$v = v_0\mathrm{e}^{-\varkappa s}, \tag{h}$$

therefore

$$\frac{\mathrm{d}s}{\mathrm{d}t} = v_0\mathrm{e}^{-\varkappa s}$$

or

$$\mathrm{e}^{\varkappa s}\,\mathrm{d}s = v_0\,\mathrm{d}t. \tag{i}$$

After integrating, we have

$$\frac{1}{\varkappa}(\mathrm{e}^{\varkappa s}-1) = v_0 t \tag{j}$$

and hence

$$s = \frac{1}{\varkappa}\ln(\varkappa v_0 t+1). \tag{k}$$

7.8 Gyroscopic and dissipative forces

In certain cases the generalized forces are linear functions of generalized velocities

$$Q_j = \varkappa_{ij}\dot{q}^i. \tag{1}$$

The matrix $\varkappa_{ij}$ can be split into a symmetrical and an antisymmetrical part according to the relationship

$$\varkappa_{ij} = \tfrac{1}{2}(\varkappa_{ij}+\varkappa_{ji})+\tfrac{1}{2}(\varkappa_{ij}-\varkappa_{ji}). \tag{2}$$

Let us calculate the power of the generalized forces

$$\frac{dW}{dt} = Q_j \dot{q}^j = \varkappa_{ij}\dot{q}^i\dot{q}^j = -h_{ij}\dot{q}^i\dot{q}^j + h^*_{ij}\dot{q}^i\dot{q}^j, \tag{3}$$

where

$$h_{ij} = -\tfrac{1}{2}(\varkappa_{ij}+\varkappa_{ji}), \quad h^*_{ij} = \tfrac{1}{2}(\varkappa_{ij}-\varkappa_{ji}). \tag{3a}$$

We shall discuss two cases:

(1) $h^*_{ij} = 0$, i.e., the matrix $\varkappa_{ij}$ is a symmetrical matrix. In the considered case,

$$\frac{dW}{dt} = -h_{ij}\dot{q}^i\dot{q}^j. \tag{4}$$

If additionally the quadratic form on the right-hand side of (4) is positive definite, then the function

$$D = \tfrac{1}{2}h_{ij}\dot{q}^i\dot{q}^j \tag{5}$$

gives energy dissipation and is called *Rayleigh's dissipation function*. In such a case

$$\frac{dW}{dt} = -2D < 0.$$

It is therefore evident that the power of dissipative forces is negative.

From (4) we can easily see that the generalized force inducing dissipation is given by

$$Q_j = -h_{ij}\dot{q}^i = -\frac{\partial D}{\partial \dot{q}^j}. \tag{6}$$

(2) $h_{ij} = 0$, i.e., the matrix $\varkappa_{ij}$ is an antisymmetrical matrix. Then

$$\frac{dW}{dt} = h^*_{ij}\dot{q}^i\dot{q}^j = 0. \tag{7}$$

Hence the conclusion that gyroscopic forces do not dissipate the energy of the system. Consequently, they may occur in a conservative system. These forces, as one can easily see, are given by

$$Q_i = h^*_{ij}\dot{q}^j. \tag{8}$$

They owe their name to the fact that they occur in gyroscopic motion. The precession of a humming top is made possible because gyroscopic forces, all equal, act on it:

$$J(\boldsymbol{\omega}\times\boldsymbol{\omega}_1) = \mathbf{M}.$$

Indeed, these generalized forces have no influence on change in the energy of the system, since

$$\mathbf{M}\cdot\boldsymbol{\omega} = J(\boldsymbol{\omega}\times\boldsymbol{\omega}_1)\cdot\boldsymbol{\omega} = 0.$$

We return now to Rayleigh's dissipation function and introduce it into the Lagrange equations of motion. In this way, equation (28), Section 7.5, is augmented by $\partial D/\partial \dot{q}^j$:

$$\frac{\mathrm{d}}{\mathrm{d}t}\left(\frac{\partial L}{\partial \dot{q}^j}\right) - \frac{\partial L}{\partial q^j} + \frac{\partial D}{\partial \dot{q}^j} = Q_j + \sum_{\mu=1}^{l_q} \lambda_\mu^{(1)} \frac{\partial \Phi_\mu}{\partial q^j} + \sum_{\nu=1}^{l_r} \lambda_\nu^{(2)} C_{\nu j}^{(2)} \qquad (9)$$

$$(j = 1, 2, \ldots, r).$$

Example 1. A centrifugal governor (Fig. 7.10) is in angular motion about a vertical axis with angular velocity $\omega = \omega(t)$. Because of the field of centrifugal forces, the masses attached at points A and B move apart simultaneously raising a weight of mass M. This raising is opposed by a spring with a constant c. Find the differential equation giving the changes in the configuration of the system in response to changes in angular velocity $\omega(t)$. Assume viscous friction between the bush T and the axis OO_1.

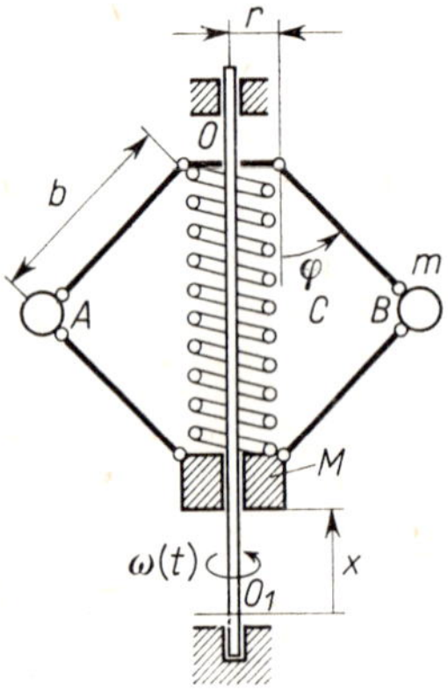

Fig. 7.10

We find first the kinetic energy of the system:

$$E = \tfrac{1}{2}M\dot{x}^2 + m(\dot{\varphi}^2 b^2 + \omega^2(r+b\sin\varphi)^2) + \tfrac{1}{2}J_T\omega^2.$$

Now

$$x = 2b(1-\cos\varphi) \quad \text{and} \quad \dot{x} = 2b\dot{\varphi}\sin\varphi,$$

therefore

$$E = \tfrac{1}{2}M\cdot 4b^2\dot{\varphi}^2\sin^2\varphi + m[\dot{\varphi}^2 b^2 + \omega^2(r+b\sin\varphi)^2] + \tfrac{1}{2}J_T\omega^2, \qquad \text{(a)}$$

where M is mass of bush T, m governor mass attached at points A and B, b length of rods, and J_r moment of inertia of the governor about the axis of rotation.

We calculate in turn the potential energy V:

$$V = 2mgb(1-\cos\varphi)+\tfrac{1}{2}cx^2+Mgx, \tag{b}$$

where c is the spring constant. The considered system is a holonomic system with one degree of freedom. For such a system the following equation of motion is valid:

$$\frac{\mathrm{d}}{\mathrm{d}t}\left(\frac{\partial L}{\partial \dot{\varphi}}\right)-\frac{\partial L}{\partial \varphi}+\frac{\partial D}{\partial \dot{\varphi}} = 0. \tag{c}$$

In the considered case apart from potential and dissipative forces, no other generalized forces are involved. The function D is as follows:

$$D = \tfrac{1}{2}\alpha\dot{x}^2 = \tfrac{1}{2}\alpha 4b^2\dot{\varphi}^2\sin^2\varphi. \tag{d}$$

Forming the function $L = E-V$ and substituting it together with (d) into (c) and also performing the indicated operations, we obtain

$$\begin{aligned}2b^2(m+2M\sin^2\varphi)\ddot{\varphi}+4Mb^2\dot{\varphi}^2\sin\varphi\cos\varphi-&\\ -2m\omega^2(r+b\sin\varphi)b\cos\varphi+2g(m+M)b\sin\varphi+&\\ +4cb^2(1-\cos\varphi)\sin\varphi+4b^2\alpha\dot{\varphi}^2\sin^2\varphi = 0.&\end{aligned} \tag{e}$$

We can find the stationary form of motion from the derived equation (e). It will suffice to put

$$\ddot{\varphi} = \dot{\varphi} = 0.$$

On this basis for an exact value of ω = const, we can find the configuration of the system, i.e., we can specify the angle φ.

7.9 The Lagrange equations for electromechanical systems

Using Lagrange's equations we can describe the dynamics of electromechanical systems [10, 16, 42]. The extension, made by Maxwell, of the equations to apply to these systems, consists of complementing the Lagrange potential with an analogue of the kinetic and potential energy, and also of finding an analogue of the generalized force.

Let $q^1, q^2, \ldots, q^r$ denote geometric coordinates and $q^{r+1}, q^{r+2}, \ldots, q^u$ electrical coordinates, strictly speaking the electric charges passing through conductors. Next, denote the matrix of electromagnetic inductance by L_{ij}; the jth magnetic flux is then expressed by

$$\Phi_j = \sum_{i=r+1}^{u} L_{ij}\dot{q}^i. \tag{1}$$

The magnetic energy

$$F_e = \tfrac{1}{2}\sum_{i=r+1}^{u}\sum_{j=r+1}^{u} L_{ij}\dot{q}^i\dot{q}^j \tag{2}$$

is the analogue of the kinetic energy of a mechanical system: $\dot{q}^j$ is the current intensity in the jth branch.

Let b_{ij} denote the matrix of electrostatic induction, by means of which the jth voltage can be expressed as follows:

$$u_j = -\sum_{i=r+1}^{u} b_{ij} q^i, \tag{3}$$

and the electrostatic energy as

$$V_e = \tfrac{1}{2} \sum_{i=r+1}^{u} \sum_{j=r+1}^{u} b_{ij} q^i q^j. \tag{4}$$

The kinetic potential of the electromechanical system is given by

$$L = E + E_e - V - V_e. \tag{5}$$

We now find the force Q_j in electric systems. Let R_{ij} denote effective resistance and e_j electromotive forces. The power of generalized forces in an electric system is

$$\sum_{j=r+1}^{u} Q_j \dot{q}^j = \sum_{j=r+1}^{u} e_j \dot{q}^j - \sum_{i=r+1}^{u} \sum_{j=r+1}^{u} R_{ij} \dot{q}^i \dot{q}^j. \tag{6}$$

Hence

$$Q_j = e_j - \sum_{i=r+1}^{u} R_{ij} \dot{q}^i \quad (j = r+1, \ldots, u). \tag{7}$$

Rayleigh's dissipation function is expressed in electric systems by

$$D = \tfrac{1}{2} \sum_{i=r+1}^{u} \sum_{j=r+1}^{u} R_{ij} \dot{q}^i \dot{q}^j \quad (i, j = r+1, \ldots, u). \tag{8}$$

And in electromechanical systems

$$D = \tfrac{1}{2} \sum_{i=1}^{r} \sum_{j=1}^{r} h_{ij} \dot{q}^i \dot{q}^j + \tfrac{1}{2} \sum_{i=r+1}^{u} \sum_{j=r+1}^{u} R_{ij} \dot{q}^i \dot{q}^j. \tag{9}$$

Example 1. An electric system consists of two coils with inductance L_{22} and L_{33}, respectively, two condensers with capacitors C_1 and C_2, two resistors R_1 and R_2, and an electromotive force $e(t)$. The diagram of the system is shown in Fig. 7.11. Give the differential equations describing current propagation in the circuits.

We begin by finding the analogue of the kinetic energy given by equation (2):

$$E_e = \tfrac{1}{2} L_{22} \dot{q}^2 \dot{q}^2 + \tfrac{1}{2} L_{33} \dot{q}^3 \dot{q}^3. \tag{a}$$

Next we determine the electrostatic energy (equation (4)):

$$V_e = \frac{1}{2C_1}(q^1-q^2)^2+\frac{1}{2C_2}(q^2-q^3)^2. \tag{b}$$

Due to the existence of effective resistance, dissipation exists:

$$D_e = \tfrac{1}{2}R_1(\dot{q}^1-\dot{q}^2)^2+\tfrac{1}{2}R_2(\dot{q}^2-\dot{q}^3)^2. \tag{c}$$

Forming a Lagrange function from (a) and (b) and then substituting it together with the dissipation function into the system of equations (9), we obtain

$$\frac{1}{C_1}(q^1-q^2)+R_1(\dot{q}^1-\dot{q}^2) = e_1(t),$$

$$L_{22}\ddot{q}^2-\frac{1}{C_1}(q^1-q^2)+\frac{1}{C_2}(q^2-q^3)-R_1(\dot{q}^1-\dot{q}^2)+R_2(\dot{q}^2-\dot{q}^3) = 0,$$

$$L_{33}\ddot{q}^3-\frac{1}{C_2}(q^2-q^3)-R_2(\dot{q}^2-\dot{q}^3) = 0.$$

This is the looked for system of equations.

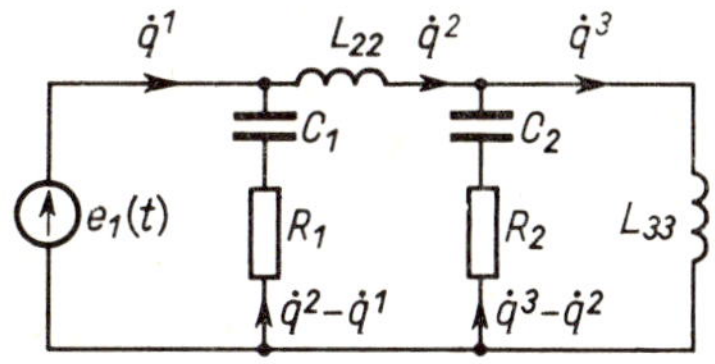

Fig. 7.11

Example 2. Give the differential equations describing current alternation in th system presented in Fig. 7.12. A mutual inductance coupling L_{12} exists between two coils with inductance L_{11} and L_{22}, respectively.

An analogous procedure to that followed in the preceding problem leads to the expressions for E_e and V_e:

$$E_e = \tfrac{1}{2}L_{11}\dot{q}^1\dot{q}^1+L_{12}\dot{q}^1\dot{q}^2+\tfrac{1}{2}L_{22}\dot{q}^2\dot{q}^2, \tag{a}$$

$$V_e = \frac{1}{2C_1}q^1q^1+\frac{1}{2C_2}q^2q^2. \tag{b}$$

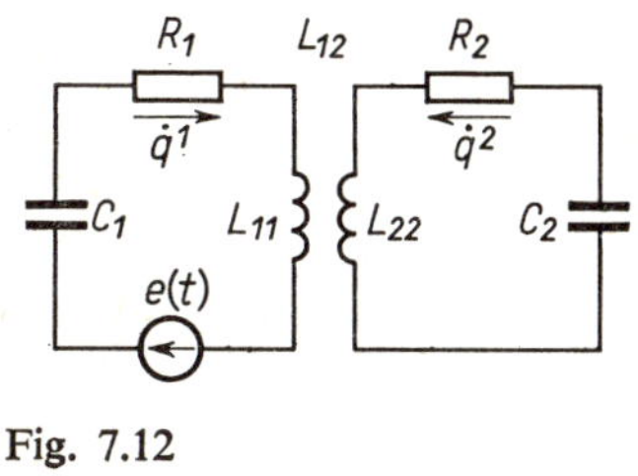

Fig. 7.12

Similarly, we find the energy dissipation of the system,

$$D_e = \tfrac{1}{2} R_1 \dot{q}^1 \dot{q}^1 + \tfrac{1}{2} R_2 \dot{q}^2 \dot{q}^2. \tag{c}$$

Forming a Lagrange function from (a) and (b) and substituting it together into Lagrange equations (9), we obtain the following system of differential equations:

$$L_{11}\ddot{q}^1 + L_{12}\ddot{q}^2 + \frac{1}{C_1} q^1 + R_1 \dot{q}^1 = e(t), \quad L_{12}\ddot{q}^1 + L_{22}\ddot{q}^2 + \frac{1}{C_2} q^2 + R_2 \dot{q}^2 = 0. \tag{d}$$

Example 3. An electromechanical system comprises two circuits: a one, stationary consists of a coil with inductance L_{11}, a resistor R_1 and an electromotive force $e_1(t)$; the other, moving, consists of a coil with inductance L_{22}, a condenser with capacitance C, a resistor R_2 and an electromotive force $e_2(t)$ (Fig. 7.13). The mutual inductance of the coils L_{12} is a function of the angle $\varphi \equiv q^3$ of the moving-circuit coil.

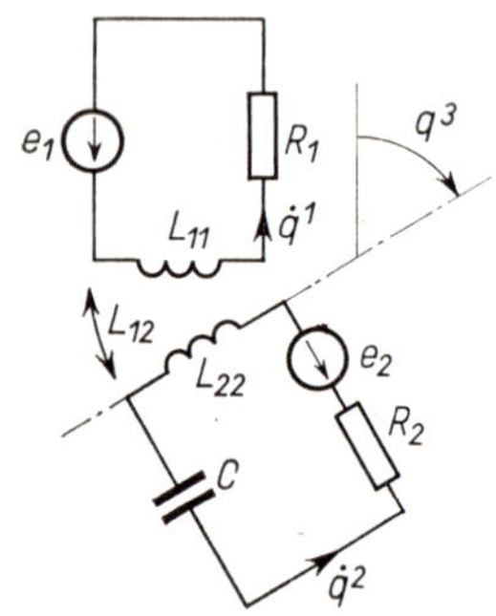

Fig. 7.13

The mechanical part, i.e., the moving circuit, has a moment of inertia about the axis of rotation equal to J. During motion a moment of friction occurs proportional to angular velocity, having an absolute value $h\dot{q}^3$; moreover the moment of a spring, equal to kq^3, acts against the angle of rotation of the coil. Find the differential equations describing the behaviour of the system.

It should first be made clear that the system has three degrees of freedom. Denoting the currents in the circuits by q^1 and q^2 and the angle of rotation of the coil in the moving part by q^3 we can determine both the kinetic and the potential energy of the system:

$$E = \tfrac{1}{2} L_{11} \dot{q}^1 \dot{q}^1 + L_{12} \dot{q}^1 \dot{q}^2 + \tfrac{1}{2} L_{22} \dot{q}^2 \dot{q}^2 + \tfrac{1}{2} J \dot{q}^3 \dot{q}^3, \tag{a}$$

$$V = \frac{1}{2C} q^2 q^2 + \tfrac{1}{2} k q^3 q^3. \tag{b}$$

The dissipation function in this case is expressed by

$$D = \tfrac{1}{2} R_1 \dot{q}^1 \dot{q}^1 + \tfrac{1}{2} R_2 \dot{q}^2 \dot{q}^2 + \tfrac{1}{2} h \dot{q}^3 \dot{q}^3. \tag{c}$$

Considering (a), (b) and (c) in the Lagrange equations, we have

$$\frac{\partial L_{12}}{\partial q^3}\dot{q}^3\dot{q}^2+L_{11}\ddot{q}^1+L_{12}\ddot{q}^2+R_1\dot{q}^1 = e_1(t),$$

$$L_{22}\ddot{q}^2+L_{12}\ddot{q}^1+\frac{\partial L_{12}}{\partial q^3}\dot{q}^1\dot{q}^3+R_2\dot{q}^2+\frac{1}{C}\ddot{q}^2 = e_2(t), \tag{d}$$

$$J\ddot{q}^3-\frac{\partial L_{12}}{\partial q^3}\dot{q}^1\dot{q}^2+h\dot{q}^3+kq^3 = 0.$$

The above equations constitute the solution to our problem.

7.10 Hamilton's canonical equations

The presentation of equations of motion in the Lagrange form is not the only one avilable. By introducing the definition of generalized moments

$$p_i \stackrel{\text{df}}{=} \frac{\partial L}{\partial \dot{q}^i} \tag{1}$$

it is possible, as we shall immediately show, to reduce a system of Lagrange equations of the second order to a normal system of equations of the first order. Note that (31), Section 7.5, can be written in the following form:

$$\frac{\partial L}{\partial q^i} = \dot{p}_i - Q_i. \tag{2}$$

The variation of the Lagrange kinetic potential, without time variation, is

$$\delta L = \frac{\partial L}{\partial q^j}\,\delta q^j+\frac{\partial L}{\partial \dot{q}^j}\,\delta\dot{q}^j. \tag{3}$$

Or after considering (2)

$$\delta L = (\dot{p}_j-Q_j)\,\delta q^j+\frac{\partial L}{\partial \dot{q}^j}\,\delta\dot{q}^j. \tag{4}$$

We note further that

$$\delta(p_j\dot{q}^j) = p_j\delta\dot{q}^j+\dot{q}^j\delta p_j. \tag{5}$$

On this basis we can write

$$\delta L = (\dot{p}_j-Q_j)\,\delta q^i+p_j\,\delta\dot{q}^j = (\dot{p}_j-Q_j)\,\delta q^j+\delta(p_j\dot{q}^j)-\dot{q}^j\delta p_j. \tag{6}$$

We form the function

$$H = p_j\dot{q}^j-L. \tag{7}$$

It is known as *Hamilton's function*. The occurrence of Hamilton's function in equation (6) is evident. Therefore

$$\delta H = \dot{q}^j\delta p_j-(\dot{p}_j-Q_j)\,\delta q^j, \tag{8}$$

where H is a function of generalized coordinates, generalized moment, and time. On the other hand,

$$\delta H = \frac{\partial H}{\partial p_j}\delta p_j + \frac{\partial H}{\partial q^j}\delta q^j. \tag{9}$$

Comparing (8) and (9), we find that

$$\dot{q}^j = \frac{\partial H}{\partial p_j}, \quad \dot{p}_j - Q_j = -\frac{\partial H}{\partial q^j}. \tag{10}$$

Equations (10) are the Hamilton's canonical equations we seek.

In the case of non-holonomic systems the following expression should replace that of the generalized force Q_j:

$$\Omega_j = Q_j + \sum_{\mu=1}^{l_q} \lambda_\mu^{(1)} \frac{\partial \Phi_\mu}{\partial q^j} + \sum_{\nu=1}^{l_r} \lambda_\nu^{(2)} C_{\nu j}^{(2)}. \tag{11}$$

7.11 The total energy of a mechanical system

The overall, total, energy of a mechanical system is the sum of the potential and the kinetic energy

$$T = E + V.$$

We shall now attempt to give on the basis of Hamilton's equations an expression by which changes in the overall, total, energy of a system can be assessed. They are assessable if we know T or the derivative $\mathrm{d}T/\mathrm{d}t$. It appears that using Hamilton's equations, it is easy to achieve this goal.

Note first that Hamilton's function can be expressed by components of the kinetic energy (equation (3), Section 7.6) and of the potential energy. We easily find that

$$p_j \dot{q}^j = \frac{\partial L}{\partial \dot{q}^j}\dot{q}^j = \frac{\partial E}{\partial \dot{q}^j}\dot{q}^j = 2E - E_1 - 2E_0. \tag{1}$$

Next, substituting this result (1) into equation (7), Section 7.10, we have

$$H = 2E - E_1 - 2E_0 - E + V = T - E_1 - 2E_0. \tag{2}$$

Differentiating the derived expression with respect to time,

$$\frac{\mathrm{d}H}{\mathrm{d}t} = \frac{\mathrm{d}T}{\mathrm{d}t} - \frac{\mathrm{d}}{\mathrm{d}t}(E_1 + 2E_0). \tag{3}$$

On the other hand,

$$\frac{\mathrm{d}H}{\mathrm{d}t} = \frac{\partial H}{\partial p_j}\dot{p}_j + \frac{\partial H}{\partial q^j}\dot{q}^j + \frac{\partial H}{\partial t}.$$

Considering Hamilton's equations, we note that

$$\frac{\mathrm{d}H}{\mathrm{d}t} = \Omega_j \dot{q}^j + \frac{\partial H}{\partial t}. \tag{4}$$

Partially differentiating (7), Section 7.10, we obtain

$$\frac{\partial H}{\partial t} = p_j \frac{\partial \dot{q}^j}{\partial t} - \frac{\partial L}{\partial \dot{q}^j}\frac{\partial \dot{q}^j}{\partial t} - \frac{\partial L}{\partial t} = -\frac{\partial L}{\partial t}. \tag{5}$$

Comparing the right-hand sides of (3) and (4), we find the expression for the derivative of the total energy of the system,

$$\frac{\mathrm{d}T}{\mathrm{d}t} = \Omega_j \dot{q}^j + \frac{\partial V}{\partial t} + \frac{\mathrm{d}}{\mathrm{d}t}(E_1 + 2E_0) - \frac{\partial E}{\partial t}. \tag{6}$$

For scleronomic systems result (6) is simplified, since

$$\left(\sum_{\mu=1}^{l_q} \lambda_\mu^{(1)} \frac{\partial \Phi_\mu}{\partial q^j} + \sum_{\nu=1}^{l_r} \lambda_\nu^{(2)} C_{\nu j}\right) \dot{q}^j = 0$$

as magnitudes proportional to the work of the reactions of ideal constraints on virtual displacements. The expression

$$\frac{\mathrm{d}}{\mathrm{d}t}(E_1 + 2E_0) - \frac{\partial E}{\partial t} = 0$$

also vanishes, which follows directly from the fact that in these systems the kinetic energy is not explicitly dependent on time $E_1 = E_0 = 0$. If, in addition, the potential energy V is independent of time, then

$$\frac{\mathrm{d}T}{\mathrm{d}t} = Q_j \dot{q}^j. \tag{7}$$

A system is termed *conservative* if

$$\frac{\mathrm{d}T}{\mathrm{d}t} = 0, \tag{8}$$

which means that the total energy of the system is constant.

7.12 Configurational space

Let us begin our considerations with a reference to the problem of transformation of acceleration (cf. Chapter 1) from a cartesian to a curvilinear frame of reference. The law of this transformation is

$$\mathbf{a}\cdot\frac{\partial \mathbf{r}}{\partial q^j} = a_j. \tag{1}$$

Note that a_j is a component of the vector $\mathbf{a}$ of the direction $\partial\mathbf{r}/\partial q^j$. Vectors $\partial\mathbf{r}/\partial q^j$ are tangential vectors to a hypersurface described by the equation

$$\mathbf{r} = \mathbf{r}(q^1, q^2, \ldots, q^r, t) \tag{2}$$

and form a vector base in the tangential space.

The metric tensor of this space can be derived from (2) as

$$g_{ij} = \frac{\partial \mathbf{r}}{\partial q^i}\cdot\frac{\partial \mathbf{r}}{\partial q^j}. \tag{3}$$

The square of the length of the vector of virtual displacement is expressed by

$$\delta s^2 = \delta\mathbf{r}\cdot\delta\mathbf{r} = g_{ij}\,\delta q^i \delta q^j. \tag{4}$$

The tensor g_{ij} also makes it possible to find the angles between vectors since the scalar product is expressed by

$$\mathbf{a}\cdot\mathbf{b} = g_{ij}a^i b^j. \tag{5}$$

Thus we can see that the tensor g_{ij} describes local properties of the space considered.

Before we return to equation (1) we should recall certain basic concepts of the tensor calculus. A set of magnitudes T_i are called components of a *covariant tensor of the first order* or components of a *covariant vector*, provided these magnitudes are transformable on transition to some other frame of reference, according to

$$T_i' = \frac{\partial q^j}{\partial q'^i} T_j. \tag{6}$$

We define the *contravariant tensor* in a similar manner. In this case, the law of transformation is decisive for the tensor character:

$$T'^i = \frac{\partial q'^i}{\partial q^j} T^j. \tag{7}$$

In equations (6) and (7) the summation convention is of course valid.

From the definition of tensors of the first order, we can introduce the concept of tensors of the nth order, covariant or contravariant. We shall now give the *law of tensor multiplication*:

$$B^i C^k = A^{ik}, \quad B^i C_k = A^i_k, \quad B_i C_k = A_{ik}. \tag{8}$$

By multiplying a tensor of the first order according to scheme (8) we obtain tensors of the second order. If the components of a stated magnitude are transformable as products of tensors of the first order, then they are components of tensors of the second order. And so the relationship

$$A'^{ik} = A^{nm}\frac{\partial q'^{i}}{\partial q^{n}}\frac{\partial q'^{k}}{\partial q^{m}} \tag{9}$$

is the law of transformation of contravariant tensors of the second order.

The laws of transformation for components A^{i}_{k} and A_{ik} are similarly derived:

$$A'^{i}_{k} = A^{n}_{m}\frac{\partial q'^{i}}{\partial q^{n}}\frac{\partial q^{m}}{\partial q'^{k}}, \tag{10}$$

$$A'_{ik} = A_{nm}\frac{\partial q^{n}}{\partial q'^{i}}\frac{\partial q^{m}}{\partial q'^{k}}. \tag{11}$$

Tensors of higher orders are formed analogously.

The point should be made that co and contravariant components are not physical components, therefore in the general case their dimensions are not compatible. To obtain physical components they should be multiplied internally by corresponding contra or covariant components of unit vectors tangential to parametric lines.

Note further that the concept of co or contravariant tensors is so conceived as to obtain a scalar by internal multiplication. Let $\boldsymbol{\tau}^{i}$ be a unit vector of the direction of a parametric line.

$$\boldsymbol{\tau}^{i} = \frac{\mathrm{d}q^{i}}{\mathrm{d}s}. \tag{12}$$

Naturally, its modulus is equal to 1, therefore

$$1 = g_{ik}\boldsymbol{\tau}^{i}\boldsymbol{\tau}^{k} = \boldsymbol{\tau}_{k}\boldsymbol{\tau}^{k},$$

or

$$g_{ik}\boldsymbol{\tau}^{i} = \boldsymbol{\tau}_{k}. \tag{13}$$

Thus we can see that the metric tensor g_{ik} enables index lowering, i.e., it enables transition from contravariant to covariant components. The elements of the inverse matrix to g_{ik} form a covariant tensor coupled to the metric tensor. The components of g_{ik} may be used for index raising. Simultaneously, the following occurs:

$$g^{ik}g_{ij} = \delta^{k}_{j}, \tag{14}$$

where δ_j^k is Kronecker's delta. In cartesian orthogonal coordinates $g_{ij} = \delta_{ij}$, thus there is no difference in this case between a covariant and contravariant tensor, a fact we have repeatedly utilized before.

Returning to equation (1), we can see that it constitutes the transformation law of a covariant vector; we transform cartesian to generalized coordinates. By this transformation we obtain

$$a_j = \frac{\mathrm{d}}{\mathrm{d}t}\left[\frac{\partial}{\partial \dot{q}^j}\left(\frac{1}{2}v^2\right)\right] - \frac{\partial}{\partial q^j}\left(\frac{1}{2}v^2\right). \tag{15}$$

Assuming that **r** does not depend on time in explicit form, the square of the velocity can be expressed by a metric tensor:

$$v^2 = g_{ik}\dot{q}^i\dot{q}^k. \tag{16}$$

Substituting (16) into (15) we have

$$\begin{aligned} a_j &= g_{ij}\frac{\mathrm{d}\dot{q}^i}{\mathrm{d}t} + \frac{\partial g_{ij}}{\partial q^k}\dot{q}^i\dot{q}^k - \frac{1}{2}\frac{\partial g_{ik}}{\partial q^j}\dot{q}^i\dot{q}^k \\ &= g_{ij}\frac{\mathrm{d}\dot{q}^i}{\mathrm{d}t} + \frac{1}{2}\left(\frac{\partial g_{ij}}{\partial q^k} + \frac{\partial g_{kj}}{\partial q^i} - \frac{\partial g_{ik}}{\partial q^j}\right)\dot{q}^i\dot{q}^k. \end{aligned} \tag{17}$$

We introduce the notation

$$\Gamma_{j,ik} = \frac{1}{2}\left(\frac{\partial g_{ij}}{\partial q^k} + \frac{\partial g_{kj}}{\partial q^i} - \frac{\partial g_{ik}}{\partial q^j}\right). \tag{18}$$

Whence

$$a_j = g_{ij}\frac{\mathrm{d}\dot{q}^i}{\mathrm{d}t} + \Gamma_{j,ik}\dot{q}^i\dot{q}^k \tag{19}$$

where $\Gamma_{j,ik}$ is *Christoffel's symbol of the first order.* This symbol is not a tensor since it is not transformable according to tensor transformation. Now

$$\dot{q}^i = v^i = \frac{\mathrm{d}q^i}{\mathrm{d}t},$$

therefore

$$a_j = g_{ij}\frac{\mathrm{d}v^i}{\mathrm{d}t} + \Gamma_{j,ik}v^i\frac{\mathrm{d}q^k}{\mathrm{d}t}. \tag{20}$$

Evidently the covariant component of acceleration has been expressed in contravariant components of velocity.

We perform a transformation to achieve a certain modification of the derived equation

$$a_j = \frac{\mathrm{d}}{\mathrm{d}t}(g_{ij}v^i) - \frac{\partial g_{ij}}{\partial q^k}\frac{\mathrm{d}q^k}{\mathrm{d}t}v^i + \Gamma_{j,ik}v^i\frac{\mathrm{d}q^k}{\mathrm{d}t} = \frac{\mathrm{d}v_j}{\mathrm{d}t} - \Gamma_{i,jk}v^i\frac{\mathrm{d}q^k}{\mathrm{d}t}. \tag{21}$$

We introduce the notation

$$\Gamma^s_{kj} = g^{si}\Gamma_{i,kj}, \tag{22}$$

where Γ^s_{jk} is *Christoffel's symbol of the second order.*

From (22) and from the properties of the metric tensor we find that the relationship

$$\Gamma_{i,kj} = \Gamma^s_{kj}g_{si} \tag{23}$$

holds, and substituting (23) into (21) we write

$$a_j = \frac{\mathrm{d}v_j}{\mathrm{d}t} - \Gamma^s_{kj}v_s\frac{\mathrm{d}q^k}{\mathrm{d}t}. \tag{24}$$

Expression (24) gives the method of differentiating the covariant velocity vector. It shows how one finds the so-called *absolute derivative* of tensor v_j, which we shall denote by the symbol $\delta/\delta t$. The acceleration is therefore the absolute derivative of velocity:

$$a_j = \frac{\delta v_j}{\delta t}. \tag{25}$$

After performing a differentiation operation according to (24) we obtain the tensor components and in this way we dispense, so to say, with the frame of reference.

Multiplying the mass m of a point by acceleration, we obtain a force Q_j which, by virtue of the method of its derivation, is a covariant tensor. The multiplication of the tensor by the scalar changes neither its order nor kind, since it is a tensor of the zero order. Newton's equation written in the following form:

$$m\frac{\delta v_j}{\delta t} = Q_j \tag{26}$$

is a tensor equation.

The importance of the tensor calculus is that tensor equations hold in all frames of reference if they hold in one of them. This is due to the fact that tensor transformations are linear and homogeneous. From equation (26) it follows that

$$m\frac{\delta v_j}{\delta t} - Q_j = T_j = 0. \tag{27}$$

The vanishing of the components of tensor T_j in one arbitrary frame of reference results in their being identically equal to zero in all frames. Multiplying (27) by g^{js} we have

$$m\frac{\delta v^s}{\delta t} - Q^s = 0. \tag{28}$$

The absolute derivative $\delta v^s/\delta t$ can be determined from (20) from which it directly follows that

$$\frac{\delta v^s}{\delta t} = \frac{dv^s}{dt} + \Gamma^s_{ik} v^i \frac{dq^k}{dt}.\ * \tag{29}$$

We shall now discuss the dynamics of a mechanical system with many degrees of freedom. The discussion of the problem will be confined to holonomic and scleronomic systems.

The Lagrange equations describing the dynamics of these systems can be written as follows:

$$\frac{d}{dt}\left(\frac{\partial E}{\partial \dot{q}^j}\right) - \frac{\partial E}{\partial q^j} = Q_j \quad (j = 1, 2, \ldots, r). \tag{30}$$

For the considered systems the kinetic energy is given by

$$E = \tfrac{1}{2} a_{ik} \dot{q}^i \dot{q}^k = \tfrac{1}{2} v^2. \tag{31}$$

Comparing (15) and (16) with (30) and (31), we notice an analogy. A difference to be observed here is that we have now a different structure of the metric tensor a_{ik} given by

$$a_{ik} = \sum_{s=1}^{n} m_s \frac{\partial \mathbf{r}_s}{\partial q^i} \cdot \frac{\partial \mathbf{r}_s}{\partial q^k}. \tag{32}$$

---

* In the case of differential equations with partial derivatives on transition to curvilinear frames of reference in place of the derivative $\partial/\partial q^k$, the so-called *covariant derivative* ∇_k should be considered. The relationship between the partial and the covariant derivative is expressed by

$$\nabla_k v_l = \frac{\partial v_l}{\partial q^k} - \Gamma^m_{lk} v_m$$

or in the case of a contravariant tensor v^j:

$$\nabla_k v^j = \frac{\partial v^j}{\partial q^k} + \Gamma^j_{ik} v^i.$$

A space in which a specified configuration corresponds to any point, this correspondence being one-to-one and continuous, is called a *configurational space*. The quadratic form positively determined

$$\mathrm{d}s^2 = a_{ik}\,\mathrm{d}q^i\,\mathrm{d}q^k \tag{33}$$

is a metric kinematical form of a kinematical square of linear element. Performing the same transformations on equations (30) and (15), replacing g_{ik} with the tensor a_{ik}, we have

$$\frac{\delta v_j}{\delta t} = Q_j. \tag{34}$$

We can therefore treat the motion of the system as the motion of a point with mass $m = 1$ in a configurational space. $\delta v_j/\delta t$ is the generalized covariant acceleration vector. The integral of equation (34) in this space gives a certain curve, a trajectory of motion, whose points correspond to different configurations of the dynamical system. It will be noticed that the generalized velocity vector $\dot{q}^j$ is tangential to the trajectory.

Indeed,

$$\dot{q}^j = \frac{\mathrm{d}q^j}{\mathrm{d}t} = \frac{\mathrm{d}q^j}{\mathrm{d}s}\frac{\mathrm{d}s}{\mathrm{d}t} = \tau^j\frac{\mathrm{d}s}{\mathrm{d}t}, \tag{35}$$

since the vector τ^j is tangential to the trajectory. The square of the length of vector v^j is

$$v^2 = a_{ik}\dot{q}^i\dot{q}^k = a_{ik}\tau^i\tau^k\left(\frac{\mathrm{d}s}{\mathrm{d}t}\right)^2 = \left(\frac{\mathrm{d}s}{\mathrm{d}t}\right)^2. \tag{36}$$

We can also prove that the generalized acceleration consists of a tangential acceleration and a normal acceleration proportional to the first curvature $\varkappa$ of the trajectory of motion:

$$\frac{\delta\dot{q}^j}{\delta t} = \frac{\delta\tau^j}{\delta t}\frac{\mathrm{d}s}{\mathrm{d}t} + \tau^j\frac{\delta}{\delta t}\left(\frac{\mathrm{d}s}{\mathrm{d}t}\right). \tag{37}$$

But

$$\frac{\delta\tau^j}{\delta t} = \frac{\delta\tau^j}{\delta s}\frac{\delta s}{\delta t} = \varkappa^j\frac{\mathrm{d}s}{\mathrm{d}t} \tag{38}$$

and

$$\frac{\delta}{\delta t}\left(\frac{\mathrm{d}s}{\mathrm{d}t}\right) = \frac{\mathrm{d}^2 s}{\mathrm{d}t^2}, \tag{39}$$

therefore

$$\frac{\delta \dot{q}^j}{\delta t} = \varkappa^j \left(\frac{\mathrm{d}s}{\mathrm{d}t}\right)^2 + \tau^j \frac{\mathrm{d}^2 s}{\mathrm{d}t^2}. \tag{40}$$

Introducing the concept of homeomorphism as a one-to-one and continuous mapping, the discussion can be focussed on the question of global properties of a configurational space. The introduction of topological concepts proves most useful here [59]. Considering the angular motion of a rigid body we notice that the characteristic feature of this motion is the periodic return of the system to its initial position. The kinetic energy is

$$E = \frac{1}{2} J \left(\frac{\mathrm{d}\varphi}{\mathrm{d}t}\right)^2, \tag{41}$$

and the square kinematical linear element

$$\mathrm{d}s^2 = J \mathrm{d}\varphi^2. \tag{42}$$

From the structure of the metric tensor, we can only infer the local properties of a space, consequently, analysis of its structure gives no possibility of exploring the global properties. The periodic return of a point in a configurational space to the initial position is a global property of a one-dimensional space, unlimited but finite. The configurational space is thus homeomorphic here with a circle. If angular motion is superimposed on translatory motion in a direction perpendicular to the direction of the axis of rotation (plane motion), then the conclusion that the configurational space is homeomorphic with a cylinder will be reached by similar analysis. A system of two non-interacting toothed wheels gives a space homeomorphic with a torus.

From these few examples we can see that by introducing configurational space into the discussion as well as basic topological concepts, possibilities open for most interesting geometrical interpretations.

So far we have considered holonomic systems with scleronomic constraints. The question arises: What will happen with holonomic systems when constraints depend on time, i.e., with rheonomic constraints?

Note first that for rheonomic systems the functions $\mathbf{r}_k = \mathbf{r}_k(q^1, q^2, \ldots$ $\ldots, q^r, t)$ are dependent on time in explicit form. For this reason, the kinetic energy consists of three members E_1, E_2, and E_0 as given by equations (5a), Section 7.6. Substituting (5), Section 7.6, into (30), we obtain

$$\frac{\mathrm{d}}{\mathrm{d}t}\left(\frac{\partial E_2}{\partial \dot{q}^j}\right) - \frac{\partial E_2}{\partial q^j} = Q_j - \frac{\mathrm{d}}{\mathrm{d}t}\left(\frac{\partial (E_1 + E_0)}{\partial \dot{q}^j}\right) + \frac{\partial}{\partial q^j}(E_1 + E_0). \tag{43}$$

But

$$\frac{\mathrm{d}}{\mathrm{d}t}\left(\frac{\partial E_2}{\partial \dot{q}^j}\right)-\frac{\partial E_2}{\partial q^j} = a_{ij}\frac{\mathrm{d}\dot{q}^i}{\mathrm{d}t}+\Gamma_{j,ik}\dot{q}^i\dot{q}^k+\frac{\partial a_{ij}}{\partial t}\dot{q}^i = \frac{\delta v_j}{\delta t}+\frac{\partial a_{ij}}{\partial t}\dot{q}^i$$

and

$$\frac{\partial E_0}{\partial \dot{q}^j} = 0,$$

therefore, in this case, the differential equation for the motion of a point in a configurational space has the form

$$\frac{\delta v_j}{\delta t} = Q_j - \frac{\mathrm{d}}{\mathrm{d}t}\left(\frac{\partial E_1}{\partial \dot{q}^j}\right)+\frac{\partial}{\partial q^j}(E_1+E_0)-\frac{\partial a_{ij}}{\partial t}\dot{q}^i. \tag{44}$$

Thus we can see that for rheonomic systems the local properties of a configurational space vary in time. This is of course due to the fact that the components of the metric tensor are dependent on time in explicit form. We should also take note of the occurrence of new members on the right-hand side of the differential equations of motion.

7.13 The stability of mechanical systems

The problem of stability is one of the major problems of the qualitative theory of differential equations. The following question needs to be answered here: What relationships occur between the given particular integral and the neighbouring solutions?

Differential equations describe the behaviour of dynamic systems, therefore stability is a feature peculiar to them. Stability is decisive of the sensitivity of a system to changes in the magnitudes which determine its reaction. For the development of this theory, we owe a debt first and foremost to Poincaré, Laplace, Lapunov, Malkin, and others, especially to Lapunov's studies which have been the most significant contribution. One can say that he has created the foundation of modern stability theory.

A stable system has the property of having little sensitivity to changes in the magnitudes determining its reactions. In other words, if the motion of a system (conditioned by the initial state and by the forces acting) reacts negligibly to changes in these magnitudes it is said to be *stable*; if not, we deal with *instability*.

The stability concept, intuitively understood, is so general that, wanting to bring it within the framework of exact mathematical terms, we must very accurately describe the characteristics which a system should maintain.

Stability is associated with the concept of differential equations. The concept of stability is commonly identified with the stability of solutions of special differential equations [35, 65].

There are two basic methods of studying stability: the first assumes that the solution to a problem is known and it therefore has limited application; the second, known as the *direct method*, on which we shall primarily focus our attention, does not require knowledge of the solution.

Let the state of a system be described by a system of Hamilton's canonical equations:

$$\dot{q}^i = \frac{\partial H}{\partial p_i}, \quad \dot{p}_i - Q_i = -\frac{\partial H}{\partial q^i}. \tag{1}$$

From the mathematical point of view, it is a system of the first order of the type

$$\dot{\mathbf{y}} = \mathbf{f}(\mathbf{y}, t), \tag{1a}$$

where

$$\dot{\mathbf{y}} = \begin{bmatrix} \dot{q}^1 \\ \vdots \\ \dot{q}^r \\ \dot{p}_1 \\ \vdots \\ \dot{p}_r \end{bmatrix}, \tag{1b}$$

$$\mathbf{f}(\mathbf{y}, t) = \begin{bmatrix} \dfrac{\partial H}{\partial p_1} \\ \vdots \\ \dfrac{\partial H}{\partial p_r} \\ -\dfrac{\partial H}{\partial q^1} + Q_1 \\ \vdots \\ -\dfrac{\partial H}{\partial q^r} + Q_r \end{bmatrix}. \tag{1c}$$

The concept of stability is commonly identified with that of the stability of special solutions of system (1) in the sense that approximating solutions for stable systems remain close with minor changes in the initial conditions and forces Q_i.

If the particular integral of (1) is the vector

$$\mathbf{y} = \boldsymbol{\varphi}(t),$$

then the basis of determining the proximity of neighbouring solutions of $\mathbf{y}$ is a function $\mathbf{x}$,

$$\mathbf{x} = \mathbf{y} - \boldsymbol{\varphi}(t). \tag{2}$$

Differentiating (2) we have

$$\dot{\mathbf{x}} = \dot{\mathbf{y}} - \dot{\boldsymbol{\varphi}}(t), \tag{3}$$

but from (1) we can write

$$\dot{\mathbf{y}} = \mathbf{f}(\mathbf{x}+\boldsymbol{\varphi}(t), t), \quad \dot{\boldsymbol{\varphi}}(t) = \mathbf{f}(\boldsymbol{\varphi}(t), t). \tag{4}$$

We form the difference of expressions (4):

$$\dot{\mathbf{x}} = \mathbf{f}(\mathbf{x}+\boldsymbol{\varphi}(t), t) - \mathbf{f}(\boldsymbol{\varphi}(t), t) \equiv \mathbf{F}(\mathbf{x}, t)$$

or

$$\dot{\mathbf{x}} = \mathbf{F}(\mathbf{x}, t). \tag{5}$$

We have arrived at a new system of differential equations [65] which are called *equations of disturbances*. Note that the particular integral for these equations is the solution $\mathbf{x} \equiv \mathbf{0}$. In this way, we have confined the examination of the stability of the solution of system (1) to the examination of the stability of the reference origin, i.e., the stability of a point.

Before we proceed with the discussion of the actual theme of this section, we shall introduce the concept of the norm, which will be used further on. In our considerations we refer to the concept of vector. We have already come across the definition of the modulus of a vector (length). This was an example of the norm of a vector. More generally, we can say that the *norm of a vector* $\mathbf{x}$ is a positive real number $||\mathbf{x}||$ obeying the following axioms:

(1) $||\mathbf{x}|| = 0 \Leftrightarrow \mathbf{x} = 0$,

(2) $||\lambda\mathbf{x}|| = |\lambda|\, ||\mathbf{x}||, \; \lambda \in R$ (R field of real numbers),

(3) $||\mathbf{x}+\mathbf{y}|| \leqslant ||\mathbf{x}||+||\mathbf{y}||$ (inequality of triangle).

The symbol $\Leftrightarrow$ stands for the range of identical equality; the logical clause $a \Leftrightarrow b$ we read: a then and only then if b. Examples of the norm of a vector $\mathbf{x}$ are, say,

$$||\mathbf{x}|| = \sum_i |x^i|, \quad ||\mathbf{x}|| = \sqrt{\sum_i (x^i)^2}, \quad ||\mathbf{x}|| = \max_i |x^i|.$$

In finite-dimensional spaces all norms are equivalent in the sense that

$$\lim_{n\to\infty} ||\mathbf{x}_n - \mathbf{x}_0||_1 = 0 \Leftrightarrow \lim_{n\to\infty} ||\mathbf{x}_n - \mathbf{x}_0||_2 = 0,$$

i.e., following from the convergence of the series $\{\mathbf{x}_n\}$ to the element $\mathbf{x}_0$ by the first norm is the convergence of the series $\{\mathbf{x}_n\}$ to the same limit by the second norm and vice versa [38]. We shall now give the definition of the stability of the solution $\mathbf{x} \equiv \mathbf{0}$.

The solution $\mathbf{x} \equiv \mathbf{0}$ to a system $\dot{\mathbf{x}} = \mathbf{F}(\mathbf{x}, t)$ we shall regard as *stable* if for any number $\varepsilon > 0$ there is such a number $\delta > 0$ that if $||\mathbf{x}(0)|| < \delta$ then, for any time $t > 0$, $||\mathbf{x}(t)|| < \varepsilon$.

This can be shortened, using symbols of mathematical logic, to: The solution $\mathbf{x} \equiv \mathbf{0}$ to a system $\dot{\mathbf{x}} = \mathbf{F}(\mathbf{x}, t)$ we shall regard as stable, provided the following logical clause

$$\bigwedge_{\varepsilon>0} \bigvee_{\delta>0} ||\mathbf{x}(0)|| < \delta \Rightarrow \bigwedge_{t>0} ||\mathbf{x}(t)|| < \varepsilon \tag{6}$$

holds. If besides (6) we have

$$\bigvee_{\Delta>0} ||\mathbf{x}(0)|| < \Delta \Rightarrow \lim_{t\to\infty} ||\mathbf{x}(t)|| = 0, \tag{7}$$

the system is said to be *asymptotically stable.*

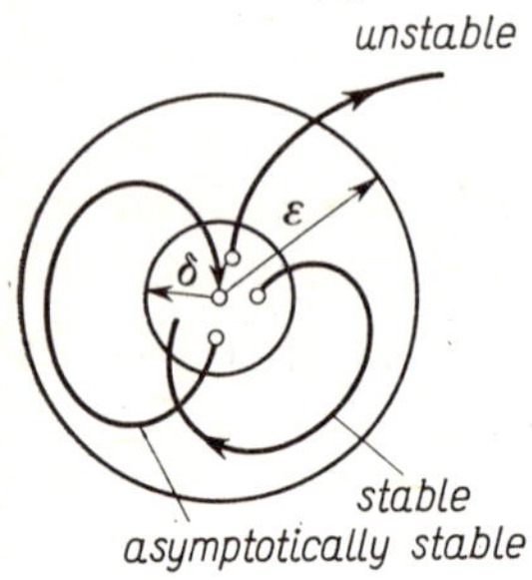

Fig. 7.14

The geometrical interpretation of the aforesaid definitions is as follows. A system is stable if for any sphere with radius ε there exists such a sphere, with radius δ, that any trajectory in a phase space originating from the sphere δ will not reach the boundary of the sphere ε (Fig. 7.14).

If additionally the phase trajectory tends to the reference origin, the system is asymptotically stable.

Now let us consider a conservative system with scleronomic constraints. For this system the sum of the kinetic and the potential energy is a constant:

$$H = E+V = \text{const} \tag{8}$$

or

$$\tfrac{1}{2}a^{ik}p_i p_k + V(q^1, \ldots, q^r) = \text{const}. \tag{8a}$$

Assume that the potential energy V at the reference origin has an extremum. With this assumption $\mathbf{y} \equiv \mathbf{0}$ is a particular solution of (1). Equation (8) specifies a certain surface in a phase space. It is precisely on this surface that the phase trajectory is placed. The stability of a system depends on the overall total properties of the surface (8).

The condition for the system now considered to be stable is that the surface (8) should be homeomorphic with the sphere.

To establish that this is so, we adopt the following reasoning. The magnitude of the area bounded by the surface (8) homeomorphic with the sphere depends on the initial conditions, i.e., on moments p_i and the position given by coordinates q^i at an instant $t = 0$. Note that with the decrease of the energy in the initial instant, this area "shrinks". Thus it is always possible to choose for such a sphere with radius ε, a sphere with radius δ, so that the former should embrace the considered surface and the latter be contained within it.

This shows that the considered system is stable in terms of the definition itself. For the system to be stable it is necessary for the function $V(q^1, \ldots, q^r)$ to be positive definite since the kinetic energy is a positive definite form ($E \geqslant 0$). A sufficient condition for the stability of a conservative system is that the potential energy should be minimum at the reference origin.

A point in a phase space is thus stable if the potential energy in it assumes the minimum.

This theorem is known in literature [37] under the name of the *Lagrange–Dirichlet criterion.*

Example 1. A mass m was fixed at one end of a rod with length l. The other end was elastically constrained at the base (Fig. 7.15). Specify the conditions for the stable position of the system while $\varphi = 0$.

The potential energy is

$$V = \tfrac{1}{2}c\varphi^2 - mgl(1-\cos\varphi), \tag{a}$$

$$\frac{\partial V}{\partial \varphi} = c\varphi - mgl\sin\varphi. \tag{b}$$

It is seen that for $\varphi = 0$ we have $\partial V/\partial\varphi = 0$ and therefore the position $\varphi = 0$ is the equilibrium position of the system.

The condition for this position to be the position of stable equilibrium is that second derivative of V be positive in it, i.e., the following inequality should occur:

$$\frac{\partial^2 V}{\partial \varphi^2} = c - mgl\cos\varphi|_{\varphi=0} > 0$$

or

$$c > mgl. \tag{c}$$

It will easily be found then that the phase trajectory is homeomorphic with a circle. In this case, vibrations of the system about the point $\varphi = 0$ are possible.

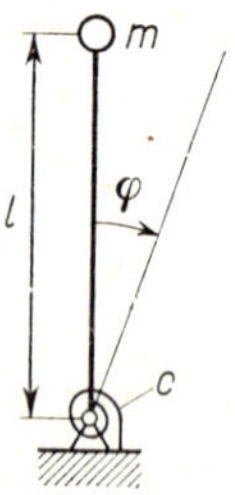

Fig. 7.15

Example 2. Derive Euler's equation (4), Section 4.12, for the critical force in the rod-stability problem from the Lagrange–Dirichlet criterion.

The potential energy of the system is given by

$$V = -P\Delta l + \int_0^l \frac{M^2}{2EJ}\,\mathrm{d}x, \tag{a}$$

$$\Delta l = \int_0^l (1-\cos\varphi)\,\mathrm{d}x \approx \frac{1}{2}\int_0^l \varphi^2 \mathrm{d}x \approx \frac{1}{2}\int_0^l \left(\frac{\mathrm{d}y}{\mathrm{d}x}\right)^2 \mathrm{d}x. \tag{b}$$

But

$$M = EJ\frac{\mathrm{d}^2 y}{\mathrm{d}x^2},$$

therefore

$$V = \frac{1}{2}\left[EJ\int_0^l \left(\frac{\mathrm{d}^2 y}{\mathrm{d}x^2}\right)^2 \mathrm{d}x - P\int_0^l \left(\frac{\mathrm{d}y}{\mathrm{d}x}\right)^2 \mathrm{d}x\right]. \tag{c}$$

Let

$$y = f\sin\frac{\pi x}{l}. \tag{d}$$

Substituting (d) into (c), integrating and equating the second derivative of V to zero ($\partial^2 V/\partial f^2 = 0$), we obtain the expression for the critical force

$$P_{\mathrm{cr}} = \frac{\pi^2 EJ}{l^2}.$$

The Lagrange–Dirichlet criterion plays an important part in applications.

Let us now consider a system with scleronomic constraints on which besides potential forces significant dissipative forces given by equation (6), Section 7.8, act. Assume that the equality $H = \text{const}$ describes a surface homeomorphic with a sphere. The energy in this system is decreasing, therefore this surface will be "shrinking" in time to a single point. Hence we infer that in this particular case, the system is asymptotically stable. In the two cases just considered, we have studied the behaviour of Hamilton's function H. To ascertain the stability of the solution $q^i = 0$ it was sufficient for the function to be determined as *significantly positive* (positive everywhere except at the reference origin of the system where it assumes zero value) and also for its derivative along the trajectory of motion to be non-positive. Lapunov [35] generalized this approach for a broader class of differential equations, expounding the following theorem:

If in any environment D of the reference origin there exists a differential function $V_L(x, t)$ (called *Lapunov's function*) such that

$$(1)\quad \bigwedge_{x\in D}\bigwedge_{t\geqslant 0} V_L(x, t) > 0, \tag{9}$$

$$(2)\quad \bigwedge_{t\geqslant 0} V_L(0, t) = 0, \tag{10}$$

$$(3)\quad \bigvee_{\substack{W(x)>0\\ W(0)=0}}\bigwedge_{x\in D}\bigwedge_{t\geqslant 0} V_L(x, t) \geqslant W(x),\ (W(x)\text{ a continuous function}), \tag{11}$$

$$(4)\quad \dot{V}_L(x, t) = \frac{\partial V_L}{\partial t} + \frac{\partial V_L}{\partial x^i} F^i(x, t) \leqslant 0, \tag{12}$$

then the solution $x \equiv 0$ to system (5) is stable.

The proof is as follows: Let the minimum value of the function $W(x)$ on the sphere $||x|| = \varepsilon$ be denoted by m or

$$m = \inf_{x\in\{x:\ ||x|| = \varepsilon\}} W(x). \tag{13}$$

Let us next choose δ so that

$$\sup_{x\in\{x:\ ||x|| < \delta\}} V_L(x, 0) < m. \tag{14}$$

By virtue of (12) the more so in following instants of time the value of the function V_L will not cross the boundaiy m, and consequently the trajectory of motion will not part from the sphere $||x|| \leqslant \varepsilon$, which is precisely what was to be proved.

For autonomous systems $\dot{x} = F(x)$; studying stability by this method we seek the function V_L independent of time t. Condition (11) can then be replaced by the condition

$$(3')\, V_L(x) \geqslant 0. \tag{15}$$

We shall now discuss Lapunov's theorem on asymptotic stability.

If, within a certain enviroment D of the reference origin, there exists a Lapunov's function V_L such that

$$(1) \bigwedge_{t>0} \bigwedge_{x\in D} \bigvee_{\substack{W(x)>0 \\ W(0)=0}} \bigvee_{\substack{W_1(x)>0 \\ W_1(0)=0}} W_1(x) \geqslant V_L(x, t) \geqslant W(x) \geqslant 0 \tag{16}$$

($W(x)$ and $W_1(x)$ are continuous functions),

$$(2)\ \dot{V}_L = \frac{\partial V_L}{\partial t} + \frac{\partial V_L}{\partial x^i} F^i(x, t) \leqslant -W_2(x) \leqslant 0 \tag{17}$$

($W_2(x)$ is a continuous function assuming zero value at the reference origin),

(3) the function $V_L(x, t)$ is uniformly convergent to zero against time t with $||x|| \to 0$, i.e.,

$$\bigwedge_{\varepsilon>0} \bigvee_{\delta(\varepsilon)>0} \bigwedge_{x\in\{x:\ ||x||<\delta\}} \bigwedge_{t>0} V_L(x, t) < \varepsilon, \tag{18}$$

then the solution $x \equiv 0$ to system (5) is asymptotically stable.

By virtue of existence of Lapunov's function $V_L(x, t)$ the system is stable in terms of definition (6). Now $\bigwedge_{\mathbf{x}} V_L(\mathbf{x}, t) > 0$ and (17) applies, therefore

$$\lim_{t\to\infty} V_L(\mathbf{x}(t), t) = k \geqslant 0.$$

If $k = 0$ then, by virtue of (16) and (18), we have

$$\bigwedge_i \lim_{t\to\infty} x^i(t) = 0.$$

The proof amounts, therefore, to showing that $k = 0$. Assume for now that $k > 0$; then

$$\bigwedge_{t>0} V_L(\mathbf{x}, t) \geqslant k > 0.$$

By virtue of (17) we get

$$\dot{V}_L \leqslant -\mu < 0,$$

where

$$\mu = \inf_{\mathbf{x}\varepsilon\{\mathbf{x}:\ \delta<||\mathbf{x}||<\varepsilon\}} W_2(\mathbf{x}).$$

Hence it follows that

$$V(\mathbf{x}(t), t) - V(\mathbf{x}(0), 0) \leqslant -\mu t,$$

which excludes the possibility of the existence of $k > 0$, thereby concluding the proof of the theorem.

Note that, when stability theorems are considered, reference is usually made to the sign of the function or derivative involved; therefore, when we refer to a function as positive and subsequently to its non-positive derivative, we may, whilst maintaining the validity of the theorem, require a negative determined function and its non-negative derivative (this would follow from formal multiplication of the function by -1). We should also appreciate the fact that the conditions for stability associated with the direct method of studying them, i.e., the method discussed here, are sufficient (adequate).

We shall discuss the problem of stability with reference to systems described by differential equations of the form

$$\dot{\mathbf{x}} = \mathbf{F}(\mathbf{x}, t) + \mathbf{R}(\mathbf{x}, t). \tag{19}$$

The system of equations (19) was formed from (5) by the addition to the right-hand side of (5) the vector $\mathbf{R}(\mathbf{x}, t)$ representing disturbance of the system. In a special case, this disturbance may be a function of time t only. Value $\mathbf{x} \equiv 0$ is the solution to (5) but not to (19), since in the general case $\mathbf{R}(\mathbf{0}, t) \neq 0$. In spite of that, we shall study the stability of the reference origin $x = 0$.

The above case was the concern of Malkin [40] and in this connection we shall cite his theorem on stability of a system with the permanent disturbances.

The solution $\mathbf{x} \equiv 0$ to system (5) with the permanent disturbances $\mathbf{R}(\mathbf{x}, t)$ is stable in the following sense:

$$\bigwedge_{\varepsilon>0} \bigvee_{\delta>0} \bigvee_{\delta_1>0} \{[||\mathbf{x}(0)|| < \delta \wedge \bigwedge_{\mathbf{x} \in \{\mathbf{x}: ||\mathbf{x}|| < \varepsilon\}} \bigwedge_{t>0} ||\mathbf{R}(\mathbf{x}, t)|| < \delta_1] \Rightarrow \bigwedge_{t>0} ||\mathbf{x}(t)|| < \varepsilon\}; \tag{20}$$

if:

(1) system (5) is stable in terms of the theorem on asymptotic stability;

(2) Lapunov's function V_L has bounded derivatives, i.e.,

$$\bigwedge_{t>0} \bigwedge_{i} \left| \frac{\partial V_L}{\partial x^i} \right| \leqslant M, \tag{21}$$

then the solution $\mathbf{x}(t)$ to system (18) has the property that $\bigwedge_{t>0} ||\mathbf{x}(t)|| < \varepsilon$.

In other words: if a trivial solution to (5) is asymptotically stable, in accord with the previous theorem, and if there exists Lapunov's function with bounded derivatives, the disturbance having a limited norm for an arbitrary time $t > 0$, and for an arbitrary sphere ε, then the solution $\mathbf{x} \equiv 0$ to (19) is stable in Lapunov's sense.

The proof is as follows. If the solution $\mathbf{x} \equiv 0$ to the equation $\dot{\mathbf{x}} = \mathbf{F}(\mathbf{x}, t)$ is asymptotically stable, in the sense that there exists Lapunov's function $V_L(\mathbf{x}, t)$ which with respect to the basic area D conforms to assumptions (16), (17), and (18), then for system (19) to be stable at the point $\mathbf{x} \equiv 0$, it is sufficient for the derivative of this function along the trajectory of motion (19) to be non-increasing, i.e.

$$\dot{V}_{LR} = \frac{\partial V_L}{\partial t} + \frac{\partial V_L}{\partial x^i} F^i(\mathbf{x}, t) + \frac{\partial V_L}{\partial x^i} R^i(\mathbf{x}, t) \leqslant 0. \tag{22}$$

By virtue of the asymptotic stability of the system without disturbances we have

$$\frac{\partial V_L}{\partial t} + \frac{\partial V_L}{\partial x^i} F^i(\mathbf{x}, t) \leqslant -W_2(\mathbf{x}), \tag{23}$$

whence

$$\dot{V}_{LR} \leqslant -W_2(\mathbf{x}) + \frac{\partial V_L}{\partial x^i} R^i(\mathbf{x}, t). \tag{24}$$

Assume that in the sphere $||\mathbf{x}|| < \varepsilon$

$$\bigvee_{\eta(\varepsilon)>0} \sup_i |R^i(\mathbf{x}, t)| \leqslant \eta(\varepsilon).$$

Considering assumptions (21) and (24)

$$\frac{\partial V_L}{\partial x^i} R^i(\mathbf{x}, t) \leqslant 2rM\eta(\varepsilon). \tag{25}$$

Taking account of (25) in (24) we write

$$\dot{V}_{LR} \leqslant -W_2(\mathbf{x}) + 2rM\eta(\varepsilon). \tag{26}$$

Let μ be the lower boundary of the function $W_2(\mathbf{x})$ in the area bounded by the spheres $||\mathbf{x}|| = \varepsilon$ and $||\mathbf{x}|| = \delta$, or

$$\mu = \inf_{\mathbf{x} \in \{\mathbf{x}:\ \delta < ||\mathbf{x}|| < \varepsilon\}} W_2(\mathbf{x}), \tag{27}$$

where δ is such that

$$\inf_{\mathbf{x} \in \{\mathbf{x}:\ ||\mathbf{x}|| = \varepsilon\}} W(\mathbf{x}) \geqslant \bigwedge_{\mathbf{x} \in \{\mathbf{x}:\ ||\mathbf{x}|| < \delta\}} W_1(\mathbf{x}). \tag{28}$$

By virtue of (20) we have

$$\bigwedge_{\mathbf{x}\in\{\mathbf{x}:\,||\mathbf{x}||=\delta\}} V_L(\mathbf{x}, t) \leqslant \inf_{\mathbf{x}\in\{\mathbf{x}:\,||\mathbf{x}||=\varepsilon\}} W(\mathbf{x}) \tag{29}$$

and

$$\bigwedge_{\mathbf{x}\in\{\mathbf{x}:\,||\mathbf{x}||=\varepsilon\}} V_L(\mathbf{x}, t) \geqslant \inf_{\mathbf{x}\in\{\mathbf{x}:\,||\mathbf{x}||=\varepsilon\}} W(\mathbf{x}). \tag{30}$$

Thus passing from the sphere $||\mathbf{x}|| = \delta$ to the sphere $||\mathbf{x}|| = \varepsilon$ we observe an increment in the function $V_L(\mathbf{x}, t)$. If the derivative V_{LR} is non-positive, the phase trajectory will not go beyond the sphere $||\mathbf{x}|| = \varepsilon$.

Let us choose $\eta(\varepsilon)$ so that

$$\eta(\varepsilon) = \frac{\beta\mu}{2rM}, \tag{31}$$

where $0 < \beta \leqslant 1$. Replacing the first and second terms on the right-hand side of (26) with μ and the right-hand side of (31), we write

$$\dot{V}_{LR} \leqslant -\mu+\beta\mu = -(1-\beta)\mu \leqslant 0. \tag{32}$$

In this way the theorem has been proved.

Returning to dynamical systems we shall discuss the stability problem with reference to stationary motion. We shall examine a conservative system, its state being described by the following equations:

$$\dot{q}^i = \frac{\partial H}{\partial p^i}, \quad \dot{p}_i = -\frac{\partial H}{\partial q^i} \quad (i = 1, 2, \ldots, r). \tag{33}$$

We define the configuration of the system with m position coordinates and $(r-m)$ cyclic coordinates. Let it be recalled that the Lagrange kinetic potential does not depend on the latter. Hamilton's function H is represented in this case by

$$H = \tfrac{1}{2}a^{ij}p_i p_j + V(q^1, q^2, \ldots, q^m). \tag{34}$$

The components of the contravariant tensor of the configurational space are dependent on m component position coordinates.

We shall call the motion of the system *stationary* if all the position coordinates remain unchanged (q^i = const).

In this case, therefore, the generalized momenta must remain the same during the motion. Consequently, equations (33) for stationary motion assume a special construction:

$$\frac{\partial H}{\partial p_i} = 0, \quad \frac{\partial H}{\partial q^i} = 0 \quad (i = 1, 2, \ldots, m). \tag{35}$$

For the cyclic coordinates the generalized velocities are constant. Indeed,

$$\dot{q}^{\alpha} = \frac{\partial H}{\partial p_{\alpha}} = a^{\alpha i} p_i = \text{const} \qquad (\alpha = m+1, m+2, \ldots, r). \tag{36}$$

Hence

$$q^{\alpha} = \dot{q}^{\alpha}(0)t + q^{\alpha}(0). \tag{37}$$

To follow up our discussion on the stability of stationary motion we consider stability with respect to position coordinates and generalized momenta; for it has to be remembered that cyclic coordinates change linearly with time, i.e., the system is always unstable in relation to these latter.

If Hamilton's function, $H(q^i, p_i, p(0))$ at point $q^i = q^i(0)$, $p_i = p_i(0)$ assumes the value of a significant extremum, then stationary motion with such initial conditions is stable.

Indeed, if we put

$$q^i = q^i(0) + \Delta q^i, \qquad p_i = p_i(0) + \Delta p_i, \qquad p_{\alpha} = p_{\alpha}(0) + \Delta p_{\alpha}$$

$$(i = 1, \ldots, m; \; \alpha = m+1, \ldots, r),$$

then Hamilton's function

$$H = H(q^i(0) + \Delta q^i, p_i(0) + \Delta p_i, p_{\alpha}(0))$$

may be regarded, with accuracy of sign, as Lapunov's function since, by virtue of its being an integral of motion, the following occurs

$$\frac{\mathrm{d}H}{\mathrm{d}t} = 0. \tag{38}$$

But we have assumed that no changes in generalized momenta, corresponding to cyclic coordinates, take place. This assumption can be dispensed. To find out that it is actually so, it is enough to construct another Lapunov function of the following form [12]:

$$V_L(\Delta q^i, \Delta p_i, \Delta p_{\alpha}) = [H(q^i(0) + \Delta q^i, p_i(0) + \Delta p_i, p_{\alpha}(0)) - \\ - H(q^i(0), p_i(0), p_{\alpha}(0))]^2 + k \sum_{\alpha=m+1}^{r} \Delta p_{\alpha}^2, \tag{39}$$

where k is an arbitrarily chosen positive coefficient ($i = 1, \ldots, m$).

Example 3. A pivotally suspended weightless rod with length l, at the end of which a point mass m had been placed, (Fig. 7.16) was set in angular motion about its vertical axis. Examine the stability of the stationary motion of the system.

We shall begin with finding Hamilton's function

$$H = \tfrac{1}{2}m(l^2\sin^2\varphi\cdot\dot{\psi}^2+l^2\dot{\varphi}^2)-mgl\cos\varphi, \tag{a}$$

where φ is the coordinate position and ψ the cyclic position. Expressing H by generalized momenta p_φ and p_ψ, we write

$$H = \frac{1}{2ml^2\sin^2\varphi}p_\psi^2+\frac{1}{2ml^2}p_\varphi^2-mgl\cos\varphi. \tag{b}$$

The stationary motion of the system is given by

$$\frac{\partial H}{\partial\varphi} = -\frac{p_\psi^2}{ml^2}\frac{\cos\varphi}{\sin^3\varphi}+mgl\sin\varphi = 0, \tag{c}$$

$$\frac{\partial H}{\partial p_\varphi} = \frac{p_\varphi}{ml^2} = 0. \tag{d}$$

Hence

$$p_\varphi = 0, \tag{e}$$

$$p_\psi = m\sin^2\varphi\sqrt{\frac{gl^3}{\cos\varphi}}. \tag{f}$$

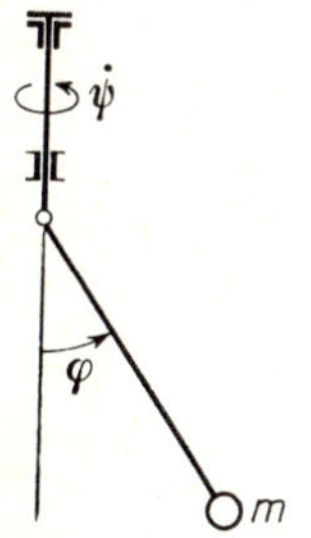

Fig. 7.16

The stability condition for stationary motion determined by conditions (c) and (d) is that

$$\frac{\partial^2 H}{\partial\varphi^2} = \frac{p_\psi^2}{ml^2}\frac{1+2\cos^2\varphi}{\sin^4\varphi}+mgl\cos\varphi > 0, \tag{g}$$

$$\frac{\partial^2 H}{\partial p_\varphi^2} = \frac{1}{ml^2} > 0. \tag{h}$$

Condition (h) is always satisfied whereas (g) will occur when

$$\frac{1+3\cos^2\varphi}{\cos\varphi} > 0. \tag{i}$$

Interpreting the derived result we say that the stationary motion conditions, (e) and (f), can be satisfied for the position coordinate φ in the interval $(0, \pi/2)$. This range is open on the right-hand side since, on the basis of (f), we must exclude $\varphi = \pi/2$.

If we assume $\dot{\psi} = \omega =$ const, then

$$p_\psi = ml^2 \sin^2 \varphi \omega . \tag{j}$$

Substituting (j) into (f) we obtain

$$l^2 \sin^2 \varphi \omega = \sin^2 \varphi \sqrt{\frac{gl^3}{\cos \varphi}}; \tag{k}$$

hence

$$\varphi = \arccos \frac{g}{l\omega^2}$$

or

$$\varphi = 0 \quad \text{for} \quad \frac{g}{l} > \omega^2.$$

In this way, we have a uniquely determined the stationary motion for an exact value ω of the angular velocity.

CHAPTER 8

Vibrations of Systems with One Degree of Freedom

8.1 Preliminary discussion

In the present chapter we shall deal with holonomic systems. For holonomic systems with one degree of freedom Hamilton's equations of the form

$$\dot{q} = \frac{\partial H}{\partial p}, \quad \dot{p}-Q = -\frac{\partial H}{\partial q} \tag{1}$$

are valid. The Lagrange function for the considered system may be expressed by

$$L = \tfrac{1}{2}A(q, t)\dot{q}^2+a_1(q, t)\dot{q}+a_0(q, t)-V(q, t). \tag{2}$$

Making use of (7) and (1), Section 7.10, we find Hamilton's function:

$$H = \frac{[p-a_1(q, t)]^2}{2A(q, t)} -a(q, t)+V(q, t). \tag{3}$$

Substituting (3) into (1) we obtain

$$\dot{q} = \frac{1}{A(q, t)}(p-a_1(q,t)), \tag{4}$$

$$\dot{p} = \frac{1}{2[(q, t)]^2}\frac{\partial A(q, t)}{\partial q}[p-a_1(q, t)]^2+\frac{p-a_1(q, t)}{A(q, t)}\frac{\partial a_1(q, t)}{\partial q}+$$
$$+Q(q, p, t)+\frac{\partial a_0(q, t)}{\partial q}-\frac{\partial V(q, t)}{\partial q}. \tag{5}$$

Equations (4) and (5) are valid in the most general case of the motion of a holonomic system with one degree of freedom. They can therefore, under certain conditions of the problem, describe vibration. The solution to the problem consists in finding the integrals of the system of differential equations (4) and (5) with given initial conditions. Every solution

$$p = p(t), \quad q = q(t)$$

gives a specified curve in a plane pq, called the *phase trajectory*. The plane pq itself is called the *phase plane*. From the structure of phase trajectories we can easily infer the character of the motion of a system. In principle, systems can be divided into two groups: *oscillatory* and *non-oscillatory*. We shall focus our attention on the first group.

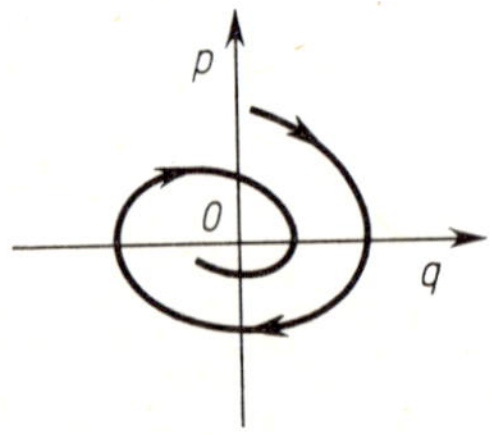

Fig. 8.1

Two phase trajectories representing an oscillatory and a non-oscillatory system are presented respectively in Fig. 8.1 and Fig. 8.2. The best method to find the phase trajectories is to divide both sides of equations (4) and (5) by each other. After division we obtain

$$\frac{\mathrm{d}p}{\mathrm{d}q} = F(q, p, t), \tag{6}$$

where

$$F(q,p,t) = \frac{p-a_1(q,t)}{2A(q,t)}\,\frac{\partial A_1(q,t)}{\partial q} + \frac{\partial a_1(q,t)}{\partial q} + \\ + \frac{A(q,t)}{p-a_1(q,t)}\left[Q(q,p,t) + \frac{\partial a_0(q,t)}{\partial q} - \frac{\partial V(q,t)}{\partial q}\right]. \tag{6a}$$

The result (6) can also easily be derived using equation (6), Section 7.11.

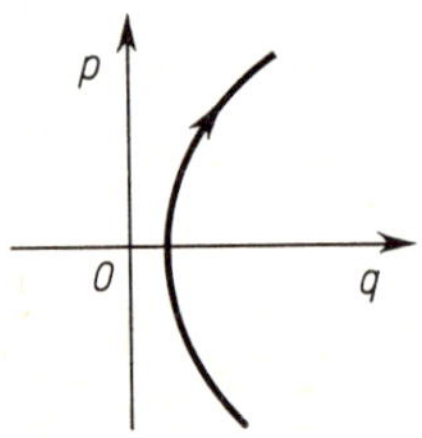

Fig. 8.2

From equation (2), Section 7.11, it follows that

$$T = H+E_1+2E_0. \tag{7}$$

Substituting the derived result into (6), Section 7.11, we have

$$\frac{\mathrm{d}H}{\mathrm{d}t} = Q\dot{q}+\frac{\partial V}{\partial t}-\frac{\partial E}{\partial t}. \tag{8}$$

Since the kinetic energy is expressed by

$$E = \tfrac{1}{2}A(q,t)\dot{q}^2+a_1(q,t)\dot{q}+a_0(q,t) \tag{9}$$

and also (3) applies, equation (8) can be written in the following form:

$$\begin{aligned}
&-\frac{1}{2[A(q,t)]^2}\left[\frac{\partial A(q,t)}{\partial q}\dot{q}+\frac{\partial A(q,t)}{\partial t}\right][p-a_1(q,t)]^2+\\
&+2[\dot{p}-\dot{a}_1(q,t)]\frac{p-a_1(q,t)}{2A(q,t)}-\frac{\partial a_0(q,t)}{\partial q}\dot{q}-\frac{\partial a_0(q,t)}{\partial t}+\\
&+\frac{\partial V(q,t)}{\partial q}\dot{q}+\frac{\partial V(q,t)}{\partial t} = Q(q,p,t)\dot{q}+\frac{\partial V(q,t)}{\partial t}-\\
&-\left[\frac{1}{2}\frac{\partial A(q,t)}{\partial t}\dot{q}^2+\frac{\partial a_1(q,t)}{\partial t}\dot{q}+\frac{\partial a_0(q,t)}{\partial t}\right].
\end{aligned} \tag{10}$$

Reducing and then dividing the result by $\dot{q}^3$, simultaneously making use of equation (4) and regrouping the terms, we obtain

$$\frac{\dot{p}}{\dot{q}} = F(q,p,t), \tag{11}$$

where $F(q,p,t)$ is described by equation (6a). The result (11) accords with (6).

For systems with constraints independent of time, i.e., scleronomic systems, the derived result becomes simplified ($a_1 = a_0 = 0$). Therefore, for such systems the following equation is valid:

$$\frac{\mathrm{d}p}{\mathrm{d}q} = \frac{p}{2A(q)}\frac{\partial A(q)}{\partial q}+\frac{A(q)}{p}\left[Q(q,p,t)-\frac{\partial V(q,t)}{\partial q}\right] \tag{12}$$

or

$$\frac{\mathrm{d}E}{\mathrm{d}q}+\frac{\partial V(q,t)}{\partial q}-Q(q,p,t) = 0, \tag{12a}$$

where

$$E = \frac{p^2}{2A(q)}.$$

The generalized momentum p can be expressed by means of the kinetic energy E. Hence

$$\frac{\mathrm{d}E}{\mathrm{d}q}+\frac{\partial V(q,t)}{\partial q}-Q(q,E,t)=0. \tag{12b}$$

Putting p or E equal to zero we find the equilibrium position of the system

$$\frac{\partial V(q,t)}{\partial q}-Q(q,0,t)=0. \tag{13}$$

Evidently, the position of equilibrium may vary with time.

Serious difficulty may be encountered in general case with solving the differential equations (12). If an exact method fails, one should resort to an approximative method, e.g. the *isocline method*, which enables the progress of phase trajectories to be estimated. The phase trajectory described by equation (12) has the property that it intersects the axis $p = 0$ right angles (for $p = 0$ we have $\mathrm{d}q/\mathrm{d}p = 0$); we assume at the same time that the point of intersection is not the equilibrium point, since

$$\left.\frac{\mathrm{d}^2q}{\mathrm{d}p^2}\right|_{p=0}=\frac{1}{A(q)\left[Q(q,0,t)-\dfrac{\partial V(q,t)}{\partial q}\right]}. \tag{14}$$

To conclude the present section, we shall consider a special form of equation (12b), namely

$$\frac{\mathrm{d}E}{\mathrm{d}q}+\frac{\partial V(q)}{\partial q}=0; \tag{15}$$

hence

$$E+V=\text{const} \tag{16}$$

giving the equation of phase trajectory and at the same time the law of conservation of energy (a conservative system).

8.2 Free vibrations of harmonic oscillators

A system for which the kinetic and the potential energy can be expressed by

$$E=\frac{p^2}{2a},\qquad V(q)=\tfrac{1}{2}cq^2, \tag{1}$$

where a and c are constants, is called a *harmonic oscillator*.

Hamilton's function for such a system is

$$H = \frac{p^2}{2a} + \tfrac{1}{2}cq^2. \tag{2}$$

Substituting (2) into (1) we have

$$\dot{q} = \frac{p}{a}, \quad \dot{p} = -cq. \tag{3}$$

Dividing both sides of the derived equations by each other, we arrive at a differential equation of phase trajectory

$$\frac{\mathrm{d}p}{\mathrm{d}q} = -\frac{acq}{q}. \tag{4}$$

Separating the variables and integrating we obtain

$$\frac{p^2}{2} = -\frac{acq^2}{2} + C. \tag{5}$$

The phase trajectories of the harmonic oscillator are ellipses (Fig. 8.3). Hamilton's function, it will be noted, assumes the least value at this point.

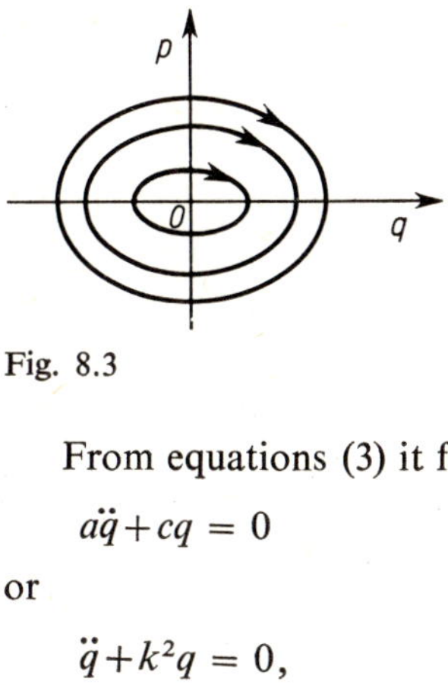

Fig. 8.3

From equations (3) it follows that

$$a\ddot{q} + cq = 0 \tag{6}$$

or

$$\ddot{q} + k^2 q = 0, \tag{6a}$$

where

$$k^2 = \frac{c}{a}. \tag{6b}$$

Evidently, the free motion of the harmonic oscillator is described by a linear differential equation of the second order with constant coefficients. Note that the equation of motion (6a) can be obtained forming the Lagrange kinetic potential:

$$L = \tfrac{1}{2}(aq^2 - cq^2), \tag{7}$$

and then substituting L into the Lagrange equation:

$$\frac{\mathrm{d}}{\mathrm{d}t}\left(\frac{\partial L}{\partial \dot{q}}\right) - \frac{\partial L}{\partial q} = 0. \tag{8}$$

Example 1. Prove that a mechanical system (Fig. 8.4), consisting of a mass m and a spring with constant c N/m, is a harmonic oscillator.

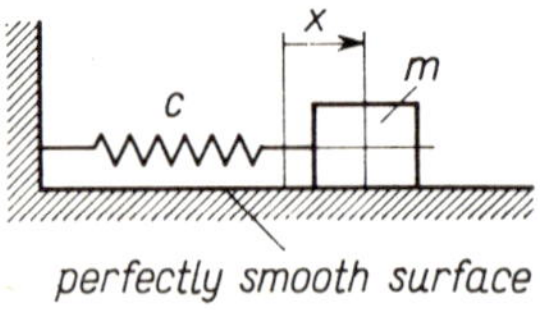

Fig. 8.4

Let q be the deflection of the system from the equilibrium position. The elastic energy accumulated in the spring with this deflection is

$$V = \tfrac{1}{2}cq^2 \tag{a}$$

and the kinetic energy

$$E = \tfrac{1}{2}m\dot{q}^2. \tag{b}$$

The generalized momentum

$$p = m\dot{q}. \tag{c}$$

Finding $\dot{q}$ from (c) and substituting it into (b) we have

$$E = \frac{p^2}{2m}. \tag{d}$$

This shows that the system considered is a harmonic oscillator, since c and m are constant.

Example 2. Find the differential equation for the motion of a system consisting of an elastic rod rigidly fixed at one end and having a mass m with a moment of inertia J attached at the other end (Fig. 8.5). The mass can move about the rod axis.

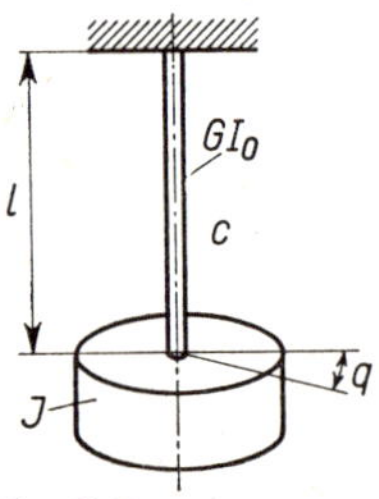

Fig. 8.5

The potential energy accumulated in a rod subjected to torsion is

$$V = c\frac{q^2}{2},$$

where q is the angle of rotation of the solid, c a constant dependent on the geometrical dimensions and material of the rod such that

$$c = \frac{GI_0}{l},$$

where G is the modulus of rigidity of the rod material, I_0 the geometric polar moment of inertia of the circular cross-section of the rod, l length of the rod.

The kinetic energy of the solid which is in angular motion is

$$E = \tfrac{1}{2}J\dot{q}^2 = \frac{1}{2J}p^2.$$

The differential equation of the motion of the solid is therefore equation (6a), in which

$$k^2 = \frac{c}{J}.$$

Example 3. An electric system consists of a coil with inductance L and a condenser with capacitance C (Fig. 8.6). Find the differential equations for oscillations of the system.

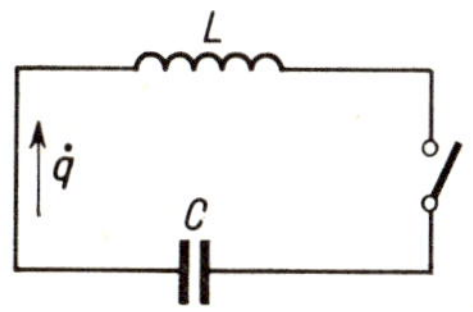

Fig. 8.6

For the given circuit we have

$$E = -\tfrac{1}{2}L\dot{q}^2 = \frac{p^2}{2L}$$

and

$$V = \frac{1}{2C}q^2.$$

The considered system is easily seen to be a harmonic oscillator; consequently, the differential equation of the changes in its state is (6a), in which

$$k^2 = \frac{1}{LC}.$$

Example 4. A mass m is placed at the end of a beam with bending rigidity EI (E the modulus of elasticity, I geometric moment of inertia of cross-section) and length l (Fig. 8.7). Find the differential equation for the motion of the mass in the vertical plane.

Let q be the deflection of the mass from the position of equilibrium. In the considered case

$$E = \tfrac{1}{2}m\dot{q}^2,$$

whereas

$$V = \tfrac{1}{2}c(q+f)^2 - mg(q+f)$$

since the mass is moving in the gravitational field (g the acceleration of gravity). Forming a Lagrange function on the lines of equation (7) and substituting the result into equation (8), we obtain

$$m\ddot{q}+cq = 0. \tag{a}$$

The constant c in the particular case is

$$c = \frac{3EI}{l}.$$

Deriving equation (a), it should be noted that in the equilibrium position

$$mg = cf$$

occurs, where f is the deflection of the beam at its end under the weight mg.

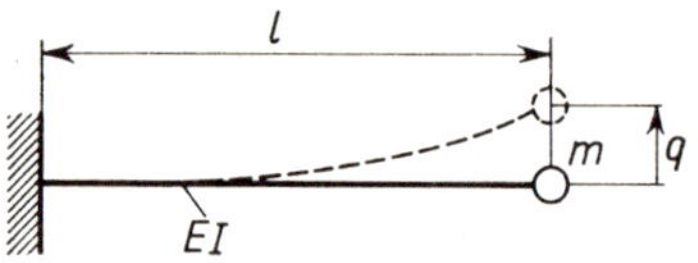

Fig. 8.7

If the scleronomic system is subjected to small vibrations about the equilibrium position while in motion, it can be regarded as a harmonic oscillator. Indeed, if the functions $A(q)$ and $V(q)$ are expanded into series in the vicinity of the equilibrium point

$$A(q) = A(0)+A(0)q+ \ldots,$$
$$V(q) = V(0)+V'(0)q+\tfrac{1}{2}V''(0)q^2+ \ldots,$$

then, assuming a minor deflection q from the position of equlibrium, we can reduce the above equations to the first non-zero terms of each expansion. Consequently, if we accept

$$A(q) \approx A(0) = a \tag{9}$$

and

$$V(q) \approx \tfrac{1}{2}V''(0)q^2 = \tfrac{1}{2}cq^2, \tag{10}$$

then the considered system will be described by the differential equation (6) for harmonic oscillators. By such limitations of the expansion of the functions $A(q)$ and $V(q)$, we achieve a linearization of the differential equation of motion. We have neglected the constant member $V(0)$ in equation (10), because the energy V is given with an accuracy up to the additive constant. The term $V'(0)q$ does not come in either, because the point $q = 0$ is the equilibrium position.

Example 5. A physical pendulum is given (Fig. 8.8). Derive the differential equation for large and small oscillations of the system.

The kinetic energy is

$$E = \tfrac{1}{2}J\dot{q}^2, \tag{a}$$

where J is the moment of inertia of the pendulum about its axis of rotation and q an angle giving its position.

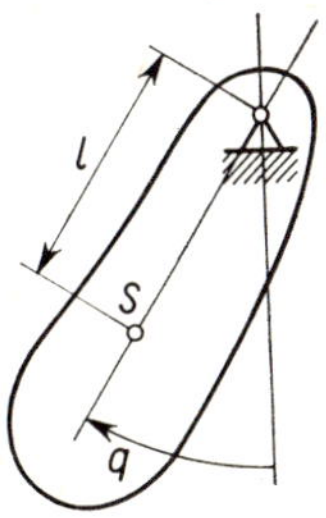

Fig. 8.8

The potential energy is expressed by

$$V = mgl(1-\cos q). \tag{b}$$

Forming the Langrange kinetic potential and substituting the derived result into equation (8), we obtain

$$J\ddot{q} - mgl\sin q = 0. \tag{c}$$

Equation (c) is obviously a non-linear equation and holds for the entire deflection range of the pendulum. Wanting to obtain its linear approximation, i.e., an equation valid for small oscillations, we expand the function $V(q)$ into a series:

$$V(q) = mgl[1-1+\tfrac{1}{2}q^2+ \ldots] \approx mgl\tfrac{1}{2}q^2. \tag{d}$$

Next, forming the Lagrange kinetic potential from (a) and (d) and substituting the result obtained into equation (8), we arrive at a linear equation

$$J\ddot{q} + mglq = 0. \tag{e}$$

Example 6. A system (Fig. 8.9) consists of a toothed wheel A with radius r and moment of inertia J, a toothed bar B with mass m, a crank $OC = a$, rigidly fixed to the wheel, having a coupler $CD = l$, and three springs with constants c_A, c_B and c_D acting on the wheel, the toothed bar and also on a slide block with mass M. In the equilibrium position the angle DOC is a right angle. Give the equation for the phase trajectory. Find the differential equation for small vibrations of the system.

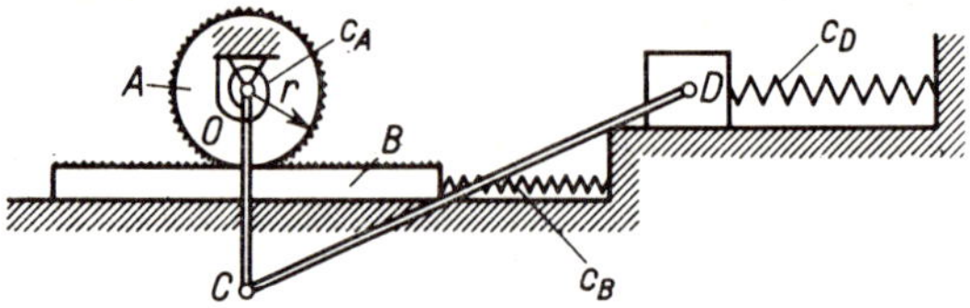

Fig. 8.9

First we determine the kinetic energy of the system:

$$E = \tfrac{1}{2}A(q)\dot{q}^2 \tag{a}$$

where q denotes the angle of rotation of the toothed wheel A from the position of equilibrium $q = 0$ (Fig. 8.10). From Fig. 8.10 we find that

$$E = \tfrac{1}{2}(J\dot{q}^2 - mr^2\dot{q}^2 + M\dot{x}^2). \tag{b}$$

The slide-block displacement x is a function of the angle q (a system with one degree of freedom)

$$x = l\cos\alpha - (l\cos\psi - a\sin q);$$

hence

$$\dot{x} = l\sin\psi\dot{\psi} + a\cos q\dot{q}.$$

But

$$l\sin\psi = a\cos q; \tag{c}$$

hence

$$\dot{x} = a\cos q(\dot{\psi} + \dot{q}).$$

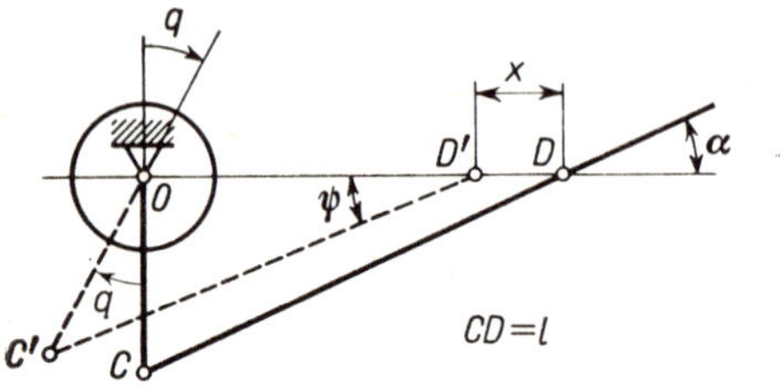

Fig. 8.10

Differentiating (c) we obtain

$$l\cos\psi\dot{\psi} = -a\sin q\dot{q},$$

or

$$\dot{\psi} = -\frac{a\sin q}{l\cos\psi}\dot{q} = -\frac{a\sin q\dot{q}}{\sqrt{l^2 - a^2\cos^2 q}},$$

therefore

$$\dot{x} = a\cos q\left(1-\frac{a\sin q}{\sqrt{l^2-a^2\cos^2 q}}\right)\dot{q}.$$

Substituting for the slide-block velocity thus calculated into equation (b), we have

$$E = \frac{1}{2}\left[J+mr^2+Ma^2\cos^2 q\left(1-\frac{a\sin q}{\sqrt{l^2-a^2\cos^2 q}}\right)^2\right]\dot{q}^2.$$

Comparing the derived equation with (a) we find that

$$A(q) = J+mr^2+Ma^2\cos^2 q\left(1-\frac{a\sin q}{\sqrt{l^2-a^2\cos^2 q}}\right)^2.$$

We now calculate the potential energy:

$$V = \tfrac{1}{2}(c_A q^2+c_B r^2 q^2+c_D x^2).$$

Making use of the law of conservation of energy, we write the equation of phase trajectory as

$$\frac{p^2}{2\left[J+mr^2+Ma^2\cos q\left(1-\frac{a\sin q}{\sqrt{l^2-a^2\cos^2 q}}\right)^2\right]}+\frac{1}{2}(c_A q^2+c_B r^2 q^2+c_D x^2) = C.$$

Wishing to derive a differential equation of motion, we may use the second order Lagrange equations. Forming the Lagrange function

$$L = E-V = \tfrac{1}{2}A(q)\dot{q}^2-V(q)$$

and substituting L into

$$\frac{d}{dt}\left(\frac{\partial L}{\partial \dot{q}}\right)-\frac{\partial L}{\partial q} = 0,$$

we arrive at

$$A(q)\ddot{q}+\frac{1}{2}\frac{\partial A}{\partial q}\dot{q}^2+\frac{\partial V}{\partial q} = 0. \tag{d}$$

As we can see the equation obtained is non-linear. Linearization, as we have mentioned earlier, consists both in adopting

$$A(q) \approx A(0),$$

where in the considered case

$$A(0) = J+mr^2+Ma^2, \tag{e}$$

and in expanding the function $V(q)$ into a series, limiting ourselves to the first non-zero terms

$$V(q) = V(0)+V'(0)q+V''(0)\frac{q^2}{2}+\ldots$$

In the considered case, as is easily seen, $V(0) = 0$ and $V'(0)q = 0$—whereas

$$V(q) = \left[c_A+c_B r^2+c_D\left(\frac{dx}{dq}\right)^2+\frac{d^2x}{dq^2}x\right]\frac{q^2}{2}+\ldots,$$

$$\frac{dx}{dq} = a\cos q\left(1-\frac{a\sin q}{\sqrt{l^2-a^2\cos^2 q}}\right).$$

Now $y = 0$ for $q = 0$, therefore

$$V(q) = (c_A + c_B r^2 + c_D a^2)\frac{q^2}{2}. \tag{f}$$

Substituting (e) and (f) into (d) we have

$$(J + mr^2 + Ma^2)\ddot{q} + (c_A + c_B r^2 + c_D a^2)q = 0. \tag{g}$$

Evidently, equation (g) is already linear.

A differential equation for the motion of harmonic oscillators presented itself in each of these examples:

$$\ddot{q} + k^2 q = 0.$$

We now come to the solution of this equation. Performing Laplace's transformation* we write

$$s^2\bar{q}(s) - sq(0) - \dot{q}(0) + k^2\bar{q}(s) = 0.$$

Solving the operator equation with respect to the transform of the looked for solution, we obtain

$$\bar{q}(s) = q(0)\frac{s}{s^2+k^2} + \frac{\dot{q}(0)}{k}\,\frac{k}{s^2+k^2}. \tag{11}$$

Performing an inverse transformation on equation (11) we find the equation of motion

$$q(t) = q(0)\cos kt + \frac{\dot{q}(0)}{k}\sin kt, \tag{12}$$

where $q(0)$ is the deflection from the position of equilibrium and $\dot{q}(0)$ is the oscillator velocity at instant $t = 0$. Equation (12) can be transformed to

$$q(t) = A_0 \sin(kt + \varepsilon), \tag{13}$$

where

$$A_0 = \sqrt{[q(0)]^2 + \left[\frac{\dot{q}(0)}{k}\right]^2} \tag{13a}$$

$$\tan\varepsilon = \frac{kq(0)}{\dot{q}(0)}. \tag{13b}$$

---

* *Laplace's transform* of the function $f(t)$ is called the integral

$$\bar{f}(s) \stackrel{\text{df}}{=} \int^{\infty} f(t)e^{-st}dt.$$

Evidently, both the amplitude A_0 of the harmonic motion and the phase depend on the initial conditions. The magnitude k is called the *vibration frequency* of a harmonic oscillator. The free vibration period, being the time required for the full cycle of motion, can be expressed by means of vibration frequency according to the relation

$$T = \frac{2\pi}{k}. \tag{14}$$

Example 7. During braking, a lift (Fig. 8.11) was moving with a deceleration $\ddot{x}_0 = 5$ m/s². When it had the speed $v = 4$ m/s, the rope suddenly jammed at the top. Find the maximum tension of the rope, knowing that its elastic constant is $c = 40\,000$ N/cm and the mass of the lift $m = 2000$ kg (the effect of rope weight should be ignored).

The system is a harmonic oscillator. Following the sudden halt of the rope, vibration of the weight will develop as described by equation (13). The maximum tension of the rope occurs when there is maximum deflection of the system from the equilibrium position

$$S = mg + A_0 c. \tag{a}$$

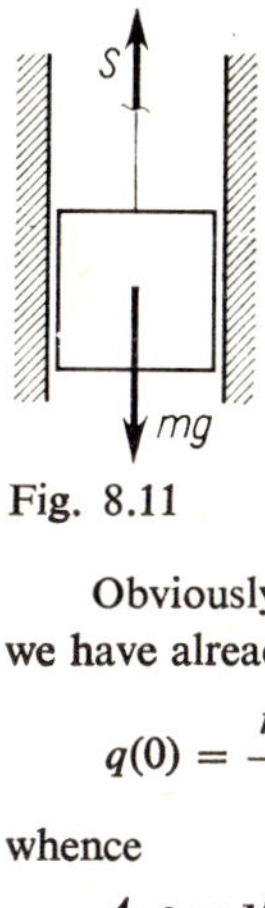

Fig. 8.11

Obviously the solution to the problem amounts to finding the amplitude A_0. As we have already noted, it depends on the initial conditions. In the considered case

$$q(0) = \frac{m\ddot{x}_0}{c} \quad \text{and} \quad \dot{q}(0) = v,$$

whence

$$A_0 c = \sqrt{m^2\ddot{x}_0^2 + v^2 cm},$$

or

$$S = mg + \sqrt{m^2\ddot{x}_0^2 + v^2 cm} = 380\,000 \text{ N}.$$

8.3 The influence of dissipative forces in the free vibrations of harmonic oscillators

Let the dissipation of the energy of a system be characterized by Rayleigh's function

$$D = \tfrac{1}{2}h\dot{q}^2. \tag{1}$$

This signifies that frictional forces are proportional to velocity. Substituting (7), Section 8.2, and (1) into the Lagrange equation

$$\frac{\mathrm{d}}{\mathrm{d}t}\left(\frac{\partial E}{\partial \dot{q}}\right)-\frac{\partial E}{\partial q}+\frac{\partial D}{\partial \dot{q}}=0,$$

we obtain

$$a\ddot{q}+h\dot{q}+cq=0. \tag{2}$$

Dividing equation (2) by a and introducing the notations

$$2n=\frac{h}{a}, \qquad k^2=\frac{c}{a},$$

we write

$$\ddot{q}+2n\dot{q}+k^2q=0. \tag{3}$$

We shall seek the solution using the method of integral transformations. Performing Laplace's transformation on equation (3), we arrive at the following operator equation:

$$s^2\bar{q}(s)-\dot{q}(0)-sq(0)+2ns\bar{q}(s)-2nq(0)+k^2\bar{q}(s)=0.$$

Hence

$$\begin{aligned}\bar{q}(s)&=\frac{q(0)s+2nq(0)+\dot{q}(0)}{s^2+2ns+k^2}\\&=q(0)\frac{s+n}{(s+n)^2+k^2-n^2}+\frac{\dot{q}(0)+q(0)n}{(s+n)^2+k^2-n^2}.\end{aligned}$$

Retransforming we obtain:

(a) for $k>n$

$$q(t)=\mathrm{e}^{-nt}\left(C_1\cos\sqrt{k^2-n^2}\,t+C_2\sin\sqrt{k^2-n^2}\,t\right), \tag{4}$$

where

$$C_1=q(0), \qquad C_2=\frac{\dot{q}(0)+q(0)n}{\sqrt{k^2-n^2}}; \tag{4a}$$

(b) for $n=k$

$$q(t)=\mathrm{e}^{-nt}(C_1+C_2t), \tag{5}$$

where

$$C_1=q(0), \qquad C_2=\dot{q}(0)+q(0)n; \tag{5a}$$

(c) for $n>k$

$$q(t)=\mathrm{e}^{-nt}\left(C_1\cosh\sqrt{n^2-k^2}\,t+C_2\sinh\sqrt{n^2-k^2}\,t\right), \tag{6}$$

where

$$C_1 = q(0), \quad C_2 = \frac{\dot{q}(0)+q(0)n}{\sqrt{n^2-k^2}}. \tag{6a}$$

As we can see, harmonic motion occurs only when $n < k$, i.e., only when the damping coefficient n is sufficiently low; if not, we are dealing with an aperiodic motion. The vibration period is

$$T = \frac{2\pi}{\sqrt{k^2-n^2}}. \tag{7}$$

Moreover, we can easily see that the vibrations have a decaying character (Fig. 8.12). We may ask: What proportion of the absolute values of deflections of the system remains after half the period T?

$$\left|\frac{q(t)}{q(t+T/2)}\right| = e^{nT/2}. \tag{8}$$

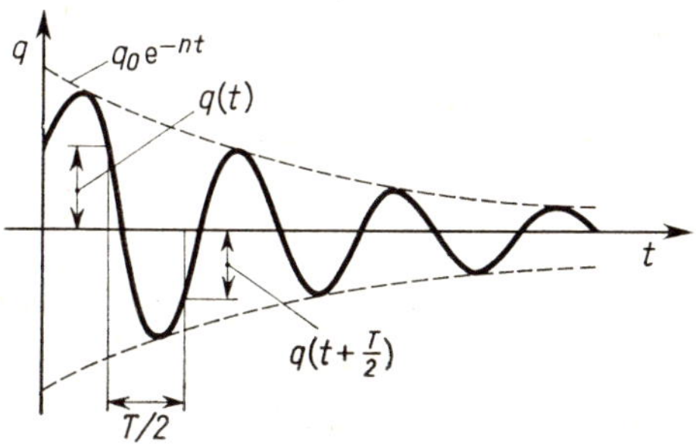

Fig. 8.12

The logarithm of this ratio is called the *logarithmic decrement*

$$\Delta = \ln\left|\frac{q(t)}{q(t+T/2)}\right| = n\frac{T}{2}. \tag{9}$$

Evidently, the logarithmic decrement characterizes the vibration damping rate.

Example 1. Deflections were found to decrease w-fold during time t_0, being the multiple of a vibration period T. Calculate the damping coefficient of vibrations n.

Using (8) we write

$$e^{nt_0} = w.$$

Therefore

$$nt_0 = \ln w,$$

whence

$$n = \frac{\ln w}{t_0}.$$

In conclusion of the present section we shall discuss the phase trajectory problem. Making use of equation (12), Section 8.1 for the considered case, we arrive at

$$\frac{dp}{dq} = \frac{a}{p}\left[-\frac{\partial D}{\partial \dot{q}} - cq\right]$$

or

$$\frac{d}{dt}\left[\frac{p^2}{2a} + \frac{cq^2}{2}\right] = -\frac{\partial D}{\partial \dot{q}}\dot{q}. \tag{10}$$

Integrating (10) we obtain

$$\frac{p^2}{2a} + \frac{cq^2}{2} = -\int_0^t \frac{\partial D}{\partial \dot{q}}\dot{q}\,dt. \tag{11}$$

The right-hand side of (11) decreases to zero, following from the fact that

$$\frac{\partial D}{\partial \dot{q}}\dot{q} = 2D > 0.$$

Hence the conclusion that a point in a phase plane is moving along a curve tending to the reference origin. The considered system is therefore an asymptotically stable system. Knowing the solution it will be easy to find the equation of phase trajectory. To analyse this fact it is more convenient to change to another frame of reference by way of a linear transformation of the form

$$x = a_{11}q + a_{12}p, \qquad y = a_{21}q + a_{22}p. \tag{12}$$

If the determinant

$$\begin{vmatrix} a_{11} & a_{12} \\ a_{21} & a_{22} \end{vmatrix} \neq 0,$$

then the transformation (12) is a one-to-one mapping of the plane qp onto the plane xy. In the present case we assume

$$a_{11} = \sqrt{k^2 - n^2}, \qquad a_{12} = 0, \qquad a_{21} = n, \qquad a_{22} = \frac{1}{a}.$$

Therefore, considering (4), we write

$$\begin{aligned} x &= A_0 e^{-nt}\sqrt{k_2 - n^2}\,\sin\left(\sqrt{k^2 - n^2}\,t + \varepsilon\right), \\ y &= A_0 e^{-nt}\sqrt{k^2 - n^2}\,\cos\left(\sqrt{k^2 - n^2}\,t + \varepsilon\right), \end{aligned} \tag{13}$$

where $A_0 = \sqrt{C_1^2+C_2^2}$, $\tan\varepsilon = C_1/C_2$, from which we obtain the equation for the phase trajectory in the plane Oxy:

$$x^2+y^2 = e^{-2nt}A_0^2(k^2-n^2). \tag{14}$$

The point under consideration evidently tends to the reference origin of the frame Oxy (Fig. 8.13). Since transformation (12) has the property that the original reference origin passes into the reference origin of the transformed frame, the transformation being one-to-one, therefore in the plane qp the point also tends to the reference origin.

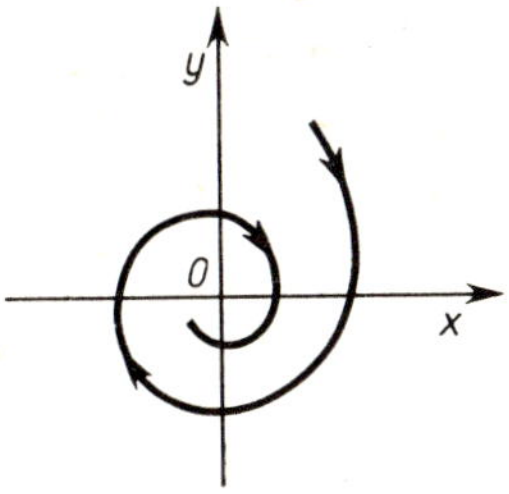

Fig. 8.13

8.4 Forced vibrations of harmonic oscillators

In the case of vibrations of harmonic oscillator under exciting forces, the following differential equation is valid:

$$a\ddot{q}+h\dot{q}+cq = Q(t). \tag{1}$$

Equation (1) differs from equation (2), Section 8.3, only in the right-hand side. Dividing (1) by a we obtain an equation analogous to (3), Section 8.3:

$$\ddot{q}+2n\dot{q}+k^2q = a_0(t), \tag{2}$$

where

$$2n = \frac{h}{a}, \quad k^2 = \frac{c}{a}, \quad a_0(t) = \frac{Q(t)}{a}. \tag{3}$$

We integrate (2) by the method of integral transformations.

Performing Laplace's transformation on equation (2) we have

$$s^2\bar{q}(s)-sq(0)-\dot{q}(0)+2ns\bar{q}(s)-2nq(0)+k^2\bar{q}(s) = \bar{a}_0(s). \tag{4}$$

The transform of the looked for $\bar{q}(s)$ is included in equation (4). Solving it with respect to this transform we obtain

$$\bar{q}(s) = q(0)\frac{s+n}{(s+n)^2+k^2-n^2}+\frac{\dot{q}(0)+q(0)n}{(s+n)^2+k^2-n^2}+$$
$$+\frac{1}{(s+n)^2+k^2-n^2}\bar{a}_0(s). \tag{5}$$

Performing Laplace's inverse transformation on equation (3) while making use of the theorem on convolution of a function, we find the looked for equation of motion:

$$q(t) = q_0(t) + \int_0^t G(t-\tau)a_0(\tau)\mathrm{d}\tau, \tag{6}$$

where $q_0(t)$ is given by the right-hand sides of equations (4), (5) and (6) from the preceding section and

$$G(t) = \frac{1}{\sqrt{k^2-n^2}} \mathrm{e}^{-nt} \sin\sqrt{k^2-n^2}\, t \quad \text{for} \quad n < k, \tag{6a}$$

$$G(t) = t\mathrm{e}^{-nt} \quad \text{for} \quad n = k, \tag{6b}$$

and finally

$$G(t) = \frac{1}{\sqrt{n^2-k^2}} \mathrm{e}^{-nt} \sinh\sqrt{n^2-k^2}\, t \quad \text{for} \quad n > k. \tag{6c}$$

The function $G(t)$ is obviously the response of the harmonic oscillator to a Dirac delta impulse, i.e., it is an *influence function* (Green's function). Its transform, $\bar{G}(s) = [(s+n)^2+k^2-n^2]^{-1}$, is called the *operator transfer function* or *transmittance*.

Note that putting $q(0) = \dot{q}(0) = 0$ we obtain

$$\bar{q}(s) = \bar{G}(s)\bar{a}_0(s).* \tag{7}$$

Thus we can see that the product of the operator function of transmittance and the transform of the input signal to the system is equal to the transform of the output signal. Diagramatically we mark the transmittance of the signal through the system in the way shown in Fig. 8.14.

We shall call the *time function of transmittance* for the system the response of the system to a unit step in the input to the system:

$$a_0(t) = H(t), \tag{8}$$

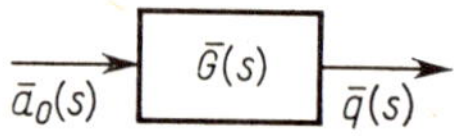

Fig. 8.14

* Generally we say that the *operator transfer function* or *transmittance* for the linear system described by the linear equation [64] $L_1(D)x(t) = L_2(D)y(t)$ is

$$\bar{G}(s) = \frac{L_1(s)}{L_2(s)},$$

when $x(t)$ is the input signal and $y(t)$ the output signal (compare (16), Section 4.10).

where $H(t)$ is Heaviside's function. It follows that this function is an integral of the influence function:

$$k(t) = \int_0^t G(t-\tau)H(\tau)\mathrm{d}\tau. \tag{9}$$

In the case of an oscillatory system ($n < k$)

$$\begin{aligned} k(t) &= \frac{1}{\sqrt{k^2-n^2}}\mathrm{e}^{-nt}\int_0^t \mathrm{e}^{n\tau}\sin\sqrt{k^2-n^2}\,(t-\tau)\,H(\tau)\mathrm{d}\tau \\ &= \frac{1}{\sqrt{k^2-n^2}}\int_0^t \mathrm{e}^{-n\tau}\sin\sqrt{k^2-n^2}\,\tau H(t-\tau)\mathrm{d}\tau \\ &= \frac{1}{k^2}\left[1-\mathrm{e}^{-nt}\left(\frac{n}{\sqrt{k^2-n^2}}\sin\sqrt{k^2-n^2}\,t+\cos\sqrt{k^2-n^2}\,t\right)\right]. \end{aligned} \tag{10}$$

Figure 8.15 presents graphically the time function of transmittance for an oscillatory system.

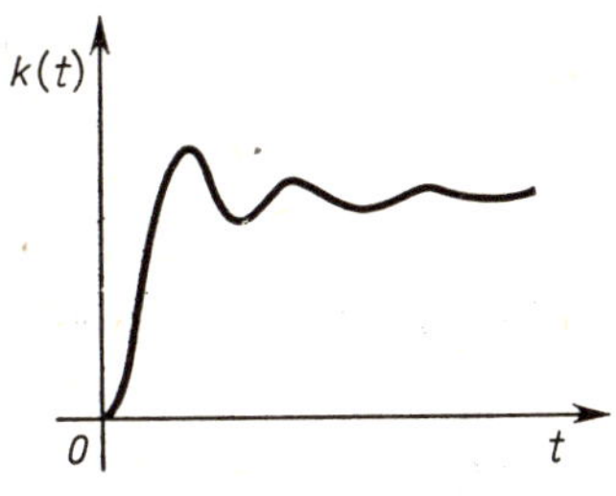

Fig. 8.15

We shall now discuss a different case. The problem of transmittance of a sinusoidal signal through the system will be considered. For $k > n$, we have

$$\begin{aligned} q(t) &= \mathrm{e}^{-nt}\left[q(0)\cos\sqrt{k^2-n^2}\,t+\frac{\dot{q}(0)+q(0)n}{\sqrt{k^2-n^2}}\sin\sqrt{k^2-n^2}\,t\right]+ \\ &+\frac{1}{\sqrt{k^2-n^2}}\int_0^t \mathrm{e}^{-n(t-\tau)}\sin\sqrt{k^2-n^2}\,(t-\tau)\,a_0(\tau)\mathrm{d}\tau. \end{aligned} \tag{11}$$

Considering the sinusoidal nature of the input signal ($a_0(\tau) = P_0 \sin(\omega\tau + \varphi)$) and calculating the integral in equation (1), we obtain

$$q(t) = e^{-nt}\left[q(0)\cos\sqrt{k^2-n^2}\,t + \frac{\dot{q}(0)+q(0)n}{\sqrt{k^2-n^2}}\sin\sqrt{k^2-n^2}\,t\right] -$$

$$-\frac{P_0 e^{-nt}}{\sqrt{(k^2-\omega^2)^2+4n^2\omega^2}}\left[\frac{n\sin(\varphi-\varepsilon)+\omega\cos(\varphi-\varepsilon)}{\sqrt{k^2-n^2}}\times\right.$$

$$\left.\times\sin\sqrt{k^2-n^2}\,t + \sin(\varphi-\varepsilon)\cos\sqrt{k_2-n^2}\,t\right] +$$

$$+\frac{P_0}{\sqrt{(k^2-\omega^2)^2+4n^2\omega^2}}\sin(\omega t+\varphi-\varepsilon), \tag{12}$$

where

$$\tan\varepsilon = \frac{2n\omega}{k^2-\omega^2}. \tag{12a}$$

Transition from expression (11) to (12) is very troublesome [64]. In this connection, the case of sinusoidal input is usually treated in literature as a separate problem of solving a linear non-homogeneous differential equation of the form

$$\ddot{q}+2n\dot{q}+k^2 q = P_0\sin(\omega t+\varphi). \tag{13}$$

We seek the solution in the form of a sum of solutions

$$q(t) = q^*(t)+q^{**}(t),$$

where $q^*(t)$ is the general integral of a homogeneous equation, i.e., without the right-hand side. For $n < k$ we have

$$q^*(t) = e^{-nt}(S_1\cos\sqrt{k^2-n^2}\,t+S_2\sin\sqrt{k^2-n^2}\,t) \tag{14}$$

(S_1 and S_2 are integration constants), $q^{**}(t)$ is a special integral of the equation with the right-hand side. We accept $q^{**}(t)$ in the form

$$q^{**}(t) = A\sin(\omega t+\varphi)+B\cos(\omega t+\varphi), \tag{15}$$

wheie A and B are constant. Substituting the sum (15) into (13) gives

$$-A\omega^2\sin(\omega t+\varphi)-B\omega^2\cos(\omega t+\varphi)+2nA\omega\cos(\omega t+\varphi)-$$
$$-2nB\omega\sin(\omega t+\varphi)+k^2A\sin(\omega t+\varphi)+k^2B\cos(\omega t+\varphi)$$
$$= P_0\sin(\omega t+\varphi).$$

The system thus obtained will be identically equal if

$$-A\omega^2-2nB\omega+k^2A = P_0, \qquad -B\omega^2+2nA\omega+k^2B = 0. \tag{16}$$

From the above we find the constants A and B:

$$A = \frac{P_0(k^2-\omega^2)}{(k^2-\omega^2)^2+4n^2\omega^2}, \qquad B = \frac{2P_0 n\omega}{(k^2-\omega^2)^2+4n^2\omega^2}. \tag{17}$$

Substituting the calculated constants into (15) and making use of relationship (12a), we obtain the following expression:

$$q^{**}(t) = \frac{P_0}{\sqrt{(k^2-\omega^2)^2+4n^2\omega^2}}\sin(\omega t+\varphi-\varepsilon), \tag{18}$$

which represents a form of steady vibrations.

Therefore, we write the general solution as follows:

$$q(t) = \mathrm{e}^{-nt}(S_1\cos\sqrt{k^2-n^2}\,t+S_2\sin\sqrt{k^2-n^2}\,t)+ \\ +\frac{P_0}{\sqrt{(k^2-\omega^2)^2+4n^2\omega^2}}\sin(\omega t+\varphi-\varepsilon). \tag{19}$$

We calculate the integration constants S_1 and S_2 from the initial conditions, i.e., from the known deflection $q(t)$ and velocity $\dot{q}(t)$ at an instant $t = 0$. Whence

$$\begin{aligned} q(0) &= S_1+\frac{P_0}{\sqrt{(k^2-\omega^2)^2+4n^2\omega^2}}\sin(\varphi-\varepsilon), \\ \dot{q}(0) &= -nS_1+S_2\sqrt{k^2-n^2}+\frac{P_0\omega}{\sqrt{(k^2-\omega^2)^2+4n^2\omega^2}}\cos(\varphi-\varepsilon). \end{aligned} \tag{20}$$

Finding therefrom S_1 and S_2 and then substituting the derived values into (19) we arrive at the result (12).

A somewhat different method leading to the same result (12), was proposed by S. Węgrzyn [77], namely the so-called method of "two opposed steady states". It consists in seeking the response of the system, being the outcome of the action on a harmonic oscillator of two opposed inputs, $P_0\sin(\omega t+\varphi)$ and $-P_0\sin(\omega t+\varphi)$ one of which has ceased at an instant $t = 0$. The sudden cessation of one of the inputs will disturb the state of equilibrium of the system. The disturbance will be characterized by the occurrence of a deflection $q_z(0)$ and a velocity $\dot{q}_z(0)$ at an instant $t = 0$. The steady response originating from input function $-P_0\sin(\omega t+\varphi)$ is given by

$$q(t) = -\frac{P_0}{\sqrt{(k^2-\omega^2)^2+4n^2\omega^2}}\sin(\omega t+\varphi-\varepsilon). \tag{21}$$

To determine the steady state we can use in this case the symbolic method. From (21) we write

$$\begin{aligned} q_z(0) &= -\frac{P_0}{\sqrt{(k^2-\omega^2)^2+4n^2\omega^2}}\sin(\varphi-\varepsilon), \\ \dot{q}_z(0) &= -\frac{P_0\omega}{\sqrt{(k^2-\omega^2)^2+4n^2\omega^2}}\cos(\varphi-\varepsilon). \end{aligned} \tag{22}$$

The cessation of input function $-P_0\sin(\omega t+\varphi)$ will induce response of the following form:

$$q_z(t) = e^{-nt}\left(q_z(0)\cos\sqrt{k^2-n^2}\,t+\frac{\dot{q}_z(0)+nq_z(0)}{\sqrt{k^2-n^2}}\sin\sqrt{k^2-n^2}\,t\right). \tag{23}$$

The steady response resulting from input $P_0\sin(\omega t+\varphi)$ is given by

$$q_s(t) = \frac{P_0}{\sqrt{(k^2-\omega^2)^2+4n^2\omega^2}}\sin(\omega t+\varphi-\varepsilon). \tag{24}$$

By the action of the input

$$a_0(t) = P_0\sin(\omega t+\varphi)-P_0H(-t)\sin(\omega t+\varphi)$$

the system will respond according to the equation

$$q(t) = q_z(t)+q_s(t),$$

which constitutes the solution to (12) for zero initial conditions. This method, being very efficatious, has been widely used in the study of instability in multiple-stage resistance-capacitance coupled amplifiers [76].

Transforming this equation we obtain

$$-\omega^2\bar{q}(j\omega)+2j\omega n\bar{q}(j\omega)+k^2\bar{q}(j\omega) = \bar{a}_0(j\omega).$$

Hence we determine $\bar{q}(j\omega)$:

$$\bar{q}(j\omega) = \frac{1}{k^2-\omega^2+2n\omega j}\bar{a}_0(j\omega). \tag{25}$$

Introducing the notation

$$\bar{G}(j\omega) = \frac{1}{k^2-\omega^2+2jn\omega}, \tag{26}$$

we write

$$\bar{q}(j\omega) = \bar{G}(j\omega)\bar{a}_0(j\omega). \tag{27}$$

Performing Fourier's inverse transformation* on (27) while making use of the theorem of convolution, we get the solution:

$$q(t) = \int_{-\infty}^{t} G(t-\tau)a_0(\tau)\mathrm{d}\tau,^{**} \tag{28}$$

where $G(t)$ is Green's function. $\bar{G}(\mathrm{j}\omega)$ is called the *frequency characteristic.* Note that it can be formally derived if we know the operator function of transmittance for the system; putting $s = \mathrm{j}\omega$ we have

$$\bar{G}(\mathrm{j}\omega) = \mathrm{Re}[\bar{G}(\mathrm{j}\omega)]+\mathrm{j}\,\mathrm{Im}[\bar{G}(\mathrm{j}\omega)] = |\bar{G}(\mathrm{j}\omega)|\mathrm{e}^{\mathrm{j}\varepsilon\omega}. \tag{29}$$

For the case of a harmonic oscillator

$$\mathrm{Re}[\bar{G}(\mathrm{j}\omega)] = \frac{k^2-\omega^2}{(k^2-\omega^2)^2+4n^2\omega^2} = \frac{1}{\sqrt{(k^2-\omega^2)^2+4n^2\omega^2}}\cos\varepsilon(\omega),$$

$$\mathrm{Im}[\bar{G}(\mathrm{j}\omega)] = -\frac{2n\omega}{(k^2-\omega^2)^2+4n^2\omega^2} \tag{30}$$

$$= -\frac{1}{\sqrt{(k^2-\omega^2)^2+4n^2\omega^2}}\sin\varepsilon(\omega).$$

See Fig. 8.16 for the graphical representation of the frequency characteristic. Obviously,

$$\tan\varepsilon(\omega) = \frac{\mathrm{Im}[\bar{G}(\mathrm{j}\omega)]}{\mathrm{Re}[\bar{G}(\mathrm{j}\omega)]},$$

---

* The *Fourier transform* of the function $f(t)$ is the integral

$$\bar{f}(\mathrm{j}\omega) \overset{\mathrm{df}}{=} \int_{-\infty}^{\infty} f(t)\mathrm{e}^{-\mathrm{j}\omega t}\mathrm{d}t.$$

** The validity of (28) can easily be proved by performing Fourier's transformation

$$\bar{q}(\mathrm{j}\omega) = \int_{-\infty}^{\infty} \mathrm{e}^{-\mathrm{j}\omega t}\Big[\int_{-\infty}^{t} G(t-\tau)a_0(\tau)\mathrm{d}\tau\Big]\mathrm{d}t$$

$$= \int_{-\infty}^{\infty} \mathrm{e}^{-\mathrm{j}\omega t}\int_{-\infty}^{\infty} G(t-\tau)H(t-\tau)a_0(\tau)\mathrm{d}\tau\,\mathrm{d}t$$

$$= \int_{-\infty}^{\infty}\int_{-\infty}^{\infty} \mathrm{e}^{-\mathrm{j}(\tau+t_1)\omega}G(t_1)H(t_1)a_0(\tau)\mathrm{d}\tau\,\mathrm{d}t_1$$

$$= \int_{-\infty}^{\infty} \mathrm{e}^{-\mathrm{j}\omega\tau}a_0(\tau)\mathrm{d}\tau \int_{-\infty}^{\infty} \mathrm{e}^{-\mathrm{j}\omega t_1}G(t_1)H(t_1)\mathrm{d}t_1 = \bar{a}_0(\mathrm{j}\omega)\bar{G}(\mathrm{j}\omega),$$

which is what was to be proved.

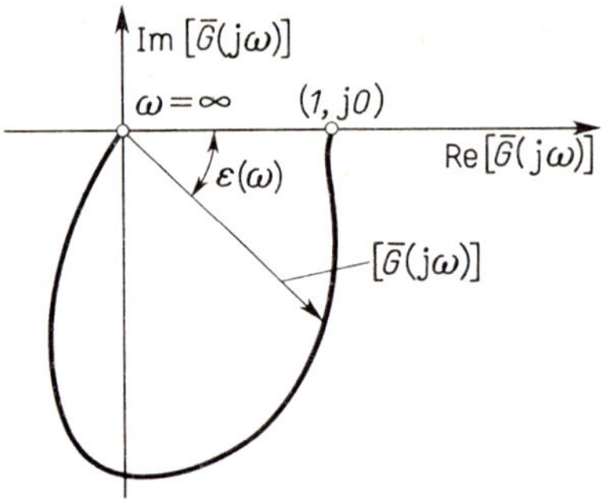

Fig. 8.16

whence

$$\varepsilon = \varepsilon(\omega) = \arctan\frac{\mathrm{Im}[\bar{G}(\mathrm{j}\omega)]}{\mathrm{Re}[\bar{G}(\mathrm{j}\omega)]}. \tag{31}$$

This is the so-called *phase characteristic* of the system (Fig. 8.17).

The *amplitude characteristic* is the modulus of the function $\bar{G}(\mathrm{j}\omega)$:

$$|\bar{G}(\mathrm{j}\omega)| = \frac{1}{\sqrt{(k^2-\omega^2)^2+4n^2\omega^2}}. \tag{32}$$

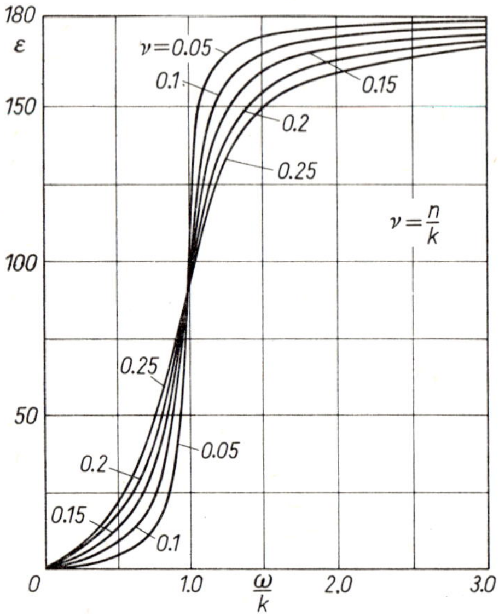

Fig. 8.17

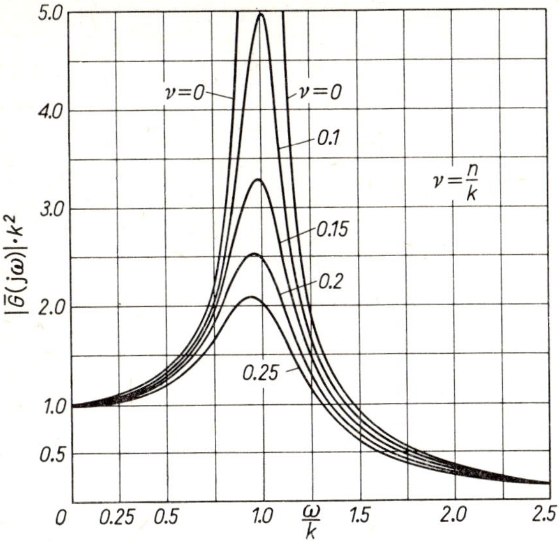

Fig. 8.18

The dependence of $|\bar{G}(j\omega)|$ on the frequency ω has been illustrated in Fig. 8.18. The curves plotted, for different values of n, show a dependence of the amplitude of forced vibrations on the input frequency ω. For $n = 0$ with $\omega = k$ the amplitude is infinitely large. We determine the vibration growth process by finding the limit of the indeterminate symbol derived from solution (12):

$$q(t) = q(0)\cos kt + \frac{\dot{q}(0)}{k}\sin kt -$$

$$-\lim_{\omega\to k}\left\{\frac{P_0}{k^2-\omega^2}\left[\frac{\omega}{k}\sin kt\cos\varphi + \cos kt)\sin\varphi - \sin(\omega t+\varphi)\right]\right\}$$

$$= q(0)\cos kt + \frac{\dot{q}(0)}{k}\sin kt + \frac{P_0}{2k^2}\left[\sin kt\cos\varphi - kt\cos(kt+\varphi)\right]. \quad (33)$$

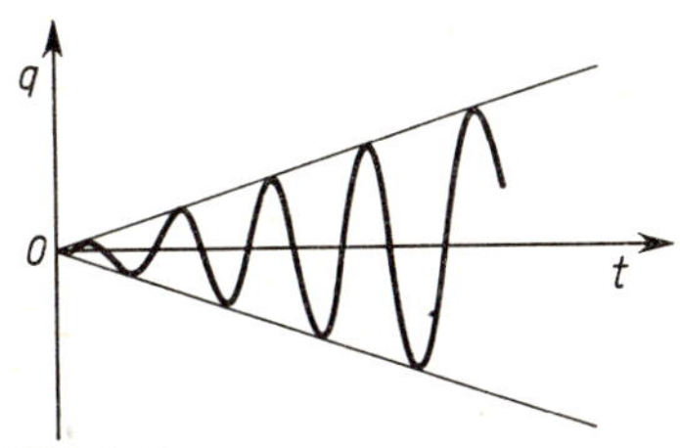

Fig. 8.19

Thus we can see that the vibration amplitude grows linearly (Fig. 8.19). This is the resonance case.

8.5 Vibrations of harmonic oscillators with kinematical input

In the preceding section we considered the case of the input of a generalized force dependent on time. But frequently vibrations due to the motion of a construction are observed. We shall first examine a vibrations recording device consisting of rigid casing A, recording drum B, and harmonic oscillator C (Fig. 8.20). Placing the device on a vibrating base, as $x = x(t)$ we

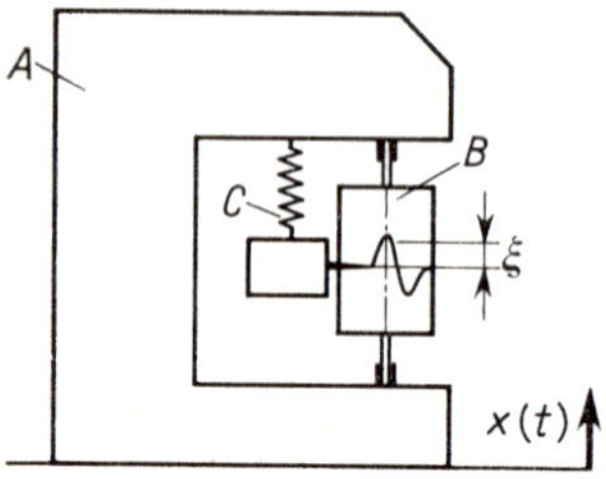

Fig. 8.20

record $\xi = \xi(t)$. The vibrations of the oscillator are due to the existence of a force of inertia in a non-inertial system, i.e., in the system ξ. We shall write the differential equation of motion in the following form:

$$m\ddot{\xi}+h\dot{\xi}+c\xi = -m\ddot{x}.$$

Dividing it by the oscillator mass m, we have

$$\ddot{\xi}+2n\dot{\xi}+k^2\xi = -\ddot{x}. \tag{1}$$

Abstracting from the cited example we say that vibrations of the harmonic oscillator with kinematical input are described by the following differential equation:

$$\ddot{q}+2n\dot{q}+k^2q = -\ddot{f}. \tag{2}$$

Performing Fourier's transformation on equation (2) we obtain

$$-\omega^2\bar{q}(\mathrm{j}\omega)+2n\mathrm{j}\omega\bar{q}(\mathrm{j}\omega)+k^2\bar{q}(\mathrm{j}\omega) = \omega^2\bar{f}(\mathrm{j}\omega). \tag{3}$$

Hence

$$\bar{q}(\mathrm{j}\omega) = \frac{\omega^2}{k^2-\omega^2+2\mathrm{j}n\omega}\bar{f}(\mathrm{j}\omega) = \bar{G}_1(\mathrm{j}\omega)\bar{f}(\mathrm{j}\omega). \tag{4}$$

After performing Fourier's inverse transformation while making use of the theorem on convolution we have

$$q(t) = \int_{-\infty}^{t} G_1(t-\tau)f(\tau)\mathrm{d}\tau, \quad (5)$$

where

$$G_1(t) = \omega^2 G(t).$$

Note that such magnitudes as the frequency characterisctic $G_1(\mathrm{j}\omega)$ and the amplitude characteristic $|\bar{G}_1(\mathrm{j}\omega)|$, in the case of kinematical input, will be obtained by multiplying (26) and (32), Section 8.4, by ω^2. Figure 8.21 graphically represents the dependence of $|\bar{G}_1(\mathrm{j}\omega)|$ on ω/k.

Problem. Give the equation for vibrations of harmonic oscillator, kinematically induced, when $f(t) = f_0 \sin(\omega t+\varphi)$ where f_0 = const.

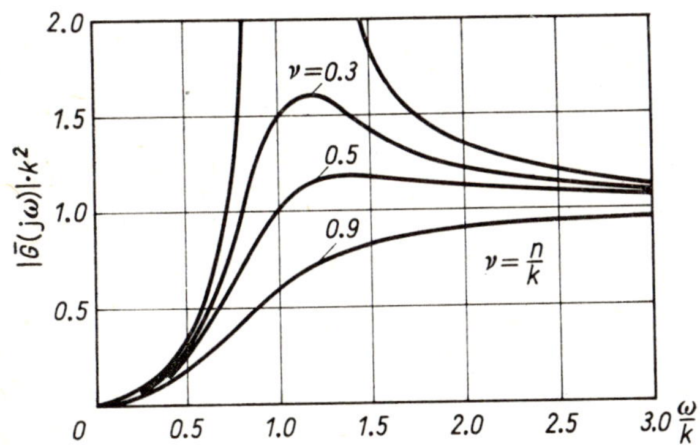

Fig. 8.21

8.6 Vibrations of harmonic oscillators under periodic input forces

We seek the solution to the differential equation

$$\ddot{q}+2n\dot{q}+k^2q = Q(t). \quad (1)$$

in the case of a periodic force $Q(t)$ $(Q(t) = Q(t+T))$, expanding the right-hand side into a Fourier's series

$$Q(t) = \frac{1}{2}H_0 + \sum_{i=1}^{\infty} H_i \sin(\omega_i t+\varphi_i), \quad (2)$$

where

$$\omega_i = \frac{2\pi i}{T}, \quad (2a)$$

$$H_0 = \frac{2}{T}\int_0^T Q(t)\,\mathrm{d}t, \tag{2b}$$

$$H_i = \frac{2}{T}\sqrt{\left[\int_0^T Q(t)\sin\omega_i t\,\mathrm{d}t\right]^2+\left[\int_0^T Q(t)\cos\omega_i t\,\mathrm{d}t\right]^2}, \tag{2c}$$

$$\tan\varphi_i = \frac{\int_0^T Q(t)\cos\omega_i t\,\mathrm{d}t}{\int_0^T Q(t)\sin\omega_i t\,\mathrm{d}t}. \tag{2d}$$

Making use of equation (6), Section 8.4, in the case $n < k$ we obtain

$$q(t) = q_0(t)+\frac{H_0}{2}k(t)+$$
$$+\int_0^t \frac{\mathrm{e}^{-n(t-\tau)}}{\sqrt{k^2-n^2}}\sin\sqrt{k^2-n^2}\,(t-\tau)\sum_{i=1}^{\infty} H_i\sin(\omega_i\tau+\varphi_i)\,\mathrm{d}\tau \tag{3}$$

where $q_0(t)$ is the integral of the homogeneous equation defined by (4), Section 8.3, and where $k(t)$ is the time function of transmittance defined by equality (10), Section 8.4.

The integral of a sum is equal to the sum of the integrals, therefore, making use of the result (12), Section 8.4, we write

$$q(t) = q_0(t)+\frac{H_0}{2}k(t)-$$
$$-\sum_{i=1}^{\infty}\frac{H_i\mathrm{e}^{-nt}}{\sqrt{(k^2-\omega_i^2)^2+4n^2\omega_i^2}}\left[\frac{n\sin(\varphi_i-\varepsilon_i)+\omega_i\cos(\varphi_i-\varepsilon_i)}{\sqrt{k^2-n^2}}\times\right.$$
$$\left.\times\sin\sqrt{k^2-n^2}\,t+\sin(\varphi_i-\varepsilon_i)\cos\sqrt{k^2-n^2}\,t\right]+$$
$$+\sum_{i=1}^{\infty}\frac{H_i}{\sqrt{(k^2-\omega_i^2)^2+4n^2\omega_i^2}}\sin(\omega_i t+\varphi_i-\varepsilon_i), \tag{4}$$

where

$$\tan\varepsilon_i = \frac{2n\omega_i}{k^2-\omega_i^2}.$$

The solution to the problem discussed can also be obtained by a somewhat different procedure. We shall discuss the method proposed by S. Węgrzyn [77]. By this method, the first step towards getting the solution is to find the steady response of the system. It is obvious that here the vibration period is equal to the input period. Counting the time from the beginning of the period we integrate equation (1). The solution thus obtained has the form of (36), Section 8.4. It contains a deflection $q(0)$ and a velocity $\dot{q}(0)$. We find these magnitudes by making use of the properties of periodic motion. For steady motion we have

$$q(0) = q(T), \tag{5}$$

$$\dot{q}(0) = \dot{q}(T). \tag{6}$$

Equations (5) and (6) are used to find $q(0)$ and $\dot{q}(0)$. The disturbed state of the system is determined by the method of two opposed steady states.

Example 1. An input indicated in Fig. 8.22 begins to act at an instant $t = 0$ on a harmonic oscillator for which $n < k$. Find the response curve with respect to time for the system assuming zero initial conditions.

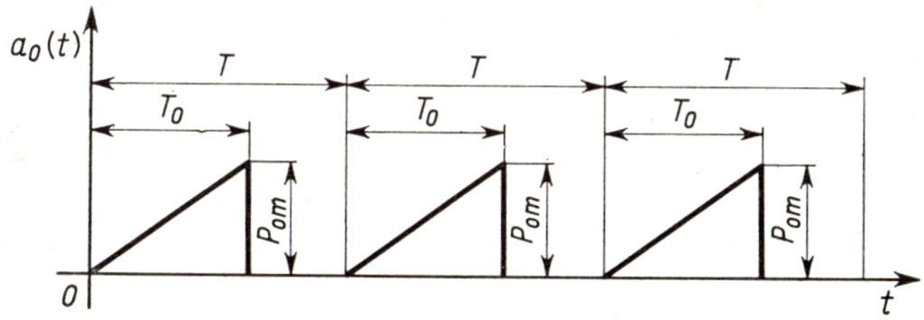

Fig. 8.22

To solve the problem we use Węgrzyn's method. For time $0 < t < T_0$, the solution is as follows:

$$q^{(1)}(t) = e^{-nt}\left[q(0)\cos\sqrt{k^2-n^2}\,t + \frac{\dot{q}(0)+nq(0)}{\sqrt{k^2-n^2}}\sin\sqrt{k^2-n^2}\,t\right]+$$

$$+\int_0^t G(t-\tau)P_{0m}\frac{\tau}{T_0}\,dt, \tag{a}$$

where

$$\int_0^t G(t-\tau)P_{0m}\frac{\tau}{T_0}\,d\tau = \frac{P_{0m}}{\nu T_0}\left\{\frac{\nu t}{k^2} - \frac{2n\nu}{k^4} + \frac{e^{-nt}}{k^4}[(n^2-\nu^2)\sin\nu t + 2n\nu\cos\nu t]\right\},$$

$$\nu = \sqrt{k^2-n^2},$$

and for time $T_0 < t < T$

$$q^{(2)}(t) = e^{-n(t-T_0)}\left[q^{(1)}(T_0)\cos\nu(t-T_0)+\frac{\dot{q}^{(1)}(T_0)+nq^{(1)}(T_0)}{\nu}\sin\nu(t-T_0)\right], \quad \text{(b)}$$

where

$$q^{(1)}(T_0) = e^{-nT_0}\left[q(0)\cos\nu T_0+\frac{\dot{q}(0)+nq(0)}{\nu}\sin\nu T_0\right]+$$
$$+\frac{P_{0m}}{\nu T_0}\left\{\frac{\nu T_0}{k^2}-\frac{2n\nu}{k^4}+\frac{e^{-nT_0}}{k^4}[(n^2-\nu^2)\sin\nu T_0+2n\nu\cos\nu T_0]\right\}$$

and

$$\dot{q}^{(1)}(T_0) = -ne^{-nT_0}\left[q(0)\cos\nu T_0+\frac{\dot{q}(0)+nq(0)}{\nu}\sin\nu T_0\right]+$$
$$+e^{-nT_0}\left[\frac{\dot{q}(0)+nq(0)}{\nu}\nu\cos\nu T_0-q(0)\nu\sin\nu T_0\right]+$$
$$+\frac{P_{0m}}{\nu T_0}\left\{\frac{\nu}{k^2}-\frac{n}{k^4}e^{-nT_0}[(n^2-\nu^2)\sin\nu T_0+2n\nu\cos\nu T_0]+\right.$$
$$\left.+\frac{\nu}{k^4}e^{-nT_0}[(n^2-\nu^2)\cos\nu T_0-2n\nu\sin\nu T_0]\right\}.$$

To calculate $q(0)$ and $\dot{q}(0)$ we make use of equations (5) and (6) which in the considered case we write as follows:

$$q^{(1)}(0) = q^{(2)}(T), \quad \dot{q}^{(1)}(0) = \dot{q}^{(2)}(T),$$

whence

$$q(0) = e^{-n(T-T_0)}\left[q^{(1)}(T_0)\cos\nu(T-T_0)+\frac{\dot{q}^{(1)}(T_0)+nq^{(1)}(T_0)}{\nu}\sin\nu(T-T_0)\right],$$

$$\dot{q}(0) = -ne^{-n(T-T_0)}\left[q^{(1)}(T_0)\cos\nu(T-T_0)+\frac{\dot{q}^{(1)}(T_0)+nq^{(1)}(T_0)}{\nu}\sin\nu(T-T_0)\right]+$$
$$+e^{-n(T-T_0)}\{[\dot{q}^{(1)}(T_0)+nq^{(1)}(T_0)]\cos\nu(T-T_0)-q^{(1)}(T_0)\sin\nu(T-T_0)\}.$$

The response of the system will be the superposition of steady vibrations (equations (a) and (b)) with the disturbance process $q_d(t)$ determined by the method of two opposed steady states:

$$q_d(t) = -e^{-nt}\left[q(0)\cos\nu t+\frac{\dot{q}(0)+nq(0)}{\nu}\sin\nu t\right].$$

8.7 Vibrations of non-linear systems

In specific cases, linearization may give rise to excessive errors. This occurs when deflections from the equilibrium position during vibration are not small in the sense that the neglect of the further terms of the expansion of functions $A(q)$ and $V(q)$ is inadmissible, i.e., the form of approximation (9) as well (10), Section 8.2, is unacceptable. The situation can be saved in a way by going over to linearization by the least squares method, but this is only a compromise solution. The evident reluctance to tackle non-linear problems by recourse to linearization should be attributed to the fact that mathematical elaborations relating to non-linear problems are still far from perfection.

The theory of non-linear vibrations poses the problem of having to solve non-linear differential equations. In many important fields, solutions in a closed form have proved simply impossible. This has given rise to a search for new ways and means, to the launching of investigations into the qualitative properties of these solutions, on the one hand, and to the development of approximating methods, on the other. Among qualitative methods, we reckon the study of the boundedness or unboundedness of solutions or the study of the stability of solutions [65], etc. So many approximating methods have been developed in recent times that only to name them would take a great deal of space, being evidence of the lack of a uniform theory for non-linear vibrations.

We shall now discuss some cases of non-linear vibrations.

(a) *Conservative systems.* This problem has already been discussed in Chapter 3 as a problem of the motion of a particle under a force dependent on position. The assumption $P = f(x)$ required the solving of a non-linear equation of the form

$$m\ddot{q} = f(q).$$

This case was discussed in Section 2.2 and consequently these considerations can be passed over here.

(b) *Free vibrations of intervally linear systems.* Consider vibrations of a body capable of moving on a horizontal plane under a spring reaction force S, and a dry friction force T occurring in the plane of contact of the body with the base (Fig. 8.23). We can call the system in question *intervally linear.*

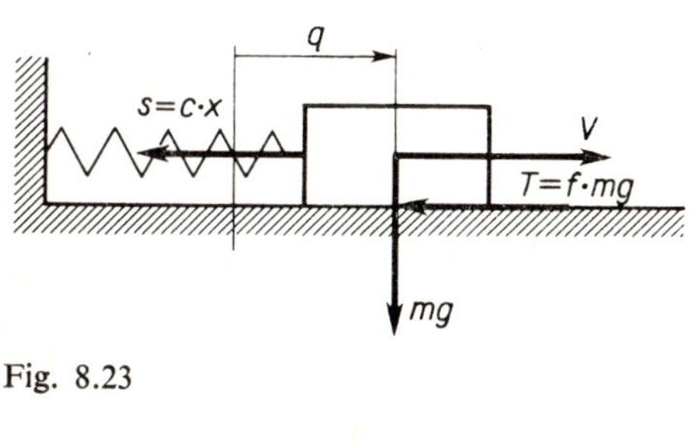

Fig. 8.23

The equation for the motion of mass m is

$$m\ddot{q} = -S - mgf\,\mathrm{sign}\,\dot{q} \tag{1}$$

where f is the coefficient of friction and

$$\mathrm{sign}\,\dot{q} = \begin{cases} 1 & \text{for} \quad \dot{q} > 0, \\ -1 & \text{for} \quad \dot{q} < 0. \end{cases} \tag{2}$$

Since $S = cq$, we write

$$m\ddot{q} + cq = -mgf\,\mathrm{sign}\,\dot{q}. \tag{3}$$

Dividing both sides of the equation by m we obtain

$$\ddot{q} - k^2 q = -gf\,\mathrm{sign}\,\dot{q} \tag{4}$$

where $k^2 = c/m$.

Let us fix the initial conditions. Let

$$q(0) > 0, \quad \dot{q}(0) = 0 \tag{5}$$

and, for the solution not be trivial, the following should also be satisfied:

$$c|q(0)| > mgf \quad \text{or} \quad |q(0)| > \frac{gf}{k^2}.$$

For these initial conditions, the motion of the body begins on the side of positive values of q, and it will be such that $\dot{q}(t) < 0$ in the first cycle. The motion is therefore described by

$$\ddot{q} + k^2 q = gf. \tag{6}$$

The solution to (6) is as follows:

$$q = \frac{fg}{k^2} + \left(q(0) - \frac{fg}{k^2}\right)\cos kt. \tag{7}$$

Hence

$$\dot{q} = -\left(q(0) - \frac{fg}{k^2}\right)k \sin kt. \tag{8}$$

Equations (7) and (8) describe the motion for intervals in which $\dot{q}(t) < 0$. This occurs when $t \in \{0 \leqslant t \leqslant \pi/k\}$. After time $t_1 = \pi/k$, the direction of motion is reversed; therefore, for interval $t_1 \leqslant t \leqslant 2\pi/k = t_2$, the motion proceeds according to the equation (velocity changes sign)

$$\ddot{q} + k^2 q = -gf \tag{9}$$

the integral of which is the function

$$q(t) = -\frac{fg}{k^2} + \left(q_1 + \frac{fg}{k^2}\right)\cos k(t - t_1) \tag{10}$$

where

$$q_1 = -q(0)+\frac{2fg}{k^2}.$$

The return motion of the body takes place only if the tensile force of the spring exceeds the static friction force, i.e.,

$$c|q_1| > mgf_1$$

where f_1 is the coefficient of static friction. We can extend this procedure obtaining the sequence of alternate solutions to equations (6) and (9). For the nth cycle, the initial conditions will be as follows:

$$q_n = (-1)^n\left(q(0)-\frac{2nfg}{k^2}\right) \tag{11}$$

and

$$\dot{q}_n = 0. \tag{11a}$$

The motion involved is a periodic motion with a period $T = 2\pi/k$. Figure 8.24 shows graphically the motion of the oscillator in the presence of solid friction.

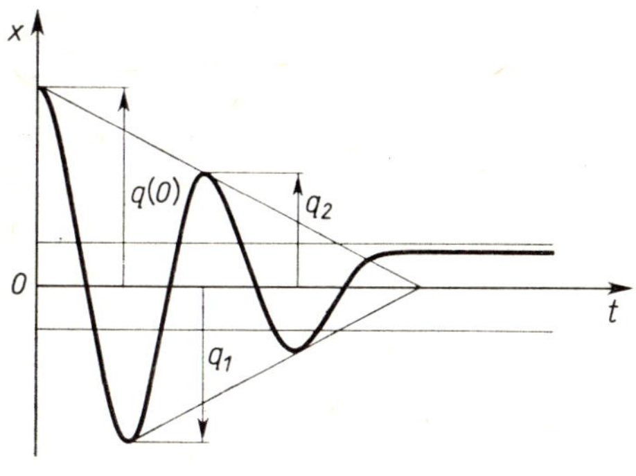

Fig. 8.24

The example of vibration, in which dry friction is considered, is a comparatively simple non-linear problem.

(c) *Asymptotic methods in the investigations of vibrations.* A somewhat more complex case represents a non-linear problem leading to

$$\ddot{q}+k^2q = \varepsilon f(\ddot{q}, \dot{q}). \tag{12}$$

Equations of this type with $\varepsilon \to 0$ describe vibrations of a harmonic oscillator. Thus the right-hand side constitutes a certain disturbance. If it is a small positive number the influence of the right-hand side is negligible and the system is said to be *quasi-harmonic*. This fact justifies the search for a solution in the form (Krylov, Bogolubov–Mitropolskii method [7, 45])

$$q = A(t)\cos\big(kt+\nu(t)\big) \tag{13}$$

where $A(t)$ and $\nu(t)$ are *slow-variable functions*, i.e. their changes during one period are small ($T = 2\pi/k$). These functions are slow-variable and their changes tend to zero along with ε. If let us say $f(\tau)$, where $\tau = \varepsilon t$, then

$$\frac{\mathrm{d}f}{\mathrm{d}t} = \frac{\mathrm{d}f}{\mathrm{d}\tau}\frac{\mathrm{d}\tau}{\mathrm{d}t} = \varepsilon\frac{\mathrm{d}f}{\mathrm{d}\tau}$$

and if $\varepsilon \to 0$, then also

$$\frac{\mathrm{d}f}{\mathrm{d}t} \to 0 \quad \text{and thereby} \quad f(\tau) \to \text{const}.$$

Due to this slow-variable nature of the function, the derivative $\dot{A}(t)$ can be neglected when differentiating (13) thus obtaining

$$\dot{q} = -A(t)k\sin\big(kt+\nu(t)\big). \tag{14}$$

After elementary transformations of equations (13) and (14) we have

$$\dot{A}(t) = \sqrt{q^2+\dot{q}^2/k^2}\,,$$

$$\dot{\nu}(t) = -\arctan\frac{\dot{q}}{qk} - \omega t.$$

Differentiating the relationship thus derived and making use of the initial equation (12) we arrive at the following system of equations of the first order:

$$\dot{A}(t) = \frac{\varepsilon\dot{q}}{k^2A}f(q,\dot{q}), \tag{15}$$

$$\dot{\nu}(t) = -\frac{\varepsilon q}{kA^2}f(q,\dot{q}). \tag{16}$$

These are called *standard equations*.

Considering (13) and (14) we obtain

$$\dot{A}(t) = -\frac{\varepsilon}{k}f(A\cos\psi,\ -kA\sin\psi)\sin\psi, \tag{17}$$

$$\dot{\nu}(t) = -\frac{\varepsilon}{kA} f(A\cos\psi, -kA\sin\psi)\cos\psi, \tag{18}$$

where

$$\psi = kt + \nu(t). \tag{19}$$

The right-hand sides of equations (17) and (18) are time functions through intermediary trigonometric functions $n\omega t$, where n is a natural number. These members give rapid changes and for this reason we shall henceforth call them *vibrational.*

We shall attempt to construct equations which, without including these rapid variable members, would be at the same time equivalent, with an arbitrary accuracy, to standard equations.

This can be achieved, in the first approximation, by averaging the right-hand sides of equations (17) and (18) (*Krylov's method*) taking advantage of the slow-variable nature of the amplitude $A(t)$ and of the phase $\nu(t)$. We then obtain

$$\begin{aligned} \dot{A}(t) &= -\frac{\varepsilon}{2\pi k}\int_0^{2\pi} f(A\cos\psi, -kA\sin\psi)\sin\psi\, d\psi, \\ \dot{\nu}(t) &= -\frac{\varepsilon}{2\pi Ak}\int_0^{2\pi} f(A\cos\psi, -kA\sin\psi)\cos\psi\, d\psi. \end{aligned} \tag{20}$$

Calculating functions $A(t)$ and $\nu(t)$ from the differential equations (20) and substituting them into (13) with suitable initial conditions, it is possible to find solutions to the problem of vibrations of a non-linear system. Solutions to the equation

$$\int_0^{2\pi} f(A\cos\psi, -kA\sin\psi)\sin\psi\, d\psi = 0 \tag{a}$$

give stationary amplitudes; obviously, we should also have

$$\int_0^{2\pi} f(A_0\cos\psi, -kA_0\sin\psi)\cos\psi\, d\psi = 0$$

where A_0 is the solution to equation (a).

The Krylov's method briefly decribed here may, in specific cases, be insufficiently accurate, and in this connection we shall proceed with its generalization which gives a technique used by N. N. Bogolubov and I. A. Mitropolskii [7, 45].

The generalization of this method consists in that solutions are looked for in the form

$$q = A\cos\psi + \sum_{k=1}^{N} \varepsilon^k u_k(A, \psi), \tag{21}$$

$$\psi = kt + \nu(t) \tag{21a}$$

and that standard equations having right-hand sides without vibration members aıe assumed, i.e.,

$$\dot{A} = \sum_{k=1}^{N} \varepsilon^k a_k(A), \tag{22}$$

$$\dot{\psi} = k + \sum_{k=1}^{N} \varepsilon^k b_k(A). \tag{23}$$

In addition, we assume that the functions $u_k(A, \psi)$ are periodic functions but missing the first harmonic, i.e., we assume that the following occurs:

$$\int_0^{2\pi} u_k(A, \psi) e^{j\psi} d\psi = 0, \quad j = \sqrt{-1} \quad (k = 1, 2, \dots). \tag{24}$$

Differentiating (21) we have

$$\dot{q} = \dot{A}\left[\cos\psi + \sum_{k=1}^{N} \varepsilon^k \frac{\partial u_k}{\partial A}\right] + \dot{\psi}\left[-A\sin\psi + \sum_{k=1}^{N} \varepsilon^k \frac{\partial u_k}{\partial \psi}\right],$$

$$\ddot{q} = \ddot{A}\left[\cos\psi + \sum_{k=1}^{N} \varepsilon^k \frac{\partial u_k}{\partial A}\right] + \ddot{\psi}\left[-A\sin\psi + \sum_{k=1}^{N} \varepsilon^k \frac{\partial u_k}{\partial \psi}\right] +$$

$$+ \dot{A}^2 \sum_{k=1}^{N} \varepsilon^k \frac{\partial^2 u_k}{\partial A^2} + 2\dot{A}\dot{\psi}\left[-\sin\psi + \sum_{k=1}^{N} \varepsilon^k \frac{\partial^2 u_k}{\partial \psi\, \partial A}\right] +$$

$$+ \dot{\psi}^2\left[-A\cos\psi + \sum_{k=1}^{N} \varepsilon^k \frac{\partial^2 u_k}{\partial \psi^2}\right].$$

Differentiating (22) and (23) we obtain

$$\ddot{A} = \sum_{k=1}^{N} \varepsilon^k \dot{a}_k(A) = \sum_{k,l=1}^{N} \varepsilon^{k+l} \frac{\partial a_k}{\partial A} a_l,$$

$$\ddot{\psi} = \sum_{k=1}^{N} \varepsilon^k \dot{b}_k(A) = \sum_{k\ l=1}^{N} \varepsilon^{k+l} \frac{\partial b_k}{\partial A} a_l.$$

We form left-hand side of equation (12) by making use of the previously derived relationships

$$\ddot{q}+k^2q = \cos\psi \sum_{l,k=1}^{N} \varepsilon^{l+k}\frac{\partial a_k}{\partial A}a_l + \sum_{l,k,m=1}^{N} \varepsilon^{l+k+m}\frac{\partial a_k}{\partial A}\frac{\partial u_m}{\partial A}a_l -$$

$$-A\sin\psi \sum_{l,k=1}^{N} \varepsilon^{l+k}\frac{\partial b_k}{\partial A}a_l + \sum_{l,k,m=1}^{N} \varepsilon^{l+k+m}\frac{\partial b_k}{\partial A}\frac{\partial u_m}{\partial \psi}a_l +$$

$$+\sum_{l,k,m=1}^{N} \varepsilon^{l+k+m}a_l a_k\frac{\partial^2 u_m}{\partial A^2} - 2\sin\psi \sum_{l,k=1}^{N} \varepsilon^{l+k}a_l b_k +$$

$$+2\sum_{l,k,m=1}^{N} \varepsilon^{l+k+m}a_l b_k\frac{\partial^2 u_m}{\partial \psi\,\partial A} - 2k\sin\psi \sum_{l=1}^{N} \varepsilon^{l} a_l +$$

$$+2k\sum_{l,k=1}^{N} \varepsilon^{l+k}a_l\frac{\partial^2 u_k}{\partial \psi\,\partial A} - Ak^2\cos\psi - 2Ak\cos\psi \sum_{l=1}^{N} \varepsilon^{l} b_l -$$

$$-A\cos\psi \sum_{l,k=1}^{N} \varepsilon^{l+k}b_l b_k + k^2\sum_{l=1}^{N} \varepsilon^{l}\frac{\partial^2 u_k}{\partial \psi^2} +$$

$$+2k\sum_{l,k=1}^{N} \varepsilon^{l+k}b_l\frac{\partial^2 u_k}{\partial \psi^2} + \sum_{l,k,m=1}^{N} \varepsilon^{l+k+m}b_l b_k\frac{\partial^2 u_m}{\partial \psi^2} +$$

$$+Ak^2\cos\psi + k^2\sum_{l=1}^{N} \varepsilon^{l}u_l. \tag{25}$$

As for the right-hand side, $\varepsilon f(q, \dot{q})$, considering that

$$\dot{q} = -kA\sin\psi - A\sin\psi \sum_{l=1}^{N} \varepsilon^{l}b_l + \sum_{l=1}^{N} \varepsilon^{l}a_l\cos\psi +$$

$$+\sum_{l=1}^{N} \varepsilon^{l}k\frac{\partial u_l}{\partial \psi} + \sum_{l,k=1}^{N} \varepsilon^{l+k}a_l\frac{\partial u_k}{\partial A} + \sum_{l,k=1}^{N} \varepsilon^{l+k}b_l\frac{\partial u_k}{\partial \psi},$$

we have, after expanding into a power series:

$$\varepsilon f(q, \dot{q}) = \varepsilon f(A\cos\psi, \, -kA\sin\psi) + \\ +\varepsilon^2\left[u_1 \frac{\partial f}{\partial q}(A\cos\psi, \, -kA\sin\psi) + \left(a_1\cos\psi - Ab_1\sin\psi + \\ +k\frac{\partial u_1}{\partial\psi}\right)\frac{\partial f}{\partial\dot{q}}(A\cos\psi, \, -kA\sin\psi)\right] + \varepsilon^3[\ldots] + \ldots \tag{26}$$

Equating the right-hand sides of (25) and (26) and comparing the expressions with the same powers of ε we obtain

$$\begin{aligned} k^2\left(\frac{\partial^2 u_1}{\partial\psi^2} + u_1\right) &= f^{(0)}(A, \psi) + 2ka_1\sin\psi + 2kAb_1\cos\psi, \\ k^2\left(\frac{\partial^2 u_2}{\partial\psi^2} + u_2\right) &= f^{(1)}(A, \psi) + 2ka_2\sin\psi + 2kAb_2\cos\psi, \\ &\ldots\ldots\ldots\ldots\ldots\ldots \\ k^2\left(\frac{\partial^2 u_N}{\partial\psi^2} + u_N\right) &= f^{(N-1)}(A, \psi) + 2kA_N\sin\psi + 2kAB_N\cos\psi \end{aligned} \tag{27}$$

where

$$\begin{aligned} f^{(0)}(A, \psi) &= f(A\cos\psi, \, -kA\sin\psi), \\ f^{(1)}(A, \psi) &= u_1\frac{\partial f}{\partial q}(A\cos\psi, \, -kA\sin\psi) + \\ &+\left(a_1\cos\psi - Ab_1\sin\psi + k\frac{\partial u_1}{\partial\psi}\right)\frac{\partial f}{\partial\dot{q}}(A\cos\psi, \, -kA\sin\psi) + \\ &+\left(Ab_1^2 - a_1\frac{\partial a_1}{\partial A}\right)\cos\psi + \left(2a_1b_1 + a_1\frac{\partial b_1}{\partial A}A\right)\sin\psi - \\ &-2ka_1\frac{\partial^2 u_1}{\partial A\,\partial\psi} - 2kb_1\frac{\partial^2 u_1}{\partial\psi^2}. \end{aligned} \tag{28}$$

The functions $f^{(i)}(A, \psi)$ are periodic functions like $\sin\psi$ and $\cos\psi$. In this connection, they can be expanded into a Fourier series:

$$\begin{aligned} f^{(0)}(A, \psi) &= h_0^{(0)}(A) + \sum_{n=1}^{\infty}\left[h_n^{(0)}(A)\cos n\psi + g_n^{(0)}(A)\sin n\psi\right], \\ u_1(A, \psi) &= w_0^{(1)}(A) + \sum_{n=1}^{\infty}\left[w_n^{(1)}(A)\cos n\psi + v_n^{(1)}(A)\sin n\psi\right]. \end{aligned} \tag{29}$$

Substituting expressions (29) into the first of equations (27), we have

$$k^2 w_0^{(1)}(A) + \sum_{n=1}^{\infty} k^2(1-n^2)\,[w_n^{(1)}(A)\cos n\psi + v_n^{(1)}(A)\sin n\psi]$$

$$= h_0^{(0)}(A) + \sum_{n=2}^{\infty} [h_n^{(0)}(A)\cos n\psi + g_n^{(0)}(A)\sin n\psi] +$$

$$+ [h_1^{(0)}(A) + 2kAb_1]\cos\psi + [g_1^{(0)}(A) + 2ka_1]\sin\psi.$$

Due to the orthogonality of the functions $\sin\psi$ and $\cos\psi$, relationships can be formally obtained by comparing the coefficients for harmonics of the same order,

$$\begin{aligned} &k^2 w_0^{(1)}(A) = h_0^{(0)}(A), \\ &h_1^{(0)}(A) + 2kAb_1 = 0, \\ &g_1^{(0)}(A) + 2ka_1 = 0, \\ &k^2(1-n^2) w_n^{(1)}(A) = h_n^{(0)}(A), \\ &k^2(1-n^2) v_n^{(1)}(A) = g_n^{(0)}(A), \qquad n = 2, 3, \ldots \end{aligned} \tag{30}$$

By virtue of assumption (24), $w_1(A) \equiv 0$ and $v_1(A) \equiv 0$. Expressing $w_0(A)$, $w_n(A)$, and $v_n(A)$ by the coefficient of expansion of the function $f^{(0)}(A, \psi)$ and then substituting them into the second equation (29) we obtain

$$u_1(A, \psi) = \frac{h_0^{(0)}(A)}{k^2} + \frac{1}{k^2} \sum_{n=2}^{\infty} \frac{h_n^{(0)}(A)\cos n\psi + g_n^{(0)}(A)\sin n\psi}{1-n^2}; \tag{31}$$

we also have the determination of the functions $a_1(A)$ and $b_1(A)$, namely

$$a_1(A) = -\frac{1}{2k} g_1^{(0)}(A), \qquad b_1(A) = -\frac{1}{2kA} h_1^{(0)}(A), \tag{32}$$

where

$$\begin{aligned} g_1^{(0)}(A) &= \frac{1}{\pi} \int_0^{2\pi} f^{(0)}(A, \psi)\sin\psi\, d\psi \\ h_1^{(0)}(A) &= \frac{1}{\pi} \int_0^{2\pi} f^{(0)}(A, \psi)\cos\psi\, d\psi. \end{aligned} \tag{32a}$$

It is therefore evident that calculating $a_1(A)$ and $b_1(A)$ according to equations (32) and then making use of the first approximation ($N = 1$) of the standard equations (22) and (23), we obtain equation (20); however, the method described here is more general and enables arbitrary accuracies to be obtained by expanding the further functions $f^{(i)}(A, \psi)$ into a Fourier series and by making use of system (27). Having determined $u_1(A, \psi)$, $a_1(A)$, and $b_1(A)$, we can expand $f^{(1)}(A, \psi)$ into a series:

$$f^{(1)}(A, \psi) = h_0^{(1)}(A) + \sum_{n=1}^{\infty} [h_n^{(1)}(A)\cos n\psi + g_n^{(1)}(A)\sin n\psi], \tag{33}$$

$$u_2(A, \psi) = w_0^{(2)}(A) + \sum_{n=1}^{\infty} [w_n^{(2)}(A)\cos n\psi + v_n^{(2)}(A)\sin n\psi]. \tag{34}$$

Substituting (33) and (34) into the second equation of system (27) and comparing the coefficients for harmonics of the same order leads to the following relationships:

$$\begin{aligned} h_1^{(1)}(A) + 2kAb_2 &= 0, \\ g_1^{(1)}(A) + 2ka_2 &= 0. \end{aligned} \tag{35}$$

Consideration has been given here to condition (24) ($w_1^{(2)}(A) \equiv 0$, $v_1^{(2)}(A) \equiv 0$). From (35) we obtain

$$\begin{aligned} a_2(A) &= -\frac{1}{2k} g_1^{(1)}(A) = -\frac{1}{2\pi k}\int_0^{2\pi} f^{(1)}(A, \psi)\sin\psi\, d\psi, \\ b_2(A) &= -\frac{1}{2kA} h_1^{(1)}(A) = -\frac{1}{2\pi kA}\int_0^{2\pi} f^{(1)}(A, \psi)\cos\psi\, d\psi. \end{aligned} \tag{36}$$

On the other hand, we obtain $u_2(A, \psi)$ in the form

$$u_2(A, \psi) = \frac{h_0^{(1)}(A)}{k^2} + \frac{1}{k^2}\sum_{n=2}^{\infty} \frac{1}{1-n^2} [h_n^{(1)}(A)\cos n\psi + g_n^{(1)}(A)\sin n\psi]. \tag{37}$$

Generalizing we write

$$\begin{aligned} a_s(A) &= -\frac{1}{2\pi k}\int_0^{2\pi} f^{(s-1)}(A, \psi)\sin\psi\, d\psi, \\ b_s(A) &= -\frac{1}{2\pi kA}\int_0^{2\pi} f^{(s-1)}(A, \psi)\cos\psi\, d\psi, \end{aligned} \tag{38}$$

and

$$\begin{aligned} u_s(A, \psi) = \frac{h_0^{(s-1)}(A)}{k^2} &+ \\ &+ \frac{1}{k^2}\sum_{n=2}^{\infty} \frac{1}{1-n^2} [h_n^{(s-1)}(A)\cos n\psi + g_n^{(s-1)}(A)\sin n\psi], \end{aligned} \tag{39}$$

where

$$h_0^{(s-1)}(A) = \frac{1}{2\pi}\int_0^{2\pi} f^{(s-1)}(A, \psi)\mathrm{d}\psi,$$

$$h_n^{(s-1)}(A) = \frac{1}{\pi}\int_0^{2\pi} f^{(s-1)}(A, \psi)\cos n\psi\,\mathrm{d}\psi, \tag{40}$$

$$g_n^{(s-1)}(A) = \frac{1}{\pi}\int_0^{2\pi} f_n^{(s-1)}(A, \psi)\sin n\psi\,\mathrm{d}\psi.$$

Standard equations (22) and (23) with arbitrary accuracy (arbitrary N) assume the following construction:

$$\dot{A}(t) = -\frac{1}{2\pi k}\sum_{s=1}^{N}\varepsilon^s\int_0^{2\pi} f^{(s-1)}(A, \psi)\sin\psi\,\mathrm{d}\psi,$$

$$\dot{\psi}(t) = k - \frac{1}{2\pi kA}\sum_{s=1}^{N}\varepsilon^s\int_0^{2\pi} f^{(s-1)}(A, \psi)\cos\psi\,\mathrm{d}\psi. \tag{41}$$

Solving the differential equations (41), i.e., deriving from them the functions $A(T)$ and $\psi(t)$, makes it possible to find a solution in the form (21) for arbitrary N:

$$q(t) = A(t)\cos\psi(t) + \sum_{s=1}^{N}\varepsilon^s u_s\big(A(t), \psi(t)\big), \tag{42}$$

where $u_s\big(A(t), \psi(t)\big)$ is given by relationship (39).

Returning to the first approximation we write

$$q \cong A\cos\psi + \varepsilon u_1(A, \psi), \tag{43}$$

$$\dot{A} \cong \varepsilon a_1(A), \tag{44}$$

$$\psi \cong k + \varepsilon b_1(A). \tag{45}$$

It appears that the term $\varepsilon u_1(A, \psi)$ on the right-hand side of (43) can be ignored, since acceptance of approximations (44) and (45) results in an error of the same order.

Indeed, in a time $t = 1/\varepsilon$, we observe an error of amplitude

$$\Delta A = \int_0^{1/\varepsilon} \big(\dot{A} - \varepsilon a_1(A)\big)\mathrm{d}t = \int_0^{1/\varepsilon} [\varepsilon^2 a_2(A) + \ldots]\mathrm{d}t \cong \varepsilon\langle a_2(A)\rangle$$

where the brackets $\langle\ \rangle$ denote the mean value in the time interval $(0, t)$. Similarly, in that same time the error of the phase in

$$\Delta\nu = \int_0^{1/\varepsilon} [\dot{\psi} - k - \varepsilon b_1(A)]\mathrm{d}t$$

$$= \int_0^{1/\varepsilon} [\varepsilon^2 b_2(A) + \ldots]\mathrm{d}t \approx \varepsilon\langle b_2(A)\rangle.$$

Consequently, the error of the first member of (43) is for that same time $t = 1/\varepsilon$, of the order of ε, and is therefore equal in order to the error due to omission of the term $\varepsilon u_1(A, \psi)$ in (43).

Thus, we shall assume in the first approximation

$$\begin{aligned} q &= A\cos(kt+\nu), \\ \dot{A} &= \varepsilon a_1(A), \\ \dot{\psi} &= k+\varepsilon b_1(A), \end{aligned} \tag{46}$$

in the second approximation,

$$\begin{aligned} q &= A\cos(kt+\nu)+\varepsilon u_1(A, \psi), \\ \dot{A} &= \varepsilon a_1(A)+\varepsilon^2 a_2(A), \\ \dot{\psi} &= k+\varepsilon b_1(A)+\varepsilon^2 b_2(A), \end{aligned} \tag{47}$$

and so forth.

Example. Analyse by the asymptotic method the solution to the differential equation

$$\ddot{q}+\alpha\dot{q}^2\operatorname{sign}\dot{q}+k^2 q = 0; \tag{a}$$

in this case, $\varepsilon f(q, \dot{q}) = -\alpha\dot{q}^2\operatorname{sign}\dot{q}$.

We successively calculate

$$g_1^{(0)}(A) = -\frac{1}{\pi}\int_0^{2\pi} f^{(0)}(-kA\sin\psi)\sin\psi\,\mathrm{d}\psi$$

$$= -\frac{1}{\pi}\int_0^{2\pi} \alpha k^2 A^2 \sin^3\varphi \operatorname{sign}(-kA\sin\psi)\mathrm{d}\psi = \frac{8\alpha k^2 A^2}{3\pi},$$

$$h_1^{(0)}(A) = \frac{1}{\pi}\int_0^{2\pi} f^{(0)}(-kA\sin\psi)\cos\psi\,\mathrm{d}\psi$$

$$= \frac{1}{\pi}\int_0^{2\pi} \alpha k^2 A^2 \sin^2\psi \cos\psi \operatorname{sign}(-kA\sin\psi)\mathrm{d}\psi = 0,$$

$$a_1(A) = -\frac{4\alpha k A^2}{3\pi},$$

$$b_1(A) = 0.$$

Hence

$$\dot{A} = -\frac{4\alpha k A^2}{3\pi}, \quad \dot{\psi} = k. \tag{b}$$

Integrating the first equation of (b) we obtain

$$\frac{1}{A(t)} - \frac{1}{A(0)} = \frac{4\alpha k t}{3\pi}$$

which gives

$$A(t) = \frac{A(0)}{1+\dfrac{4\alpha k A(0)}{3\pi}t} \tag{c}$$

whereas

$$\psi = kt+\nu(0).$$

This means that the first approximation gives

$$q(t) = \frac{A(0)}{1+\dfrac{4\alpha k A(0)}{3\pi}t}\sin(kt+\nu(0)). \tag{d}$$

The second approximation [45] leads to

$$q(t) = A(t)\cos\psi - \frac{8[A(t)]^2\alpha}{\pi}\sum_{n=1}^{\infty}\frac{\sin(2n+1)\psi}{(2n+1)\,[(2n+1)^2-1]\,[(2n+1)^2-4]}. \tag{e}$$

At the same time $A(t)$ is expressed by equation (c) whereas

$$\psi(t) = kt - \frac{3C\alpha A(0)}{\pi}\left[1-\left(1+\frac{4\alpha k[A(0)]^2}{3\pi}t\right)^{-1}\right]+\nu(0) \tag{f}$$

where

$$C = 8\sum_{n=1}^{\infty}\frac{1}{[(2n+1)^2-1]\,[(2n+1)^2-4]} = 0.0407.$$

Problem. Analyse the problem of vibrations of a system subject to friction, considered at the beginning of this section, using the asymptotic method.

We shall conclude the present section with remarks on the question of response of non-linear systems induced by periodic forces. The Krylov–Bogolubov–Mitropolskii method [7, 45] is also applicable here. Let the response of the system considered be described by the following differential equation:

$$\ddot{q}+k^2q = \varepsilon f(\omega t, q, \dot{q}) \tag{48}$$

where the right-hand side is dependent on time in explicit form and contains periodic terms. Thus, this side can be presented in the form of a trigonometric polynomial

$$f(\omega t, q, \dot{q}) = \sum_{n=-N}^{N} \mathrm{e}^{in\omega t} f_n(q, \dot{q}). \tag{49}$$

The Krylov–Bogolubov–Mitropolskii method is analogous in application to the study of the considered problem. Note is first taken of solution $q = A\cos\psi$, which gives terms containing the functions $\sin(n\omega+mk)t$ and $\cos(n\omega+mk)t$ where n and m are constant. If one of the frequencies $(n\omega+mk)$ now approximates to the frequency of the system described by the equation with $\varepsilon = 0$, i.e., without the right-hand side, then one can expect an increase in the amplitude, as in the case of resonance in linear systems.

The case of increasing amplitude is observed: (i) when $\omega \cong k$; this is the so-called *base resonance*, (ii) when $\omega \cong pk$; this is the *subharmonic resonance*, (iii) when $k \cong \omega q$; the case of *hyperharmonic resonance.*

The crucial fact here is that under specified conditions a minor input $\varepsilon f(\omega t, q, \dot{q})$ may produce considerable change in response solutions, such solutions being said to be *unstable* in relation to input $\varepsilon f(\omega t, q, \dot{q})$. The verification of this fact is of relevance in the case of the neglect of small terms of differential equations for the purpose of linearization.

When one of the named conditions (i)–(iii) is satisfied, we are dealing with so-called *resonant vibrations.* In all other cases, non-resonant vibrations are involved.

The interested reader will find in [27] and [45] a detailed treatment of both cases.

CHAPTER 9

Vibrations of Systems with Many Degrees of Freedom

9.1 Preliminary discussion

We shall now deal with vibrations of holonomic systems with many degrees of freedom. For such systems, Hamilton's equations are valid written in the form

$$\dot{q}^j = \frac{\partial H}{\partial p_j}, \qquad \dot{p}_j - Q_j = -\frac{\partial H}{\partial q^j}, \tag{1}$$

where

$$H = p_j \dot{q}^j - L. \tag{1a}$$

We obviously apply the summation convention here and we shall be making use of it further on.

We can write the Lagrange function L as follows:

$$L = \tfrac{1}{2} a_{ik} \dot{q}^i \dot{q}^k + a_j \dot{q}^j + a_0 - V, \tag{2}$$

where a_{ik}, a_j, a_0 are given by equations (4), Section 7.6. We write the generalized momentum according to (1), Section 7.10, in the form

$$p_j = a_{ij} \dot{q}^i + a_j. \tag{3}$$

Hence

$$\dot{q}^i = a^{ik}(p_k - a_k), \tag{4}$$

where a^{ik} are elements of a tensor coupled to a_{ik} (equation (32), Section 7.12). Considering (4) in (2) we have

$$L = \tfrac{1}{2} a^{ik}(p_i - a_i)(p_k - a_k) + a_j a^{ji}(p_i - a_i) + a_0 - V. \tag{5}$$

Substituting the obtained result into (1a) we find Hamilton's function H dependent on generalized momenta, coordinates, and time:

$$H = \tfrac{1}{2} a^{ik}(p_i - a_i)(p_k - a_k) - a_0 + V. \tag{1b}$$

Considering Hamilton's function thus derived in equations (1) we write

$$\dot{q}_j = a^{jk}(p_k - a_k), \tag{6}$$

$$\dot{p}_j - Q_j = -\frac{1}{2}\frac{\partial a^{ik}}{\partial q^j}(p_i - a_i)(p_k - a_k) + a^{ik}(p_i - a_i)\frac{\partial a_k}{\partial q^j} + \frac{\partial a_0}{\partial q^j} - \frac{\partial V}{\partial q^j}. \tag{7}$$

Equations (6) and (7) hold in the most general case of the motion of a holonomic system with many degrees of freedom and they can therefore, under certain conditions of the problem, decribe vibration. Solving differential equations of the motion of a system with many degrees of freedom, in the general case, is reckoned among difficult problems. The solution amounts to finding, from (6) and (7) and from the initial conditions, the coordinates and momenta as a function of time. Each solution

$$q = q(t) = \begin{bmatrix} q^1(t) \\ q^2(t) \\ \vdots \\ q^r(t) \end{bmatrix}, \quad p = p(t) = \begin{bmatrix} p_1(t) \\ p_2(t) \\ \vdots \\ p_r(t) \end{bmatrix}$$

gives specified curve in a multidimensional space, called the *phase space*. This curve we shall call the *phase trajectory*.

For scleronomic systems equations (6) and (7) take the form

$$\dot{q}^j = a^{jk}p_k, \tag{6a}$$

$$\dot{p}_j - Q_j = -\frac{1}{2}\frac{\partial a^{ik}}{\partial q^j}p_i p_k - \frac{\partial V}{\partial q^j}, \tag{7a}$$

due to functions a_i and a_0 decaying to zero.

9.2 Problems of linearization of the equations

Let us assume that the considered systems are scleronomic, consequently, their motion is described by equations (6a) and (7a) from the preceding section. We shall consider vibrations of systems in the vicinity of the stable equilibrium position (min V), therefore certain simplifications can be introduced for elements of the matrix a^{ik} and the kinetic potential function V.

Assume that a^{ik} are constants calculated at a point $q^1 = q^2 = \ldots = q^N = 0$, which we shall regard as the equilibrium point. Expanding V into a series we write

$$V = \left(\frac{\partial V}{\partial q^i}\right)_0 q^i + \frac{1}{2}\left(\frac{\partial^2 V}{\partial q^i \partial q^k}\right)_0 q^i q^k + \ldots. \tag{1}$$

We have assumed here that $V(0, \ldots, 0) = 0$. Due to the fact that the expansion has been performed in the vicinity of a point at which the function V assumes the least value, the first derivatives vanish while the quadratic form

$$\left(\frac{\partial^2 V}{\partial q^i \partial q^k}\right)_0 q^i q^k$$

is positive definite.

We denote the second derivatives calculated at the reference origin by c_{ik}. Bearing in mind the assumed small deflections from the equilibrium position, we shall confine ourselves to the first terms of the expansion (1). Thus we express the potential energy of the system by

$$V = \tfrac{1}{2} c_{ik} q^i q^k. \tag{2}$$

It should be noted that the approximate quadratic form should continue to be significantly positive definite. That this is not a trivial demand is attested to by the following example.

Let the potential energy be

$$V = \alpha[(q^1)^4 + (q^2)^4] + \beta(q^1 - q^2)^2.$$

As one will easily notice, it is significantly positive definite but after being expanded into a Maclaurin series and limited to the form (2) it looses this property. By assuming constancy of the elements of the matrix a^{ik} and defining the function V by the form (2), equations (6a) and (7a) from the preceding section may be writen as follows:

$$\dot{q}^j = a^{ik} p_k, \tag{3a}$$

$$\dot{p}_j - Q_j = -c_{jk} q^k. \tag{3b}$$

Calculating the generalized momentum p_j from the first equation, we write

$$p_j = a_{jk} \dot{q}^k. \tag{4}$$

In equation (4) the elements of the matrix a_{jk} are constant, just like the elements of the inverse matrix a^{ik}. Differentiating (4) and substituting the result into (3b) we obtain

$$\begin{aligned} a_{jk}\ddot{q}^k - Q_j &= -c_{jk} q^k \quad \text{or} \\ a_{jk}\ddot{q}^k + c_{jk} q^k &= Q_j. \end{aligned} \tag{5}$$

In the case of systems with energy dissipation determined by the dissipation function

$$D = \tfrac{1}{2} h_{ik} \dot{q}^i \dot{q}^k,$$

the right-hand side of (5) is augmented by an extra member

$$\frac{\partial D}{\partial \dot{q}^j} = -h_{jk}\dot{q}^k.$$

Thus, for holonomic systems with dissipation, the differential equations of motion in the vicinity of the stable equilibrum point of the system are as follows:

$$a_{jk}\ddot{q}^k + h_{jk}\dot{q}^k + c_{jk}q^k = Q_j \qquad (j = 1, 2, \ldots, N). \tag{6}$$

As we can see, it is a linear system of differential equations.

With linearization thus performed, small deflections from the equilibrium position should be assumed. Under certain circumstances, when greater deflections are likely to be involved, the least squares method appears to be more satisfactory. The method consists in finding a matrix c_{ik} with constant coefficients under conditions of minimization of an integral which is the measure of the approximation errror

$$\varepsilon_i = \int\limits_{1^{q^1}}^{2^{q^1}} \ldots \int\limits_{1^{q^r}}^{2^{q^r}} \left(\frac{\partial V}{\partial q^i} - c_{ik}q^k\right)^2 dq^1 \ldots dq^r. \tag{7}$$

The extremum condition demands that

$$\frac{\partial \varepsilon_i}{\partial c_{mn}} = 0.$$

By differentiating (7) we obtain

$$\int\limits_{1^{q^1}}^{2^{q^1}} \ldots \int\limits_{1^{q^r}}^{2^{q^r}} \left(\frac{\partial V}{\partial q^i} - c_{ik}q^k\right) \frac{\partial c_{ik}}{\partial c_{mn}} q^k \mathrm{d}q^1 \mathrm{d}q^2 \ldots \mathrm{d}q^r = 0.$$

But

$$\frac{\partial c_{ik}}{\partial c_{mn}} = \delta_{im}\delta_{kn},$$

therefore

$$\int\limits_{1^{q^1}}^{2^{q^1}} \ldots \int\limits_{1^{q^r}}^{2^{q^r}} \frac{\partial V}{\partial q^m} q^n \mathrm{d}q^1 \mathrm{d}q^2 \ldots \mathrm{d}q^r - c_{mk} \int\limits_{1^{q^1}}^{2^{q^1}} \ldots \int\limits_{1^{q^r}}^{2^{q^r}} q^k q^n \mathrm{d}q^1 \mathrm{d}q^2 \ldots \mathrm{d}q^r = 0.$$

These equations can be written differently:

$$D^n_m - c_{mk}R^{nk} = 0. \tag{8}$$

Evidently, we have N^2 linear equations with N^2 unknows c_{mk}. Solving this system of equations, we write

$$c_{mk} = D^n_m R_{nk}, \tag{9}$$

where R_{nk} are elements of the matrix inverse to R^{nk}.

Example 1. For the system shown in Fig. 7.15 the potential energy is expressed by

$$V(\varphi) = \tfrac{1}{2}k\varphi^2 - mgl(1-\cos\varphi). \tag{a}$$

Using the method of expansion into a series (the tangents method), we obtain

$$V \approx \tfrac{1}{2}(k-mgl)\varphi^2. \tag{a1}$$

We shall attempt to compare the above method with the least squares method. In the considered case

$$D = \int_{-\varphi_k}^{\varphi_k} \frac{\partial V}{\partial\varphi}\varphi\,\mathrm{d}\varphi = \int_{-\varphi_k}^{\varphi_k} (k\varphi - mgl\sin\varphi)\varphi\,\mathrm{d}\varphi$$

$$= 2\left(\int_0^{\varphi_k} k\varphi^2\,\mathrm{d}\varphi - mgl\int_0^{\varphi_k} \varphi\sin\varphi\,\mathrm{d}\varphi\right) = 2\left(\tfrac{1}{3}k\varphi_k^3 - mgl(\sin\varphi_k - \varphi_k\cos\varphi_k)\right),$$

$$R = 2\int_0^{\varphi_k} \varphi^2\,\mathrm{d}\varphi = \tfrac{2}{3}\varphi_k^3,$$

whence

$$c = \frac{D}{R} = k - 3mgl\frac{\sin\varphi_k - \varphi_k\cos\varphi_k}{\varphi_k^3}.$$

Thus, the approximating form of the potential energy is

$$V = \frac{1}{2}\left(k - 3mgl\frac{\sin\varphi_k - \varphi_k\cos\varphi_k}{\varphi_k^3}\right)\varphi^2, \tag{a2}$$

where φ_k gives the area in which we have performed linearization by the least squares method. Comparing equations (a1) and (a2), we notice that when $\varphi_k \to 0$, these equations become identical, since

$$\lim_{\varphi_k\to 0} \frac{\sin\varphi_k - \varphi_k\cos\varphi_k}{\varphi_k^3} = \frac{1}{3}.$$

It is evident from the cited example that the least squares method is an extension of the tangents method; it is superior in being more general and also in being suitable for determining the approximation of non-analytical functions.

The linearization just performed applies only to potential forces. Obviously linearization can also be performed in an analogous manner in the case of forces which are functions of position and velocity. If vibrations were taking place not about the reference origin but elsewhere, it would be necessary to seek an approximating linear form of the type $c_{0i}-c_{ik}q^k$. In order to find the additional constants c_{0i}, it would be necessary to bring in additional conditions for the extremum:

$$\frac{\partial \varepsilon_i}{\partial c_{0m}} = 0 \quad (m = 1, \dots, N).$$

The possibility of entering a correction for a_{ik} may also be explored. The elements of a_{ik} are assumed to be constant equal values of the function a_{ik} at a point $q^1 = q^2 = \dots = q^N = 0$.

Surveying the linearization methods we have tacitly assumed that the generalized forces Q_j are functions of time, t, only. In more general cases, i.e., $Q_j = Q_j(q, \dot{q}, t)$, one may work towards linearizing these expression or, on assuming that these forces do not induce great changes in the amplitude and phase during one period, one may find the reactions of a system to have a quasi-harmonic character and apply to solve the equations of motion the Bogolubov–Mitropolskii asymptotic method [7].

9.3 Free vibrations of conservative systems

The differential equations for vibrations of conservative systems are a special case of equations (6), Section 9.2. Assuming that no external forces act on the system and that there is no energy dissipation ($D = 0$), we have

$$a_{ik}\ddot{q}^k + c_{ik}q^k = 0 \quad (i = 1, 2, \dots, N). \tag{1}$$

We shall seek a solution in the form

$$q^k = u^k \sin(\omega t + \varphi). \tag{2}$$

Substituting (2) into (1) and simplifying by $\sin(\omega t + \varphi)$, we obtain

$$(c_{ik} - \omega^2 a_{ik})u^k = 0. \tag{3}$$

A non-trivial solution exists only when the characteristic determinant of system (3) vanishes, i.e.,

$$\det[c_{ik} - \omega^2 a_{ik}] = 0. \tag{4}$$

From the system of equations (4) we can find the free vibration frequencies ω of the system. The roots $\lambda = \omega^2$, as we intend to show, are real and positive. The number of these roots in the general case is N and each of them determines a different amplitude vector. Thus $\lambda_{(j)}$ gives (from equation (3)) the vector

$$\mathbf{u}_j = \begin{bmatrix} u_j^1 \\ u_j^2 \\ \vdots \\ u_j^r \end{bmatrix}. \tag{5}$$

This vector is determined with an accuracy up to the constant factor, which follows from the homogeneity of system (3). Therefore, the following system of equations is satisfied:

$$(c_{ik} - \lambda_{(j)} a_{ik}) u_j^k = 0. \tag{6}$$

For $l \neq j$, it is also valid that

$$(c_{ik} - \lambda_{(l)} a_{ik}) u_l^k = 0. \tag{7}$$

Multiplying (6) by u_l^i and (7) by u_j^i we obtain

$$c_{ik} u_l^i u_j^k = \lambda_{(j)} a_{ik} u_l^i u_j^k, \tag{8}$$

$$c_{ik} u_j^i u_l^k = \lambda_{(l)} a_{ik} u_j^i u_l^k. \tag{9}$$

The left-hand sides of equations (8) and (9) are equal, whence

$$\lambda_{(j)} a_{ik} u_j^i u_l^k = \lambda_{(l)} a_{ik} u_j^i u_l^k. \tag{10}$$

Hence

$$(\lambda_{(j)} - \lambda_{(l)}) a_{ik} u_j^i u_l^k = 0. \tag{11}$$

Obviously, if $j \neq l$, then

$$a_{ik} u_j^i u_l^k = 0, \tag{12}$$

therefore the amplitude vectors are orthogonal in a configurational space. Since they are determined with an accuracy up to a constant factor, we can write

$$a_{ik} u_j^i u_l^k = \delta_{jl} \alpha^2, \tag{13}$$

where δ_{jl} is Kronecker's symbol, α an arbitrary constant factor.

We shall now prove that the roots of the characteristic equation (4) are real and positive. To this end, we make use of equation (8) putting $l = j$:

$$c_{ik} u_{(j)}^i u_{(j)}^k = \lambda_{(j)} a_{ik} u_{(j)}^i u_{(j)}^k; \tag{14}$$

hence

$$\lambda_{(j)} = \frac{c_{ik} u_{(j)}^i u_{(j)}^k}{a_{ik} u_{(j)}^i u_{(j)}^k}. \tag{15}$$

Assume that $\underline{\lambda}_{(j)}$ is complex. Complex amplitude vector $\underline{\mathbf{u}}_j = \underset{1}{u}_j + \mathrm{j}\underset{2}{u}_j$ corresponds to this value. Therefore

$$\underline{\lambda}_{(j)} = \frac{C_1 + \mathrm{j}C_2}{A_1 + \mathrm{j}A_2}, \tag{16}$$

where

$$C_1 = \sum_{i,k} (c_{ik}\underset{1}{u}_j^i \underset{1}{u}_j^k - c_{ik}\underset{2}{u}_j^i \underset{2}{u}_j^k), \tag{17}$$

$$C_2 = 2\sum_{i,k} c_{ik}\underset{1}{u}_j^i \underset{2}{u}_j^k, \tag{18}$$

$$A_1 = \sum_{i,k} (a_{ik}\underset{1}{u}_j^i \underset{1}{u}_j^k - a_{ik}\underset{2}{u}_j^i \underset{2}{u}_j^k), \tag{19}$$

$$A_2 = 2\sum_{i,k} a_{ik}\underset{1}{u}_j^i \underset{2}{u}_j^k. \tag{20}$$

The root $\underline{\lambda}^*_{(j)}$, conjugate to $\underline{\lambda}_{(j)}$, of the characteristic equation corresponds to the value conjugate to $\underline{\mathbf{u}}_j$ since substituting the vector $\underline{\mathbf{u}}_j$ into (15), we obtain

$$\underline{\lambda}^*_{(j)} = \frac{C_1 - \mathrm{j}C_2}{A_1 - \mathrm{j}A_2} \quad * \tag{21}$$

which confirms our supposition.

Considering two amplitude vectors conjugate to each other and corresponding to two conjugate roots of the characteristic equation, we write

$$\sum_{i,k} a_{ik}(\underset{1}{u}_j^i + \mathrm{j}\underset{2}{u}_j^i)(\underset{1}{u}_j^k - \mathrm{j}\underset{2}{u}_j^k) = \sum_{i,k} a_{ik}\underset{1}{u}_j^i \underset{1}{u}_j^k + \sum_{i,k} a_{ik}\underset{2}{u}_j^i \underset{2}{u}_j^k > 0.$$

Consequently, the assumed existence of complex roots leads to a contradiction, for we have previously proved that amplitude vectors are orthogonal, i.e., the quadratic form just written should vanish; in this way we preclude the existence of complex roots.

It still remains to be proved that the just named roots are positive. The proof is immediate if we consider that the quadratic forms occurring in expression (15) are positively determined. It can also be shown that the amplitude vectors are linearly independent. If they vere linearly dependent, such $\beta^j \neq 0$ would then exist that

$$\beta^j \mathbf{u}_j = 0. \tag{22}$$

---

* Distinction should be made between the index j and the imaginary unit $\mathrm{j} = \sqrt{-1}$.

We shall now construct the quadratic form

$$0 = a_{ik} u_l^i \beta^j u_j^k = \beta^j \alpha^2 \delta_{lj} = \beta_l \alpha^2.$$

It follows that for any l there is $\beta_l = 0$; this is proof that amplitude vectors are linearly independent, since no such $\beta^j \neq 0$ exists that would satisfy equation (22).

Returning to the integrals of the initial system of differential equations we find that we have obtained N solutions for each degree of freedom of the system:

$$q_j^k = u_j^k \sin(\omega_j t + \varphi_j). \tag{23}$$

By virtue of the linearity of differential equations, the linear combination of solutions of type (23) is also a solution, therefore

$$q^k = C^j q_j^k. \tag{24}$$

We calculate the constants C^j from the initial conditions, i.e., from known deflections $q^k(0)$ and velocities $\dot{q}^k(0)$. Evidently, we have at our disposal $2N$ equations with $2N$ unknows C^j and φ_j. Note also that there is complete freedom in the choice of the constant factor in equation (13); it can be assumed to be equal, let us say, to 1. Then

$$a_{ik} u_j^i u_l^k = \delta_{jl}. \tag{25}$$

By adding equation (25) to $N-1$ equations from system (3) it is possible to determine uniquely the amplitude vectors.

Example 1. Find the free vibration frequencies of a mechanical system consisting of two masses m_1 and m_2, capable of moving on a smooth plane surface, if it is known that the springs have constants c_1 and c_2 (Fig. 9.1).

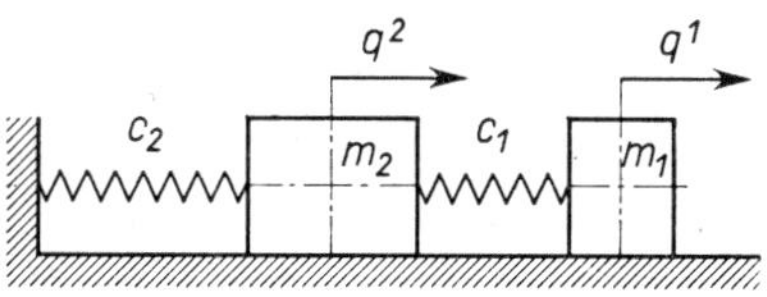

Fig. 9.1

In the considered case, the kinetic energy is expressed by

$$E = \tfrac{1}{2} m_1 (\dot{q}^1)^2 + \tfrac{1}{2} m_2 (\dot{q}^2)^2.$$

The system is easily seen to have two degrees of freedom. The matrix a_{ik} is diagonal:

$$[a_{ik}] = \begin{bmatrix} m_1 & 0 \\ 0 & m_2 \end{bmatrix}.$$

We now find the potential energy:

$$V = \tfrac{1}{2}c_1(q^1-q^2)^2+\tfrac{1}{2}c_2 q_2^2 = \tfrac{1}{2}[c_1(q^1)^2-2c_1 q^1 q^2+c_1(q^2)^2+c_2(q^2)^2].$$

We identify the matrix c_{ik}

$$[c_{ik}] = \begin{bmatrix} c_1 & -c_1 \\ -c_1 & c_1+c_2 \end{bmatrix}.$$

The characteristic equation in the considered case is

$$\det[c_{ik}-\lambda a_{ik}] = \begin{vmatrix} c_1-\lambda m_1 & -c_1 \\ -c_1 & c_1+c_2-\lambda m_2 \end{vmatrix} = 0$$

or after expanding the determinant

$$\lambda^2 m_1 m_2-(c_1 m_2+c_1 m_1+c_2 m_1)\lambda+c_1 c_2 = 0.$$

Hence

$$\lambda_{1,2} = \frac{c_1 m_2+c_1 m_1+c_2 m_1 \mp \sqrt{(c_1 m_2+c_1 m_1+c_2 m_1)^2-4c_1 c_2 m_1 m_2}}{2m_1 m_2},$$

$$\omega_{1,2} = \sqrt{\lambda_{1,2}}\,.$$

Example 2. A simple pendulum with length l and mass m_2 was attached to mass m_1, capable of moving on smooth horizontal surface. The motion of m_1 is limited by a spring with constant c. Find the equation of motion for the system assuming small deflections and that, in the initial instant, the mass m_1 is deflected from the equilibrium position, the pendulum is in a vertical position and the system remains at rest (Fig. 9.2).

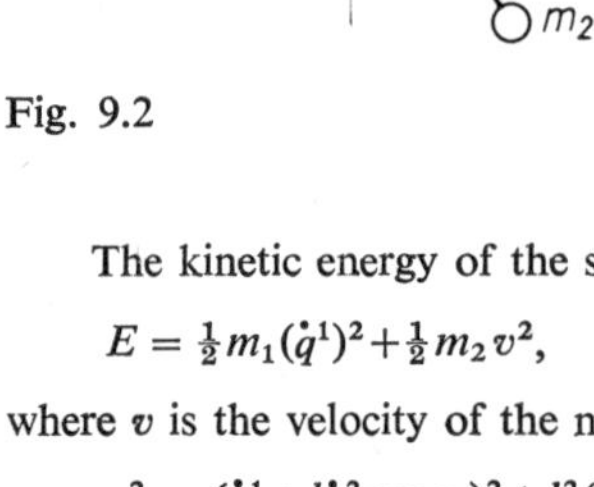

Fig. 9.2

The kinetic energy of the system is

$$E = \tfrac{1}{2}m_1(\dot{q}^1)^2+\tfrac{1}{2}m_2 v^2, \tag{a}$$

where v is the velocity of the mass m_2

$$v^2 = (\dot{q}^1+l\dot{q}^2\cos q_2)^2+l^2(\dot{q}^2)^2\sin^2 q^2 = (\dot{q}^1)^2+2\dot{q}^1 l\dot{q}^2\cos q^2+l^2(\dot{q}^2)^2. \tag{b}$$

Assuming that the deflections from the equilibrium position are small, we replace $\cos q^2$ with 1. In this way a_{ik} will be constant:

$$E = \tfrac{1}{2}[(m_1+m_2)(\dot{q}^1)^2+2m_2 l\dot{q}^1\dot{q}^2+m_2 l^2(\dot{q}^2)^2]. \tag{c}$$

Identifying the matrix a_{ik}, we write

$$a_{ik} = \begin{bmatrix} m_1+m_2 & m_2 l \\ m_2 l & m_2 l^2 \end{bmatrix}. \tag{d}$$

We shall now find the potential energy of the system:

$$V = \tfrac{1}{2}c(q^1)^2+m_2 gl(1-\cos q^2) = \tfrac{1}{2}c(q^1)^2+\tfrac{1}{2}m_2 gl(q^2)^2+ \ldots, \tag{e}$$

or

$$[c_{ik}] = \begin{bmatrix} c & 0 \\ 0 & m_2 gl \end{bmatrix}. \tag{f}$$

From (d) and (f) we build the characteristic determinant (4) which is expanded as follows:

$$\lambda^2 - \left[\frac{c}{m_1}+(m_1+m_2)\frac{g}{m_1 l}\right]\lambda+\frac{cg}{m_1 l} = 0, \tag{g}$$

from which we find the free vibration frequencies of the system:

$$\omega_1 = \sqrt{\lambda_1} \quad \text{and} \quad \omega_2 = \sqrt{\lambda_2}\,.$$

Now, we shall find the amplitude vectors. For this purpose, we make use of system (6). In the considered case, we have two degrees of freedom ($N = 2$). We use one of equations (6) to which we add equation (13):

$$\begin{aligned} &(c_{11}-\lambda_j a_{11})u_j^1+(c_{12}-\lambda_j a_{12})u_j^2 = 0, \\ &a_{11}(u_j^1)^2+2a_{12}u_j^1 u_j^2+a_{22}(u_j^2)^2 = \alpha^2 \qquad (j = 1, 2). \end{aligned} \tag{h}$$

Finding u_j^2 from the first equation and then substituting the derived expression into the second equation of the system (h) we have

$$\left[a_{11}\left(\frac{c_{12}-\lambda_j a_{12}}{c_{11}-\lambda_j a_{11}}\right)^2+2a_{12}\frac{c_{12}-\lambda_j a_{12}}{c_{11}-\lambda_j a_{11}}+a_{22}\right](u_j^2)^2 = \alpha^2.$$

It is convenient to accept α as being equal to the root of the multiple of $(u_j^2)^2$. In this way

$$u_j^2 = 1, \tag{i}$$

whereas

$$u_j^1 = -\frac{c_{12}-\lambda_j a_{12}}{c_{11}-\lambda_j a_{11}} = \frac{\lambda_j m_2 l}{c-\lambda_j(m_1+m_2)}. \tag{j}$$

Making use of equations (24) we write

$$\begin{aligned} q^1 &= C^1u_1^1\sin(\omega_1 t+\varphi_1)+C^2u_2^1\sin(\omega_2 t+\varphi_2), \\ q^2 &= C^1u_1^2\sin(\omega_1 t+\varphi_1)+C^2u_2^2\sin(\omega_2 t+\varphi_2). \end{aligned} \tag{k}$$

In the initial instant, the mass m_1 was deflected from the equilibrium position by a value $q^1(0)$ and the entire system was at rest, i.e., $q^2(0) = 0$, $\dot{q}^1(0) = \dot{q}^2(0) = 0$. On this basis we can calculate the integration constants. For this purpose we differentiate equations (k), add the result obtained to the initial system and set $t = 0$; we arrive at

$$\begin{aligned} q^1(0) &= C^1u_1^1\sin\varphi_1+C^2u_2^1\sin\varphi_2, \\ 0 &= C^1u_1^2\sin\varphi_1+C^2u_2^2\sin\varphi_2, \\ 0 &= C^1u_1^1\omega_1\cos\varphi_1+C^2u_2^1\omega_2\cos\varphi_2, \\ 0 &= C_1u_1^2\omega_1\cos\varphi_1+C^2u_2^2\omega_2\cos\varphi_2. \end{aligned} \tag{l}$$

Solving system (1) we obtain

$$\varphi_1 = \varphi_2 = \pi/2 \tag{m}$$

and

$$C^1 = -C^2 = \frac{q^1(0)}{u_1^1 - u_2^1}. \tag{n}$$

Knowing now the integration constants we return to system (k):

$$\begin{aligned} q^1 &= \frac{q^1(0)}{u_1^1 - u_2^1}\left[u_1^1 \sin\left(\omega_1 t + \frac{\pi}{2}\right) - u_2^1 \sin\left(\omega_2 t + \frac{\pi}{2}\right)\right], \\ q^2 &= \frac{q^1(0)}{u_1^1 - u_2^1}\left[u_1^2 \sin\left(\omega_1 t + \frac{\pi}{2}\right) - u_2^2 \sin\left(\omega_2 t + \frac{\pi}{2}\right)\right], \end{aligned} \tag{o}$$

where the amplitude vectors are given by equations (i) and (j).

9.4 Normal coordinates

We shall seek these coordinates for which the differential equations of motion assume the simplest form. The structure of these equations depends on the structure of the quadratic forms which give the kinetic and the potential energy, therefore the energies for these coordinates should be expressed by the simplest formulae. The simplest of quadratic forms is the sum of squares, hence we are confronted with the problem of the simultaneous reduction of two quadratic forms to a sum of squares. We find that linear transformations of the coordinates lead to this objective. By θ^j we denote the normal coordinates. By transforming:

$$q^k = u_j^k \theta^j. \quad \det[u_j^k] \neq 0,$$

it is possible to reduce E and V to the form of a sum of squares. Indeed,

$$E = \tfrac{1}{2} a_{ik} \dot{q}^i \dot{q}^k = \tfrac{1}{2} a_{ik} u_l^i u_j^k \dot{\theta}^l \dot{\theta}^j = \tfrac{1}{2} \delta_{lj} \dot{\theta}^l \dot{\theta}^j. \tag{1}$$

Similarly,

$$V = \tfrac{1}{2} c_{ik} q^i q^k = \tfrac{1}{2} c_{ik} u_l^i u_j^k \theta^l \theta^j. \tag{2}$$

Considering equation (8), Section 9.3, we write

$$V = \tfrac{1}{2} \lambda_{(j)} a_{ik} u_l^i u_j^k \theta^l \theta^j = \tfrac{1}{2} \lambda_{(j)} \delta_{lj} \theta^l \theta^j. \tag{3}$$

The Lagrange function is expressed by

$$L = \tfrac{1}{2} \delta_{lj} (\dot{\theta}^l \dot{\theta}^j - \lambda_{(j)} \theta^l \theta^j). \tag{4}$$

Putting

$$\frac{\mathrm{d}}{\mathrm{d}t}\left(\frac{\partial L}{\partial \dot{\theta}^j}\right) - \frac{\partial L}{\partial \theta^j} = 0,$$

into the Lagrange equations, we obtain

$$\ddot{\theta}^j + \lambda_{(j)}\theta^j = 0\,^* \qquad (j = 1, 2, \ldots, N). \tag{5}$$

The integrals of these equations are

$$\theta^j = C^j \sin(\omega_j t + \varphi_j), \quad \text{where} \quad \omega_j = \sqrt{\lambda_j}\,. \tag{6}$$

Returning to the former system we obtain

$$q^k = C^j u_j^k \sin(\omega_j t + \varphi_j). \tag{7}$$

Evidently the result coincides with (24), Section 9.3.

9.5 Forced vibrations of a system

Vibrations develop when a system is subjected to the action of exciting forces Q_i. If dissipation is disregarded, i.e., if we assume that $D = 0$, then the differential equations of motion take the form

$$a_{ik}\ddot{q}^k + c_{ik}q^k = Q_i. \tag{1}$$

We now transform system (1) to normal coordinates. For this purpose we multiply the equations belonging to this system by virtual displacements and then add both sides. In this way we obtain

$$a_{ik}\ddot{q}^k \delta q^i + c_{ik}q^k \delta q^i = Q_i \delta q^i. \tag{2}$$

But $q^i = u_j^i \theta^j$, therefore $\delta q^i = u_j^i\, \delta\theta^j$; similarly, $q^k = u_l^k \theta^l$, $\ddot{q}^k = u_l^k \ddot{\theta}^l$ and therefore

$$a_{ik} u_j^i u_l^k \ddot{\theta}^l \delta\theta^j + c_{ik} u_j^i \theta^l \delta\theta^j = Q_i u_j^i\, \delta\theta^j. \tag{3}$$

Taking advantage of the fact that the amplitude vectors are orthogonal, we reduce the system to the form

$$\delta_{lj}\ddot{\theta}^l \delta\theta^j + \omega_{(j)}^2 \delta_{lj}\theta^l \delta\theta^j = Q_i u_j^i\, \delta\theta^j. \tag{4}$$

The virtual displacements $\delta\theta^j$ are independent, therefore

$$\delta_{lj}(\ddot{\theta}^l + \omega_{(j)}^2 \theta^l) = Q_i u_j^i. \tag{5}$$

The sum $Q_i u_j^i$ is a generalized force in normal coordinates; denoting it by P_j we write

$$\ddot{\theta}_j + \omega_{(j)}^2 \theta_j = P_j. \tag{6}$$

Let the force P_j be a time function. To arrive at the solution we perform Laplace's transformation on equation (6):

$$s^2\bar{\theta}_j(s) - s\theta_j(0) - \dot{\theta}_j(0) + \omega_j^2 \bar{\theta}_j(s) = \bar{P}_j(s); \tag{7}$$

* No summation should be done over j.

hence

$$\bar{\theta}_j(s) = \theta_j(0)\frac{s}{s^2+\omega_j^2} + \frac{\dot{\theta}_j(0)}{\omega_j}\frac{\omega_j}{s^2+\omega_j^2} + \frac{\bar{P}_j(s)}{\omega_j}\frac{\omega_j}{s^2+\omega_j^2}. \tag{8}$$

Performing Laplace's inverse transformation on operator equation (8) while making use of the theorem of convolution of function, we write:

$$\theta_j(t) = \theta_j(0)\cos\omega_j t + \frac{\dot{\theta}_j(0)}{\omega_j}\sin\omega_j t + \frac{1}{\omega_j}\int_0^t P_j(t-\tau)\sin\omega_j\tau\,d\tau$$

or

$$\theta_j(t) = C_j\sin(\omega_j t+\varphi_j) + \frac{1}{\omega_j}\int_0^t P_j(t-\tau)\sin\omega_j\tau\,d\tau, \tag{9}$$

where

$$C_j = \sqrt{[\theta_j(0)]^2 + \left[\frac{\dot{\theta}_j(0)}{\omega_j}\right]^2}, \qquad \tan\varphi_j = \frac{\theta_j(0)\omega_j}{\dot{\theta}_j(0)}.$$

Knowing $\theta_j(t)$, we can go back to the former coordinates by linear transformation: $q^k(t) = u_j^k\theta^j(t)$ $(\theta^j \equiv \theta_j)$. This shows that forced vibration of systems without damping does not belong to the category of complex problems.

9.6 Free vibrations of dissipative systems

The differential equations of motion for free vibrations in dissipative systems are as follows:

$$a_{ik}\ddot{q}^k + h_{ik}\dot{q}^k + c_{ik}q^k = 0. \tag{1}$$

We shall seek the solution to the system of equations (1) in the form

$$q^k = u^k e^{st}. \tag{2}$$

Substituting (2) into (1), and simplifying by $e^{st} \neq 0$ we obtain

$$(a_{ik}s^2 + h_{ik}s + c_{ik})u^k = 0. \tag{3}$$

A non-trivial solution exists only when the characteristic determinant of system (3) is equal to zero:

$$\det[a_{ik}s^2 + h_{ik}s + c_{ik}] = 0. \tag{4}$$

From equation (4) we can determine s. The roots of equation (4) can be estimated using system (3). Multiplying this system by the value u^{i*} conjugated to u^k we have

$$s^2 a_{ik} u^k u^{i*} + s h_{ik} u^k u^{i*} + c_{ik} u^k u^{i*} = 0. \tag{5}$$

The quadratic forms occurring in (5) are positive. Let us denote them by the letters a, b and c:

$$as^2 + bs + c = 0; \tag{6}$$

hence

$$s_{1,2} = \frac{1}{2a}(-b \pm \sqrt{b^2 - 4ac}),$$

Note further that

$$s_1 + s_2 = -b/a \quad \text{and} \quad s_1 s_2 = c/a.$$

We can see that the sum of the roots is negative while the product is positive. This is proof that if the roots are conjugated, then the real parts are negative and if they are real, then $|s_1| > |s_2|$. In either case, deflections from the equilibrium position decrease with time.

One could reach the same conclusion in a somewhat different way too. Consider equation (7), Section 7.11, expressing a derivative of the total energy of the system. In the considered case completely dissipative forces are in action, therefore

$$Q_j \dot{q}^j = -h_{jk} \dot{q}^k \dot{q}^j,$$

and this proves that the overall total energy of the system decreases with time.

If the total energy decreases, then the system tends to the equilibrium position. Evidently, the system is in this case asymptotically stable (the matrix h_{ik} is significantly positive finite).

We may present the solution to system (1) in the form

$$q^k = C^j u_j^k e^{s_j t} \tag{7}$$

where s_j are the roots of the characteristic equation (4), whereas u_j^k—the components of amplitude vectors derived from equation (3) with an accuracy up to the constant factor.

The roots s_j may be complex. Each pair of complex conjugate roots $s = -n + j\omega$ and $s^* = -n - j\omega$ has a corresponding pair of amplitude vectors:

$$u = \underset{1}{u} + j\underset{2}{u} \quad \text{and} \quad u^* = \underset{1}{u} - j\underset{2}{u}.$$

In this case we consider each sum of two such solutions separately. Considering that the constants occurring with the special integrals are also complex and conjugate, we write

$$Cue^{st}+C^*u^*e^{s^*t} = e^{-nt}[(\underset{1}{C}\underset{1}{u}-\underset{2}{C}\underset{2}{u})\cos\omega t-(\underset{1}{C}\underset{2}{u}+\underset{2}{C}\underset{1}{u})\sin\omega t], \tag{8}$$

where $\underset{1}{C}$ and $\underset{2}{C}$ are the real and the imaginary parts of the double values of the constants C and C^*.

Evidently, conjugate roots occur in the case of vibrations. These are in the nature of damped vibrations.

9.7 Forced vibrations in dissipative systems

The system of differential equations (6), Section 9.2, defines forced vibrations of systems in which energy dissipation occurs. Let the generalized forces Q_i be time functions.

To arrive at a general solution we shall use the integral transformation method. Performing Laplace's transformation on system (6), Section 9.2, we obtain

$$\begin{aligned} &a_{ik}s^2\bar{q}^k(s)+h_{ik}s\bar{q}^k(s)+c_{ik}\bar{q}^k(s)\\ &\qquad = a_{ik}sq^k(0)+a_{ik}\dot{q}^k(0)+h_{ik}q^k(0)+\bar{Q}_i(s), \end{aligned} \tag{1}$$

where $q^k(0)$ is a deflection, $\dot{q}^k(0)$ generalized velocity at instant $t = 0$. Denoting the inverse matrix to the matrix

$$[a_{ik}s^2+h_{ik}s+c_{ik}]$$

by $\bar{\mathbf{G}}(s)$ and substituting

$$W_i^{(1)} = a_{ik}q^k(0), \qquad W_i^{(2)} = a_{ik}\dot{q}^k(0)+h_{ik}q^k(0),$$

we write

$$\bar{q}^k(s) = s\bar{G}^{ki}(s)\,W_i^{(1)}+\bar{G}^{ki}(s)\,W_i^{(2)}+\bar{G}^{ki}(s)\bar{Q}_i(s) \tag{2}$$

or in the matrix form

$$\bar{\mathbf{q}}(s) = s\bar{\mathbf{G}}(s)\mathbf{W}^{(1)}+\bar{\mathbf{G}}(s)\mathbf{W}^{(2)}+\bar{\mathbf{G}}(s)\bar{\mathbf{Q}}(s). \tag{2a}$$

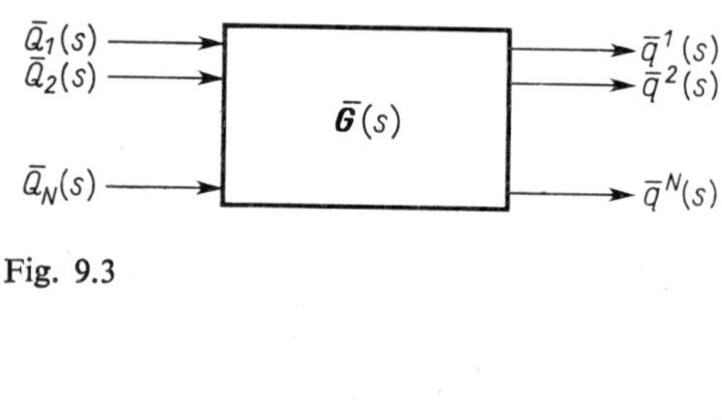

Fig. 9.3

The matrix $\overline{\mathbf{G}}(s)$ is called the *matrix of transmittance* of a system with N inputs and N outputs. The symbol of such a system has been presented in Fig. 9.3. Returning to the operator equation (2) or (2a) we find that the first two terms on the right-hand side are transforms of the solutions to the problem of free vibrations of dissipative systems. Performing Laplace's inverse transformation on equation (2) while making use of the theorem on convolution of function, we obtain

$$\mathbf{q}(t) = \dot{\mathbf{G}}(t)\mathbf{W}^{(1)}+\mathbf{G}(t)\mathbf{W}^{(2)}+\int_0^t \mathbf{G}(t-\tau)\mathbf{Q}(\tau)\mathrm{d}\tau. \tag{3}$$

The result (3) is the most general solution to the problem posed in this section.

9.8 Vibrations of Clapeyron's systems *

So far we have discussed in our considerations problems of vibrations of systems about their equilibrium positions. We have assumed at the same time that the deflections are sufficiently small for the discussion to be confined to the first terms of expansion (1), Section 9.2, and elements of the matrix a_{ik} are accepted as constant. Clapeyron's systems, being elastic systems for which the displacements q^i are linear functions of loads Q_k ($q^i = \sum c^{ik}Q_k$) and whose potential energy we reckon as half the products of displacements q^i and the corresponding loads Q_i, are systems for which the cited theory holds. Thus the solutions (3), Section 9.7, describe vibrations of all kinds of frameworks, trusses, beams, plates, and shells, and the like, with masses concentrated at different points of their systems.

In order to establish this, we shall derive the differential equations of motion for such systems. The kinetic and the potential energy are expressed by

$$E = \tfrac{1}{2}\sum_{i=1}^{n} m_i(\dot{q}^i)^2, \tag{1}$$

$$V = \tfrac{1}{2}\sum_{i=1}^{n} q^iQ_i, \tag{2}$$

where m_i are the masses or moments of inertia of the masses at different points of the system.

* The summation convention will not be used in this section.

Wanting to express the potential energy by displacement of points q^i of the system the following system of equations needs to be solved:

$$q^i = \sum_{k=1}^{n} c^{ik} Q_k \tag{3}$$

with respect to the generalized forces Q_k. In the above equation, c^{ik} are the influence coefficients and signify the influence of a force at a point k on a displacement at a point i. By Betti's theorem of reciprocity, the matrix formed from the influence coefficients c^{ik} is a symmetrical matrix. Solving (3) we write

$$Q_i = \sum_{k=1}^{n} c_{ik} q^k, \tag{4}$$

where c_{ik} are elements of the inverse matrix to c^{ik}. Substituting (4) into (2) we have

$$V = \tfrac{1}{2} \sum_{i,k} c_{ik} q^i q^k. \tag{5}$$

The determination of the matrix c_{ik} amounts here to finding the inverse matrix to the influence matrix. Evidently, for Clapeyron's systems the potential energy is a positive definite homogeneous quadratic form.

Note that by virtue of (1) the matrix a_{ik} is a diagonal matrix. We write differential equations (6), Section 9.2, in the form

$$m_i \ddot{q}^i + \sum_{k=1}^{n} h_{ik} \dot{q}^k + \sum_{k=1}^{n} c_{ik} q^k = Q_i \quad (i = 1, 2, \ldots, n). \tag{6}$$

If there is no dissipation of the energy, the sum $\sum_{k=1}^{n} h_{ik} \dot{q}^k$ disappears and then

$$m_i \ddot{q}^i + \sum_{k=1}^{n} c_{ik} q^k = Q_i \quad (i = 1, 2, \ldots, n). \tag{6a}$$

We ascertain from the derived differential equation that solution (3), Section 9.7, is a general solution to the problem of vibrations of Clapeyron's systems.

Solving system (6a) in relation to q^k we obtain

$$q^k = \sum_{i=1}^{n} c^{ki} (Q_i - m_i \ddot{q}_i) \quad (k = 1, 2, \ldots, n). \tag{7}$$

Comparing the obtained result, i.e., (7), with expression (3), we find that a problem of dynamics can be treated as a problem of statics, all that needs to be done is to add forces of inertia $B_i = -m_i q^i$ to the external forces Q_i. The observed analogy between equations (3) and (7) is called the *statical-dynamical analogy*. This analogy is most helpful in setting up differential equations of motion. It should be appreciated that equation (7) is also analogous to equation (1), Section 9.5. On comparing the two systems of equations we can easily see that the equaivalent of the matrix c_{ik} is reduced to a unit matrix in equations (7).

Example 1. Find the free vibration frequencies of a beam with bending rigidity EJ, shown in Fig. 9.4, with two concentrated masses m_1 and m_2. The beam's own mass could be ignored.

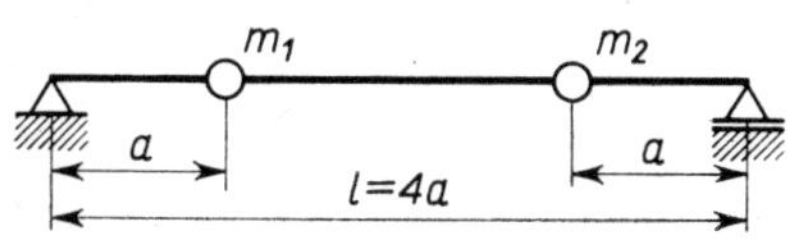

Fig. 9.4

First we shall write the differential equations of free vibrations for a system with two degrees of freedom, such as is the considered beam with the two masses. From (7) we write

$$q_1 + c^{11} m_1 \ddot{q}_1 + c^{12} m_2 \ddot{q}_2 = 0, \quad q_2 + c^{21} m_1 \ddot{q}_1 + c^{22} m_2 \ddot{q}_2 = 0, \tag{a}$$

where the influence matrix c^{ik} is

$$\begin{bmatrix} \dfrac{3a^3}{4EJ} & \dfrac{7a^3}{12EJ} \\ \dfrac{7a^3}{12EJ} & \dfrac{3a^3}{4EJ} \end{bmatrix}. \tag{b}$$

Using equation (4), Section 9.3, we write

$$\det[\delta^{kl} - c^{kl} m_l \lambda] = 0$$

or in an expanded form

$$\begin{vmatrix} (1 - c^{11} m_1 \lambda) & -c^{12} m_2 \lambda \\ -c^{21} m_1 \lambda & (1 - c^{22} m_2 \lambda) \end{vmatrix} = 0. \tag{c}$$

After expanding the determinant we obtain a quadratic equation with respect to $\omega^2 = \lambda$:

$$m_1 m_2 \left(c^{11} c^{22} - (c^{12})^2\right) \lambda^2 - (m_1 c^{11} + m_2 c^{22}) \lambda + 1 = 0. \tag{d}$$

The derived equation gives two free vibration frequencies of the system:

$$\omega_1 = \sqrt{\frac{c^{11}m_1 + c^{22}m_2 - \sqrt{(m_1 c^{11} - m_2 c^{22})^2 + 4m_1 m_2 (c^{12})^2}}{2m_1 m_2 [c^{11}c^{22} - (c^{12})^2]}}, \tag{e}$$

$$\omega_2 = \sqrt{\frac{c^{11}m_1 + c^{22}m_2 + \sqrt{(m_1 c^{11} - m_2 c^{22})^2 + 4m_1 m_2 (c^{12})^2}}{2m_1 m_2 [c^{11}c^{22} - (c^{12})^2]}}. \tag{f}$$

The calculated vibration frequencies are general solutions in so far as they constitute expressions for free vibration frequencies for all systems with two degrees of freedom. Naturally, a different matrix c^{ik} corresponds to any of the systems in question.

CHAPTER 10

Some Methods of Describing Random Phenomena in Mechanics

10.1 Basic concepts

Up to the present we have considered determined problems in the sense that both the systems and forces involved had a strictly defined character. Engineering practice brings to mind the fact of the frequent occurrence of phenomena of a random nature. The motion of a system is often undetermined due to a random occurrence of forces. Examples of random loads may be kinematical forcings of vibrations of vehicles by rough road surfaces, of vibrations of buildings by earthquakes, of movements of the frames of masts or trusses or vibrations of factory chimneys by wind force, of vibrations of shells induced by acoustic forces, and the like.

In all these examples, causal description is absolutely inadequate and some other must be used in its place. These cases should be viewed as probability problems demanding different mathematical methods whose foundations we owe first and foremost to A. Y. Khinchin, A. N. Kolmogorov, A. A. Markov, N. Wiener, and others. Following the ideas of R. L. Stratonovich [69] we shall present here a distributive approach of a more intuitive character. Before discussing the problem of random phenomena taking place in time, we shall discuss the fundamentals of the calculus of probability.

Suppose that we perform an experiment as a result of which a certain event A of interest to us is likely to occur. Performing it n times we find that A occurred n_A times. The ratio n_A/n we shall call the *frequency* (*relative frequency*).

Underlying the concept of probability is an experimentally observable regularity, consisting in that the frequency of events of concern to us, occurring in a series of experiments carried under identical conditions, varies about a certain constant value. For a large number of experiments n we assume the relative frequency n_A/n to be a probability $P(A)$ being a number from the closed interval [0, 1]:

$$P(A) \approx \frac{n_A}{n} \tag{1}$$

or, if n were unlimited,

$$P(A) = \lim_{n\to\infty} \frac{n_A}{n}. \tag{1a}$$

When the event A is impossible, then $P(A) = 0$, and when certain, $P(A) = 1$.

Let us imagine an experiment consisting in the random sampling of numbered balls from a drum (like a draw). Let there be N balls numbered from 1 to N. In addition let us agree that after each sampling the ball is replaced into the drum to ensured identical conditions for each test.

Consider the frequency of drawing a ball with a number less than k $(1 \leqslant k \leqslant N)$.

Let the number of tests be n. We calculate the number of hits according to $\sum_{i=1}^{n} H(k-\xi_i)$ where ξ_i is the number of the ball drawn in the ith sampling and $H(k-\xi_i)$ is Heaviside's function. The frequency of events $k > \xi_i$ in n tests is therefore given by

$$S_n(k) = \frac{1}{n} \sum_{i=1}^{n} H(k-\xi_i). \tag{2}$$

Note that we have introduced a certain subordination, namely that to each event corresponds a number. If ξ is a unique function determined for a set of elementary events, we shall call it the *random variable*. To be more specific [8], we should say that the random variable is a certain series of numerical sequences, $\xi_{1\alpha}, \xi_{2\alpha}, \ldots$, such that for any $x \in \overline{\boldsymbol{R}} = \boldsymbol{R} \cup \{-\infty\} \cup \{+\infty\}$ there exists a limit, in the distributive sense, $\lim_{n\to\infty} S_n(x)$ which does not depend on the choice of the sequence $\{\xi_{i\alpha}\}$, $i \in N$, $\alpha \in \Lambda$ (independent of $\alpha \in \Lambda$).

Equation (2) also remains valid in the case of a random variable of continuous type. To emphasise this generality, let us assume that both

k and ξ_i may vary in a continuous manner. We have exchanged the symbol k for x. Therefore

$$S_n(x) = \frac{1}{n}\sum_{i=1}^{n} H(x-\xi_i). \tag{2a}$$

Experience shows that with fixed x the frequency of events $\xi < x$ given by equation (2a) varies about a stated value. Assume hypotetically that it is the limit of expression (2a) with $n \to \infty$:

$$\lim_{n\to\infty} S_n(x) = \langle H(x-\xi)\rangle, \tag{2b}$$

which denotes the mean value of Heaviside's function.

The concept of the *mean* which will play a most essential part in our considerations, can also be introduced for an arbitrary function $f(\xi)$, which we shall define as follows:

$$\langle f(\xi)\rangle \stackrel{\text{df}}{=} \lim_{n\to\infty} \frac{1}{n}\sum_{i=1}^{n} f(\xi_i), \tag{3}$$

provided this limit exists and is independent of the choice of ξ_i.

Returning to our experiment we notice that the limit $\lim\limits_{n\to\infty} S_n(k)$ will give the frequency of the occurrence of a ball with a number below k. Note that if $k > N$, each event (sampling of a ball with a number below k) is *certain*. Hence the calculated mean from Heaviside's function will be equal to 1 in this case. In the other extreme case, if k were zero or a number below it, each sampling would be a failure in the sense that $\bigwedge\limits_i k < \xi_i$, and the occurrence of the sampling of such a ball would be said to be *impossible*. So we can see that 1 corresponds to an event which is certain and 0 to an event which is impossible. The mean $\langle H(x-\xi)\rangle$ is easily seen to assume values from the closed interval $[0, 1]$.

We shall call the mean, $\langle H(x-\xi)\rangle$, the *cumulative distribution function* of the random variable ξ. It expresses the probability of a random variable ξ occurring below the value of x. It is said to have the character of the standard measure of the set. Probability can be treated therefore as a *measure of a set from a space of elementary events*.

As noted at the begining we shall not develop the set-theory approach [59] here, pursuing instead the more intuitive distributive aspect.

Let us denote by $F_\xi(x)$ the cumulative distribution function:

$$F_\xi(x) = \langle H(x-\xi)\rangle = P(\xi < x). \tag{4}$$

It is a non-decreasing function (Fig. 10.1).

We shall now show that probability is a denumerable additive function. Assume that the random variable ξ may assume values from the interval $[-\infty, +\infty]$. We shall be interested in the probability of the occurrence of a random variable from the interval $[x_1, x_2]$. In this connection we construct such a function $f(\xi)$ that assumes the value 1 when ξ belongs to the interval $[x_1, x_2]$ and zero value when ξ does not belong to it.

$$f(\xi) = \begin{cases} 0 & \text{for} \quad \xi < x_1, \\ 1 & \text{for} \quad x_1 < \xi < x_2, \\ 0 & \text{for} \quad x_2 < \xi. \end{cases} \tag{5}$$

This function can be presented in the form

$$f(\xi) = H(x_2-\xi)-H(x_1-\xi). \tag{5a}$$

Performing the averaging operation according to definition (3), we obtain

$$\langle f(\xi)\rangle = \langle H(x_2-\xi)\rangle-\langle H(x_1-\xi)\rangle. \tag{6}$$

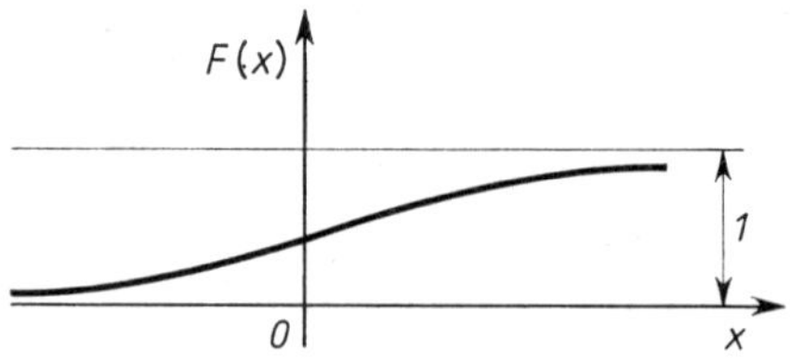

Fig. 10.1

The mean $\langle f(\xi)\rangle$ is the probability $P(x_1 < \xi < x_2)$ of random variable ξ occurring in the interval $[x_1, x_2]$, whence

$$P(x_1 < \xi < x_2) = \langle H(x_2-\xi)\rangle-\langle H(x_1-\xi)\rangle = P(\xi < x_2)-P(\xi < x_1). \tag{6a}$$

Consider now an arbitrary set, finite or denumerable, of disjoint intervals $[x_{Ai}, x_{Bi}]$ $(i = 1, \ldots, p)$. The probability of random variable ξ occurring in an arbitrary interval will be the mean of the function $f(\xi)$, defined in the following manner:

$$f(\xi) = \sum_i [H(x_{Bi}-\xi)-H(x_{Ai}-\xi)].$$

It follows that this probability will be expressed by

$$P(x_{A1} < \xi < x_{B1} \vee x_{A2} < \xi < x_{B2} \vee \ldots) = \sum_i P(x_{Ai} < \xi < x_{Bi}), \tag{6b}$$

The symbol $\vee$ is the alternative sign ($a \vee b$ should be read: a or b). If the intervals (x_{Ai}, x_{Bi}) are not disjoint intervals, then (6) is easily seen to become an inequality

$$P(x_{A1} < \xi < x_{B1} \vee x_{A2} < \xi < x_{B2} \vee \ldots) \leqslant \sum_i P(x_{Ai} < \xi < x_{Bi}). \tag{7}$$

Evidently, the probability has certain properties similar to those of measure.

With the occurrence of random phenomena we observe some regularities consisting in that that many events under observations, some occurrences are more likely than others, that is to say, some are more frequent than others. We observe here a certain density distribution of probability analogous to the distribution of masses of a continuous medium. We define the *probability density* $g_\xi(x)$ as follows:

$$g_\xi(x) \overset{\mathrm{df}}{=} \lim_{\Delta x \to 0} \frac{P(x < \xi < x+\Delta x)}{\Delta x}. \tag{8}$$

But

$$\begin{aligned} P(x < \xi < x+\Delta x) &= \langle H(x+\Delta x-\xi)\rangle - \langle H(x-\xi)\rangle \\ &= \langle H(x+\Delta x-\xi) - H(x-\xi)\rangle, \end{aligned}$$

therefore

$$\begin{aligned} g_\xi(x) &= \lim_{\Delta x \to 0} \frac{\langle H(x+\Delta x-\xi) - H(x-\xi)\rangle}{\Delta x} \\ &= \left\langle \lim_{\Delta x \to 0} \frac{H(x+\Delta x-\xi) - H(x-\xi)}{\Delta x} \right\rangle = \langle \delta(x-\xi)\rangle, \end{aligned}$$

where $\delta(x)$ is Dirac's cumulative distribution function. Thus the density, $g_\xi(x)$, is the mean of the Dirac delta function

$$g_\xi(x) = \langle \delta(x-\xi)\rangle. \tag{9}$$

It follows that, between the cumulative distribution function and the density of the probability distribution, there exists a relationship similar to that between a function and its derivative:

$$g_\xi(x) = \frac{\mathrm{d}F_\xi(x)}{\mathrm{d}x}. \tag{10}$$

Note that, given the function $g_\xi(x)$, the cumulative distribution function can be determined according to

$$F_\xi(x) = \int_{-\infty}^{x} g_\xi(x)\,\mathrm{d}x. \tag{11}$$

Naturally, the following occurs:

$$\int_{-\infty}^{\infty} g_\xi(x)\mathrm{d}x = 1. \tag{12}$$

In the language of mechanics equation (12) is a limitation to systems which have mass equal to 1 over the whole area.

In some cases where the random variable cannot assume all values from the interval $[-\infty, \infty]$, the function $g_\xi(x)$ has a distributive character. When, let us say, ξ occurs at points x_i with a probability P_i, then

$$g_\xi(x) = \sum_i P_i\,\delta(x-x_i). \tag{13}$$

For the considered case, the cumulative distribution function has the character of a step function (Fig. 10.2).

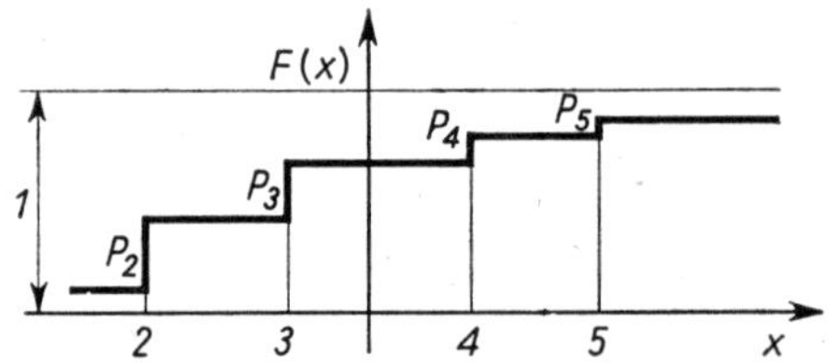

Fig. 10.2

We shall now proceed with finding the relationship between the function $g_\xi(x)$ and the mean value of the function $f(\xi)$. For this purpose we write

$$f(\xi) = \int_{-\infty}^{\infty} f(x)\,\delta(x-\xi)\mathrm{d}x.$$

Averaging we obtain

$$\langle f(\xi)\rangle = \int_{-\infty}^{\infty} f(x)\langle\delta(x-\xi)\rangle\mathrm{d}x = \int_{-\infty}^{\infty} f(x)g_\xi(x)\mathrm{d}x. \tag{14}$$

Thus we have arrived at the sought relationship giving the so-called *average* or *expected value* of the function $f(\xi)$.

Let $f(\xi) \equiv \xi$; then

$$\langle\xi\rangle = \int_{-\infty}^{\infty} xg_\xi(x)\mathrm{d}x = \frac{\int_{-\infty}^{\infty} xg_\xi(x)\mathrm{d}x}{\int_{-\infty}^{\infty} g_\xi(x)\mathrm{d}x}. \tag{15}$$

Equation (15) gives the average or expected values of the random variable ξ. We have deliberately written expression (15) in the form of a fraction to bring out the analogy between the mean value and the coordinate of the centre of mass of a system concentrated on a straight line (on the axis x). Accepting

$$f(\xi) = (\xi - \langle\xi\rangle)^2,$$

we obtain

$$\langle f(\xi)\rangle = \langle(\xi-\langle\xi\rangle)^2\rangle = \langle\xi^2 - 2\xi\langle\xi\rangle + \langle\xi\rangle^2\rangle = \langle\xi^2\rangle - \langle\xi\rangle^2 \tag{16}$$

called the *mean quadratic deviation, dispersion* or *variance*. The derived equation is easily seen to be the equivalent of Steiner's equation for moments of inertia.

The expression

$$\sigma = \sqrt{\langle(\xi-\langle\xi\rangle)^2\rangle} \tag{17}$$

is called the *standard deviation*. As we can see, the basis for calculating the mean is the knowledge of the density of probability $g_\xi(x)$. One of the distributions of major practical importance is called the *Gauss normal distribution* which is characterized by the function

$$g_\xi(x) = \frac{1}{\sigma\sqrt{2\pi}}\, \mathrm{e}^{-\frac{(x-a)^2}{2\sigma^2}}. \tag{18}$$

In this case,

$$\langle\xi\rangle = a \quad \text{and} \quad \langle\xi^2\rangle - \langle\xi\rangle^2 = \sigma^2.$$

The graph of the function $g_\xi(x)$ is presented in Fig. 10.3. The curve $g_\xi(x)$ is symmetrical about the straight line $x = a$. The cumulative distribution function of this distribution is expressed by

$$F_\xi(x) = \int_{-\infty}^{x} g_\xi(x)\,\mathrm{d}x = \int_{-\infty}^{a} g_\xi(x)\,\mathrm{d}x + \int_{-\infty}^{x} g_\xi(x)\,\mathrm{d}x.$$

By virtue of the noted symmetry, $\int_{-\infty}^{a} g_\xi(x)\,\mathrm{d}x = \frac{1}{2}$, whereas

$$\int_{a}^{x} g_\xi(x)\,\mathrm{d}x = \frac{1}{2}\Phi\left(\frac{x-a}{\sigma}\right)$$

is the so-called *error integral*. Hence

$$F_\xi(x) = \frac{1}{2} + \frac{1}{2}\Phi\left(\frac{x-a}{\sigma}\right), \tag{19}$$

where

$$\Phi\left(\frac{x-a}{\sigma}\right) = \Phi(\xi) = \frac{2}{\sqrt{2\pi}}\int_0^{\xi} e^{-u^2/2}du.$$

The respective values of the error integral $\Phi((x-a)/\sigma)$ will be found in tables.

One of the advantages of a normal distribution is that the sum of several normal distributions is also a normal distribution with a mean equal to the sum of means and with a quadratic deviation equal to the sum of quadratic deviations. Where not one random variable but a whole set $\xi^1, \xi^2, \ldots, \xi^N$ occurs, the density of probability is defined analogously to the definition in the preceding case:

$$g_{\xi^1,\ldots\xi^N}(x^1, x^2, \ldots, x^N) \overset{\mathrm{df}}{=} \langle\delta(\xi^1-x^1)\ldots\delta(\xi^N-x^N)\rangle. \tag{20}$$

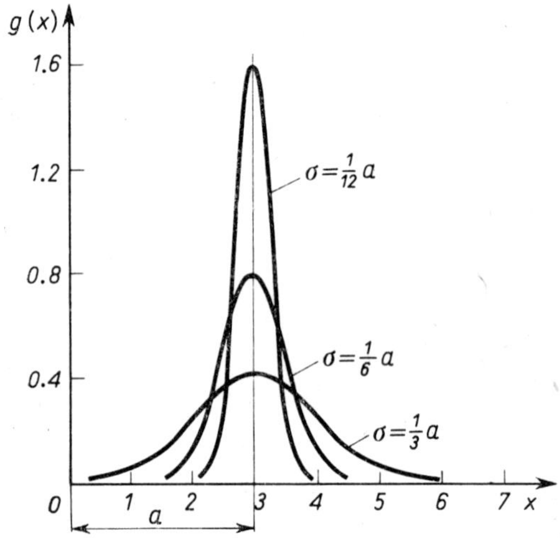

Fig. 10.3

The relationship between the cumulative distribution function $F_{\xi^1,\ldots\xi^N}(x_1, x_2, \ldots, x_n)$ and the density is also analogous to the case of a single variable:

$$g_{\xi^1,\ldots,\xi^N}(x^1, x^2, \ldots, x^N) = \frac{\partial^N F_{\xi^1,\ldots,\xi^N}(x^1, x^2, \ldots, x^N)}{\partial x^1\,\partial x^2\ldots\partial x^N}. \tag{21}$$

For we have

$$F_{\xi^1,\ldots,\xi^N}(x^1, \ldots, x^N) = \langle H(x^1-\xi^1)H(x^2-\xi^2)\ldots H(x^N-\xi^N)\rangle, \tag{22}$$

and further more we define the mean of a function with many variables in the same way as for one variable

$$\langle f(\xi^1, \ldots, \xi^N)\rangle \stackrel{\text{df}}{=} \lim_{n\to\infty} \frac{1}{n} \sum_{i=1}^{n} f(\xi_{i\alpha}^1, \ldots, \xi_{i\alpha}^N), \tag{23}$$

provided that this limit exists and does not depend on the choice of sequences (from $\alpha \in \Lambda$).

The validity of equations analogous to (6) can also be proved for a spatial case.

Also note that the following relationships occur:

$$\bigwedge_{0<k<N} F_{\xi^1,\ldots,\xi^N}(x^1, \ldots, x^k, \infty, \ldots, \infty) = F_{\xi^1,\ldots,\xi^N}(x^1, \ldots, x^k), \tag{24}$$

$$F_{\xi^1,\ldots,\xi^N}(\infty, \ldots, \infty) = 1, \tag{25}$$

$$\bigwedge_{ij} F_{\xi^1,\ldots,\xi^i,\ldots,\xi^j,\ldots,\xi^N}(x^1, \ldots, x^i, \ldots, x^j, \ldots, x^N)$$

$$= F_{\xi^1,\ldots\xi^j,\ldots\xi^i,\ldots\xi^N}(x^1, \ldots, x^j, \ldots, x^i, \ldots, x^N), \tag{26}$$

which are the so-called *conditions of consistency.*

In order to characterize the statistical link between random variables, the concept of correlation is introduced. In the case of two variables ξ^1 and ξ^2, the *correlation* is defined as follows:

$$K[\xi^1, \xi^2] \stackrel{\text{df}}{=} \langle \xi^1\xi^2\rangle - \langle\xi^1\rangle\langle\xi^2\rangle. \tag{27}$$

Variables will be called *uncorrelated* when

$$K[\xi^1, \xi^2] = 0, \tag{28}$$

which is not equivalent to the independence of variables when

$$g_{\xi^1,\,\xi^2}(x^1, x^2) = g_{\xi^1}(x^1)g_{\xi^2}(x^2). \tag{28a}$$

We also notice that for $\xi^1 = \xi^2 = \xi$ the correlation (27) becomes the *mean quadratic deviation* (*variance*):

$$K[\xi, \xi] = \sigma^2.$$

We find the correlation for three variables ξ^1, ξ^2, ξ^3 in the same way as for two random variables:

$$K[\xi^1, \xi^2, \xi^3] \stackrel{\text{df}}{=} \langle\xi^1\xi^2\xi^3\rangle - K[\xi^1, \xi^2]\langle\xi^3\rangle -$$

$$-\langle\xi^1\rangle K[\xi^2, \xi^3] - \langle\xi^2\rangle K[\xi^1, \xi^3] - \langle\xi^1\rangle\langle\xi^2\rangle\langle\xi^3\rangle. \tag{29}$$

It is possible to give a general formula [66] for the relationship between correlation functions and moments. For this purpose, it is necessary to define the correlation functions. Functions $K[\xi^{i_1}, \ldots, \xi^{i_s}] = k^{i_1,\ldots,i_s}$ of the sth order we define by the following relationship:

$$\exp \sum_{s=1}^{\infty} \frac{j^s}{s!} \sum_{i_1,\ldots i_s=1}^{N} k^{i_1,\ldots,i_s} u_{i_1} \ldots u_{i_s} = 1 + \sum_{s=1}^{\infty} \frac{j^s}{s!} \sum_{i_1,\ldots i_s=1}^{N} m^{i_1,\ldots,i_s} u_{i_1} \ldots u_{i_s} \tag{30}$$

with

$$m^{i_1,\ldots,i_s} \overset{\mathrm{df}}{=} \left\langle \prod_{k=1}^{s} \xi^{i_k} \right\rangle$$

and $k^{i_1,\ldots,i_s}$ not changing value on transposition of any two indices.

Both the right and left-hand sides of (30) are characteristic functions of random vectors, which will be defined in the following section.

Now if we put both sides of (30) into logarithms and next expand the right-hand side into a power series, then, by comparing the terms with the same powers j, we obtain a most important relationship between correlations and moments

$$k^{i_1\ldots,i_s} = m^{i_1,\ldots,i_s} + \\ + \sum_{n=2}^{s} \frac{(-1)^{n+1}}{n} \sum_{\nu_1=1}^{s} \cdots \sum_{\nu_{n-1}=1}^{s} \frac{s!H(c_n - \frac{1}{2}) D_{(n)}^{(i_1,\ldots,i_s)}(\nu_1, \ldots, \nu_{n-1})}{c_n! \nu_1! \ldots \nu_{n-1}!} \tag{31}$$

where $c_n = s - \sum_{i=1}^{n-1} \nu_i$, $H(\cdot)$ is Heaviside's function.

A symbol has been introduced for the symmetrical part of the multiple-indexed magnitude $A^{i_1,\ldots,i_s}$ denoting it by the symbol $A^{(i_1,\ldots i_s)}$. We form it according to

$$A^{(i_1,\ldots,i_s)} = \frac{1}{s!}(A^{(i_1,i_2,\ldots,i_s)} + A^{i_2,i_1,\ldots,i_s} + \ldots). \tag{32}$$

The summation embraces all permutations of the sequence $i_1, \ldots, i_s$. Note further that

$$k^{(i_1,\ldots,i_s)} = k^{i_1,\ldots,i_s} \quad \text{and} \quad m^{(i_1,\ldots,i_s)} = m^{i_1,\ldots,i_s}.$$

The symbol $D_{(n)}^{i_1,\ldots,i_s}$ is expressed by

$$D_{(n)}^{i_1,\ldots,i_s}(\nu_1, \ldots, \nu_{n-1}) = m^{i_1,\ldots,i_{s-b_1^{(n)}}} m^{i_{s-b_1^{(n)}+1},\ldots,i_{s-b_2^{(n)}}} m^{i_{s-b_{n-1}^{(n)}+1},\ldots,i_s} \tag{33}$$

where $b_k^{(n)} = \sum_{i=k}^{n-1} \nu_i$. Specifically, for $s = 2$ we obtain equation (27) and for $s = 3$, equation (29).

10.2 Methods of describing stochastic processes

We now come to the discussion of random phenomena taking place in time. The value of a function in a fixed instant of time t is a random magnitude. Functions endowed with this property are called *stochastic processes.*

Let $\xi(t)$ be the random function. We fix the times $t_1, t_2, \ldots, t_r$. The magnitudes $\xi(t_1), \xi(t_2), \ldots, \xi(t_r)$ form an r-dimensional random vector with an overall total distribution of probability characterized by the density function

$$g_r(x^1, x^2, \ldots, x^r, t_1, t_2, \ldots, t_r) = \langle \delta(x^1 - \xi(t_1)) \ldots \delta(x^r - \xi(t_r)) \rangle, \quad (1)$$

satisfying the conditions of consistency (24)–(26), Section 10.1. A stochastic process can be regarded as definite when the compatible distributions of the form (1) are known for each r.

The full characterization of the process is also given by the characteristic functional defined by

$$\Theta(u(t)) = \left\langle \exp\left(\mathrm{j} \int u(t)\xi(t)\,\mathrm{d}t \right) \right\rangle, \quad (2)$$

for $u(t)$ belonging to a certain class of functions, for which the integral on the right-hand side is definite.

For an r-dimensional process we should put

$$u(t) = \sum_{\nu=1}^{r} u_\nu \delta(t - t_\nu). \quad (3)$$

Substituting (3) into (2) we obtain a characteristic function

$$\Theta_r(u_1, \ldots, u_r, t_1, \ldots, t_r) = \left\langle \exp\left(\mathrm{j} \sum_{\nu=1}^{r} u_\nu \xi(t_\nu) \right) \right\rangle. \quad (4)$$

Note that characteristic function (4) is Fourier's transform of the distribution density (1). Thus, knowing the form of the characteristic function, we can calculate the distribution (1) by way of an inverse transformation.

We calculate the mean value of the process by making use of equation (15), Section 10.1, and (1):

$$\langle \xi(t) \rangle = \int_{-\infty}^{\infty} x g_1(x, t)\,\mathrm{d}x. \quad (5)$$

Similarly the mean can be derived from the product $\xi(t_1)\xi(t_2)$:

$$\langle\xi(t_1)\xi(t_2)\rangle = \int_{-\infty}^{\infty}\int_{-\infty}^{\infty} x^1 x^2 g_2(x^1, x^2, t_1, t_2)\mathrm{d}x^1\,\mathrm{d}x^2. \tag{6}$$

To characterize the statistical link between random variables $\xi(t_1)$ and $\xi(t_2)$, the *correlation function* concept is introduced and is given by

$$K[\xi(t_1), \xi(t_2)] \overset{\mathrm{df}}{=} \langle\xi(t_1)\xi(t_2)\rangle - \langle\xi(t_1)\rangle\langle\xi(t_2)\rangle. \tag{7}$$

Note that for $t_1 = t_2 = t$, the correlation function (7) becomes the mean quadratic deviation calculated for the random function at instant t.

Correlation functions of an order higher than the second can be built and this is possible when all the compatible density distributions (1) are known. The process is fully characterized in the same way as with conformant density distributions, by an unlimited sequence of correlation functions of the form

$$k_1(t_1),\ k_2(t_1, t_2),\ k_3(t_1, t_2, t_3),\ \ldots, \tag{8}$$

where

$$k_r(t_1, \ldots, t_r) = K[\xi(t_1), \ldots, \xi(t_r)]. \tag{9}$$

In the case of a determinate process, the terms of sequence (8) are all zero excepting $k_1(t) = \langle\xi(t)\rangle = \xi(t)$ since in this case

$$g_r(x^1, \ldots, x^r, t_1, \ldots, t_r) = \delta(x^1 - \xi(t_1))\,\delta(x^2 - \xi(t_2))\ldots\delta(x^r - \xi(t_r)). \tag{10}$$

The inverse also holds: If

$$\bigwedge_{r\neq 1} k_r(t_1, \ldots, t_r) \equiv 0,$$

then the process is determinate. Conventionally, determinate processes could be called *processes of the first order.*

Let us now consider the case where the terms of sequence (37), Section 10.1, from $k_3(t_1, t_2, t_3)$ onwards are all zero ($k_1(t_1)$ and $k_2(t_1, t_2)$ remain non-zero). Such a process is defined as *normal* or *gaussian*, since the assumption of the vanishing of the correlation functions as from the third order onward predetermines normal distributions of probability.

Indeed, expressing the characteristic function by means of correlation functions we have

$$\Theta_r(u_1, \ldots, u_r, t_1, \ldots, t_r)$$

$$= \exp\left\{\sum_{s=1}^{\infty}\frac{\mathrm{j}^s}{s!}\sum_{i_1, \ldots, i_s=1}^{r} k_s(t_{i_1}, \ldots, t_{i_s})\,u_{i_1}\ldots u_{i_s}\right\}. \tag{11}$$

Next, assuming that the process is normal, we obtain

$$\Theta_r(u_1, \ldots, u_r, t_1, \ldots, t_r)$$

$$= \exp\left\{ j \sum_{\alpha=1}^{r} k_1(t_\alpha) u_\alpha - \frac{1}{2} \sum_{\alpha=1}^{r} \sum_{\beta=1}^{r} k_2(t_\alpha, t_\beta) u_\alpha u_\beta \right\}. \quad (12)$$

Performing Fourier's inverse transformation on (12), we arrive at

$$g_r(x^1, x^2, \ldots, x^r, t_1, \ldots, t_r) = \frac{1}{\sqrt{(2\pi)^r \det[k_2(t_\alpha, t_\beta)]}} \times$$

$$\times \exp\left[-\frac{1}{2} \sum_{\alpha=1}^{r} \sum_{\beta=1}^{r} k_2^{\alpha\beta} (x^\alpha - k_1(t_\alpha))(x^\beta - k_1(t_\beta)) \right], \quad (13)$$

where the elements $k_2^{\alpha\beta}$ form the inverse matrix to $k_2(t_\alpha, t_\beta)$. The distribution obtained is a normal distribution.

Correlation functions of higher orders contribute less information about the process and for this reason we often in practical problems confine ourselves to correlation functions of the second order, in spite of the fact that the processes involved lack the character of normal processes.

A *process of the second order* is one for which the following occurs:

$$\bigwedge_{t \in R} k_2(t, t) < \infty. \quad (14)$$

Let a system of random processes $\xi^1(t)$, $\xi^2(t)$, ..., $\xi^N(t)$ be given. Means for this system form a vector with components $\langle \xi^i(t) \rangle$ whereas the correlation functions of the second order form a matrix $N \times N$ with the terms

$$K[\xi^i(t_1), \xi^j(t_2)] = \langle \xi^i(t_1) \xi^j(t_2) \rangle - \langle \xi^i(t_1) \rangle \langle \xi^j(t_2) \rangle. \quad (15)$$

We shall also use other notations to express the elements of the correlation matrix:

$$K_\xi^{ij}(t_1, t_2) \equiv K[\xi^i(t_1), \xi^j(t_2)], \quad (16)$$

for vectors with superscripts and

$$K_{\xi ij}(t_1, t_2) = K[\xi_i(t_1), \xi_j(t_2)], \quad (16a)$$

for vectors with subscripts. In the case where the processes $\xi^i(t)$ are statistically independent the correlation matrix is diagonal.

The *stationary process*, in a broad sense, is one in which the correlation function depends only on time difference, i.e., the following occurs:

$$K_\xi^{ij}(t_1, t_2) = K_\xi^{ij}(t_1 - t_2) = K_\xi^{ij}(\tau), \quad (17)$$

the mean value and the mean quadratic deviation being constant.

Note that the functions $K_\xi^{ij}(\tau)$ are even, i.e.,

$$K_\xi^{ij}(\tau) = K_\xi^{ij}(-\tau). \tag{18}$$

However, probability distributions which enable correlation functions to be found are not always known. As a result, it is often assumed that the process is ergodic.

A process is *ergodic* when the means calculated over the space of elementary events and the means reckoned over the time are equal. We write therefore

$$\langle \xi^i(t) \rangle = \lim_{T\to\infty} \frac{1}{2T} \int_{-T}^{T} \xi^i(t)\,\mathrm{d}t \tag{19}$$

and

$$K_\xi^{ij}(\tau) = \lim_{T\to\infty} \frac{1}{2T} \int_{-T}^{T} \xi^i(t+\tau)\xi^j(t)\,\mathrm{d}t - \\ -\left(\lim_{T\to\infty} \frac{1}{2T} \int_{-T-\tau}^{T-\tau} \xi^i(t+\tau)\,\mathrm{d}t\right)\left(\lim_{T\to\infty} \frac{1}{2T} \int_{-T}^{T} \xi^j(t)\,\mathrm{d}t\right). \tag{20}$$

Observation time equal to infinity is obviously unavailable to us, therefore in practical calculations, we confine ourselves to finite time T, making use of an approximating equation:

$$K_\xi^{ij}(\tau) \approx \frac{1}{2T} \int_{-T}^{T} \xi^i(t+\tau)\xi^j(t)\,\mathrm{d}t - \\ -\left(\frac{1}{2T} \int_{-T-\tau}^{T-\tau} \xi^i(t+\tau)\,\mathrm{d}t\right)\left(\frac{1}{2T} \int_{-T}^{T} \xi^j(t)\,\mathrm{d}t\right). \tag{21}$$

To find the correlation function from the graph of a random function we use an approximating equation resulting from the exchange of an integral for a finite sum.

By assuming that the magnitude $\xi(t)$ is a random stationary ergodic function, the correlation function can be calculated using a single realization of sufficiently long duration of the process $x(t)$. For this purpose, the following formula [61] may be chosen:

$$K_\xi\left(\frac{mT}{n}\right) = \frac{1}{n-m} \sum_{s=1}^{n-m} \left(x(t_{s+m}) - \langle \xi(t) \rangle\right)\left(x(t_s) - \langle \xi(t) \rangle\right), \tag{22}$$

where T is a observed duration of a realization, n number of intervals, m discrete time assuming the values 0, 1, ..., $n/4$, and

$$\langle \xi(t) \rangle = \frac{1}{n} \sum_{s=1}^{n} x(t_s). \tag{23}$$

The correlation equations in the case of two different realizations $x^i(t)$ and $x^j(t)$ are derived in like manner:

$$K_{\xi}^{ij}\left(\frac{mT}{n}\right) = \frac{1}{n-m} \sum_{s=1}^{n-m} (x^i(t_{s+m}) - \langle \xi^i(t) \rangle)(x^j(t_s) - \langle \xi^j(t) \rangle). \tag{24}$$

When use is made of approximating equations, it has to be remembered that the observational time T must be sufficiently long, so that any further increase of it should not essentially affect the result.

The determination of correlation functions from experimental data by use of equations (22) and (24) is highly labour consuming; the development of a device known as the *correlator* has substantially facilitated its calculation. Various types of this instrument, chiefly electronic [68], are now manufactured. As an example of devices of this type we shall take the mechanical correlator presented diagrammatically in Fig. 10.4.

The motor *3*, drives two components via transmission gear, *2*: roller *19* and disc *1*, respectively effecting the travel of the tape and the recording

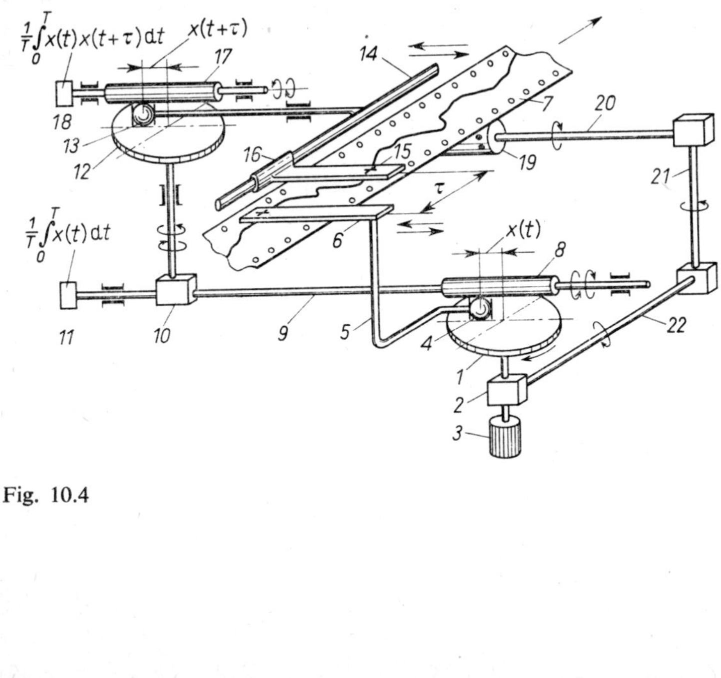

Fig. 10.4

of the realization of the process. Steel ball *4*, which has its position linked with the graph ordinate $x_1(t')$, sets shaft *8* (diameter $2r_1$) in angular motion with angular velocity

$$\omega_1 = \frac{\omega}{r_1} x_1(t'), \tag{25}$$

where ω is the angular velocity of disc *1*. For the present we shall assume $\omega = \text{const}$, consequently the angle of rotation φ_1 of shaft *8* is given by

$$\varphi_1 = \frac{\omega}{r_1} \int_0^{T'} x_1(t')\mathrm{d}t', \tag{26}$$

where t' is the time taken to reproduce the realization of the process, and T' total travelling time of tape. The real time t is $t = \mu_t t'$, where μ_t is the coefficient of time scale. Similarly, the ordinates of the realization $x_1(t)$ in the scale μ_x give the corresponding values of the process $x(t)$, i.e.,

$$x(t) = \mu_x x_1(t).$$

In the light of the above remarks and due to (26), the mean value of the process under examination is expressed by

$$\langle \xi(t) \rangle = \frac{\mu_x \mu_t}{\mu_t T'} \int_0^{T'} x_1(t')\mathrm{d}t' = \frac{\varphi_1 r_1}{\omega T'} \mu_x. \tag{27}$$

Evidently, the measurement of the angle of rotation φ_1 of shaft *8* will suffice in finding the expected value of the process; counter *11* serves this purpose. Note further that the angular motion of shaft *8* is transmitted through gear *10* to disc *12* which via ball *13* drives shaft *17* (diameter $2r_2$) with angular velocity

$$\omega_2 = \frac{\omega_1}{r_2} x_1(t'+\tau') = \frac{\omega}{r_1 r_2} x_1(t'+\tau') x_1(t'). \tag{28}$$

The angle of rotation φ_2 of shaft *17* is expressed by

$$\varphi_2 = \frac{\omega}{r_1 r_2} \int_0^{T'} x_1(t'+\tau') x_1(t')\mathrm{d}t'. \tag{29}$$

Consequently,

$$\langle \xi(t+\tau)\xi(t) \rangle \approx \mu_x^2 \frac{\varphi_2 r_1 r_2}{\omega T'}. \tag{30}$$

Counter *18* gives values of the mean as specified by equation (30). The correlation function is determined making use of (27) and (30),

$$K[\xi(t+\tau), \xi(t)] = \mu_x^2 \frac{r_1}{\omega T'}\left(r_2\varphi_2 - \frac{r_1\varphi_1^2}{\omega T'}\right). \tag{31}$$

Although we have carried out our reasoning with ω = const, this assumption can be dispensed with considering that the system has one degree of freedom.

The final position of the system defined by, say, the angle of rotation of disc *1* ($\omega T'$), defines angles φ_1 and φ_2, thus the correlation function given by equation (31) is independent of the way in which the system (the correlation) is brought to the end state. The described mechanical correlator is simple in construction but is less efficient than correlators of other types, e.g. photoelectric, magnetic or electronic, suited to quick determination of correlation functions. A reader interested in these problems is referred to the book of V. V. Solodovnikov [68].

10.3 Stochastic linearization

Correlation methods as applied to the study of the dynamics of random processes in linear systems consist in finding relationships between means and correlation functions of the second order for input and output signals. The application of these methods to non-linear systems requires prior linearization of the initial equations. This linearization can be carried out either in the deterministic stage, as has been shown in the foregoing chapters, or in the stochastic stage [60].

Stochastic linearization enables, like the least squares method, linearization of non-analytical functions. To linearize a non-linear function $\varphi_i(q^1, \ldots, q^r)$ with random terms q^i, we find its equivalent linear approximation with respect to the expected value and variance. Thus

$$\varphi_i(q^1, \ldots, q^r) \approx \varphi_{0i} + c_{ik} q^{0k}, \tag{1}$$

where

$$\varphi_{0i} = \langle \varphi_i(q^1, \ldots, q^r) \rangle \tag{2}$$

is the expected value of the function $\varphi_i(q^1, \ldots, q^r)$, q^{0k}—deviation of the kth ordinate from the mean.

Stochastic linearization can be carried out assuming the mean values and variances of the non-linear function and its linear approximation to be the same. On this basis it is possible to find φ_{0i} and the constants c_{ik}.

Another linearization technique consists in the choice of φ_{0i} and of the elements of the matrix c_{ik} such that the mean value of the quadratic deviation assumes the minimum.

If

$$x_i = \langle [\varphi_i(q^1, \ldots, q^r) - \varphi_{0i} - c_{ik} q^{0k}]^2 \rangle,$$

then the conditions

$$\frac{\partial x_i}{\partial c_{mn}} = 0, \tag{3}$$

$$\frac{\partial x_i}{\partial \varphi_{0i}} = 0 \tag{4}$$

will allow the determination of the unknowns.

From equations (4) we obtain

$$\varphi_{0i} = \langle \varphi_i(q^1, \ldots, q)^r \rangle. \tag{5}$$

From equations (3) it follows that

$$\langle \varphi_i q^{0n} \rangle = c_{ik} \langle q^{0k} q^{0n} \rangle. \tag{6}$$

As we can see, the present result corresponds to the result obtained when linearizing by the least squares method.

Assuming

$$\langle \varphi_i q^{0n} \rangle = \overline{D}_i^n \quad \text{and} \quad \langle q^{0k} q^{0n} \rangle = \overline{R}_k{}^n,$$

we rewrite equations (6):

$$\overline{D}_i^n = c_{ik} \overline{R}^{kn};$$

hence

$$c_{ik} = \overline{D}_i^n \overline{R}_{kn}, \tag{7}$$

where $\overline{R}_{kn}$ is the inverse matrix to $\overline{R}^{kn}$. Assuming the multidimensional distribution to be normal and using the ergodicity hypothesis, the linearization coefficients are easily calculated. In the present case

$$g(q^1, q^2, \ldots, q^N) = \frac{1}{\sqrt{2^N \pi^N \Delta}} \exp\left(\frac{\Delta^*}{2\Delta}\right), \tag{8}$$

where

$$\Delta = \det[\overline{R}^{kn}],$$

$$\Delta^* = \begin{vmatrix} R^{11} & \ldots & R^{1N} & q^1 - \langle q^1 \rangle \\ \multicolumn{4}{c}{\dotfill} \\ R^{N1} & \ldots & R^{NN} & q^N - \langle q^N \rangle \\ q^1 - \langle q^1 \rangle & \ldots & q^N - \langle q^N \rangle & 0 \end{vmatrix}.$$

Substituting (8) into (6), differentiating with respect to $\langle q^k \rangle$, and comparing the result from (7) we find that

$$c_{ik} = \frac{\partial \varphi_{0i}}{\partial \langle q^k \rangle}. \tag{9}$$

In presenting the linearization method, attention should be called to a certain inconvenience. It consists in that in introducing the ergodicity hypothesis (by which averaging against time can be replaced by averaging over a probabilistic space) and assuming, let us say, a normal distribution, we introduce unknown correlations of the first and second order. The determination of these magnitude has to be deferred until the differential equation thus linearized has first been solved. This greatly complicates the calculation problem.

10.4 Random vibrations of linear systems with one degree of freedom

The motion of harmonic oscillators was examined in Chapter 8. Making use of the methods of Fourier's transformation, we have arrived at the result:

$$q(t) = \int_{-\infty}^{t} G(t-\tau) a_0(\tau) \mathrm{d}\tau. \tag{1}$$

This result holds for a wider class of linear systems than harmonic oscillators.

Assume that $a_0(t)$ is a random process. Let us take its ith realization, $\underset{(i)}{a_0}(t)$. This realization is a determinate process giving, according to equation (1), a determinate response of the system.

Let $\underset{(i)}{q}(t)$ be the response (output signal) to the signal $\underset{(i)}{a_0}(t)$. We write

$$\underset{(i)}{q}(t) = \int_{-\infty}^{t} \underset{(i)}{G}(t-\tau) \underset{(i)}{a_0}(\tau) \mathrm{d}\tau, \tag{2}$$

where $\underset{(i)}{G}(t)$ denotes the impulse function during the ith realization. All the realizations (2) constitute a set (process). With these given we can concentrate on the technique of finding the means and the correlations. We shall make use of equation (23), Section 10.1, i.e.,

$$\langle q(t) \rangle = \lim_{n\to\infty} \frac{1}{n} \sum_{i=1}^{n} \underset{(i)}{q}(t) = \lim_{n\to\infty} \frac{1}{n} \sum_{i=1}^{n} \int_{-\infty}^{t} \underset{(i)}{G}(t-\tau) \underset{(i)}{a_0}(\tau) \mathrm{d}\tau$$

$$= \int_{-\infty}^{t} \lim_{n\to\infty} \frac{1}{n} \sum_{i=1}^{n} \underset{(i)}{G}(t-\tau) \underset{(i)}{a_0}(\tau) \mathrm{d}\tau. \tag{3}$$

If we assume that the structural parameters of a dynamical system are determinate, i.e.,

$$\bigwedge_i \underset{(i)}{G}(t) = G(t) = \text{idem}, \tag{4}$$

then

$$\lim_{n\to\infty} \frac{1}{n} \sum_{i=1}^{n} \underset{(i)}{G}(t-\tau)\underset{(i)}{a_0}(\tau) = G(t-\tau) \lim_{n\to\infty} \frac{1}{n} \sum_{i=1}^{n} \underset{(i)}{a_0}(\tau)$$
$$= G(t-\tau)\langle a_0(\tau)\rangle. \tag{5}$$

Substituting (5) into (3) we obtain the relationship between the means of the input signal $a_0(t)$ and the output signal $q(t)$, with determinate structural parameters of the system:

$$\langle q(t)\rangle = \int_{-\infty}^{t} G(t-\tau)\langle a_0(\tau)\rangle \mathrm{d}\tau. \tag{6}$$

We now calculate the correlation function of the process $q(t)$. By the definition of the correlation function, we have

$$K_q(t_1, t_2) = \langle q(t_1)q(t_2)\rangle - \langle q(t_1)\rangle\langle q(t_2)\rangle$$
$$= \langle [q(t_1)-\langle q(t_1)\rangle][q(t_2)-\langle q(t_2)\rangle]\rangle. \tag{7}$$

Denoting the deviation of the mean input signal $a_0(t)$ by $h(t)$, we write

$$K_q(t_1, t_2) = \left\langle \left[\int_{-\infty}^{t_1} G(t_1-\tau_1)h(\tau_1)\mathrm{d}\tau_1 \right] \left[\int_{-\infty}^{t_2} G(t_2-\tau_2)h(\tau_2)\mathrm{d}\tau_2 \right] \right\rangle$$
$$= \left\langle \int_{-\infty}^{t_1} \int_{-\infty}^{t_2} G(t_1-\tau_1)G(t_2-\tau_2)h(\tau_1)h(\tau_2)\mathrm{d}\tau_1\,\mathrm{d}\tau_2 \right\rangle. \tag{7a}$$

If $\underset{(i)}{h}(t)$ is the ith realization of the process

$$h(t) = a_0(t) - \langle a_0(t)\rangle,$$

then, using equation (23), Section 10.1, we obtain

$$K_q(t_1, t_2) = \lim_{n\to\infty} \frac{1}{n} \sum_{i=1}^{n} \int_{-\infty}^{t_1} \int_{-\infty}^{t_2} \underset{(i)}{G}(t_1-\tau_1)\underset{(i)}{G}(t_2-\tau_2)\underset{(i)}{h}(\tau_1)\underset{(i)}{h}(\tau_2)\mathrm{d}\tau_1\,\mathrm{d}\tau_2$$
$$= \int_{-\infty}^{t_1} \int_{-\infty}^{t_2} \left(\lim_{n\to\infty} \frac{1}{n} \sum_{i=1}^{n} \underset{(i)}{G}(t_1-\tau_1)\underset{(i)}{G}(t_2-\tau_2)\underset{(i)}{h}(\tau_1)\underset{(i)}{h}(\tau_2) \right) \mathrm{d}\tau_1\,\mathrm{d}\tau_2. \tag{8}$$

By virtue of (4) we have

$$\lim_{n\to\infty}\frac{1}{n}\sum_{i=1}^{n} \underset{(i)}{G(t_1-\tau_1)}\underset{(i)}{G(t_2-\tau_2)}\underset{(i)}{h(\tau_1)}\underset{(i)}{h(\tau_2)}$$

$$= G(t_1-\tau_1)G(t_2-\tau_2)\lim_{n\to\infty}\frac{1}{n}\sum_{i=1}^{n}\underset{(i)}{h(\tau_1)}\underset{(i)}{h(\tau_2)}$$

$$= G(t_1-\tau_1)G(t_2-\tau_2)\langle h(\tau_1)h(\tau_2)\rangle. \tag{9}$$

Note that $\langle h(\tau_1)h(\tau_2)\rangle$ is the correlation function $K_h(t_1, t_2)$ of the process $h(t)$. Substituting (9) into (8) we arrive at

$$K_q(t_1, t_2) = \int_{-\infty}^{t_1}\int_{-\infty}^{t_2} G(t_1-\tau_1)G(t_2-\tau_2)K_h(\tau_1, \tau_2)\mathrm{d}\tau_1\mathrm{d}\tau_2. \tag{10}$$

Equation (10) gives the relationship between $K_q(t_1, t_2)$ and $K_h(t_1, t_2)$. This is delineated in Fig. 10.5. Equation (10), backed by relationship (31), Section 10.1, associating moments with correlations can be generalized:

$$K[q(t_1), \ldots, q(t_s)] = G(t_1, \ldots, t_s)*K[h(t_1),\ldots, h(t_s)] \tag{10a}$$

where

$$G(t_1, \ldots, t_s) = \prod_{k=1}^{s} G(t_k)$$

and $*$ is the convolution symbol (in this case a multidimensional convolution).

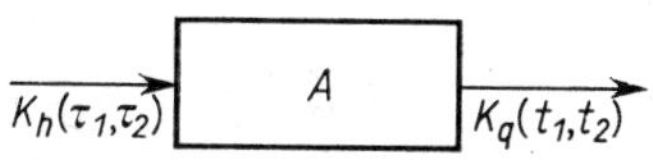

Fig. 10.5

The calculation of the integral in (10) is, in the general case, a difficult problem. The situation is considerably simplified on assuming that the considered process $h(t)$ is stationary. Then

$$K_h(t_1, \tau_2) = K_h(\tau_2-\tau_1).$$

Substituting

$$\tau_1-\tau_2 = \tau, \quad t_1-\tau_1 = \tau_1', \quad t_2-\tau_2 = \tau_2',$$

we have

$$\mathrm{d}\tau_1 = -\mathrm{d}\tau_1', \quad \mathrm{d}\tau_2 = -\mathrm{d}\tau_2',$$
$$\tau_2-\tau_1 = t_2-t_1-\tau_2'+\tau_1' = \tau'-\tau_2'+\tau_1'.$$

Consequently, when $a_0(t)$ acts infinitely long on a system, the process $q(t)$ is also stationary since

$$K_q(t_1, t_2) = K_q(t_2-t_1) = K_q(\tau')$$
$$= \int_0^\infty \int_0^\infty G(\tau_1')G(\tau_2')K_h(\tau'-\tau_2'+\tau_1')\mathrm{d}\tau_1'\mathrm{d}\tau_2'. \tag{11}$$

Describing stationary processes, the concept of spectral density is often introduced. *Spectral density* is defined as Fourier's transformation of the correlation function:

$$S_q(\omega) \stackrel{\mathrm{df}}{=} \int_{-\infty}^{\infty} K_q(\tau)\mathrm{e}^{-\mathrm{j}\omega\tau}\mathrm{d}\tau. \tag{12}$$

Knowing the spectral density, it is possible in turn to find the correlation function through Fourier's inverse transformation:

$$K_q(\tau) = \frac{1}{2\pi}\int_{-\infty}^{\infty} S_q(\omega)\mathrm{e}^{\mathrm{j}\omega\tau}\mathrm{d}\omega. \tag{13}$$

Substituting (11) into (12) we obtain

$$S_q(\omega) = \iiint_{-\infty}^{\infty} G(\tau_1')G(\tau_2')\mathrm{e}^{-\mathrm{j}\omega\tau'}K_h(\tau'-\tau_2'+\tau_1')\mathrm{d}\tau'\mathrm{d}\tau_1'\mathrm{d}\tau_2'$$
$$= \int_{-\infty}^{\infty} G(\tau_1')\mathrm{e}^{\mathrm{j}\omega\tau_1'}\mathrm{d}\tau_1 \int_{-\infty}^{\infty} G(\tau_2')\mathrm{e}^{-\mathrm{j}\omega\tau_2'}\mathrm{d}\tau_2' \int_{-\infty}^{\infty} K_h(\tau'-\tau_2'+\tau_1') \times$$
$$\times\, \mathrm{e}^{-\mathrm{j}\omega(\tau'-\tau_2'+\tau_1')}\mathrm{d}\tau' = \bar{G}(-\mathrm{j}\omega)\bar{G}(\mathrm{j}\omega)S_h(\omega)\ *$$

or

$$S_q(\omega) = |\bar{G}(\mathrm{j}\omega)|^2 S_h(\omega). \tag{14}$$

$S_h(\omega)$ is obviously the spectral density of input $h(t)$. Equation (14) gives the law of transmittance of the spectral density of a stationary signal through a linear system (Fig. 10.6).

* The limits of integration from $-\infty$ to ∞ are justified by the fact that $G(\tau) = 0$ for negative values of argument τ.

The derived equations apply to all linear systems. For the harmonic oscillator with damping n it will be

$$\bar{G}(j\omega) = \frac{1}{k^2-\omega^2+3nj\omega}.$$

But

$$|\bar{G}(j\omega)|^2 = \frac{1}{(k^2-\omega^2)^2+4n^2\omega^2},$$

or

$$S_q(\omega) = \frac{S_h(\omega)}{(k^2-\omega^2)^2+4n^2\omega^2}. \tag{15}$$

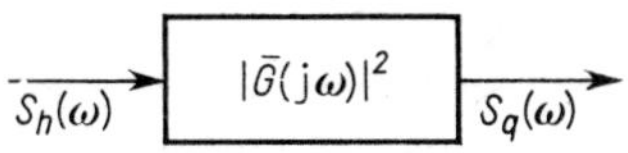

Fig. 10.6

Making use of equation (13) we calculate the correlation function as

$$K_q(\tau) = \frac{1}{2\pi}\int_{-\infty}^{\infty} \frac{S_h(\omega)e^{j\omega\tau}}{(k^2-\omega^2)^2+4n^2\omega^2}\,d\omega. \tag{16}$$

In considering stationary problems it is often assumed, bearing in mind the facility of analysis, that the input has the character of the so-called *white noise* (by analogy with white light in optics). The correlation function for random system thus defined is given by the Dirac distribution:

$$K_h(\tau) = C\delta(\tau),$$

where C = const.

The spectral density for white noise is of constant magnitude:

$$S_h(\omega) = \int_{-\infty}^{\infty} C\delta(\tau)e^{-j\omega\tau}d\tau = C.$$

Now

$$\int_{-\infty}^{\infty} S_h(\omega)d\omega = \langle[h(t)]^2\rangle \tag{17}$$

is a magnitude proportional to the power of the process $h(t)$, and hence $\int_{-\infty}^{\infty} S(\omega)d\omega$ for white noise is infinitely large and for this reason certain limitations should in this case be put on $S(\omega)$. This is achieved by, say, "clipping" the spectral density while assuming $S_h(\omega) = CH(\omega_c - |\omega|)$ where $H(\omega_c - |(\omega)|)$ is Heaviside's function, and ω_c "cut-off" frequency (Fig. 10.7). In this connection,

$$K_q(\tau) = \frac{C}{2\pi} \int_{-\omega_c}^{\omega_c} \frac{e^{j\omega\tau}}{(k^2-\omega^2)^2+4n^2\omega^2} d\omega. \tag{18}$$

ω_c is determinable if the right-hand side of (17) is known. In the general case we calculate the mean quadratic deviation of process $q(t)$ from (16) with $\tau = 0$:

$$\langle q(t)\rangle^2 = K_q(0) = \frac{1}{2\pi} \int_{-\infty}^{\infty} \frac{S_h(\omega)d\omega}{(k^2-\omega^2)^2+4n^2\omega^2}. \tag{19}$$

Note that, by virtue of $K_q(\tau)$ being odd, it is possible to confine oneself to cosine transformations.

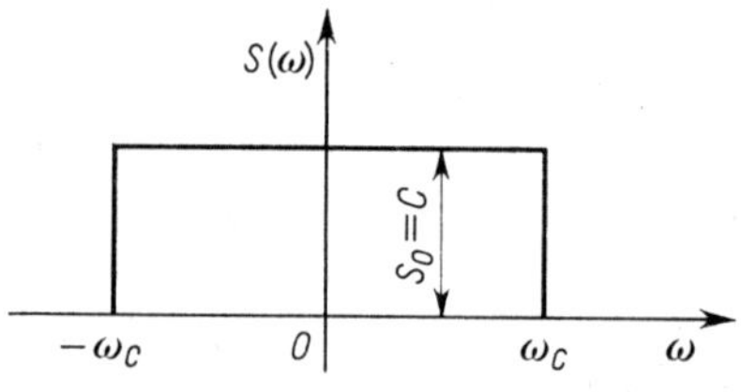

Fig. 10.7

A wide class of stationary processes can be characterized by correlation functions of the following form:

$$K_h = Ce^{-\alpha|\tau|}\cos\beta\tau \tag{20}$$

or by the sum of terms of this type (C, α, and β constant). For the correlation function $K_h(\tau)$ thus defined, we calculate the spectral density:

$$S_h(\omega) = 2\int_0^{\infty} K_h(\tau)\cos\omega\tau d\tau = 2C\int_0^{\infty} e^{-\alpha|\tau|}\cos\beta\tau\cos\omega\tau d\tau.$$

After integrating we obtain

$$S_h(\omega) = C\left[\frac{1}{\alpha^2+(\beta+\omega)^2} + \frac{1}{\alpha^2+(\beta-\omega)^2}\right]. \tag{21}$$

Using equation (14) we find the spectral density of the input signal. For harmonic oscillators it will be

$$S_h(\omega) = \frac{\alpha C}{(k^2-\omega^2)^2+4n^2\omega^2}\left[\frac{1}{\alpha^2+(\beta+\omega)^2}+\frac{1}{\alpha^2+(\beta-\omega)^2}\right]. \tag{22}$$

Finally, the mean quadratic deviation will be

$$\langle [q(t)]^2 \rangle = \frac{1}{\pi}\int_0^\infty S_q(\omega)\,d\omega = \frac{\alpha C}{\pi}\int_0^\infty \frac{1}{(k^2-\omega^2)^2+4n^2\omega^2}\times$$

$$\times\left[\frac{1}{\alpha^2+(\beta+\omega)^2}+\frac{1}{\alpha^2+(\beta-\omega)^2}\right]d\omega. \tag{23}$$

Table 3 specifies spectral densities for different correlation functions $K(\tau)$.

TABLE 3

$K(\tau)$	$S(\omega)$
$C\delta(\tau)$	C constant
$\sum_{i=1}^{n} C_i\delta^i(\tau)$	$\sum_{i=1}^{n} C_i(j\omega)^i$
$Ce^{-\alpha\|\tau\|}$	$\frac{2\alpha C}{\alpha^2+\omega^2}$
$Ce^{-\alpha\|\tau\|}\cos\beta\tau$	$\alpha C\left[\frac{1}{\alpha^2+(\beta+\omega)^2}+\frac{1}{\alpha^2+(\beta-\omega)^2}\right]$
$Ce^{-\alpha\|\tau\|}\left(\cos\beta\tau-\frac{\alpha}{\beta}\sin\beta\|\tau\|\right)$	$\frac{C\omega^2}{(\omega^2-\alpha^2-\beta^2)+4\alpha^2\omega^2}$
$Ce^{-\alpha\|\tau\|}\left(\cos\beta\tau+\frac{\alpha}{\beta}\sin\beta\|\tau\|\right)$	$\frac{4C(\alpha^2+\omega^2)\alpha}{(\omega^2-\alpha^2-\beta^2)^2+4\alpha^2\omega^2}$
$Ce^{-(\alpha\tau)^2}\cos\beta\tau$	$\frac{C\sqrt{\pi}}{2\alpha^2}\left\{\exp\left[-\frac{(\beta+\omega)^2}{4\alpha^2}\right]+\exp\left[-\frac{(\omega-\beta)^2}{4\alpha^2}\right]\right\}$
$K(\tau)=\begin{cases}\sum_{i=0}^{n} C_i\cos\frac{i\pi}{T}\tau & \text{for } \|\tau\|\leqslant T\\ 0 & \text{for } \|\tau\|>T\end{cases}$	$2T\pi\sum_{i=0}^{n}(-1)^i C_i\frac{\omega T\sin\omega T}{(\omega T)^2-(i\pi)^2}$
$K(\tau)=\begin{cases}C\left(1-\frac{\|\tau\|}{T}\right) & \text{for } \|\tau\|\leqslant T\\ 0 & \text{for } \|\tau\|>T\end{cases}$	$CT\left(\frac{\sin\frac{1}{2}\omega T}{\frac{1}{2}\omega T}\right)^2$

In the general case, the concept of *spectral density* $S(\omega_1, \ldots, \omega_s)$ can also be introduced as Fourier's multiple transform performed on the correlation function:

$$S_{(q)}(\omega_1, \ldots, \omega_s) \stackrel{\text{df}}{=} \mathscr{F}\{K[q(t_1), \ldots, q(t_s)]\}(\omega_1, \ldots, \omega_s). \tag{24}$$

To find the correlation function of the sth order from the known spectral density it is necessary to perform Fourier's inverse transform

$$\begin{aligned} K_{(q)}(t_1, \ldots, t_s) &= K[q(t_1), \ldots, q(t_s)] \\ &= \mathscr{F}^{-1}\{S_{(q)}(\omega_1, \ldots, \omega_s)\}(t_1, \ldots, t_s); \end{aligned} \tag{25}$$

specifically for $s = 2$ we obtain

$$S_{(q)}(\omega_1, \omega_2) = \int_{-\infty}^{\infty} \int_{-\infty}^{\infty} K_{(q)}(t_1, t_2) e^{-j(\omega_1 t_1 + \omega_2 t_2)} dt_1 dt_2 \tag{26}$$

and

$$K_{(q)}(t_1, t_2) = \frac{1}{4\pi^2} \int_{-\infty}^{\infty} \int_{-\infty}^{\infty} S_{(q)}(\omega_1, \omega_2) e^{j(\omega_1 t_1 + \omega_2 t_2)} d\omega_1 d\omega_2. \tag{27}$$

Definition (24) will be found more convenient than the conventional one here. Performing Fourier's multiple transformation on (10a) we obtain the general relationship between spectral density of the sth order for input and for output

$$S_{(q)}(\omega_1, \ldots, \omega_s) = \bar{G}(\omega_1, \ldots, \omega_s) S_{(h)}(\omega_1, \ldots, \omega_s)$$

where

$$\bar{G}(\omega_1, \ldots, \omega_s) = \prod_{k=1}^{s} \bar{G}(j\omega_k).$$

We shall now deal with the determination of the relationship between the density $S(\omega_1, \omega_2)$ of the input signal $h(t)$ and of the output signal $q(t)$. Considering (10) in expression (26) we obtain

$$\begin{aligned} S_{(q)}(\omega_1, \omega_2) = \int_{-\infty}^{\infty} \int_{-\infty}^{\infty} \int_{-\infty}^{t_1} \int_{-\infty}^{t_2} & G(t_1 - \tau_1) G(t_2 - \tau_2) \times \\ & \times K_h(\tau_1, \tau_2) e^{-j(\omega_1 t_1 + \omega_2 t_2)} dt_1 dt_2 d\tau_1 d\tau_2. \end{aligned}$$

Substituting

$$t_1 - \tau_1 = \xi_1, \quad t_2 - \tau_2 = \xi_2, \quad -d\tau_1 = d\xi_1, \quad -d\tau_2 = d\xi_2,$$

we arrive at

$$S_q(\omega_1, \omega_2) = \iiiint_{-\infty}^{\infty} G(\xi_1)e^{-j\omega_1\xi_1}G(\xi_2)e^{-j\omega_2\xi_2} \times$$
$$\times K_h(t_1-\xi_1, t_2-\xi_2)e^{-j[\omega_1(t_1-\xi_1)+\omega_2(t_2-\xi_2)]}d\xi_1\, d\xi_2\, dt_1\, dt_2$$
$$= \bar{G}(j\omega_1)\bar{G}(j\omega_2)S_h(\omega_1, \omega_2).$$

Finally

$$S_q(\omega_1, \omega_2) = \bar{G}(j\omega_1)\bar{G}(j\omega_2)S_h(\omega_1, \omega_2). \tag{28}$$

The equation obtained generalizes equation (14) for the stationary problem.

For the calculation of the correlation function for input process $K_q(t_1, t_2)$, it is sufficient to make use of relationship (26). Knowing this function, the mean quadratic deviation is easily calculated from the mean. For this purpose it will suffice to put $t_1 = t_2 = t$.

Displacements of a system are accompanied by stresses which for linear systems are linear functions of displacements. We determine random stresses using the presented method of testing random displacements. For correctly designed construction, there is a certain predetermined probability with which these stresses will not overstep admissible values. This probability can easily be determined when assuming that the process is normal, for we know that in this case the cumulative distribution function is

$$F_\xi(x) = P(\xi < x) = \tfrac{1}{2}[1+\Phi(x)]. \tag{29}$$

Here x denotes the ratio of the deviation from the mean to the value of the standard deviation of the stresses.

Equation (29) gives the probability that the deviation of the stresses from the mean value is likely to be below $x\sigma$ where σ is the standard deviation. If $\xi(t)$ is not a process with a normal distribution, then use can be made of Chebyshev's inequality

$$P\{|\xi-\langle\xi\rangle| \geqslant k\sigma\} < \frac{1}{k^2}. \tag{30}$$

Problems of departures will be considered in greater detail in Section 10.6.

10.5 Random vibrations of systems with many degrees of freedom

Performing Fourier's transformation on differential equations (6), Section 9.2, describing the vibrations of systems with many degrees of freedom we obtain

$$-a_{lk}\omega^2\bar{q}^k(j\omega)+h_{lk}\,j\omega\bar{q}^k(j\omega)+c_{lk}\bar{q}^k(j\omega) = \bar{Q}_l(j\omega). \tag{1}$$

Using $\overline{G}^{kl}(j\omega)$ to denote the inverse matrix to

$$[-a_{lk}\omega^2 + h_{lk}j\omega + c_{lk}],$$

we write

$$\overline{q}^k(j\omega) = \overline{G}^{kl}(j\omega)\overline{Q}_l(j\omega), \tag{2}$$

where $\overline{G}^{kl}(j\omega)$ is the characteristic of frequency for a system with many degrees of freedom. Note that this characteristic can be derived knowing the operator function of transmittance for a system with N inputs and N outputs. It is enough for this purpose to put formally $s = j\omega$:

$$\overline{G}^{kl}(j\omega) = |\overline{G}^{kl}(j\omega)|\, e^{\varepsilon_{kl}(\omega)}. \; * \tag{3}$$

As in the case of a system with one degree of freedom, $|\overline{G}^{kl}(j\omega)|$ is called the *amplitude characteristic* and $\varepsilon_{kl}(\omega)$—the *phase characteristic.*

The solution is Fourier's inverse transformation performed on equation (2). For this purpose we shall use the theorem of convolutions for Fourier's transform; we write

$$q^k(t) = \int_{-\infty}^{t} G^{kl}(t-\tau)Q_l(\tau)\,d\tau \tag{4}$$

or in matrix notation

$$\boldsymbol{q}(t) = \int_{-\infty}^{t} \boldsymbol{G}(t-\tau)\boldsymbol{Q}(\tau)\,d\tau. \tag{4a}$$

Let the forces $Q_l(t)$ be stochasitc processes of the second order. For the ith realization we have

$$\underset{(i)}{q^k}(t) = \int_{-\infty}^{t} \underset{(i)}{G^{kl}}(t-\tau)\underset{(i)}{Q_l}(\tau)\,d\tau. \tag{5}$$

Summing over i and averaging, we obtain

$$\langle q^k(t)\rangle = \lim_{n\to\infty}\frac{1}{n}\sum_{i=1}^{n}\underset{(i)}{q^k}(t)$$

$$= \lim_{n\to\infty}\frac{1}{n}\sum_{i=1}^{n}\int_{-\infty}^{t}\underset{(i)}{G^{kl}}(t-\tau)\underset{(i)}{Q_l}(\tau)\,d\tau$$

$$= \int_{-\infty}^{t}\lim_{n\to\infty}\frac{1}{n}\sum_{i=1}^{n}\underset{(i)}{G^{kl}}(t-\tau)\underset{(i)}{Q_l}(\tau)\,d\tau. \tag{6}$$

* Summation should not be performed over k and l.

If the system has determinate structural parameters, then the matrix $G^{kl}(t)$ is determinate $(\bigwedge_{i} \underset{(i)}{G}(t) = G(t))$, whence

$$\lim_{n\to\infty} \frac{1}{n} \sum_{i=1}^{n} \underset{(i)}{G^{kl}}(t-\tau)\underset{(i)}{Q_l}(\tau) = G^{kl}(t-\tau) \lim_{n\to\infty} \frac{1}{n} \sum_{i=1}^{n} \underset{(i)}{Q_l}(\tau)$$

$$= G^{kl}(t-\tau)\langle Q_l(\tau)\rangle. \tag{7}$$

Thus the relationship between the mean input-signal vector $\langle Q(t)\rangle$ and the mean output-signal vector is given by

$$\langle q^k(t)\rangle = \int_{-\infty}^{t} G^{kl}(t-\tau)\langle Q_l(\tau)\rangle \mathrm{d}\tau. \tag{8}$$

We shall now find the relationship between the correlation functions of the input and output-signal vectors. We form a valid expression for the ith realization:

$$\underset{(i)}{q^k}(t_1)\underset{(i)}{q^m}(t_2) = \int_{-\infty}^{t_1}\int_{-\infty}^{t_2} \underset{(i)}{G^{kl}}(t_1-\tau_1)\underset{(i)}{G^{ms}}(t_2-\tau_2)\underset{(i)}{Q_l}(\tau_1)\underset{(i)}{Q_s}(\tau_2)\mathrm{d}t_1\,\mathrm{d}t_2 \tag{9}$$

and then calculate the mean

$$\langle q^k(t_1)q^m(t_2)\rangle = \lim_{n\to\infty} \frac{1}{n} \sum_{i=1}^{n} \underset{(i)}{q^k}(t_1)\underset{(i)}{q^m}(t_2)$$

$$= \lim_{n\to\infty} \frac{1}{n} \sum_{i=1}^{n} \int_{-\infty}^{t_1}\int_{-\infty}^{t_2} \underset{(i)}{G^{kl}}(t_1-\tau_1)\underset{(i)}{G^{ms}}(t_2-\tau_2)\underset{(l)}{Q_l}(\tau_1)\underset{(i)}{Q_s}(\tau_2)\mathrm{d}\tau_1\,\mathrm{d}\tau_2$$

$$= \int_{-\infty}^{t_1}\int_{-\infty}^{t_2} \lim_{n\to\infty} \frac{1}{n} \sum_{i=1}^{n} \underset{(i)}{G^{kl}}(t_1-\tau_1)\underset{(i)}{G^{ms}}(t_2-\tau_2)\underset{(i)}{Q_l}(\tau_1)\underset{(i)}{Q_s}(\tau_2)\mathrm{d}\tau_1\,\mathrm{d}\tau_2. \tag{10}$$

By virtue of the deterministic character of the structural parameters of the system the following occurs:

$$\lim_{n\to\infty} \frac{1}{n} \sum_{i=1}^{n} \underset{(i)}{G^{kl}}(t_1-\tau_1)\underset{(i)}{G^{ms}}(t_2-\tau_2)\underset{(i)}{Q_l}(\tau_1)\underset{(i)}{Q_s}(\tau_2)$$

$$= G^{kl}(t_1-\tau_1)G^{ms}(t_2-\tau_2) \lim_{n\to\infty} \frac{1}{n} \sum_{i=1}^{n} \underset{(i)}{Q_l}(\tau_1)\underset{(i)}{Q_s}(\tau_2)$$

$$= G^{kl}(t_1-\tau_1)G^{ms}(t_2-\tau_2)\langle Q_l(\tau_1)Q_s(\tau_2)\rangle. \tag{11}$$

Considering (11) in (10) we obtain the relationship

$$\langle q^k(t_1)q^m(t_2)\rangle = \int_{-\infty}^{t_1}\int_{-\infty}^{t_2} G^{kl}(t_1-\tau_1)G^{ms}(t_2-\tau_2)\langle Q_l(\tau_1)Q_s(\tau_2)\rangle \mathrm{d}\tau_1\,\mathrm{d}\tau_2. \tag{12}$$

Bearing in mind definition (15), Section 10.2, giving the elements of the correlation matrix of a system with N inputs and N outputs, as well as relationships (8) and (12), we write

$$K_q^{km}(t_1, t_2) = \langle q^k(t_1)q^m(t_2)\rangle - \langle q^k(t_1)\rangle\langle q^m(t_2)\rangle$$

$$= \int_{-\infty}^{t_1}\int_{-\infty}^{t_2} G^{kl}(t_1-\tau_1)G^{ms}(t_2-\tau_2)[\langle Q_l(\tau_1)Q_s(\tau_2)\rangle -$$

$$-\langle Q_l(\tau_1)\rangle\langle Q_s(\tau_2)\rangle]\mathrm{d}\tau_1\,\mathrm{d}\tau_2.$$

Now $\langle Q_l(\tau_1)Q_s(\tau_2)\rangle - \langle Q_l(\tau_1)\rangle\langle Q_s(\tau_2)\rangle$ gives an element of the correlation matrix of a random input signal, therefore

$$K_q^{km}(t_1, t_2) = \int_{-\infty}^{t_1}\int_{-\infty}^{t_2} G^{kl}(t_1-\tau_1)G^{ms}(t_2-\tau_2)K[Q_l(\tau_1)Q_s(\tau_1)]\mathrm{d}\tau_1\,\mathrm{d}\tau_2. \tag{13}$$

The result obtained is the law of transformation of the correlation matrix of a process with transmittance through a linear system.

The introduced relationships are easily generalized. Making use of the relationship between correlation and moments (31), Section 10.2, we write

$$K[q^{i_1}(t_1), \ldots, q^{i_s}(t_s)]$$

$$= G^{i_1,\ldots,i_s,l_1,\ldots,l_s}(t_1, \ldots t_s) * K[Q_{l_1}(t_1), \ldots, Q_{l_s}(t_s)] \tag{14}$$

where $G^{i_1,\ldots,i_s,l_1,\ldots,l_s}(t_1, \ldots, t_s) = \prod_{k=1}^{s} G^{i_k l_k}(t_k)$.

10.6 The problem of departures [65]

We shall now concern ourselves with occurrences consisting in the departure from the limits of a specified area by the trajectory of the motion of a system in a configurational space. The problem is important from the point of view of reliability. The departure frequency, the time of stay outside or within the area and the probability of one, two, or three, etc., departures from the area are essential problems in assessing the safety of a system.

a. Number of departures of the area limit

An area, D, of the configurational space of a dynamical system is given. Let the limits of the area be hypersurfaces described by the equations

$$\Phi_r(q^1, \dots, q^N, t) = 0 \quad (r = 1, \dots, L) \tag{1}$$

of class C_1. Assume that dependence on t is also of class C_1. We define area D itself by the inequalities

$$\Phi_r(q^1, \dots, q^N, t) \leqslant 0^* \quad (r = 1, \dots, L). \tag{2}$$

A differentiable [52, 71] multidimensional process $\xi^i(t)$ is also given. We can identify process $\xi^i(t)$ with the random motion of a point in the configurational space, i.e., the coordinates $q^i(t)$ will be, like $\xi^i(t)$, random. The coordinates $q^i(t)$ may, for example, define the random vibration of a system with N degrees of freedom. Naturally, this motion may be induced by random factors (processes) with known stochastic characteristics as described in the earlier sections. We shall now consider how to construct such a process $\eta^+(t_0, T)$ that would represent a random number of departures of the process $\xi^i(t)$ from area D.

The construction of such a process will be achieved first by introducing a process $\zeta_r(t, \omega)$ constructed as follows:

$$\zeta_r(t, \omega) \overset{\text{df}}{=} \Phi_r(\xi^1(t, \omega), \dots, \xi^N(t, \omega), t).^{**} \tag{3}$$

Evidently, the construction of this new process consists in the formal introduction of coordinates of the N-dimensional process $\xi^i(t, \omega)$ into equation (1) in place of the coordinates q^i.

We shall now analyse the properties of the process $\zeta_r(t, \omega)$. Note that the components of process $\zeta_r(t, \omega)$ assume, depending on the process $\xi^i(t, \omega)$, either a posistive, zero or negative value; and so:

$$\begin{aligned} &\text{(i) } \xi^i(t, \omega) \in D \Rightarrow \bigwedge_{i \leqslant r \leqslant L} \zeta_r(t, \omega) < 0, \\ &\text{(ii) } \xi^i(t, \omega) \in \partial D \Rightarrow \bigvee_{i \leqslant r \leqslant L} \zeta_r(t, \omega) = 0, \\ &\text{(iii) } \xi^i(t, \omega) \notin D \Rightarrow \bigvee_{i \leqslant r \leqslant L} \zeta_r(t, \omega) > 0. \end{aligned} \tag{4}$$

* There exists the possibility of using W. F. Rvachev's R-function method what gives quite coherent results.

** In all expressions in the sequel, the functions which contain the symbol ω are of a probabilistic nature.

The important fact here is that, as process $\xi^i(t, \omega)$ crosses the limit—boundary ∂D of area D, process $\zeta_r(t, \omega)$ changes its sign (to the opposite) for a certain r.

We are now confronted with the problem of counting the departures of process $\xi^i(t, \omega)$ from area D in a stated time interval $[t_0, t_0+T]$. Each time the process $\xi^i(t, \omega)$ leaves area D there is an accompanying sign change, i.e. the event is accompanied by a positive derivative $\dot{\zeta}_r(t, \omega)$. Therefore, if we want to count the departures, we must consider the positive derivative of process $\dot{\zeta}_r(t, \omega)$ for a specified r corresponding to a hypersurface from which the process departs. The number of crossings will be the sum of departures calculated on each hypersurface $\Phi_r(q^1, \ldots, q^N, t) = 0$, being the boundary of area D.

Let us use $\eta_r^+(t_0, T)$ to denote the random number of crossings by the process $\xi^i(t, \omega)$ of the boundary of area D having an element of hypersurface $\Phi_r(q, \ldots, q^N, t) = 0$. The process $\eta^+(t_0, T)$ will be the sum

$$\eta^+(t_0, T) = \sum_{r=1}^{L} \eta_r^+(t_0, T). \tag{5}$$

We shall use the sequence $t_{1r}, t_{2r}, \ldots, t_{kr}$ to denote the instants of departures by the process $\xi^i(t, \omega)$ from an element of hypersurface $\Phi_r(q^1, \ldots, q^N, t) = 0$ constituting the boundary of area D.

In this situation the number of departures by the process $\xi^i(t, \omega)$ from this element of the limits of area D will be

$$\eta_r(t_0, T) = \sum_{k=1}^{N(t_0, T)} \int_{t_0}^{t_0+T} \delta(t-t_{kr})\mathrm{d}t = \int_{t_0}^{t_0+T} \sum_{k=1}^{N(t_0, T)} \delta(t-t_{kr})\mathrm{d}t \tag{6}$$

where $\delta(t)$ is the Dirac-delta distribution.

This would give the overall total number of departures from one of the walls constituting the boundary of area D.

But our concern is to count only events corresponding to departures of process $\xi^i(t, \omega)$ from area D. To achieve this it is necessary to eliminate from sum (6) cases corresponding to the entrance of process $\xi^i(t, \omega)$ into area D.

We remember that process $\zeta_r(t, \omega)$ has the property that, at the instants at which process $\xi^i(t, \omega)$ departs the limit of area D for a specified r, $\dot{\zeta}_r(t, \omega) \neq 0$ taking a positive value when process $\xi^i(t, \omega)$ leaves area D and a negative value when it enters the area. Thus we form a function which assumes zero value for $\dot{\zeta}_r(t, \omega) < 0$, and value 1 for $\dot{\zeta}_r(t, \omega) = 0$. This function can be built, e.g., like this

$$\psi_r(t, \omega) = \frac{1}{2}\left(1 + \frac{\dot{\zeta}_r(t, \omega)}{|\dot{\zeta}_r(t, \omega)|}\right). \tag{7}$$

Considering (7) in (6) we obtain

$$\eta_r^+(t_0, T) = \int_{t_0}^{t_0+T} \psi_r(t, \omega) \sum_{k=1}^{N(t_0, T)} \delta(t - t_{kr})\,\mathrm{d}t. \tag{8}$$

On the other hand, we should take note of the fact that instants $t_{1r}, t_{2r}, \ldots, t_{kr}$ can be determined by observation of the process $\zeta_r(t, \omega)$ which, at these instants of time, changes sign, i.e., it assumes zero value. Therefore, making use of the theorem on distributions we find that

$$\sum_{k=1}^{N(t_0, T)} \delta(t - t_{kr}) = |\dot{\zeta}_r(t, \omega)|\, \delta(\zeta_r(t, \omega))$$

or

$$\eta_r^+(t_0, T) = \int_{t_0}^{t_0+T} \psi_r(t, \omega) |\dot{\zeta}_r(t, \omega)|\, \delta(\zeta_r(t, \omega))\,\mathrm{d}t. \tag{9}$$

Note further that the sign of the absolute value under the integral can be neglected since we are interested only in those cases for which $\dot{\zeta}_r(t, \omega) > 0$ (departure of the process $\xi^i(t, \omega)$ from area D). Bearing the last remark in mind and considering relationship (5), we shall write an equation defining process $\eta^+(t_0, T)$, constituting the random number of departures of process $\xi^i(t, \omega)$ from area D in the configurational space:

$$\eta^+(t_0, T) = \int_{t_0}^{t_0+T} \sum_{r=1}^{L} \frac{\mathrm{d}\zeta_r(t, \omega)}{\mathrm{d}t} \psi_r(t, \omega)\, \delta(\xi_r(t, \omega))\,\mathrm{d}t. \tag{10}$$

Analogously, one can construct an expression for a random number of entrances $\eta^-(t_0, T)$ through the limit of area D by an external process:

$$\eta^-(t_0, T) = -\int_{t_0}^{t_0+T} \sum_{r=1}^{L} \frac{\mathrm{d}\zeta_r(t, \omega)}{\mathrm{d}t} \varphi_r(t, \omega)\, \delta(\zeta_r(t, \omega))\,\mathrm{d}t \tag{11}$$

where

$$\varphi_r(t, \omega) = \frac{1}{2}\left(1 - \frac{\dot{\zeta}_r(t, \omega)}{|\dot{\zeta}_r(t, \omega)|}\right).$$

The functions $\psi_r(t,\omega)$ and $\varphi_r(t,\omega)$ may be constructed in another way, making use of the Heaviside distribution: for we have

$$\psi_r(t,\omega) = H(\dot{\zeta}_r(t,\omega)) \quad \text{whereas} \quad \varphi_r(t,\omega) = H(-\dot{\zeta}_r(t,\omega)).$$

This is precisely the construction of functions $\psi_r(t,\omega)$ and $\varphi_r(t,\omega)$ adopted in examples.

Note that

$$\frac{d\zeta_r(t,\omega)}{dt} = \frac{\partial \Phi_r}{\partial q^i}\bigg|_{q^i=\xi^i(t,\omega)} \dot{\xi}^1(t,\omega) + \frac{\partial \Phi_r}{\partial t}\bigg|_{q^i=\xi^i(t,\omega)} \tag{12}$$

Thus $\eta^+(t_0, T)$ and $\eta^-(t_0, T)$ depend on the function $\zeta_r(t,\omega)$ given by equation (3) and on the derivative of process $\dot{\xi}^i(t,\omega)$. Therefore, to find the value of the mean number of crossings, it is necessary to know the overall total density of the distribution

$$g_{\xi^1,\ldots,\xi^N,\dot{\xi}^1,\ldots,\dot{\xi}^N}(x^1,\ldots,x^N,\dot{x}^1,\ldots,\dot{x}^N,t).$$

Knowing this density, it is possible to calculate the mean values and thus

$$\langle\eta^+(t_0,T)\rangle = \int_{t_0}^{t_0+T}\int_{-\infty}^{\infty}\cdots\int_{-\infty}^{\infty} g_{\xi^1,\ldots,\xi^N,\dot{\xi}^1,\ldots,\dot{\xi}^N}(x^1,\ldots,x^N,\dot{x}^1,\ldots,\dot{x}^N,t)\times$$

$$\times\sum_{r=1}^{L}\left[\frac{d\zeta_r(t,\omega)}{dt}\psi_r(t,\omega)\,\delta(\zeta_r(t,\omega))\right]_{\substack{\xi^i(t,\omega)=x^i,\\ \dot{\xi}^i(t,\omega)=\dot{x}^i,}}\times$$

$$\times\, dx^1,\ldots,dx^N, d\dot{x}^1\ldots, d\dot{x}^N dt, \tag{13}$$

and analogously for the process $\eta^-(t_0, T)$:

$$\langle\eta^-(t_0,T)\rangle = \int_{t_0}^{t_0+T}\int_{-\infty}^{\infty}\cdots\int_{-\infty}^{\infty} g_{\xi^1,\ldots,\xi^N,\dot{\xi}^1,\ldots\dot{\xi}^N}(x^1,\ldots,x^N,\dot{x}^1,\ldots,\dot{x}^N,t)\times$$

$$\times\sum_{r=1}^{L}\left[\frac{d\zeta_r(t,\omega)}{dt}\varphi_r(t,\omega)\,\delta(\zeta_r(t,\omega))\right]_{\substack{\xi^i(t,\omega)=x^i\\ \dot{\xi}^i(t,\omega)=\dot{x}^i}}\times$$

$$\times\, dx^1\ldots dx^N d\dot{x}^1\ldots d\dot{x}^N dt \tag{14}$$

We shall now discuss several special cases in examples.

Example 1. Specify the equation for $\langle\eta^+(_0, T)\rangle$ in the one-dimensional case (Fig. 10.8).

The areas will be rays obtained by dividing the numerical axis by a point $q = f(t)$. In this way area D is defined by the following inequality:

$$\Phi_1 = \Phi_1(q,t) = q - f(t) \leqslant 0. \tag{a}$$

Next, we form the expression (see equation (9))

$$\zeta(t,\omega) = \xi(t,\omega) - f(t) \tag{b}$$

and then we build the derivatives

$$\dot{\zeta}(t,\omega) = \dot{\xi}(t,\omega) - \dot{f}(t). \tag{c}$$

We now give the formula for the mean number of crossings, from below to above function $f(t)$, by process $\xi(t)$ in the interval $[t_0, t_0+T]$ (see equation (13)):

$$\langle \eta^+(t_0, T)\rangle = \int_{t_0}^{t_0+T} \int_{-\infty}^{\infty} \int_{-\infty}^{\infty} g_{\xi\dot{\xi}}(x, \dot{x}|t)\,(\dot{x}-\dot{f}(t))\,H(\dot{x}-\dot{f}(t))\,\delta(x-f(t))\,\mathrm{d}t\,\mathrm{d}x\,\mathrm{d}\dot{x}. \tag{d}$$

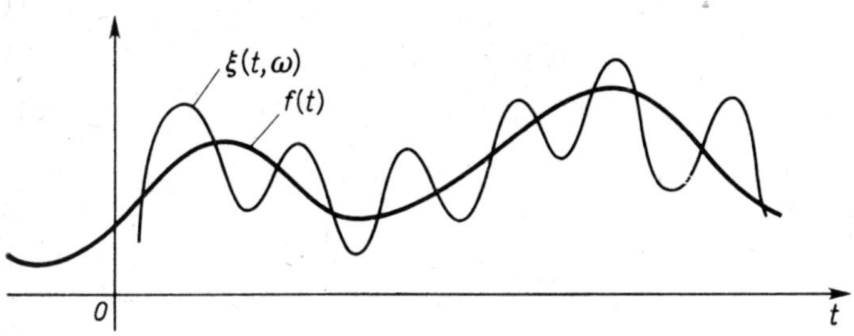

Fig. 10.8

Making use of the properties of the Dirac delta function, we have

$$\langle \eta^+(t_0, T)\rangle = \int_{t_0}^{t_0+T} \int_{-\infty}^{\infty} g_{\xi\dot{\xi}}(f(t), \dot{x}|t)\,(\dot{x}-\dot{f}(t))\,H(\dot{x}-\dot{f}(t))\,\mathrm{d}t\,\mathrm{d}\dot{x}$$

$$= \int_{t_0}^{t_0+T} \int_{\dot{f}(t)}^{\infty} g_{\xi\dot{\xi}}(f(t), \dot{x}|t)\,(\dot{x}-\dot{f}(t))\,\mathrm{d}t\,\mathrm{d}\dot{x}. \tag{e}$$

An interesting problem is the departure of process $\xi(t)$ from an area limited by two curves $f_1(t)$ and $f_2(t)$ such that

$$\bigwedge_{t>0} f_1(t) > f_2(t). \tag{f}$$

Area D is defined by the following inequalities:

$$\Phi_1 = q - f_1(t) \leqslant 0,$$
$$\Phi_2 = -q + f_2(t) \leqslant 0.$$

Making use of equation (9) we write

$$\langle \eta^+(t_0, T)\rangle = \left\langle \int_{t_0}^{t_0+T} [(\dot{\xi}(t)-\dot{f}_1(t))\,H(\dot{\xi}(t)-\dot{f}_1(t))\,\delta(\xi(t)-f_1(t)) + \right.$$

$$\left. +(-\dot{\xi}(t)+\dot{f}_2(t))\,H(-\dot{\xi}(t)+\dot{f}_2(t))\,\delta(-\xi(t)+f_2(t))]\,\mathrm{d}t \right\rangle. \tag{g}$$

Evidently, the mean number of departures from the area $\langle \eta^+(t_0, T)\rangle$ will be equal to the sum of the mean number of crossings, from below to above the curve $y = f_1(t)$, by the process $\xi(t)$ and the mean number of crossing, from above to below the function $y = f_2(t)$, by the same process $\xi(t)$.

Example 2. We now find the number of departures of a two-dimensional process $\xi^i(t)$ from a circle with a radius which varies with time, $r(t)$ (Fig. 10.9).

In the considered case, area D is defined by the inequality

$$\Phi_1 = (q^1)^2+(q^2)^2-(r(t))^2 \leqslant 0. \tag{a}$$

Making use of equation (13) we write

$$\langle\eta^+(t_0, T)\rangle = \int_{t_0}^{t_0+T} \iiiint_{-\infty}^{\infty} g_{\xi^1,\dots,\xi^N,\dot{\xi}^1,\dots,\dot{\xi}^N}(x^1,\dots,x^N,\dot{x}^1,\dots,\dot{x}^N|t)\times$$
$$\times 2(x^1\dot{x}^1-x^2\dot{x}^2-r(t)\dot{r}(t))H(x^1\dot{x}^1+x^2\dot{x}^2-r(t)\dot{r}(t))\times$$
$$\times \delta\big((x^1)^2+(x^2)^2-(r(t))^2\big)\,dx^1dx^2d\dot{x}^1d\dot{x}^2dt. \tag{b}$$

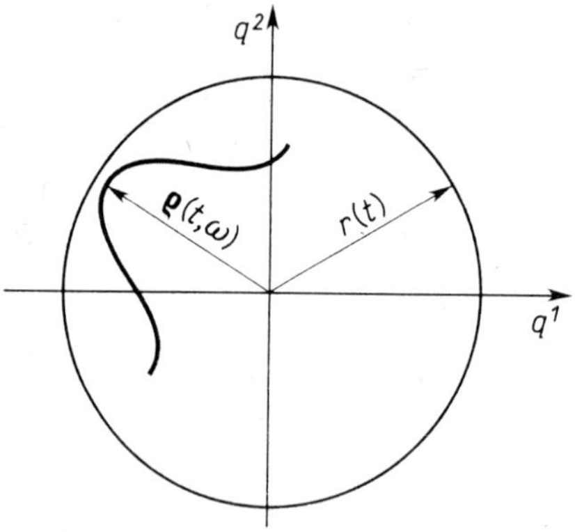

Fig. 10.9

The function under the Dirac delta becomes zero in two cases:

(a) when $x^2 = \sqrt{(r(t))^2-(x^1)^2}$, and

(b) when $x^2 = -\sqrt{(r(t))^2-(x^1)^2}$; therefore we obtain

$$\langle\eta^+(t_0, T)\rangle = 2\int_{t_0}^{t_0+T}\int_{-\infty}^{\infty}\int_{-\infty}^{\infty}\int_{-\infty}^{\infty} g_{\xi^1,\xi^2,\dot{\xi}^1,\dot{\xi}^2}\Big(x^1, \sqrt{(r(t))^2-(x^1)^2}, \dot{x}^1, \dot{x}^2\Big|t\Big)\times$$
$$\times\frac{1}{2\sqrt{(r(t))^2-(x^1)^2}}\Big(x^1\dot{x}^1+\dot{x}^2\sqrt{(r(t))^2-(x^1)^2}-r(t)\dot{r}(t)\Big)\times$$
$$\times H\Big(x^1\dot{x}^1+x^2\sqrt{(r(t))^2-(x^1)^2}-r(t)\dot{r}(t)\Big)\,dx^1d\dot{x}^1d\dot{x}^2dt+$$
$$+2\int_{t_0}^{t_0+T}\int_{-\infty}^{\infty}\int_{-\infty}^{\infty}\int_{-\infty}^{\infty} g_{\xi^1,\xi^2,\dot{\xi}^1,\dot{\xi}^2}\Big(x^1, -\sqrt{(r(t))^2-(x^1)^2}, \dot{x}^1, \dot{x}^2\Big|t\Big)\times$$

$$\times \frac{1}{2\sqrt{(r(t))^2-(x^1)^2}}\left(x^1\dot{x}^1-\dot{x}^2\sqrt{(r(t))^2-(x^1)^2}-r(t)\dot{r}(t)\right)\times$$
$$\times H\left(x^1\dot{x}^1-\dot{x}^2\sqrt{(r(t))^2-(x^1)^2}-r^1(t)\dot{r}(t)\right)\mathrm{d}x^1\mathrm{d}\dot{x}^1\mathrm{d}\dot{x}^2\mathrm{d}t. \tag{c}$$

Thus we can see that x^1 is contained in the interval $[-r(t), r(t)]$ while the integration with respect to the remaining variables proceeds over areas for which the value of Heaviside's function in both integrals of equation (c), is equal to 1.

Equation (c) will be simplified if we assume $r(t) = r = \text{const}$, and that the process is stationary; then

$$\langle\eta^+(t_0, T)\rangle = 2T\int_{-r}^{r}\int_{-\infty}^{\infty}\int_{-\infty}^{\infty} g_{\xi^1,\xi^2,\dot{\xi}^1,\dot{\xi}^2}(x^1, \sqrt{r^2-(x^1)^2}, \dot{x}^1, \dot{x}^2)\times$$
$$\times\frac{1}{2\sqrt{r^2-(x^1)^2}}\cdot(x^1\dot{x}^1+\dot{x}^2\sqrt{r^2-(x^1)^2})\,H(x^1\dot{x}^1+\dot{x}^2\sqrt{r^2-(x^1)^2})\times$$
$$\times\mathrm{d}x^1\mathrm{d}\dot{x}^1\mathrm{d}\dot{x}^2+2T\int_{-r}^{r}\int_{-\infty}^{\infty}\int_{-\infty}^{\infty} g_{\xi^1,\xi^2,\dot{\xi}^1,\dot{\xi}^2}(x^1, -\sqrt{r^2-(x^1)^2}, \dot{x}^1, \dot{x}^2)\times$$
$$\times\frac{1}{2\sqrt{r^2-(x^1)^2}}\cdot(x^1\dot{x}^1-\dot{x}^2\sqrt{r^2-(x^1)^2})\,H(x^1\dot{x}^1-\dot{x}^2\sqrt{r^2-(x^1)^2})\times$$
$$\times\mathrm{d}x^1\mathrm{d}\dot{x}^1\mathrm{d}\dot{x}^2. \tag{c1}$$

We can construct an expression for the mean number of entrances of a process into a circle in an analogous manner.

Example 3. Find the mean number of departures, by a two-dimensional process $\xi^i(t)$, from a rectangle with height h and width b (Fig. 10.10). In this case, area D is defined by the following inequalities:

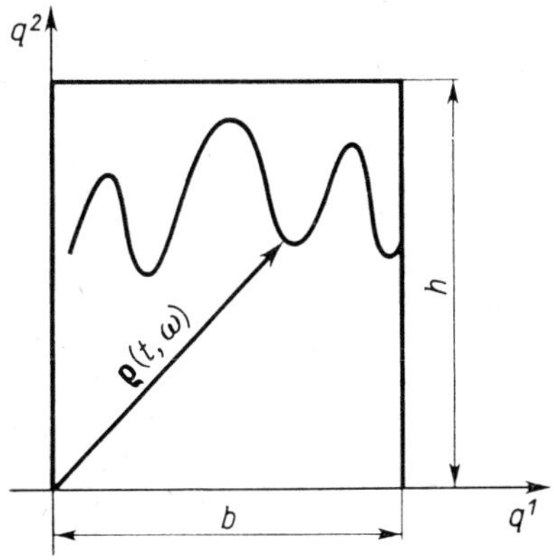

Fig. 10.10

$$\begin{aligned} \Phi_1 &= -q^1 \leqslant 0, \\ \Phi_2 &= q^1 - b \leqslant 0, \\ \Phi_3 &= -q^2 \leqslant 0, \\ \Phi_4 &= q^2 - h \leqslant 0. \end{aligned} \tag{a}$$

Making use of equation (13) we write

$$\begin{aligned} \langle \eta^+(t_0, T) \rangle = & \int_{t_0}^{t_0+T} \int_{-\infty}^{\infty} \int_{-\infty}^{\infty} \int_{-\infty}^{\infty} \int_{-\infty}^{\infty} g_{\xi^1, \xi^2, \dot{\xi}^1, \dot{\xi}^2}(x^1, x^2, \dot{x}^1, \dot{x}^2 | t) \times \\ & \times [-\dot{x}^1 H(-\dot{x}^1)\, \delta(x^1)\, (H(x^2) - H(x^2 - h)) + \\ & + \dot{x}^1 H(\dot{x}^1)\, \delta(x^1 - b)\, (H(x^2) - H(x^2 - h)) - \\ & - \dot{x}^2 H(-\dot{x}^2)\, \delta(x^2)\, (H(x^1) - H(x^1 - b)) + \\ & + \dot{x}^2 H(\dot{x}^2)\, \delta(x^2 - h)\, (H(x^1) - H(x^1 - b))] \mathrm{d}x^1 \mathrm{d}x^2 \mathrm{d}\dot{x}^1 \mathrm{d}\dot{x}^2 \mathrm{d}t \\ = & \int_{t_0}^{t_0+T} \int_{0}^{h} \int_{-\infty}^{0} \int_{-\infty}^{\infty} g_{\xi^1, \xi^2, \dot{\xi}^1, \dot{\xi}^2}(0, x^2, \dot{x}^1, \dot{x}^2 | t)\, (-\dot{x}^1) \mathrm{d}x^2 \mathrm{d}\dot{x}^1 \mathrm{d}\dot{x}^2 \mathrm{d}t + \\ & + \int_{t_0}^{t_0+T} \int_{0}^{h} \int_{0}^{\infty} \int_{-\infty}^{\infty} g_{\xi^1, \xi^2, \dot{\xi}^1, \dot{\xi}^2}(b, x^2, \dot{x}^1, \dot{x}^2 | t) \dot{x}^1 \mathrm{d}x^2 \mathrm{d}\dot{x}^1 \mathrm{d}\dot{x}^2 \mathrm{d}t + \\ & + \int_{t_0}^{t_0+T} \int_{0}^{b} \int_{0}^{\infty} \int_{-\infty}^{\infty} g_{\xi^1, \xi^2, \dot{\xi}^1, \dot{\xi}^2}(x^1, 0, x^1, x^2 | t)\, (-\dot{x}^2) \mathrm{d}x^1 \mathrm{d}\dot{x}^1 \mathrm{d}\dot{x}^2 \mathrm{d}t + \\ & + \int_{t_0}^{t_0+T} \int_{0}^{b} \int_{-\infty}^{\infty} \int_{0}^{\infty} g_{\xi^1, \xi^2, \dot{\xi}^1, \dot{\xi}^2}(x^1, h, x^1, x^2 | t) \dot{x}^2 \mathrm{d}x^1 \mathrm{d}\dot{x}^1 \mathrm{d}\dot{x}^2 \mathrm{d}t. \end{aligned} \tag{b}$$

Heaviside's functions occurring in the first part of equation (b) have provided a safeguard against counting the intersection of the straight lines described by equations (a) beyond the edges of the rectangle.

One could cite a great many similar problems. However, significant here are the general equations (13) and (14) which serve as the basis for solving such problems. In the theory of dynamical systems we may find of interest the density distribution of processes $\eta^+(t_0, T)$ or $\eta^-(t_0, T)$.

For, knowing the distribution of probability of these processes, it is possible to determine the probability of an departure or entrance of a process from area D in a stated time interval. Exact solving of this problem is possible but rather complicated. It is greatly simplified when we deal with a special case where the mean number of entrances into area D in a given time interval is sufficiently small for the successive departures or entrances to be regarded as independent occurrences. These assumptions make it legitimate in turn to accept that the number of departures or entrances has the *Poisson distribution*, i.e.,

$$P\{\eta^+(t_0, T) = m\} = \frac{1}{m!} \langle \eta^+(t_0, T) \rangle^m \exp\left[-\langle \eta^+(t_0, T) \rangle\right]. \tag{15}$$

b. The mean time of stay of the process $\xi^i(t)$ outside or inside area D in the time interval $[t_0, t_0+T]$ and during one departure from or entrance into area D

In the preceding point we have assumed that the distributions

$$g_{\xi^1,\ldots,\xi^N,\dot{\xi}^1,\ldots,\dot{\xi}^N}(x^1, \ldots, x^N, \dot{x}^1, \ldots, \dot{x}^N|t)$$

are known. On this basis we can find the density $g_{\xi^1,\ldots,\xi^N}(x^1, \ldots, x^N|t)$ since these distributions are compatible:

$$g_{\xi^1,\ldots,\xi^N}(x^1, \ldots, x^N|t)$$
$$= \int_{-\infty}^{\infty} \ldots \int_{-\infty}^{\infty} g_{\xi^1,\ldots,\xi^N,\dot{\xi}^1,\ldots,\dot{\xi}^N}(x^1, \ldots, x^N, \dot{x}^1, \ldots, \dot{x}^N|t)\,\mathrm{d}\dot{x}^1 \ldots \mathrm{d}\dot{x}^N. \tag{16}$$

Distributions (16) make it possible to determine the probability of the stay of the process at a given instant t in an arbitrary area of the space.

Thus the probability that the process is outside area D is

$$P\{\xi^i(t) \notin D\} = \int_{(x^1,\ldots,x^N)\notin D} \ldots \int g_{\xi^1,\ldots,\xi^N}(x^1, \ldots, x^N|t)\,\mathrm{d}x^1, \ldots, \mathrm{d}x^N. \tag{17}$$

For example, for the one-dimensional case the probability that the value of process $\xi(t)$ at a given instant is likely to be above the level $y = f(t)$ is given by

$$P\{\xi(t) > f(t)\} = \int_{f(t)}^{\infty} g_\xi(x|t)\,\mathrm{d}x. \tag{17a}$$

We now go back to the general case. To find the mean time of stay of the process $\xi^i(t, \omega)$ outside area D we make use of property (i) (equation (4)) of process $\zeta_r(t, \omega)$.

First we form an auxilliary process

$$\chi(t, \omega) = 1 - \frac{1}{2^L} \prod_{r=1}^{L} \left(1 - \frac{\xi_r(t, \omega)}{|\xi_r(t, \omega)|}\right). \tag{18}$$

Note that process $\chi(t, \omega)$ has the property:

$$\chi(t, \omega) = \begin{cases} 1 & \text{when} \quad \xi^i(t, \omega) \notin D, \\ 0 & \text{when} \quad \xi^i(t, \omega) \in D, \end{cases}$$

in connection with which

$$\tau^+ = \int_{t_0}^{t_0+T} \chi(t, \omega)\,\mathrm{d}t; \tag{19}$$

this connotes the random crossing duration of the process.

We find the mean value of the random variable τ^+ in the following manner:

$$\langle\tau^+\rangle = \langle \int_{t_0}^{t_0+T} \chi(t, \omega)\mathrm{d}t\rangle = \int_{t_0}^{t_0+T} \langle\chi(t, \omega)\rangle \mathrm{d}t$$

$$= \int_{t_0}^{t_0+T} \int_{(x^1, \ldots, x^n)\notin D} \cdots \int g_{\xi^1, \ldots, \xi^N}(x^1, \ldots, x^N|t)\mathrm{d}x^1 \ldots \mathrm{d}x^N \mathrm{d}t. \tag{20}$$

For a stationary process

$$\langle\tau^+\rangle = T \int_{(x^1, \ldots, x^N)\notin D} \cdots \int g_{\xi^1, \ldots, \xi^N}(x^1, \ldots, x^N)\mathrm{d}x^1 \ldots \mathrm{d}x^N. \tag{20a}$$

It is evident from equation (20a) that the mean time of stay of the stationary process $\xi^i(t, \omega)$ outside area D is proportional to observational time T. Knowing the mean time $\langle\tau^+\rangle$ and the mean number of departures $\langle\eta^+(t_0, T)\rangle$ from the limits of area D by process $\xi^i(t, \omega)$, it is possible to determine the mean time of stay of the process outside area D during one departure from the limits of area D by the process $\xi^i(t, \omega)$

$$\langle\Delta\tau\rangle = \frac{\langle\tau^+\rangle}{\langle\eta^+(t_0, T)\rangle}.$$

For a stationary process and for limitations independent of time, it will become

$$\langle\Delta\tau\rangle = \frac{\int_{(x^1, \ldots, x^N)\notin D} \cdots \int g_{\xi^1, \ldots, \xi^N}(x^1., \ldots, x^N)\mathrm{d}x^1, \ldots, \mathrm{d}x^N}{\int_{-\infty}^{\infty} \cdots \int_{-\infty}^{\infty} g_{\xi^1, \ldots, \xi^N, \dot{\xi}^1, \ldots, \dot{\xi}^N}(x^1, \ldots, x^N, \dot{x}^1, \ldots, \dot{x}^N) \times R(x^1 \ldots x^N, \dot{x}^1 \ldots \dot{x}^N)\mathrm{d}x^1 \ldots \mathrm{d}x^N \mathrm{d}\dot{x}^1 \ldots \mathrm{d}\dot{x}^N} \tag{21}$$

where

$$R(x^1, \ldots, x^N, \dot{x}^1, \ldots, \dot{x}^N) = \sum_{r=1}^{L} \frac{\mathrm{d}\dot{\zeta}_r}{\mathrm{d}t} H\left(\frac{\mathrm{d}\xi_r}{\mathrm{d}t}\right) \delta(\zeta_r). \tag{22}$$

The derived equations, as we could see, make it possible to find the mean number of departures, the mean time of stay in a stated area in the required time interval and the mean time of stay in a given area with a single departure of the limits.

In all these cases, knowledge of the function

$$g_{\xi^1, \ldots, \xi^N, \dot{\xi}^1, \ldots, \dot{\xi}^N}(x^1, \ldots, x^N, \dot{x}^1, \ldots, \dot{x}^N|t)$$

is postulated. This is an definite inconvenience which can in some cases be overcome. If we make use, for example, of correlation methods leading to the determination of the correlation of a process constituting the dynamical reaction of a linear system to stochastic excitation, then, assuming that the process is normal, the problem is comparatively simple.

For it is then possible to find [66] the sought distribution function as (equation (13), Section 10.2)

$$g_{\xi^1,\dots,\xi^N,\dot{\xi}^1,\dots,\dot{\xi}^N}(x^1,\dots,x^N,\dot{x}^1,\dots,\dot{x}^N|t) = g_{\xi^1,\dots,\xi^{2N}}(x^1,\dots,x^{2N}|t)$$

$$= \frac{1}{(2\pi)^N \sqrt{\det[k^{ik}_{(\xi)}(t,t)]}} \times$$

$$\times \exp\left[-\frac{1}{2}\sum_{\alpha,\beta=1}^{2N} k_{(\xi)\alpha\beta}(t,t)(x^\alpha - \langle\xi^\alpha(t)\rangle)(x^\beta - \langle\xi^\beta(t)\rangle)\right] \tag{23}$$

where the elements $k_{(\xi)\alpha\beta}(t,t)$ form the inverse matrix to $k^{ik}_{(\xi)}(t,t)$ with $4N^2$ elements.

Matrix $k^{ik}_{(\xi)}(t_1, t_2)$ is constructed as follows:

$$k^{ik}_{(\xi)}(t_1, t_2) = \langle\xi^i(t_1)\,\xi^k(t_2)\rangle - \langle\xi^i(t_1)\rangle\langle\xi^k(t_2)\rangle \tag{24}$$

and

$$k^{i,k+N}_{(\xi)}(t_1, t_2) = \frac{\partial k^{ik}_{(\xi)}(t_1, t_2)}{\partial t_2}, \tag{25}$$

$$k^{i+N,k+N}_{(\xi)}(t_1, t_2) = \frac{\partial^2 k^{ik}_{(\xi)}(t_1, t_2)}{\partial t_1\,\partial t_2}. \tag{26}$$

In equations (24) to (26), "i" and "k" vary between limits 1 and N.

We shall now discuss the case where an arbitrary stochastic process is observed at the output of a linear system.

If we can find the correlation function of the output signal (equation (14), Section 10.5), for an arbitrary order, then we can also find its characteristic function and then its density. In the theory of departures the problem is to find the overall total density of the probability of positions and velocities.

Renumbering the coordinates in such a way that

$$\dot{q}^{i_k} = q^{i_k+N},$$

we shall seek the density distribution of probability in the form

$$g_{q^1,\dots,q^{2N}}(x^1,\dots,x^{2N}|t).$$

We construct correlation functions of order s formed from all the combinations of processes q^1 to q^{2N} by adding all the possible combinations of derivatives with respect to times $t_1, t_2, \dots, t_s$ to correlations $k_{(q)}^{i_1,\dots,i_s}(t_1, \dots, t_s)$. For example,

$$k_{(q)}^{i_1,\dots,i_k+N,\dots,i_s}(t_1, \dots, t_s) = \frac{\partial k_{(q)}^{i_1,\dots,i_s}(t_1, \dots, t_s)}{\partial t_k}. \tag{27}$$

By successive differentiation operations we obtain correlations $k_q^{i_1,\dots,i_s}(t_1, \dots, t_s)$ where i_k for arbitrary k will assume values not from 1 to N as previously, but from 1 to $2N$. Due to this change the characteristic function of the output signal will have the following construction:

$$\Theta_{(q)}(u_1, \dots, u_{2N}, t_1, \dots, t_s) = \exp \sum_{s=1}^{\infty} \frac{j^s}{s!} \sum_{i_1,\dots,i_s=1}^{2N} k_{(q)}^{i_1,\dots,i_s}(t_1, \dots, t_s) u_{i_1} \dots u_{i_s}. \tag{28}$$

The looked for density distribution of probability depends on a single instant t, it is therefore Fourier's transform of the characteristic function for $t_1 = t_2 = \dots = t_s = t$, i.e., the function

$$\Theta_{(q)}(u_1, \dots, u_{2N}|t) = \exp \sum_{s=1}^{\infty} \frac{j^s}{s!} \sum_{i_1,\dots,i_s=1}^{2N} k_{(q)}^{i_1,\dots,i_s}(t) u_{i_1} \dots u_{i_s}. \tag{28a}$$

For the sake of brevity, functions $k(t_1, \dots, t_s)$ have been denoted by the symbol $k(t)$.

To define the density as Fourier's inverse transformation performed on (28a), we use a technique [66] consisting in introducing the concept of quasimoments $b_{(q)}^{i_1,\dots,i_s}(t)$ associated with the correlation functions by the relationship

$$\exp \sum_{s=3}^{\infty} \frac{j^s}{s!} \sum_{i_1,\dots,i_s=1}^{2N} k^{i_1,\dots,i_s}(t) u_{i_1} \dots u_{i_s} = 1 + \sum_{s=3}^{\infty} \frac{j^s}{s!} \sum_{i_1,\dots,i_s=1}^{2N} b_q^{i_1,\dots,i_s}(t) u_{i_1} \dots u_{i_s}. \tag{29}$$

If the left-hand side of equation (29) is now expanded into a power series, then, by comparing the terms with the same powers i, we obtain the relationship between correlations and quasimoments. This relationship is analogous to relationship (31), Section 10.1, between correlations and moments:

$$b^{i_1,\ldots,i_s}(t) = k^{i_1,\ldots,i_s}(t) + \\ + \sum_{n=2}^{s} \frac{1}{n!} \sum_{\nu_1=1}^{s} \cdots \sum_{\nu_{n-1}=1}^{s} \frac{s!\, H(c_n - \frac{1}{2})\, B_{(n)}^{(i_1,\ldots,i_s)}(t, \nu_1, \ldots, \nu_{n-1})}{c_n!\, \nu_1! \ldots \nu_{n-1}!} \tag{30}$$

where $c_n = s - \sum_{i=1}^{n-1} \nu_i$, $H(\cdot)$ is Heaviside's function,

$$B_{(n)}^{(i_1,\ldots,i_s)}(t, \nu_1, \ldots, \nu_{n-1})$$

is the symmetrical part of the expression

$$B_{(n)}^{(i_1,\ldots,i_s)}(t, \nu_1, \ldots, \nu_{n-1}) \\ = k^{i_1,\ldots,i_{s-b_1^{(n)}}}(t)\, k^{i_{s-b_1^{(n)}+1},\ldots,i_{s-b_2^{(n)}}}(t) \ldots k^{i_{s-b_{n-1}^{(n)}},\ldots,i_s}(t),$$

where

$$b_k^{(n)} = \sum_{i=k}^{n-1} \nu_i .$$

Equation (30), for the case where we also consider the first two correlations k^{i_1} and k^{i_1,i_2} on the right-hand side, gives moments $m_{(q)}^{i_1,\ldots,i_s}$ of arbitrary order. To find the quasimoments, the first two correlations on the right-hand side should be assumed to be equal to zero. Knowing correlation functions $k^{i_1,\ldots,i_s}(t)$, we can find all the quasimoments.

Returning to the characteristic function (28a) and considering equality (29) we write

$$\Theta_{(q)}(u_1, \ldots, u_{2N}|t) \\ = \exp\left[\mathrm{j} \sum_{i_1=1}^{2N} k_q^{i_1}(t) u_{i_1} - \tfrac{1}{2} \sum_{i_1,i_2=1}^{2N} k_q^{i_1,i_2}(t) u_{i_1} \cdot u_{i_2} \right] \times \\ \times \left[1 + \sum_{s=3}^{\infty} \frac{\mathrm{j}^s}{s!} \sum_{i_1,\ldots,i_s=1}^{2N} b_q^{i_1,\ldots,i_s}(t) u_{i_1} \ldots u_{i_s} \right]. \tag{31}$$

The distribution density of probability we find performing on (31) Fourier's inverse transformation

$$g_{q^1,\ldots,q^{2N}}(x^1, \ldots, x^{2N}|t)$$

$$= \frac{1}{(2\pi)^{2N}} \int_{-\infty}^{\infty} \ldots \int_{-\infty}^{\infty} \exp\left(-\mathrm{j} \sum_{\alpha=1}^{2N} x^\alpha u_\alpha\right) \times$$

$$\times \left[1 + \sum_{s=1}^{\infty} \frac{\mathrm{j}^s}{s!} \sum_{i_1,\ldots,i_s=1}^{2N} b_q^{i_1,\ldots,i_s}(t) u_{i_1} \ldots u_{i_s}\right] \mathrm{d}u_1 \ldots \mathrm{d}u_{2N}. \tag{32}$$

Performing an inverse transformation we obtain

$$g_{q^1,\ldots,q^{2N}}(x^1, \ldots, x^{2N}|t)$$

$$= \left[1 + \sum_{s=3}^{\infty} \frac{1}{s!} \sum_{i_1,\ldots,i_s=1}^{2N} b_q^{i_1,\ldots,i_s}(t) \prod_{k=1}^{s} \left(-\frac{\partial}{\partial x^{i_k}}\right)\right] \times$$

$$\times \tilde{g}_{q^1,\ldots,q^{2N}}(x^1, \ldots x^{2N}|t), \tag{33}$$

where

$$\tilde{g}_{q^1,\ldots,q^{2N}}(x^1, \ldots, x^{2N}|t) = \frac{1}{(2\pi)^N \sqrt{\det\,[k_{(q)}^{\alpha\beta}(t, t)]}} \times$$

$$\times \exp\left\{-\sum_{\alpha,\beta=1}^{2N} k_{(q)\alpha\beta}^{-1}(t, t)(x^\alpha - k_{(q)}^\alpha(t))(x^\beta - k_{(q)}^\beta(t))\right\}. \tag{33a}$$

The presented procedure to find the density (33) proved rather complicated. Owing to the generality of the considerations, it is at the same time the solution to a whole class of linear problems with a significant practical aspect. The calculation of correlations has proceeded on the assumption of determined structural parameters of systems.

Moreover, the coefficients of differential equations describing the dynamics of systems were assumed to be constant; hence the impulsive matrix of transmittance is a system of functions dependent on the time difference $t-\tau(G^{ki}(t-\tau))$. Naturally, widening the class of problems considered, it was possible to assume a response of the system in the form

$$q^k(t) = \int_{-\infty}^{t} \sum_{i=1}^{N} G^{ki}(t, \tau) Q_i(\tau) \mathrm{d}\tau. \tag{34}$$

Following the same line of reasoning, it was also possible to assume a random character of the structural parameters of a system, which would correspond to a random character of the function $G^{ik}(t, \tau)$. Then the departure of the stochastic process $q^k(t)$ from area D with a determinate input signal could be interpreted as, for example, a random breakdown of the system. The presented theoretical scheme is just half the solution to a specific problem. The next difficulty is the numerical problem with its difficulty growing faster than the number of dimensions (degrees of freedom) of the system.

10.7 Fokker–Planck–Kolmogorov equations

The technique is briefly outlined here [67; 69] of deriving diffusion equations for non-linear equations of the form

$$\dot{q}^i = \psi^i\big(q^1, \ldots, q^N, \xi^1(t, \omega), \ldots, \xi^N(t, \omega), t\big) \quad (i = 1, \ldots, 2N). \tag{1}$$

In equations (1) the superscipt "i" progresses from 1 to $2N$, which may appear peculiar. This is so because the coordinates have been renumbered.

We shorten equations (1) to

$$\dot{q}^i = \tilde{\psi}^i(q, t, \omega) \tag{1a}$$

where the right-hand sides are stochastic processes dependent on the multi-dimensional input process $\xi^i(t, \omega)$.

We perform a formal transformation on (1a) consisting in dividing and multiplying the right-hand sides by a number ε which we shall regard as small,

$$\dot{q}^i = \varepsilon F^i(q, t, \omega), \tag{2}$$

where

$$F^i(q, t, \omega) = \frac{1}{\varepsilon}\tilde{\psi}^i(q, t, \omega).$$

Integrating (2) we obtain

$$\Delta q^i = q^i - q^i(0) = \int_0^t F^i(q, t_1, \omega)\,\mathrm{d}t_1. \tag{3}$$

Assume that the function $F^i(q, t_1, \omega)$ is expandable into a power series, i.e., it can be rendered in the form

$$F^i(q, t_1, \omega) = F^i\big(q(0), t_1, \omega\big) + \frac{\partial F^i\big(q(0), t_1, \omega\big)}{\partial q^m}\Delta q^m +$$

$$+ \frac{1}{2}\frac{\partial^2 F^i\big(q(0), t_1, \omega)\big)}{\partial q^m \partial q^n}\Delta q^m \Delta q^n + \ldots \tag{4}$$

Substituting (4) into (3) we have

$$\Delta q^i = \varepsilon \int_0^t F^i(q(0), t_1, \omega)\,\mathrm{d}t_1 + \varepsilon \int_0^t \frac{\partial F^i(q(0), t_1, \omega)}{\partial q^m} \Delta q^m \mathrm{d}t +$$

$$+\frac{1}{2} \int_0^t \frac{\partial^2 F^i(q(0), t_1, \omega)}{\partial q^m \partial q^n} \Delta q^m \Delta q^n \mathrm{d}t_1 + \ldots \tag{5}$$

Substituting into (5) the first approximation of Δq^m,

$$\Delta q^m = \varepsilon \int_0^t F^m(q(0), t_2, \omega)\,\mathrm{d}t_2,$$

we obtain the second approximation of Δq^i:

$$\Delta q^i = \varepsilon \int_0^t F(q(0), t_1, \omega)\,\mathrm{d}t_1 +$$

$$+ \varepsilon^2 \int_0^t \frac{\partial F^i(q(0), t_1, \omega)}{\partial q^m} \left[\int_0^{t_1} F^m(q(0), t_2, \omega)\,\mathrm{d}t_2\right] \mathrm{d}t_1 . \tag{6}$$

If we now substitute (6) into (5) again and limit ourselves to terms up to ε^3, then

$$\Delta q^i = \varepsilon a_1^i + \varepsilon^2 a_2^i + \varepsilon^3 a_3^i + \ldots = \sum_{l=1}^{\infty} \varepsilon^l a_l^i, \tag{7}$$

where

$$a_1^i = \int_0^t F^i(q(0), t_1, \omega)\,\mathrm{d}t_1; \tag{7a}$$

$$a_2^i = \int_0^t \frac{\partial F^i(q(0), t_1, \omega)}{\partial q^m} \left[\int_0^{t_1} F^m(q(0), t_2, \omega)\,\mathrm{d}t_2\right] \mathrm{d}t_1; \tag{7b}$$

$$a_3^i = \int_0^t \frac{\partial F^i(q(0), t_1, \omega)}{\partial q^m} \left\{\int_0^{t_1} \frac{\partial F^m(q(0), t_2, \omega)}{\partial q^n} \times\right.$$

$$\left. \times \left[\int_0^{t^2} F^n(q(0), t_2, \omega)\,\mathrm{d}t_3\right] \mathrm{d}t_2\right\} \mathrm{d}t_1 +$$

$$+\frac{1}{2}\int_0^t \frac{\partial^2 F^i(q(0), t_1, \omega)}{\partial q^m \partial q^n}\left[\int_0^{t_1} F^m(q(0), t_2, \omega)\,dt_2\right]\times$$

$$\times\left[\int_0^{t_1} F^n q((0), t_3, \omega)\,dt_3\right]dt_1. \tag{7c}$$

If zero initial conditions are assumed, $q^i(0) = 0$, then $\Delta q^i = q^i$. Assume that the motion of the system starts at the reference origin; this in no way reduces the generality of the solution, since the reference frame can always be so assumed as to realize this assumption. The characteristic function of process q^i (increment of the "output" signal) is

$$\Theta_{(q)}(u_1, \ldots, u_{2N}, t) = 1 + \sum_{s=1}^{\infty} \frac{j^s}{s!} \sum_{i_1 \ldots, i_s=1}^{2N} m_{(q)}^{i_1, \ldots, i_s}(t)\, u_{i_1}, \ldots, u_{i_s}. \tag{8}$$

Performing Fourier's inverse transformation on (8), we obtain the distribution density in the form

$$g_{q^1, \ldots, q^{2N}}(x^1, \ldots, x^{2N}|t) = (1+R)\prod_{k=1}^{2N} \delta(x^k) \tag{9}$$

where R is an operator of the following construction:

$$R = \sum_{s=1}^{\infty} \frac{1}{s!} \sum_{i_1, \ldots i_s=1}^{2N} \prod_{k=1}^{s} \left(-\frac{\partial}{\partial x^{i_k}}\right) m_{(q)}^{i_1, \ldots, i_s}(x^{i_1}, \ldots, x^{2N}, t)(\cdot). \tag{10}$$

In finding operator R we use the theorem on delta distributions since relationships of the type

$$\frac{\partial}{\partial x}(f(x)\,\delta(x)) = f(0)\frac{\partial}{\partial x}\delta(x)$$

are repeatedly used. In this connection, the correlation functions in (10) depend on the coordinates x^i which have been formally entered in place of $q^i(0)$ which we have assumed to be equal to zero (zero initial conditions). In other words, to calculate the function of moments $m_q{}^{i_1, \ldots, i_s}(t)$ it is necessary to use the function $F^i(x, t, \omega)$, being here a parameter (not a time function), in place of the function $F^i(q(0), t, \omega) = F^i(0, t, \omega)$.

To shorten (9) we denote density by g.

Thus

$$g = (1+R)\prod_{k=1}^{2N} \delta(x^k). \tag{11}$$

Differentiating with respect to time t gives

$$\frac{\partial g}{\partial t} = \frac{\partial R}{\partial t} \prod_{k=1}^{2N} \delta(x^k). \tag{12}$$

From (11) we find

$$\prod_{k=1}^{2N} \delta(x^k) = (1+R)^{-1} g. \tag{13}$$

Substituting (13) into (11) we obtain

$$\frac{\partial g}{\partial t} = \frac{\partial R}{\partial t} (1+R)^{-1} g. \tag{14}$$

The operator

$$(1+R)^{-1} = 1 - R + R^2 - \dots$$

or

$$\frac{\partial g}{\partial t} = \frac{\partial R}{\partial t} (1 - R + R^2 - \dots) g. \tag{15}$$

Evidently, the right-hand side of (15) has an infinite number of terms. Consequently their number should be limited, thus necessitating approximation of the density g.

Confining ourselves to the terms containing ε^2 on the right-hand side of (15), we obtain the *Fokker–Planck–Kolmogorov equation*. After transformations we have

$$\begin{aligned}\frac{\partial g}{\partial t} = & -\varepsilon \sum_{l=i}^{2N} \frac{\partial}{\partial x^l} \Bigg\{ [\langle F^l(x, t, \omega)\rangle g] + \\ & + \varepsilon \sum_{m=1}^{2N} \Bigg[\int_0^t K\Bigg[\frac{\partial F^l(x, t, \omega)}{\partial x^m}, F^m(x, t_1, \omega) \Bigg] \mathrm{d}t_1 \Bigg] g \Bigg\} + \\ & + \varepsilon^2 \sum_{m,l=1}^{2N} \frac{\partial^2}{\partial x^m \partial x^l} \Bigg\{ \Bigg[\int_0^t K[F^m(x, t, \omega), F^l(x, t_1, \omega)] \mathrm{d}t_1 \Bigg] g \Bigg\}. *\end{aligned} \tag{16}$$

* The symbol $K[\cdot, \cdot]$ has been adopted here to denote the correlaton function.

But

$$F^i(x, t, \omega) = \frac{1}{\varepsilon}\tilde{\psi}^i(x, t, \omega),$$

therefore equation (16) can be rewritten in the form

$$\frac{\partial g}{\partial t} = -\sum_{l=1}^{2N} \frac{\partial}{\partial x^l}\Bigg\{[\langle\tilde{\psi}^l(x, t, \omega)\rangle g] + \sum_{m=1}^{2N}\Bigg[\int_0^t K\Bigg[\frac{\partial\tilde{\psi}^l(x, t, \omega)}{\partial x^m}, \tilde{\psi}^m(x, t_1, \omega)\Bigg]\mathrm{d}t_1\Bigg]g\Bigg\} + \sum_{m,l=1}^{2N} \frac{\partial^2}{\partial x^m\,\partial x^l}\Bigg\{\Bigg[\int_0^t K[\tilde{\psi}^m(x, t, \omega), \tilde{\psi}^l(x, t_1, \omega)]\mathrm{d}t_1\Bigg]g\Bigg\}. \quad (16a)$$

In the case where the right-hand sides of system (1) are, in a broader sense, stationary processes, the equations become slightly simplified. If

$$K[\tilde{\psi}^m(x, t, \omega), \tilde{\psi}^l(x, t_1, \omega)] = \psi_1^{ml}(x, t-t_1)$$

and

$$\sum_{m=1}^{2N} K\Bigg[\frac{\partial\tilde{\psi}^l(x, t, \omega)}{\partial x^m}, \tilde{\psi}^m(x, t_1, \omega)\Bigg] = \varphi_2^l(x, t-t_1),$$

then substituting $t-t_1 = \tau$ we obtain

$$\frac{\partial g}{\partial t} = -\sum_{l=1}^{2N} \frac{\partial}{\partial x^l}\Bigg\{\langle\tilde{\psi}^l(x, t, \omega)\rangle g + \Bigg[\int_0^t \varphi_2^l(x, \tau)\mathrm{d}\tau\Bigg]g\Bigg\} + \sum_{m,l=1}^{2N} \frac{\partial^2}{\partial x^m\,\partial x^l}\Bigg\{\Bigg[\int_0^t \varphi_1^{ml}(x, \tau)\mathrm{d}\tau\Bigg]g\Bigg\}. \quad (17)$$

We must bear in mind the character of function g determined by integrating equations (16) or (17). Function $g(x^1, \dots, x^{2N}|t)$ is valid for increments $q^i = \Delta q^i$. The smaller the increment Δq^i the more exactly $g(x^1, \dots, x^{2N}|t)$ approximates to the proper distribution.

Large increments of processes can be characterized by smaller increments, their sum constituting large increments. Here it should be assumed that a multidimensional process is a homogeneous Markov's process. If small increments are statistically independent, then the following occurs (*Smoluchowski's equation*):

$$g(\underset{2}{x^1}, \ldots, \underset{2}{x^{2N}}|t_2, t_2)$$

$$= \int_{-\infty}^{\infty} \ldots \int_{-\infty}^{\infty} g(\underset{1}{x^1}, \ldots, \underset{1}{x^{2N}}|t_1) \times$$

$$\times\, g(\underset{2}{x^1}, \ldots, \underset{2}{x^{2N}}|\underset{1}{x^1}, \ldots, \underset{1}{x^{2N}}, t_2, t_1)\mathrm{d}\underset{1}{x^1}, \ldots, \mathrm{d}\underset{1}{x^{2N}} \qquad (18)$$

where

$$g(\underset{2}{x^1}, \ldots, \underset{2}{x^{2N}}|\underset{1}{x^1}, \ldots, \underset{1}{x^{2N}}, t_2, t_1) = g(\underset{2}{x^1}-\underset{1}{x^1}, \ldots, \underset{2}{x^{2N}}-\underset{1}{x^{2N}}|t_2-t_1).$$

It is obvious therefore that an assumption concerning homogeneity has been made here. It should suffice to define $g(x^1, \ldots, x^{2N}|t) = g_{\xi^1, \ldots, \xi^{2N}}(x^1, \ldots, x^{2N}|t)$ for arbitrary initial conditions. Equation (18) allows for the division of increment $\underset{2}{q^i}$ into two parts. It also allows division into an arbitrary number of increments.

To find the density g, differential equations with partial derivatives need to be integrated. This is by no means a simple problem. It is somewhat less complicated to solve a stationary problem, i.e., the case where $\partial g/\partial t = 0$. Examples of solving a stationary problem will be found, e.g., in [55, 69].

10.8 Proposal for a method of direct determination of probability density

The discussion of Fokker–Planck–Kolmogorov equations for non-linear problems suggests that other roads leading to the determination of the density of distributions should be looked for. Density determination involves the determination of a characteristic function built on moments or correlations

$$Q_{(q)}(u_1, \ldots, u_{2N}, t) = \exp\left\{\sum_{s=1}^{\infty} \frac{\mathrm{j}^s}{s!} \sum_{i_1, \ldots, i_s=1}^{2N} k_{(q)}^{i_1, \ldots, i_s}(t)\, u_{i_1} \ldots u_{i_s}\right\}. \qquad (1)$$

The correlations occurring on the right-hand side of relation (1) have the property

$$K[\xi^{i_1}, \ldots, \xi^{i_k}+\eta^{i_k}, \ldots, \eta^{i_s}] = K[\xi^{i_1}, \ldots, \xi^{i_k}, \ldots, \xi^{i_s}]+$$
$$+K[\xi^{i_1}, \ldots, \eta^{i_k}, \ldots, \xi^{i_s}]\,* \qquad (2)$$

for arbitrary k, which follows directly from the equation (31), p. 380.

* The notation $k^{i_1, \ldots, i_s} \equiv K[\xi^{i_1}, \ldots, \xi^{i_s}]$ has been introduced.

Property (2) is significant in the case where processes are expressed through intermediary sums.

Correlations have yet another property, namely they occur for an arbitrary number α

$$K[\xi^{i_s}, \ldots, \alpha\xi^{i_k}, \ldots, \xi^{i_s}] = \alpha K[\xi^{i_1}, \ldots, \xi^{i_k}, \ldots, \xi^{i_s}]. \tag{3}$$

If, therefore,

$$\Delta q^{i_k} = q^{i_k} = \sum_{l=1}^{\infty} \varepsilon^l a_l^{i_k},$$

then from (2) and (3) we infer that

$$K[q^{i_k}, \ldots, q^{i_s}] = \sum_{l_1, \ldots, l_s = 1}^{\infty} \varepsilon^{l_1 + \ldots + l_s} K[a_{l_1}^{i_1}, \ldots, a_{l_s}^{i_s}]. \tag{4}$$

This observation (4) plays an important part in finding the density since from this relationship it is possible to calculate the characteristic function

$$\Theta_{(q)}(u_1, \ldots, u_{2N}|t) = \exp\left\{\sum_{s=1}^{\infty} \frac{j^s}{s!} \sum_{i_1, \ldots, i_s = 1}^{2N} \sum_{l_1, \ldots, l_s = 1}^{\infty} \varepsilon^{l_1 + \ldots + l_s} K[a_{l_1}^{i_1}, \ldots, a_{l_s}^{i_s}] u_{i_1} \ldots u_{i_s}\right\}. \tag{5}$$

The determination of the sought density $g_{q^1, \ldots, q^{2N}}(x^1, \ldots, x^{2N}|t)$ consists in performing Fourier's inverse transformation. In this operation we avail ourselves of the method, presented in the preceding section, in which quasimoments were introduced. Through quasimoments it is possible to transform equation (5)

$$\Theta_{(q)}(u_1, \ldots, u_{2N}|t) = \exp\left[j \sum_{i_1=1}^{2N} k_q^{i_1}(t) u_{i_1} - \frac{1}{2} \sum_{i_1, i_2 = 1}^{2N} k_q^{i_1, i_2}(t) u_{i_1} u_{i_2}\right] \times$$
$$\times \left[1 + \sum_{s=3}^{\infty} \frac{j^s}{s!} \sum_{i_1, \ldots, i_s = 1}^{2N} b_g^{i_1, \ldots, i_s}(t) u_{i_1} \ldots u_{i_s}\right]. \tag{6}$$

In the considered case

$$k_{(q)}^{i_1}(t) = \sum_{l_1=1}^{\infty} \varepsilon^{l_1} K[a_i^{i_1}] = \sum_{l_1=1}^{\infty} \varepsilon^{l_1} \langle a_{l_1}^{i_1} \rangle, \tag{7}$$

$$k_{(q)}^{i_1 i_2}(t) = \sum_{l_1, l_2=1}^{\infty} \varepsilon^{l_1+l_2} K[a_{l_1}^{i_1}, a_{l_2}^{i_2}], \tag{8}$$

$$b_q^{i_1, i_2, i_3}(t) = \sum_{l_1, l_2, l_3=1}^{\infty} \varepsilon^{l_1+l_2+l_3} K[a_{l_1}^{i_1}, a_{l_2}^{i_2}, a_{l_3}^{i_3}], \tag{9}$$

$$b_q^{i_1, i_2, i_3, i_4}(t) = \sum_{l_1, \ldots, l_4=1}^{\infty} \varepsilon^{l_1+\ldots+l_4} K[a_{l_1}^{i_1}, \ldots, a_{l_4}^{i_4}], \tag{10}$$

$$b_q^{i_1, \ldots, i_5}(t) = \sum_{l_1, \ldots, l_5=1}^{\infty} \varepsilon^{l_1+\ldots+l_5} K[a_{l_1}^{i_1}, \ldots, a_{l_5}^{i_5}], \tag{11}$$

$$b_q^{i_1, \ldots, i_6}(t) = \sum_{l_1, \ldots, l_6=1}^{\infty} \varepsilon^{l_1+\ldots+l_6} K[a_{l_1}^{i_1}, \ldots, a_{l_6}^{i_6}] + 10 B_{(2)}^{(i_1, \ldots, i_6)}(t, 3) \tag{12}$$

where

$$B_{(2)}^{(i_1, \ldots, i_6)}(t, 3) = K[a_{l_1}^{i_1}, \ldots, a_{l_3}^{i_3}] K[a_{l_4}^{i_4}, \ldots, a_{l_6}^{i_6}].$$

Further quasimoments are determined in a similar manner from the general equation. We find the density of the distribution of probability by performing Fourier's inverse transformation on (6):

$$\begin{aligned} & g_{q^1, \ldots, q^{2N}}(x^1, \ldots, x^{2N} | t) \\ & = \left\{ 1 + \sum_{s=3}^{\infty} \frac{1}{s!} \sum_{i_1, \ldots, i_s=1}^{2N} b_q^{i_1, \ldots, i_s}(t) \prod_{k=1}^{s} \left(-\frac{\partial}{\partial x^{i_k}} \right) \right\} \times \\ & \times \tilde{g}_{q^1, \ldots, q^{2N}}(x^1, \ldots, x^{2N} | t) \end{aligned} \tag{13}$$

where

$$\begin{aligned} \tilde{g}_{q^1, \ldots, q^{2N}}(x^1, \ldots, x^{2N} | t) &= \frac{1}{(2\pi)^N \sqrt{\det [k_{(q)}^{i_1 i_2}(t)]}} \times \\ & \times \exp \left[- \sum_{i_1, i_2=1}^{2N} k_{(q) i_1, i_1}(t) \, (x^{i_1} - k_{(q)}^{i_1}(t) \, (x^{i_2} - k_{(q)}^{i_2}(t) \right] \end{aligned} \tag{14}$$

where

$$[k_{(q) i_1 i_2}(t)] = [k_{(q)}^{i_1 i_2}(t)]^{-1}.$$

This shows how the distribution density of increments Δq^i can be found. Depending on the specific case considered (on the degree of accuracy of the approximation of (7), Section 10.6) the distribution given by (13) can be regarded as good in close and less close proximity to the starting point (initial conditions). An extended validity of the distribution is possible by assuming that the process involved is of the Markov type with independent increments and by making use of Smoluchowski's equation (18), Section 10.9.

Finally, an essential conclusion should be presented. The result of our considerations points to the advisability of seeking the density by "direct" method, not by integration of the Fokker–Planck–Kolmogorov equations. What is more, the proposition that a solution to partial equations of the Fokker–Planck–Kolmogorov type can be sought in the class of functions defined by equation (13) may be put forward.

The presented approach to non-linear problems is founded on the adopted method of search after a solution. Any method other than the one used here can serve as the basis of an analogous determination of the characteristic function, followed by the determination of the density.

Bibliography

1 P. Antosik, J. Mikusiński, R. Sikorski, *Theory of Distributions. The Sequential Approach*, Elsevier, Amsterdam–PWN, Warszawa, 1973.

2 I. I. Artobolevskii, *Theory of Mechanisms and Machines*, (in Russian), Gos. Izd. Tekh.-Teor. Lit., Moskva, 1951.

3 I. M. Babakov, *Theory of Vibrations*, (in Russian), Gos. Izd. Tekh.-Teor. Lit., Moskva, 1958.

4 S. Banach, *Mechanics*, PWN, Warszawa, 1956.

5 N. M. Belayev, *Strength of Materials*, (in Russian), Gos. Izd. Tekh.-Teor. Lit., Moskva, 1956.

6 N. M. Belayev, *Collection of Problems from Strength of Materials*, (in Russian), Gos. Izd. Tekh.-Teor. Lit., Moskva, Leningrad, 1951.

7 N. N. Bogolubov, Yu. A. Mitropol'skii, *Asymptotic Methods in the Theory of Nonlinear Vibrations*, (in Russian), Gos. Izd. Fiz. Mat. Lit., Moskva, 1958.

8 D. Brückner, M. Podhorodyński, B. Skalmierski, *Fundamentals of Probability Theory in the Distri butiveAspect*, (in Polish), Zeszyty Politechn. Śląskiej, Mat. –Fiz. 29 (1978).

9 N. N. Bukhgolc, *Fundamental Course of Theoretical Mechanics*, (in Russian), Nauka, Moskva, 1969.

10 B. W. Bulgakov, *Vibrations*, (in Russian), Gos. Izd. Tekh.-Teor. Lit., Moskva, 1954.

11 R. H. Cannon, *Dynamics of Physical Systems*, McGraw-Hill Book Co., New York, 1967.

12 E. Czogała, K. Respondek, B. Skalmierski, A. Tylikowski, *Collection of Problems, from Mechanics and Strength of Materials for Automation Engineers*, (in Polish), PWN, Warszawa, 1974.

13 G. Devore, J. Rivaud, *Cours de physique. Mécanique*, I, Librairie Vuibert, Paris, 1965.

14 G. Devore, J. Rivaud, *Cours de physique. Mécanique*, II, Librairie Vuibert, Paris, 1966.

15 *Dynamics of Electromechanical Systems*, (in Polish), Collective Work, PWN, Warszawa–Kraków, 1963.

16 V. I. Feodos'ev, *Strength of Materials*, (in Russian), Izd. "Nauka", Moskva, 1970.

17 V. I. Feodos'ev, *Selected Problems and Questions from the Strength of Materials*, (in Russian), Gos. Izd. Tekh.-Teor. Lit., Moskva, Leningrad, 1950.

18 A. P. Filippov, *Variations of Elastic Systems*, (in Russian), Izd. Akademii Nauk USSR, Kiev, 1956.

19 P. Fleury, J. P. Mathieu, *Mécanique Physique*, Editions Eyrolles, Paris, 1965.

20 F. Gantmakher, *Lectures in Analytical Mechanics*, (in Russian), Mir Publishers, Moscow, 1970.

21 R. Gutowski, *Analytical Mechanics*, (in Polish), PWN, Warszawa, 1971.
22 M. T. Huber, *Engineering Stereomechanics*, (in Polish), PZWS, Warszawa, 1951.
23 A. A. Il'ushin, V. S. Lenski, *Strength of Materials*, (in Russian), Gos. Izd. Fiz.-Mat. Lit., Moskva, 1959.
24 N. I. Ivanov, *Collections of Problems from Strength of Materials*, (in Russian), Gos. Izd. Tekh.-Teor. Lit., Moskva, 1957.
25 J. A. Jovlev, *Theory of Mechanical Vibrations. Course of Lectures*, (in Russian), Leningradski politekhn. inst. im. M. I. Kalinina, Leningrad, 1965.
26 S. Kaliski (editor), *Vibrations and Waves*, (in Polish), PWN, Warszawa, 1966.
27 S. Kaliski, *Introduction to Tensor Calculus and Applications in Continuous Field Theory*, (in Polish), WAT, Warszawa, 1961.
28 R. Z. Khas'minskii, *Stability of the Systems of Differential Equations with Accidental Perturbations of their Parameters*, (in Russian), Nauka, Moskva, 1969.
29 Ch. Kittel, W. D. Knight, Ruderman A. Malvin, *Mechanics*, McGraw-Hill Book Co., New York, 1962.
30 G. Korn, T. Korn, *Mathematical Handbook*, McGraw-Hill Book Co., New York 1969.
31 J. Kožešnik. *Machine Dynamics*, (in Czech), Stătni Nakladatelstvi Téchnické Literatury, Praha, 1958.
32 R. Kurowski, M. E. Niegodziński, *Strength of Materials*, (in Polish), PWN, Warszawa, 1972.
33 L. Landau, E. Lifshitz, *Theoretical Physics*, (in Russian), Gos. Izd. Tekh.-Teor. Lit., Moskva, 1958.
34 J. La Salle, S. Lefschetz, *Stability by Lapunov's Direct Method with Applications*, Academic Press, New York, London, 1961.
35 M. Lavrentev, L. A. Lusternik, *Variation Calculus*, (in Russian), Gos. Izd. Tekh.-Teor. Lit., Moskva, 1950.
36 *Lectures of Strength of Materials*, (in Russian), edited by M. M. Filonko-Borodich, Gos. Izd. Tekh.-Teor. Lit., Moskva, 1956.
37 M. J. Lighthill, *Introduction to Fourier Analysis and Generalized Functions*, University Press, Cambridge, 1959.
38 L. G. Loycyanski, A. I. Lur'e, *Course of Theoretical Mechanics*, (in Russian), Gos. Izd. Tekh.-Teor. Lit., Moskva, 1955.
39 L. A. Lusternik, W. I. Sobolev, *Elements of Functional Analysis*, (in Russian), Gos. Izd. Tekh.-Teor. Lit., Moskva–Leningrad, 1951.
40 I. G. Malkin, *Theory of Stability of Motion*, (in Russian), 2nd ed., Izd. "Nauka", Moskva, 1966.
41 H. Margenau, G. M. Murphy, *The Mathematics of Physics and Chemistry*, New York, 1950.
42 C. Maxwell, *A Treatise on Electricity and Magnetism*, Oxford, 1873.
43 J. N. McDuff, J. R. Curreri, *Vibration Control*, McGraw-Hill Book Co., New York, Toronto, London, 1958.
44 I. W. Mieshcherskii, *Collection of Problems from Mechanics*, (in Russian), Izd. "Nauka", Moskwa, 1968.

45 N. Minorsky, *Nonlinear Oscillations*, D. Van Nostrand Co., Princeton, N. J., Toronto, New York, London, 1962.
46 C. Mira, *Cours de systémes asservis non linéaires*, Dunod, Paris, 1969.
47 *Modern Mathematics for the Engineers*, Vols. 1, 2, Editor E. F. Beckenbach, McGraw-Hill Book Co., New York, Toronto, London, 1956.
48 J. Naleszkiewicz, *Engineering Mechanics*, (in Polish), PWN, Warszawa, 1953.
49 W. Nowacki, *Building Dynamics*, (in Polish), Arkady, Warszawa, 1961.
50 W. Nowacki, *Group Theory*, (in Polish), Arkady, Warszawa, 1963.
51 W. S. Novoselov, *Variation Methods in Mechanics*, (in Russian), Izd. Leningradskogo Universiteta, Leningrad, 1966.
52 A. Papoulis, *Probability, Random Variables and Stochastic Processes*, McGraw-Hill Book Co., 1965.
53 J. W. Prokhorov, J. A. Rozanov, *Probability Theory*, (in Russian), Izd. "Nauka", Moskva, 1967.
54 W. S. Pugachev, *Theory of Stochastic Functions and its Application to Problems of Automatic Management*, (in Russian), 3rd ed., Fizmatgiz, 1962.
55 *Random Vibration*, Vol. 2, Editor S. H. Crandall, The M.I.T. Press Massachusetts Inst. Techn., Cambridge, Mass., 1963.
56 M. Reiner, *Rheologie théoretique*, Dunod, Paris, 1955.
57 *Rheology Theory and Applications*, Editor F. R. Eirich, Vol. 1, Academic Press Inc. Publishers, New York, 1956.
58 B. Skalmierski, *Some Problems of Dynamical Systems, Non-Linear Vibration Problems*, (in Polish), PWN, Warszawa, 1971.
59 B. Skalmierski, A. Tylikowski, *Stochastic Processes in Dynamics*, (in Polish), PWN, Warszawa, 1972.
60 B. Skalmierski, *Theoretical Mechanics*, (in Polish), Wyd. Uniw. Wrocławskiego, Katowice, 1972.
61 B. Skalmierski, *Mechanics and Strength of Materials for Automation Engineers*, (in Polish), Warszawa, 1973.
62 B. Skalmierski, *Dynamics of Mechanical Systems*, (in Polish), 2nd ed., Silesian Technical University, Gliwice, 1973, (mimeographed lectures).
63 B. Skalmierski, A. Tylikowski, *Stability of Dynamical Systems*, (in Polish), PWN, Warszawa, 1973.
64 B. Skalmierski, *Mechanics*, (in Polish), PWN, Warszawa, 1975.
65 B. Skalmierski, *Multidimensional Departure Problems*, (in Polish), Zeszyty Nauk. Polit. Śląskiej, Fasc. devoted to studies of Institute of Theoretical Mechanics, Mat. Fiz. 28 (1977).
66 B. Skalmierski, *Departure Problems in Dynamical Systems*, (in Polish), Zeszyty Nauk. Politechn. Śląskiej.
67 B. Skalmierski, *Problems of Departness in the Dynamics of Non-Linear Systems*, Zeszyty Nauk. Politechn. Śląskiej.
68 V. V. Solodovnikov, *Statistical Dynamics of Linear Automatic Control Systems*, (in Russian), Fizmatgiz, Moskva, 1960.
69 R. L. Stratonovich, *Selected Questions of the Theory of Fluctuations in Radio Engineering*, (in Russian), Izd. "Sovetskoe radyo", Moskva, 1961.

70 G. K. Suslov, *Theoretical Mechanics*, (in Russian), Gostekhizdat, Moskva, 1946.
71 A. A. Sveshnikov, *Basic Methods of Random Functions*, (in Russian).
72 J. L. Synge, A. Schild, *Tensor Calculus*, University of Toronto Press, Toronto, 1959.
73 W. I. Tikhonov, *Questions of Random Processes*, (in Russian), Izd. "Nauka", Moskva, 1970.
74 S. Timoshenko, *Théorie des vibrations*, Beranger, Paris et Liege, 1954.
75 C. Truesdell, *Six Lectures on Modern Natural Philosophy*, Springer-Verlag, Berlin–Heidelberg, 1966.
76 S. Węgrzyn, *Archiwum Elektroniki*, 1. 2 (1952), (in Polish).
77 S. Węgrzyn, *Calcul operationnel en electrotéchnique*, Gauthier-Villars, Paris—PWN, Editions Sci. de Pologne, Warszawa, 1967.
78 S. Węgrzyn, *Fundamentals of Automatics*, (in Polish), PWN, Warszawa, 1972.
79 E. T. Whittaker, *A Treatise on the Analytical Dynamics of Particles and Rigid Bodies with an Introduction to the Problem of the Bodies*, University Press, Cambridge, 1952.

Subject Index

Absolute derivative of a tensor, 289
Acceleration, 16
  in general motion, 38
Addition of vectors, 2
d'Alembert's principle, 257
Amplitude characteristic, 330, 398
Angular momentum, 92, 227
Angular velocity vector, 32
Articulated joint, 106
Assur's point, 62
Asymptotically stable system, 296
Axode, 37
Average value, 376

Base resonance, 350
Bearing friction, 140
Betti's reciprocal theorem, 185, 186
Bilateral constraints, 253
Bilinear product, 6
Binet's equation, 78
Boltzmann's bodies, 182

Castigliano's theorem, 187
Central axis of a system, 98
Central axis of inertia, 164
Centre of impact, 234
Centrifugal moment, 159
Chebyshev's inequality, 397
Christoffel's symbol of the first order, 288
Christoffel's symbol of the second order, 289
Clapeyron's system, 185, 190
Coaxial vectors, 2
Coefficient of friction, 137
Coefficient of restriction, 91
Collinear vectors, 4
Conditions of inseparability, 151
Cone of friction, 138
Configuration of a body, 11
Configurational space, 291
Conservative system, 71, 75, 285
Constant of spring, 215
Constrained mechanical system, 250
Construction scale, 101
Continuous vector function, 9
Contravariant tensor, 286
Convergence of a vector-function, 9
Correlaton, 379
Correlation function, 382
Correlator, 385
Couple, 96
Covariant derivative, 290
Covariant tensor, 286
Crank-slide system, 55
Crank-yoke system, 59
Creep, 182
Cremona's method, 118
Cremona's polygon, 118
Culmann's line, 121
Culmann's method, 120
Cumulative distribution function, 373
Cyclic coordinates, 269

Degrees of freedom, 249
Derivative of a vector function, 10
Derivative of the Dirac delta function, 123

Differentiable vector function, 10
Differentiable constraints, 253
Dirac delta function, 122
Dispersion, 377
Dynamically balanced body, 232

Effort hypotheses, 217
Elastic limit, 145
Elastic buckling, 199
Energy, 70
Equations of circular motions, 231
Equations of disturbances, 295
Equations of equilibrium, 95
Ergodic process, 384
Error integral, 377
Euler's angles, 33
Euler's equation, 231
Expected value, 376
Externally determinate static system, 114

Fokker–Planck–Kolmogorov equation, 418
Fourier transform, 329
Framework, 195
Free-ends beam, 114
Free vector, 2
Frequency, 371
Frequency characteristic, 329
Friction, 136
  of a cylinder rolling, 142
  of rope against wheel, 141
  on an inclined plane, 138
Friction angle, 138
Funicular polygon, 103

Gauss normal distribution, 377
Gaussian process, 382
Generalized coordinates, 20
Geometric constraints, 252
Geometrical statical moment, 81, 82

Hamilton function, 283
Hamilton's canonical equations, 284
Hamilton's principle, 258
Harmonic oscillator, 310
Heaviside distribution, 121
Helical motion, 40
Hodograph, 15
Holonomic system, 252
Homeomorphism, 11
Homogeneous body, 81
Hooke's law, 147
Horse power, 72
Huber-Mises-Hencky hypothesis, 217
Hyperharmonic resonance, 350

Ideal constraints, 256
Inertia ellipsoid, 226
Influence function, 324
Influence matrix, 185
Instantaneous centre of acceleration, 52
Instantaneous centre of velocities, 49
Instantaneous helical axis, 39
Instantaneous power of a force, 71
Instantaneous rotational axis, 36
Internal magnitudes, 129
Intervally linear system, 337
Isotropic space, 11

Joule, 70

k-fold statically indeterminate system, 151
Kinetic energy, 70, 71
Kronecker's symbol, 6
Krylov's method, 341

Lagrange equations
  of impulsive motion, 275
  of the first order, 262
  of the second order, 268
Lagrange kinetic function, 269

Lagrange kinetic potential function, 260
Lagrange multipliers, 262
Lagrange-Dirichlet criterion, 297
Laplace's transform, 318
Lapunov function, 299
Lapunov's theorem, 299
Law, 255
  of action and reaction, 70, 274
  of conservation of mechanical energy, 71, 75
  of conservation of angular momentum, 93
  of variation of angular momentum, 92
  of variation of momentum, 87
Linear dependence of vectors, 6
Logarithmic decrement, 321

Matrix of transmittance, 367
Matrix of velocity gradient, 27
Menabrea's theorem, 190
Mean quadratic deviation, 377, 379
Modulus, 2
  of rigidity, 149
  of section, 166
  of a vector, 9
Mohr's analogy, 173
Mohr–Maxwell canonical system of equations, 192
Mohr–Maxwell equations, 212
Moment of deviation, 159
Moment of momentum, 92
Moment of a vector, 8
Momentum of a system, 86
Motion, 11

Newton, 69
Newton's second law, 69
Newton's third law, 69
Non-dilatation strain, 148
Non-holonomic system, 253
Non-oscillatory system, 308
Norm of a vector, 295
Normal stochastic process, 382
Node separation method, 118

Oblique bending, 159
Operator transfer function, 324
Orientation of a space, 8
Orthogonal coordinate system, 24
Orthogonal vectors, 4
Oscillatory system, 308

Parallel vectors, 2
Permutation symbol, 7
Phase characteristic, 330, 398
Phase plane, 308
Phase space, 352
Phase trajectory, 308, 352
Phase trajectory problem, 322
Plane truss, 117
Plastic limit, 145
Poisson number, 148
Polygon of forces, 101
Position of a body, 11
Position vector, 2
  of the centre of mass, 80
Postelastic buckling, 199, 201
Potential, 70
Potential energy, 75
Principal axes of inertia, 164
Principal central axes of inertia, 226
Principal moment, 98
Principle, 255
Probability density, 375

Quantificator, 9
Quasi-harmonic system, 340

Radial velocity, 22
Random variable, 372
Radius of gyration, 225
Rayleigh's dissipation function, 277
Reduced moment, 218
Relative frequency, 371

Relative motion, 45
Relaxation, 182
Resultant vector, 98
Rheonomic constraints, 252
Right-hand screw rule, 32

Safety factor, 149
Scalar product, 5
Scleronomic constraints, 252
Self-locking inclined plane, 139
Sensitivity of a thermal bi-metal, 181
Simple bending, 159
Slenderness ratio of a rod, 201
Smoluchowski's equation, 420
Solidification hypothesis, 145
Spectral density, 392, 396
Stable solution, 296
Stable system, 293, 296
Standard equations, 340
Standard deviation, 377
Statical moment, 80
Statical-dynamical analogy, 369
Statically balanced body, 232
Statically indeterminate system, 151
Stationary motion, 303
Stationary process, 383
Steiner's equation, 162, 223
Stochastic process, 381
  of the first order, 382
  of the second order, 383
Strain of a rod, 147
Stress, 146
Subharmonic resonance, 350
Subtraction of vectors, 3

Tapering spring, 213
Tensile rigidity, 146
Tensor multiplication, 286
Tensor of velocity gradient, 27
Tetmajer's line, 202
Theorem on statical moments, 80
Time function of transmittance, 324
Torsional rigidity, 210
Transmittance, 324
Transportation, 45
Transverse component, 22
Triple-jointed system, 106
Truss, 117

Uniform space, 11
Unilateral constraints, 253
Unit vector, 4
Uncorrelated variables, 379

Variance 377, 379
Vector, 1
Vector function, 8
Vector product, 6
Velocity, 14
  of a moving space, 41
  of transportation, 41
Velocity tensor of a strain, 29
Velocity vector, 14
Vibration frequency of a harmonic oscillator, 319
Virtual displacement, 254
Voight's model, 184

Watt, 72
White noise, 393
Work, 70
Wrench, 98